Cities on the Plains

This book is presented to
L.W. Nixon Library
in memory of
Victor Bland
by Butler Community College
May 15, 2006

Cities on the Plains
The Evolution of Urban Kansas

James R. Shortridge

University Press of Kansas

© 2004 by the University Press of Kansas
All rights reserved

Published by the University Press of Kansas (Lawrence, Kansas 66049), which was organized by the Kansas Board of Regents and is operated and funded by Emporia State University, Fort Hays State University, Kansas State University, Pittsburg State University, the University of Kansas, and Wichita State University

Library of Congress Cataloging-in-Publication Data

Shortridge, James R., 1944–
 Cities on the plains : the evolution of urban Kansas / James R. Shortridge.
   p. cm.
 Includes bibliographical references and index.
   ISBN 0-7006-1312-9 (cloth : alk. paper)
  1. Cities and towns—Kansas.   I. Title.
HT123.5.K2S56 2004
307.76′09781—dc22   2003022759

British Library Cataloguing-in-Publication Data is available.

Printed in the United States of America

10 9 8 7 6 5 4 3 2 1

The paper used in this publication meets the minimum requirements of the American National Standard for Permanence of Paper for Printed Library Materials Z39.48-1984.

307.76 SHO 2004

Shortridge, James R., 1944-
Cities on the plains : the evolution of urban Kansas.

For Amy and Kate Shortridge—
*Independent, resourceful, and fair-minded
Kansans who, like Dorothy, have seen Oz
but remain true to prairie values*

Contents

List of Illustrations   ix

Preface   xiii

1. Systems of Cities on the Plains   1
2. The River Towns to 1860   41
3. Lawrence, Topeka, and the Placement of Public Institutions, 1854–1866   68
4. Railroad Promotion and a Reconceptualization of Urban Kansas, 1863–1880   96
5. Later Railroads and Railroad Towns, 1877–1910   149
6. Mining, Irrigation, and Newer Institutional Towns, 1876–1950   173
7. Urban Consolidation in a Railroad Mode, 1880–1950   208
8. Postindustrial Kansas I: The Interstate Cities since 1950   282
9. Postindustrial Kansas II: Life beyond the Exit Ramps   336
10. Conclusion   369

Appendix   381

Notes   389

Bibliography   431

Index   471

Illustrations

Charts

Theoretical Linkages between a Frontier Center and Its National Urban System   8
Populations of Selected Cities in Kansas, 1870–1950   210
Populations of Selected Cities in Kansas, 1950–2000   285
Populations of Selected Cities in Southeastern Kansas, 1950–2000   286
A Model of Major Mercantile Linkages within Kansas, 1890–1950   371

Maps

The Elbow Region, 1850–1853   13
Indian Reservations, 1830s   14
Indian Holdings in the Summer of 1854   15
Eastern Kansas and the Elbow Region, 1860   16
Rivers and Uplands   17
Consensus Plan of the Topeka Railroad Convention of October 1860   18
Routeways Authorized by the Pacific Railroad Act of 1862   19
Railroads and Cities, 1869–1870   20
Railroad Visions of Six Cities, Late 1860s   21–22
Railroads and Cities, 1872   23
Railroads and Cities, 1877–1910   24
Minerals and Cities   25
Railroad Passenger Service, 1878   26
Railroad Passenger Service, 1892   27
Railroad Passenger Service, 1948   28
Urban Populations and Economic Specializations, 1880   29
Urban Populations and Economic Specializations, 1890   30
Urban Populations and Economic Specializations, 1910   31
Urban Populations and Economic Specializations, 1930   32
Urban Populations and Economic Specializations, 1950   33
The Trade Area of Hutchinson, Kansas, 1951   34
Urban Populations and Economic Specializations, 2000   35–36
U.S. Department of War's Strategic Highway Network, 1941   37

The Kansas Turnpike and Interstate Highway System   38
Commercial Television Markets, 1982   39
Basic Trading Areas Projected for 2010   40

Tables

A Functional Classification of Cities in Kansas, 1870–1950   216
A Functional Classification of Cities in Kansas, 2000   287

Photographs

The Steamboat *Mary McDonald* at Wyandotte, 1867   51
Cathedral of the Immaculate Conception in Leavenworth, 1867   57
Advertisement for the Town of Pawnee, 1855   77
The State Capitol in Topeka, 1880   83
Administration Building of the Kansas State Penitentiary in Lansing, circa 1910   86
The University of Kansas in Lawrence, 1867   89
Osawatomie State Hospital, circa 1917   93
Headquarters of the Union Pacific Railway Company, Eastern Division, at Armstrong, 1867   105
Bridge of the Leavenworth, Lawrence and Galveston Railroad in Lawrence, 1867   111
Depot Scene at Parsons, 1935   124
Building the Union Pacific Railroad in Ellis County, 1867   133
Loading Texas Cattle onto the Kansas Pacific Railroad at Abilene, 1871   135
Santa Fe Depot in Newton, 1878   139
Railroad Yards at Dodge City, Early 1900s   146
Box Cars at the Santa Fe Yards in Kansas City, circa 1949   151
Cavalry on the Parade Ground at Fort Riley, 1897   157
Nazareth Motherhouse of the Sisters of St. Joseph in Concordia, circa 1992   159
Railroad Shops and Yards at Horton, 1936   166
Rock Island Depot in Liberal, 1972   171
St. Louis and Pittsburg Zinc Company, 1909   176
Strip Mining in Crawford County, circa 1925   178
Standard Oil Refinery in Neodesha, circa 1900   185
El Dorado Oil Field, circa 1917   193
Factory of the United States Sugar Company in Garden City, 1910   202
The Fred Harvey Dining Room in Hutchinson, 1926   212

Kansas City Stock Yards Company, 1934   223
Armour Packing Company in Kansas City, circa 1939   224
Abernathy Furniture Company in Leavenworth, circa 1949   233
Railroad Yards and Associated Businesses in Atchison, circa 1960   236
Locomotive Finished Material Company in Atchison, 1930   239
Locomotive Shops of the Atchison, Topeka and Santa Fe Railroad in Topeka, Early 1900s   244
The Capper Building in Topeka, circa 1930   246
Production Line at Beech Aircraft Corporation in Wichita, 1942   252
Plant II of Boeing Airplane Company in Wichita, circa 1944   253
Watson Wholesale Grocery Company in Salina, circa 1930   256
Officials and Main Building of Southwest Kansas Methodist Episcopal College in Winfield, circa 1897   267
Workers at Lehigh Portland Cement Company in Iola, circa 1925   278
West Lawrence Toll Station on the Kansas Turnpike, circa 1960   292
Buick-Oldsmobile-Pontiac Plant in Kansas City, 1960   297
The Great Flood of July 1951 in Kansas City   298
Aerial View of Downtown Wichita, 1961   307
Natural History Museum at Fort Hays State University in Hays, circa 1958   329
Line Workers at the Iowa Beef Processors Plant in Emporia, circa 1975   334
Corn and Watermelon Exhibit at the Kansas State Fair in Hutchinson, circa 1965   340
Re-creation of Front Street in Dodge City, 1958   347
Automobile Racing at the North Central Kansas Free Fair in Belleville, 1958   352
Reliance Manufacturing Company in Cherryvale, circa 1941   359
Farmland Industries Refinery in Coffeyville, circa 1986   363
Machine Shop at the McNally Pittsburg Manufacturing Company in Pittsburg, circa 1965   367

Preface

Friends who know my love for rural America will be surprised that I have written a book about cities. Let me try to explain. The immediate stimulation was a previous study, *Peopling the Plains,* in which I traced the settlement process in Kansas. As I read and thought about who migrated and why, it became obvious that many early residents were urban speculators. I had heard stories about county-seat "wars," of course, and about once-grand ambitions to make my hometown a railroad center, but gradually came to appreciate that city rivalries existed across a much wider range of fronts and scales. Entrepreneurs in every town, it seemed, had strategies for making their community the future hub of the state's political, industrial, or mercantile network. As I completed the settlement book, I knew that I wanted to probe deeper into the evolution of the Kansas urban system.

The realization that urban rivalries are as active today as they were in the past gave this subject additional appeal. My study could stretch over a period of a century and a half. I also was intrigued to learn that case studies of the type I contemplated were rare. Many people have written histories of individual cities, of course, but only a few address larger-scale urban systems. As a result, theories about how such networks evolve are mostly speculative. I hope that the empirical work here will inspire similar study in other places. Documentation to reconstruct the interlocking sets of metropolitan dreams is abundant, and such efforts are needed to properly test and reformulate our general models. The subject matter also is fascinating, with planning schemes that range from naive to sophisticated and from benevolent to diabolic. To follow these collective aspirations, triumphs, and failures is to see deeply into human nature. Such striving is basic to who we are.

I did most of my archival research in the Kansas Collection at Spencer Research Library on the campus of the University of Kansas. Sheryl Williams, the curator there, created a much-appreciated atmosphere of professionalism blended with informality. She and her associates Bryan Culp, Deborah Dandridge, Kristin Eshelman, Becky Schulte, and others made the place a second home for me, and I am saddened by the recent decision to collapse their facility into the larger Spencer holdings. The other half of this research—site visits to the various cities of the state—was made much more efficient and enjoyable by the presence and assistance of Barbara Shortridge.

The University of Kansas provided release time in the form of a sabbatical leave in 2001, which allowed me to begin the writing process. The final product benefits greatly from the help of four experts: Darin Grauberger and Matthew Stratton, who created quality maps from my sketches, Nancy Sherbert, who located excellent photographs in the collection of the Kansas State Historical Society, and Barbara Shortridge, again, who read and improved the entire manuscript. I thank them all. Earlier versions of selected materials from chapter 4 appear in Michael H. Hoeflich, Gayle R. Davis, and Jim Hoy, eds., *Tallgrass Essays: Papers from the Symposium in Honor of Dr. Ramon Powers* (Topeka: Kansas State Historical Society, 2003), and in *Kansas History* 26 (2003): 186–205.

1

Systems of Cities on the Plains

Kansas City is the metropolis of the central plains. Everybody knows this. Similarly, area residents all recognize Wichita as one of the air capitals of the country, Salina as the trading center for a huge section of north-central and northwestern Kansas, and Lawrence as a mellow college town. People know these things because the character and functional interaction of a region's cities are among the most significant facts of modern economic and social life. What rarely is considered, however, is that the particular pattern of urban growth and function so glibly accepted in Kansas, or anywhere else, is anything but inevitable.

A look at popular opinion from earlier times dispels any notion that specific sites were somehow preordained for urban success. Kansans in the 1850s and 1860s, for example, foresaw none of the economic roles just noted. Were these new immigrants so preoccupied with survival that they had no dreams about city growth? Not at all. Even a brief perusal of the records reveals rampant speculation in urban real estate and multiple theories for development. Rather, the lesson from such study is that the process of city growth is complicated. Many visions of the future competed with one another, and people often accorded wildly differing importance and meaning to a particular routeway, type of transportation, mineral deposit, or congressional vote. Some of the forces involved operated purely on a local scale, but others were regional in scope, and still others national or international. The rules of the game, if you will, also changed considerably over time. Early theory spoke about "natural" locations such as river junctions or the boundaries between differing agricultural regimes. A technological change, however, could transform the value of one of these sites. A political whim could relocate a key public facility or render unprofitable an important manufacturing operation. The only predictable trend seemed to be that the wishes of outside capitalists grew steadily more important, while those of local entrepreneurs declined.

As decisions were made and money invested, the existing infrastructure of a region became another important factor for planners and dreamers to consider. In Kansas, the success of the river-port cities of Atchison and Leavenworth in the 1850s produced a set of transportation arteries, manufacturing establishments, and vested interests that itself had important impacts on the decisions of the following decade. As time moved forward and the inertia of these existing systems commingled with new realities, the strategies for continued growth became ever more complex. New cities emerged, and older ones declined. Nearly every community saw its economic role and social character modified in ways its earlier citizens never could have predicted.

This book presents the story of how urban development unfolded in Kansas. I will write about Kansas City, Lawrence, Salina, and Wichita, of course, but will not limit myself to such scenes of obvious success. Chetopa and Ellis interest me just as much. These two communities are small today, but their residents and other promoters once envisioned bright economic futures. Chetopa, along the Oklahoma border in Labette County, was an incipient gateway city for the cattle trade and other dealings with the Southwest, just as Kansas City had been at an earlier time. Ellis, in west-central Kansas, was bolstered by its choice as a division point on the state's first major railroad. What caused the urban dreams in these towns and those in many similar communities to falter while others came to fruition? How did the current hierarchy of cities emerge in the state? These are the core themes I want to address.

To begin thinking about urban structure in Kansas, look at a current map. Given the prominent climatic transition across the state from east to west and the general advancement of the Euro-American settlement frontier in the same direction, the pattern of more and bigger cities in eastern Kansas makes sense. The concentration of cities along the Missouri and Kansas Rivers seems logical as well, because railroads were not yet well established when Kansas Territory was opened for general settlement in 1854. It is therefore not surprising that a river port would emerge as the metropolis. The question remains, however, why Kansas City became this center of development instead of Leavenworth or some other place. Leavenworth in particular was a larger community than Kansas City at the census of 1860 and home to an important military installation. Being in Kansas instead of Missouri, this city also was on the "right" side of the Civil War and therefore presumably in a position to gain federal favor. Suddenly the emergence of Kansas City no longer seems inevitable.

The dominance of Wichita in southern Kansas provides another good illustration. Railroad entrepreneurs, especially officials of the Atchison, Topeka and Santa Fe, heavily influenced urban development in this part of the state. Cities along their main line, especially those selected as division points or the sites of repair shops, had the best access to the outside world and therefore tremendous advantages for growth. Newton was the place so designated in south-central Kansas, however, not Wichita. In fact, Wichita originally was on only a branch line of the railroad. Another obstacle to that city's survival and prosperity emerges from a second business factor in early southern Kansas. As the example of Chetopa illustrates, entrepreneurs saw substantial trade potential with the seemingly permanent Indian Territory located on the state's border. Arkansas City and Caldwell exploited this niche locally. With these two places only fifty miles south of Wichita, and Newton less than thirty miles to the north, not many smart investors in the 1870s would have purchased urban real estate in Sedgwick County.

Mysteries of growth or the lack thereof spring up in every section of the state once a person begins to look closely. To stay with regions influenced by the Santa Fe railroad for a moment, these officials selected Dodge City for their next major repair shops and stockyard west from Newton, and that town immediately boomed. Because of this decision, no one in 1900 would ever have predicted that Garden City would eclipse the "queen of cow towns" in population by 1970 and that Liberal would rival it by 1990.[1] In northwestern Kansas, where the Union Pacific Railroad played a role as kingmaker analogous to that of the Santa Fe in the south, city growth seems even more perplexing. West of Hays, the communities of Colby and Goodland currently vie for leadership, and neither is on the old main line of that premier transportation system.

Finally, consider southeastern Kansas. This well-watered region filled quickly just after the Civil War, and town sites along railroads building south to the Gulf of Mexico were touted for rapid growth. When major deposits of lead and zinc, coal, and natural gas were discovered shortly thereafter, expectations rose to a fever pitch. The scene today belies all such dreams. No metropolis emerged, and indeed no one urban place contains as many as twenty thousand people. What caused growth in this area to disperse among many places and then to stagnate? People seemingly had coalesced so easily around Wichita to the west and Kansas City to the north. Why not here?

How did entrepreneurs in nineteenth- and early twentieth-century Kansas gauge prospects for city growth? Their principles are apparent from their actions. First, it is clear that trade was the dominant consideration in the territorial years and those of early statehood. To the long-existing interchange with Spanish settlers over the Santa Fe Trail, several new destinations had been added by the 1850s. Gold-mining camps had sprung up in California and Colorado, and silver ones in Nevada. Provisioners to these places could get rich as easily as a lucky miner. Similarly, new army bases built to protect traders also needed regular shipments of goods, as did the increasingly large number of settlers in Oregon and the Salt Lake oasis. Port cities on the Missouri and Kansas Rivers thus were much on people's minds in the 1850s, along with existing and planned trails from these supply centers and possible new locations for forts in the western section of the state. After the Civil War, railroads rapidly supplanted rivers and trails in businessmen's thinking, but the logic of trade with the West and the construction of immense warehouses continued.

Three aspects of western trade held special promise for early Kansas entrepreneurs. Two of these already have been noted: shipments into and across the large Indian Territory on the state's southern border, and the acquisition of division points and repair facilities on the emerging railroad system. The third, better known in popular history, is the cattle trade. City officials did not universally seek this business of linking trails from Texas with railroad lines to the East because everybody realized it would be both short-lived and boisterous. Still, success in this arena was seen by many people as a means to other, more permanent developments for their cities.

In the efforts to secure trading business, city boosters often turned to geographic and other arguments about various transportation advantages. Enthusiasts in Leavenworth and Kansas City touted their radiating networks of roads, for example, the only major ones in the territory in 1854. Other places had to be more imaginative in their promotion. Atchison people always reminded visitors that their community was farther west on the Missouri River than either Kansas City or Leavenworth and therefore a better departure point for a western journey. Promoters in Council Grove trotted out a climatic argument, that theirs was the westernmost humid-land location along the Santa Fe route. And leaders in Fort Scott tried to convince shippers that their rail connections across Missouri and southward to the Gulf of Mexico would lend speed and profit to any business operation.

Beyond trade and transportation, the securing of a major public or private institution was another much discussed means of fostering city growth. This competition was different from the battles over trade routes in many ways, but it was every bit as spirited. It also was concentrated in time, with the major allocations obviously being made shortly after statehood in 1861. Ostensibly, the decisions regarding where to establish a new state capital, university, prison, or similar facility were straightforward, made either by the legislature or by public vote. Since the procurement of such a public institution could do much for the immediate and long-term prosperity of a community, however, the maneuverings were intense and the suspicions many. Several cases of outright briberies have come to light, and many others almost certainly took place. Once the larger public facilities had been fixed, entrepreneurs often turned their lobbying skills to the similar but usually smaller-scale recruitment of private and federal operations. A church-sponsored college was one common goal, but the examples were many, including railroad-sponsored hospitals and military installations.

Once a basic network of railroads existed in Kansas and the public institutions were located, many people expected stability for the urban system. This turned out not to be the case. Mineral discoveries in the late nineteenth and early twentieth centuries altered the scene considerably. Later the process of suburbanization and other developments associated with the automobile created significant change. So, too, did the rise of footloose retirees, a globalized economy, and additional phenomena unimagined in 1900. In fact, the process of selective city growth and its attendant promotional activity is very much ongoing at the beginning of another new century.

Theory

The various strategies for promoting city growth in Kansas need to be understood in a series of contexts. As suggested earlier, conditions and means have changed through time. I will address these modifications through a generally chronological arrangement of text. Kansas realities also usefully can be evaluated by reference to several theoretical models and the findings of other case studies. Such work allows us to see general patterns and to assess the degree of uniqueness exhibited on the central plains during the last century and a half.

Scholarly thinking about the historical aspects of regional urban growth is not as extensive as one might suppose. We have many biographies of cities, of course. But before the 1940s and 1950s, American historians showed little concern for their patterns of distribution and roles in regional development. Strongly influenced by the frontier writings of Frederick Jackson Turner, they assumed urban places to be late and rather unremarkable occurrences in a region. Subsistence farmers were the true pioneers. Cities emerged slowly in this view, only after the farmers had accumulated capital enough to become commercial operators and then to have surpluses for investment in urban enterprises. Countrysides, in other words, created their own systems of cities.[2]

Geographers, observing the contemporary pattern of cities in the Midwest and elsewhere as it existed in the 1950s and 1960s, offered a model to explain its form and hierarchy. Their inspiration was an empirical study done by Walter Christaller in Germany in the 1930s. Called central-place theory, this model assumed that cities were primarily market centers. If the terrain were uniform and the natural resources and rural population evenly distributed, economic logic would dictate that nucleated settlements should be spaced regularly. Central-place theorists also envisioned several "orders" of city size, reasoning that some categories of goods and services required more people for their support than did others. The smallest order, hamlets, would have an average spacing of about six miles. There, merchants would sell only basic items such as fuel and groceries. Higher-order places, including villages with hardware stores and elementary schools, and towns with jewelers and dentists, also would exhibit regular spacing but at greater distances one from another. At the top of the hierarchy, of course, would be a single regional or national metropolis.[3]

The first influential suggestion that cities might play an active rather than a passive role in frontier development came in 1959 when Richard C. Wade published *The Urban Frontier: The Rise of Western Cities, 1790–1830*. Professor Wade's argument that cities were "spearheads" of progress initiated a wave of scholarly reassessment. For geographers and others interested in regional models, the new thinking revealed the ahistoric nature of central-place theory.

Clearly the standard assumption of an even distribution of population and economic resources did not apply to the frontier. Perhaps cities at that time assumed economic roles considerably more complex than just marketing centers.[4]

Charles F. J. Whebell's observation that cities in southern Ontario tended to occur in linear sequences initiated the new discussion. He attributed this pattern partly to initial cost advantages that certain locations and routeways offered to merchants and traders. This early superiority led to differentials in capital accumulation and subsequently to the continued growth of these places. Newer forms of transportation tended to follow the old routes as well, creating corridors of development.[5]

Working independently from Whebell, James E. Vance in 1970 took many of the same ideas about initial advantage, capital, and transportation and incorporated them into a generalized theory. Known as the mercantile model, it was anchored solidly in historical process and has become the basis for most contemporary thinking on the evolution of urban systems. Vance argued that trade constituted the foundation for city growth and that wholesaling, not retailing, became the most important aspect of this activity once long-distance transportation was practicable with large ships and then railroads. Zones of economic influence could be huge.[6]

The initial stage of the mercantile model assumes the pairing of an established economy in one place with a frontier elsewhere. Merchants in the older center have capital and seek to invest it in frontier resources through a process of economic colonization. Initially this exploitation may consist only of sending out people to produce or select staple goods for shipment back to the older center. As the process develops, however, populations and profits increase to the point where "points of attachment" are needed, nucleated settlements at the edge of the frontier on or near the best routes for transportation. Some of these towns will develop further, becoming "entrepôts" where it becomes profitable for the merchants to erect processing centers for the staple goods as well as distribution centers for manufactured and other trade items shipped in from older centers. Later, each of these entrepôts (or gateway cities) will develop its own series of dependent "depot" sites deeper into the developing frontier. These will be along the transportation lanes, producing the linear patterns of cities observed by Whebell. Finally, as the frontier becomes fully developed, its appearance will become increasingly similar to the region that originally exploited it. Its merchants will themselves now look for new frontiers to develop, and local economic and social conditions will approximate more and more those described by central-place theory.

The Vance model has been modified by a series of studies over the past thirty years, but it generally has withstood the scrutiny of time and several intensive case studies. One early addition to the theory suggested that each frontier experience can lead to a distinctive regional culture and that the emerging political boundaries in such places will approximate those of the trading zones.[7] Another corollary, provided by Edward K. Muller, addresses the process of selective growth of smaller cities within a developing region. Muller identified three stages based on his study of conditions in the Ohio Valley. Initially, he argued, transportation is poor and the towns few. These few would be in the larger valleys or along other natural routeways. In the second stage, increased population leads to a denser network of transportation, including nodal locations where key routeways intersect. Settlements at these nodes grow relatively fast, but a change in transportation technology can easily disrupt this nodality and thus patterns of growth. Finally comes integration of the regional system into the national one. Transportation costs decline, and manufacturing establishments begin to find the area profitable. These factories, too, locate in nodal positions, especially at sites having multiple connections to the hinterland. Big places will get bigger, overwhelming nearby cities and isolated towns alike.[8]

The linkages that connect a growing city on the frontier with the national urban system have been the subject of considerable work since 1970. Andrew F. Burghardt has focused on the gateway concept; John R. Borchert, Michael P. Conzen, and Allan R. Pred on integration at a national scale; and David R. Meyer on the creation of a dynamic model for the process. This model, especially as modified in 1988 by William K. Wyckoff via a case study of Denver, stresses the likelihood of change in the nodality of places and the potential importance of forceful individual entrepreneurs. Although case studies to test its findings are still scarce, it fits the experience in the upper Mississippi Valley and has even been applied successfully in the Ecuadorian Amazon.[9]

The Meyer-Wyckoff model has three stages (Chart 1). In the early years of development, poor transportation limits the interaction of a new frontier center (G) with major urban cities in the older section of the nation (A, B). Smaller cities adjacent to the frontier thus act as its primary suppliers (E, F). As time passes, the frontier city grows and transportation linkages improve. At this second stage, the long-distance contacts with entrepreneurs in the major cities become dominant. Superior capitalization, economies of scale, and transportation linkages drive this process. Suppliers in the smaller cities that formerly controlled the frontier trade find they have to specialize if they are to survive. Some make this transformation successfully (F), and some less so (E). New specialty suppliers also are possible if local entrepreneurs are aggressive enough (D). The third stage reflects both continued growth in the frontier center and dynamic changes in the national urban system. City A increases its influence, but that of city B declines. City D, possibly via improving transportation linkages, economic complementarity, or other reasons, increases its trade and general influence. New linkages may continue to emerge at this stage, too (C), perhaps the result of changes in the economy of the frontier city itself.

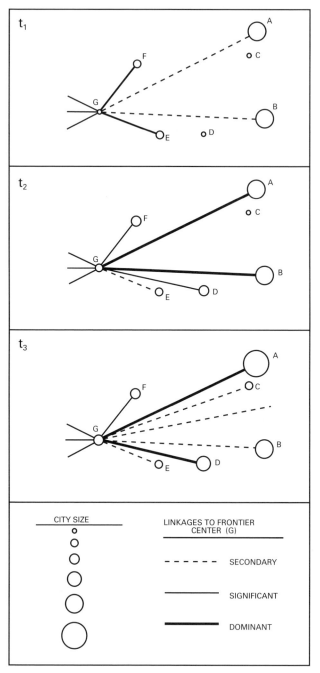

1. Theoretical Linkages between a Frontier Center and Its National Urban System. Modified from a diagram in William K. Wyckoff, "Revising the Meyer Model: Denver and the National Urban system, 1859–1879," *Urban Geography* 9 (1988): 14.

Approach

More than four thousand communities in Kansas have established post offices since the territory was opened for settlement in 1854. The majority of these places had aspirations to become important cities, but only a minority survived in any form at all. How should a person best evaluate this process of striving, success, and disappointment? To study the dreams of several thousand towns is an impossibly large task. To focus on only a few examples seems wrong as well. Histories of Emporia, Hays, and Topeka would suggest that city building was an easy process, since most of the decisions made by leaders in these communities turned out to be fruitful. Such successes need to be balanced by the much more common stories of frustration and failure.

My solution to the dilemma of city selection is arbitrary but fairly inclusive. I will focus on 118 cities, the ones among the 4,000 that have been successful enough to have attracted at least twenty-five hundred residents at some point in their histories (Appendix). Census officials use this number as a threshold size for urban places, and because of this, a wide range of data is available. One hundred eighteen communities also are enough to provide a good sampling of strategies and locations.

A chronology of maps showing urban populations as a series of proportional circles will serve as a vehicle for organizing the text, each depiction revealing the realities of geography and size that entrepreneurs and government officials of the time had to confront as they prepared their ongoing proposals for local and regional economic development. Conditions in the nearby Missouri communities of Joplin, Kansas City, and St. Joseph, although not the primary concern here, also will enter the discussion regularly, since their fortunes have always been closely intertwined with those of Kansas. My goal is to see Kansas and the broader region through the eyes of various promoters, how they evaluated and reevaluated their own cities and businesses and those of their rivals. The composite picture of their perceptions and actions is the story of an evolving urban Kansas.

I begin in the next chapter with the territorial period, 1854 to 1860, and its focus on river towns. This material actually reaches back into the 1820s, however, with the establishment of Fort Leavenworth, the instigation of trade with Santa Fe, and the resettlement of eastern Indians to Kansas. A series of new border settlements arose in Missouri eager to profit from the new commerce. Everybody agreed that the Missouri River provided the logical way to import goods from St. Louis, but they contested the best points for redistribution. Weston was an early leader, and St. Joseph became important after 1848 when it became a point of departure for emigrants to Oregon and California. Kansas settlements, prohibited until 1854, necessarily were late arrivals in the competition, but Atchison and Leavenworth emerged as major players by 1860. By the end of this period, a host of would-be metropolises, including Weston, had fallen by the wayside (a few of them literally into the river). Atchison and Leavenworth vied with the Missouri towns of Kansas City and St. Joseph to assume the classic role of entrepôt.

Chapter 3 moves the discussion forward to the early statehood years of 1861 to 1866 and focuses on the competition for public institutions. Such pursuit represents an important alternate strategy for growth, although one largely ignored by theorists. Some leaders took this approach because their cities were not on major rivers and thus were poorly situated to become trade emporiums. Others adopted a combined strategy, questing after trade and institutions simultaneously. The allocation of the state capital and other public institutions is especially interesting to study in Kansas because the group of politicians who dominated the new state (free-state Republicans) was entirely different from the one that had been in charge of the territory (proslavery Democrats).

The first state legislators reexamined all the old discussions about location, but none of the decisions were obvious. Although the potential power of railroads to influence city growth was known, for example, track locations still were speculative. The biggest existing cities were on the Missouri River, but many people argued that public facilities in a westward-growing state should be more centrally located. Differences in power and attitude among settler groups were important, too, with groups of New Englanders in Lawrence, Manhattan, and Topeka craving institutions, and former Missourians living in the eastern and southern border counties biding their time quietly in the face of wartime raids and occasional arrogance by the Yankee ruling establishment. Finally, the wishes of powerful individuals, particularly those of the two new United States senators, obviously were critical.

Railroads were the single most important determinant of city growth in nineteenth-century Kansas. These are the subject of chapters 4 and 5. Although their initial promoters usually were businessmen from the older river and institutional towns, control soon passed to moneyed investors from Boston, New York, and other northeastern places. Loss of local control meant almost immediate modifications in routes and interconnections, and, with these, the relative growth potentials for local cities. Decisions by railroad men led to the stagnation of Atchison and Leavenworth, for example, and to booms for Wyandotte, Topeka, and, to a lesser degree, for Ottawa, Fort Scott, and Junction City.

West and south of the old territorial core of cities, railroad entrepreneurs often built ahead of settlement, and thereby were even more influential. They laid out and promoted many towns and could guarantee rapid growth by naming particular sites as division headquarters, switching points, and repair centers. In at least two cases (Hutchinson and Newton), railroad people even manipulated county boundaries to ensure county-seat designations. Besides a description of general strategies, I will detail the urban hierarchies envisioned along the main lines of the six principal track systems in the state: the Atchison, Topeka and Santa Fe; the Chicago, Rock Island and Pacific; the Kansas Pacific (Union Pacific); the Missouri, Kansas and Texas; the Missouri Pacific; and the St. Louis–San Francisco. These visions focus on Arkansas City, Dodge City, Emporia, Goodland, Herington, Horton, Hutchinson, Kansas City, Liberal, Newton, Osawatomie, Parsons, Pratt, Salina, Topeka, Wellington, and Wichita.

Chapter 6 introduces a series of circumstances and strategies that, starting in the 1870s, modified the railroad model for urban growth. Mining activity is the chief of these. Although the route of the Santa Fe between Topeka and Emporia was influenced by the presence of coal in Osage County, most main lines had been constructed before development of the state's extensive mineral reserves. Rich coal deposits in Cherokee and Crawford Counties were not mined intensively until 1876, for example, and the first big lead and zinc strike at Galena happened a year later. Salt became important only in 1888, natural gas and oil in southeastern Kansas in the 1890s, and irrigation water from the Arkansas River at Garden City about 1900. The oil and gas fields in western counties were not discovered until the 1920s and 1930s.

In some cases, as with salt at Hutchinson and natural gas at Liberal, existing railroad centers were well enough positioned to accommodate the new industrial developments. Elsewhere, at Garden City and particularly in southeastern Kansas, new industrial towns vied with older railroad locations to produce situations where no one city was able to dominate a regional economy. Cities stressed in this chapter include Chanute, Coffeyville, El Dorado, Garden City, Great Bend, Hugoton, Hutchinson, Independence, Iola, Liberal, Lyons, McPherson, Neodesha, Phillipsburg, Pittsburg, and Russell.

Chapter 7 focuses on the urban network as it had evolved by 1950. I change my perspective from the historical forces that created growth to how these cities actually functioned during the last decades of railroad dominance in the national transportation system. In many ways the pattern in 1950 was the culmination of a series of urban rivalries that had been under way since the first Union Pacific tracks were laid in the 1860s. A functional classification of cities based on the number of people employed in different sectors of the economy helps to organize this chapter. The identification of almost pure "railroad towns" such as Hoisington, Newton, and Parsons, and their separation from other specialist cities in education, oil extraction, or heavy manufacturing, reveals important subsystems within the overall urban network. Such a classification also helps to explain contrasts between places of apparent economic stability and those in flux. The steadily increasing influence of Hutchinson and Salina over the trade from western Kansas becomes obvious, for example, as does the rise of manufacturing in southeastern Kansas in response to its short-lived energy boom at the turn of the century. I look at the state region by region with the goal of presenting a Kansas that people (unaware of the transforming possibilities of interstate highways, suburbanization, and instant communication) thought had settled into its permanent pattern.

Kansas since 1950 has been characterized by rapid growth at a few urban centers, with stagnation or decline elsewhere. In chapters 8 and 9 I explore causes for this phenomenon and demonstrate its effects on the preexisting urban structure. One key, of course, has been new economic opportunity created when a system of superhighways challenged and then overwhelmed the older railroad hegemony. I begin, therefore, with discussion of the routeway selections for the Kansas Turnpike in the 1950s and for Interstate Highways 35, 70, and 135 a decade later. These decisions have proved more influential than originally imagined. They explain, for example, why Salina now has many more people than does Hutchinson, its longtime rival that was bypassed by the interstate planners. Similarly, I-70 communities in western Kansas dwarf their counterparts along U.S. Highways 24 and 36. Increased ease of transportation is part of an ongoing transformation in the United States to what scholars term a postindustrial society. This concept, which implies the possibility of industrial decentralization and a greater reliance on the service sector of the economy, is central to understanding recent city growth. Chapter 8 examines its range of effects along the state's interstate corridors, including the suburbanization of Kansas City and Wichita and the growing popular appeal of Lawrence's university atmosphere.

Chapter 9 changes the perspective to focus on Kansas communities that lie beyond the new transportation routes and how they have coped with a changed and changing world. I begin with an exception to the often-made assertion that superhighways are necessary for population growth—spectacular increases in Dodge City, Garden City, and Liberal associated with irrigation and the meatpacking industry. Discussion of the highly varied fates of former railroad towns and small industrial cities in far-northern and southeastern Kansas concludes the book. Individual cases include Atchison, Beloit, Concordia, Fort Scott, Independence, Iola, and Pittsburg.

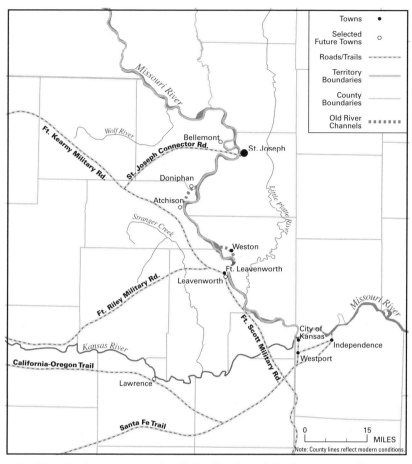

1. The Elbow Region, 1850–1853. Population data from federal census records.

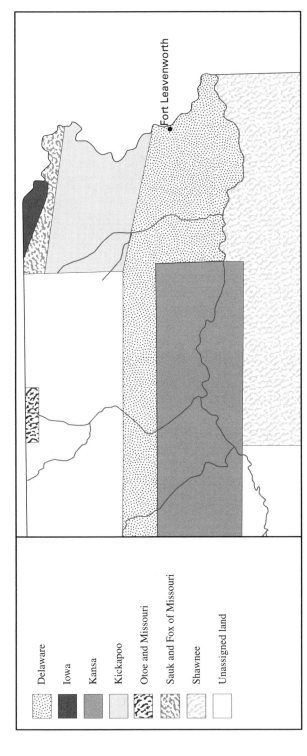

2. Indian Reservations, 1830s. Modified from a map in James R. Shortridge and Barbara G. Shortridge, "Yankee Town on the Kaw: A Geographical and Historical Perspective on Lawrence and Its Environs," in *Embattled Lawrence: Conflict and Community*, ed. Dennis Domer and Barbara Watkins (Lawrence: University of Kansas Continuing Education, 2001), p. 9.

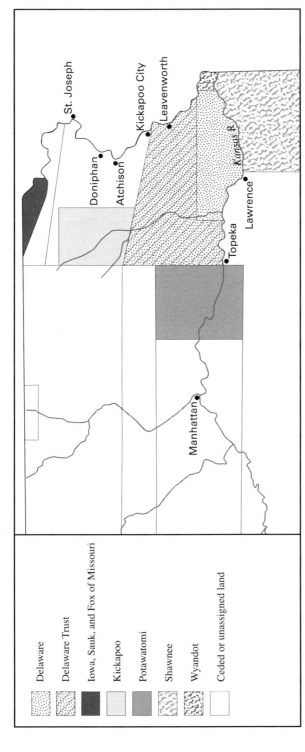

3. Indian Holdings in the Summer of 1854. Modified from a map in James R. Shortridge and Barbara G. Shortridge, "Yankee Town on the Kaw: A Geographical and Historical Perspective on Lawrence and Its Environs," in *Embattled Lawrence: Conflict and Community*, ed. Dennis Domer and Barbara Watkins (Lawrence: University of Kansas Continuing Education, 2001), p. 9.

4. Eastern Kansas and the Elbow Region, 1860. Population data from federal census records.

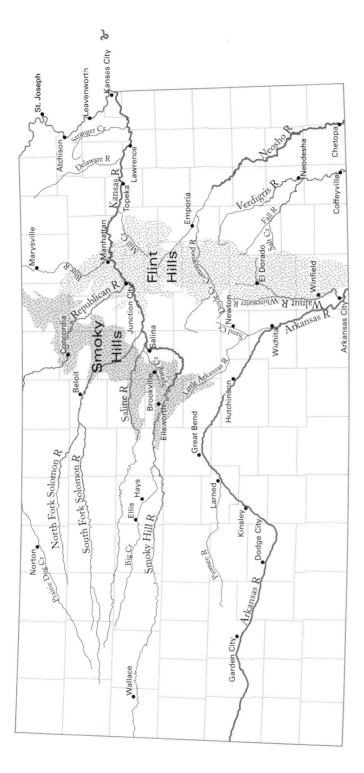

5. Rivers and Uplands.

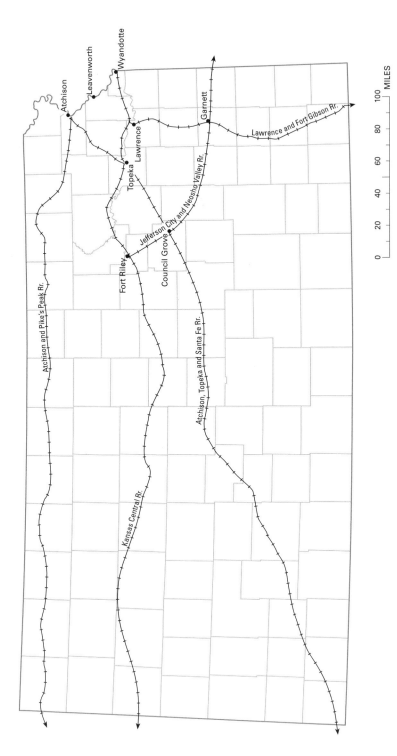

6. Consensus Plan of the Topeka Railroad Convention of October 1860. Modified from a map in George W. Glick, "The Railroad Convention of 1860," *Transactions of the Kansas State Historical Society, 1905–1906* 9 (1906): 477.

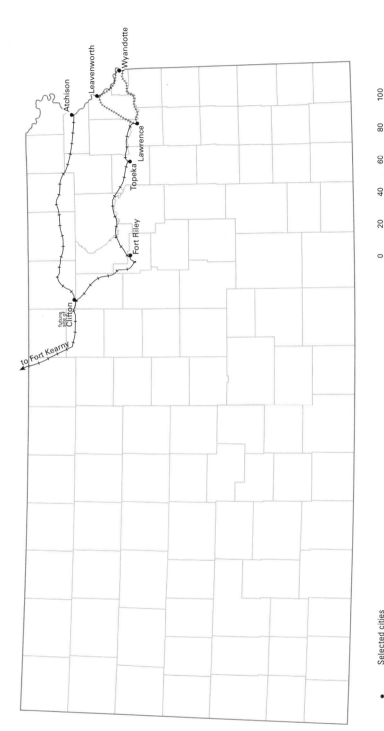

7. Routeways Authorized by the Pacific Railroad Act of 1862. Information from William R. Petrowski, *The Kansas Pacific: A Study in Railroad Promotion* (New York: Arno Press, 1981), pp. 52–54.

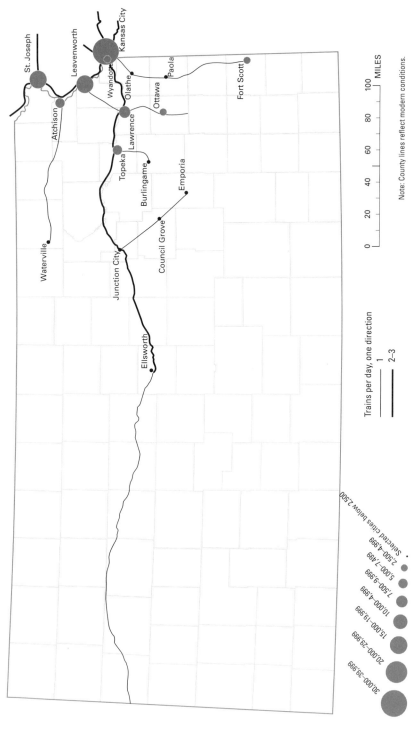

8. Railroads and Cities, 1869–1870. Population data from federal census records; railroad data from Edward Vernon, ed., *Travelers' Official Railway Guide* (New York: J. W. Pratt and Co., 1869), and other sources.

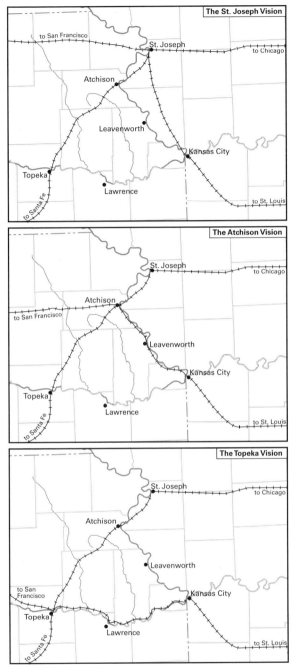

9. Railroad Visions of Six Cities, Late 1860s. *(Continued on next page)*

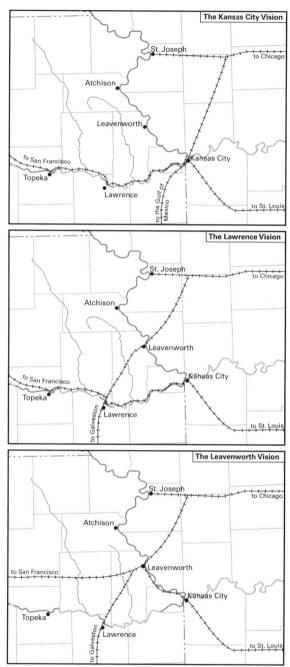

Note: County lines reflect modern conditions.

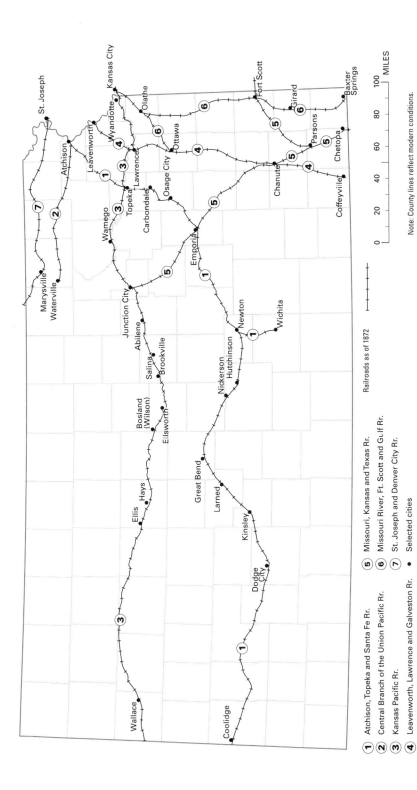

10. Railroads and Cities, 1872.

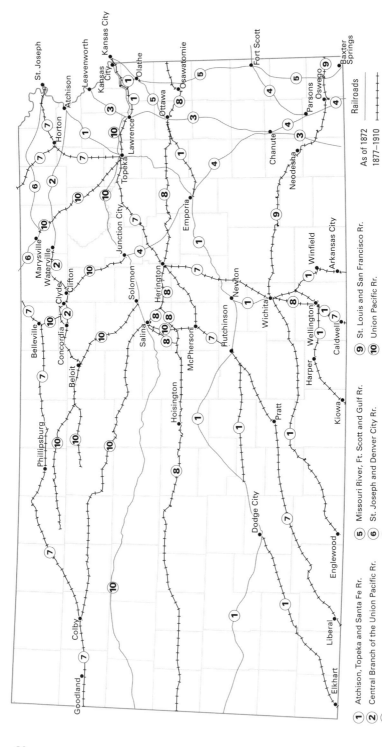

11. Railroads and Cities, 1877–1910.

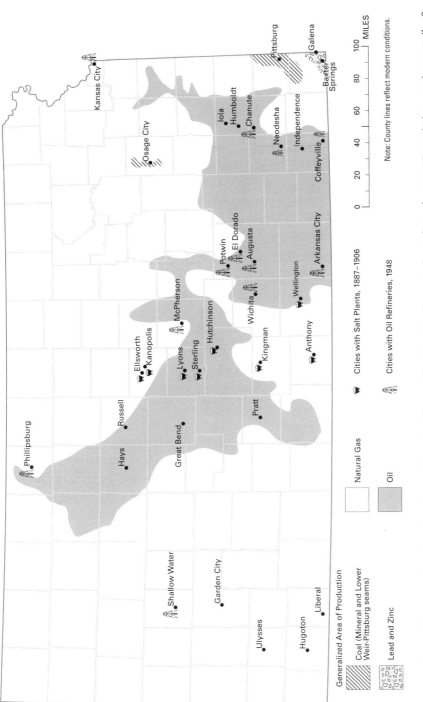

12. Minerals and Cities. Boundary data from George F. Jenks, *A Kansas Atlas* (Topeka: Kansas Industrial Development Commission, 1952), p. 10; oil refinery and salt plant data from, respectively, Earl K. Nixon, "The Petroleum Industry of Kansas," *Transactions of the Kansas Academy of Science* 51 (1948): 414; and Frank Vincent, "History of Salt Discovery and Production in Kansas, 1887–1915," *Kansas Historical Collections, 1915–1918* 14 (1918): 358–78.

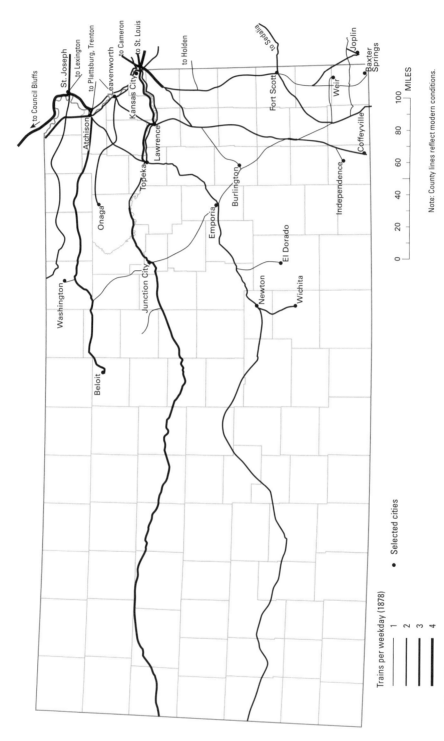

13. Railroad Passenger Service, 1878. Data from W. F. Allen, ed., *Travelers' Official Railway Guide for the United States and Canada* (Philadelphia: National Railway Publication Co., 1878).

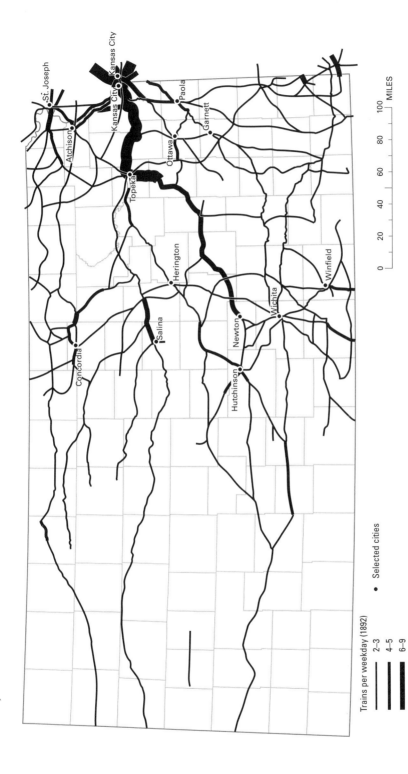

14. Railroad Passenger Service, 1892. Data from W. F. Allen, ed., *Travelers' Official Railway Guide for the United States and Canada* (New York: National Railway Publication Co., 1892).

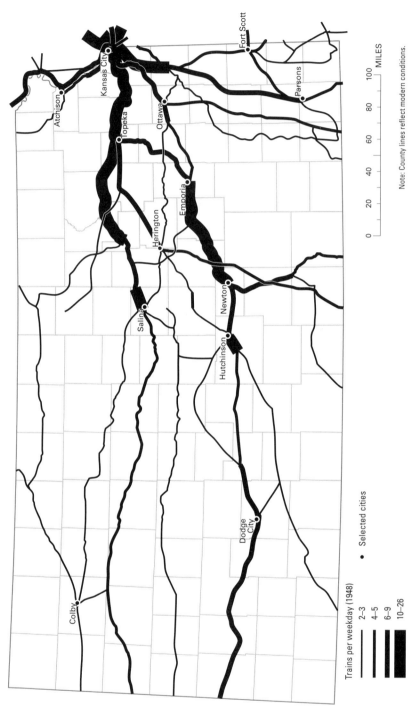

15. Railroad Passenger Service, 1948. Data from *The Official Guide of the Railways and Steam Navigation Lines of the United States, Porto Rico, Canada, Mexico, and Cuba* (New York: National Railway Publication Co., 1948).

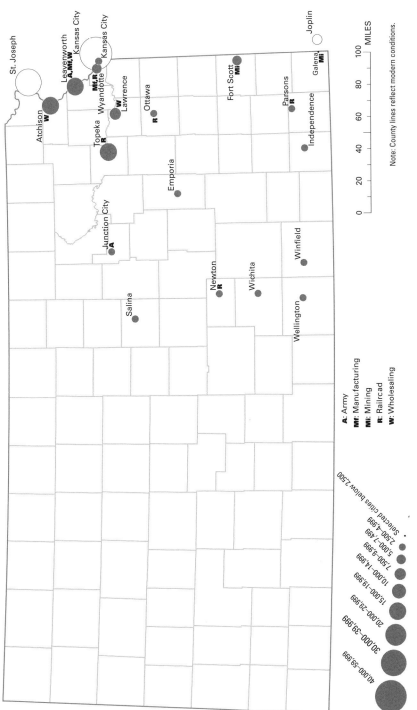

16. Urban Populations and Economic Specializations, 1880. Population data from federal census records; calculation procedures for economic specializations are described in the text.

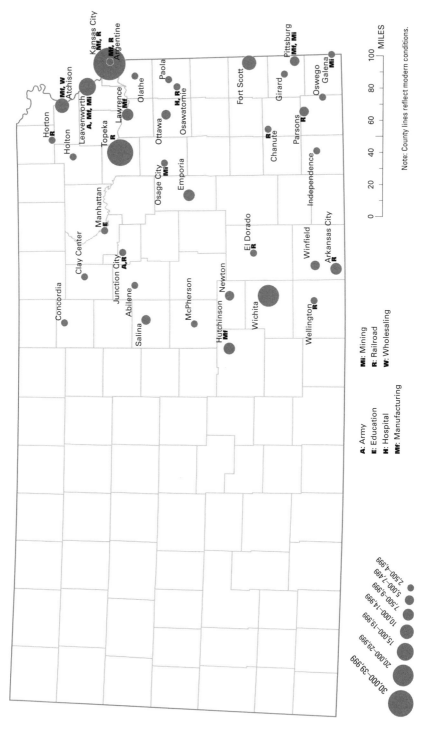

17. Urban Populations and Economic Specializations, 1890. Population data from federal census records; calculation procedures for economic specializations are described in the text.

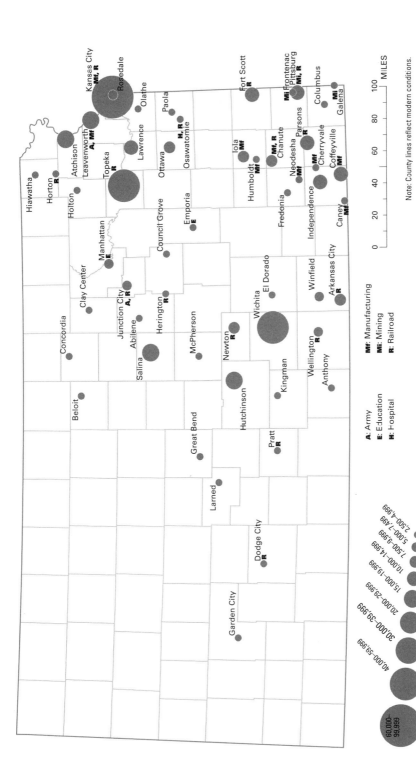

18. Urban Populations and Economic Specializations, 1910. Population data from federal census records; calculation procedures for economic specializations are described in the text.

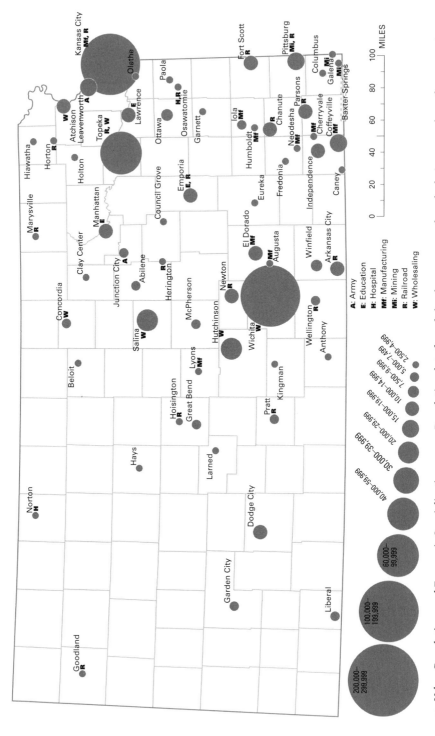

19. Urban Populations and Economic Specializations, 1930. Population data from federal census records; calculation procedures for economic specializations are described in the text.

32

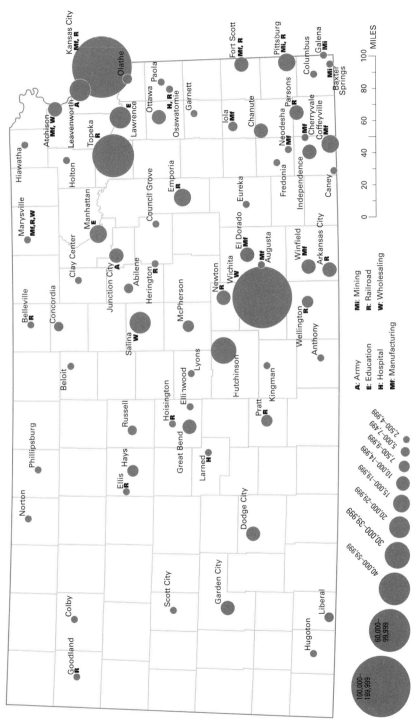

20. Urban Populations and Economic Specializations, 1950. Population data from federal census records; calculation procedures for economic specializations are described in the text.

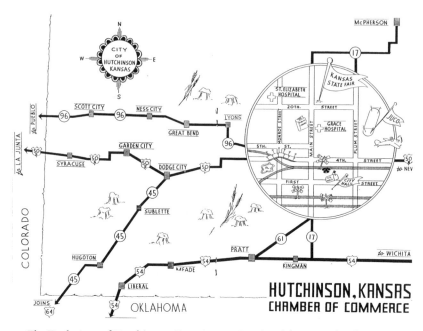

21. The Trade Area of Hutchinson, Kansas, 1951. Reprinted from Beach Advertising Co., comp., *Hutchinson, Kansas* (Hutchinson, Kans.: Hutchinson Chamber of Commerce, 1951), pp. 14–15.

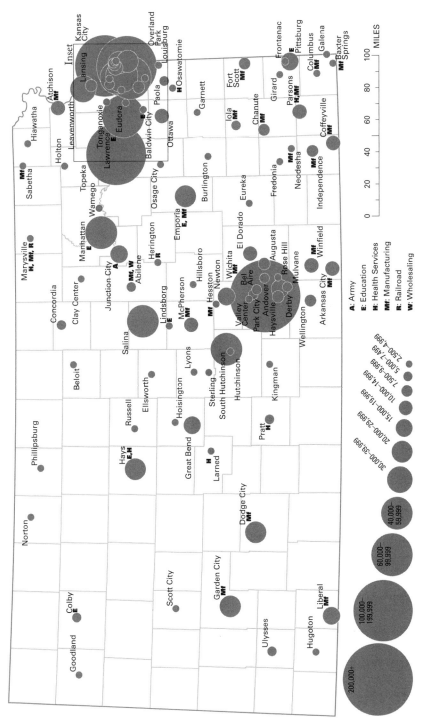

22. Urban Populations and Economic Specializations, 2000. Population data from federal census records; calculation procedures for economic specializations are described in the text.

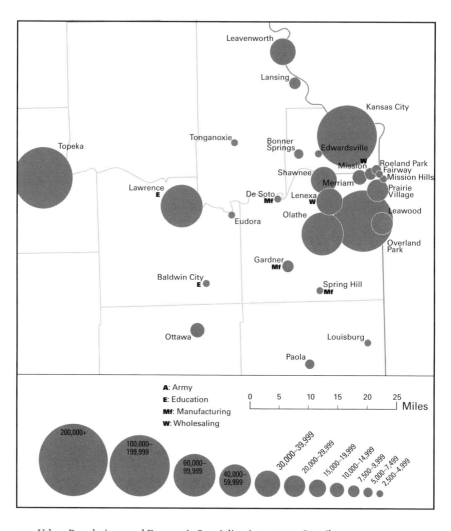

22. Urban Populations and Economic Specializations, 2000. Detail.

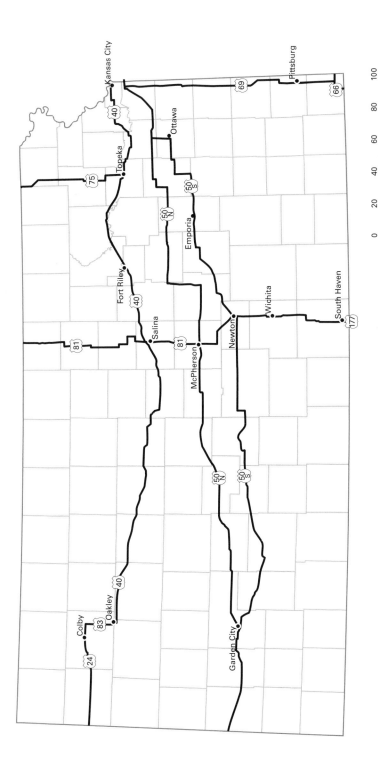

23. U.S. Department of War's Strategic Highway Network, 1941. Information from Sherry L. Schirmer and Theodore A. Wilson, *Milestones: A History of the Kansas Highway Commission and the Department of Transportation* (Topeka: Kansas Department of Transportation, 1986), p. 4:4.

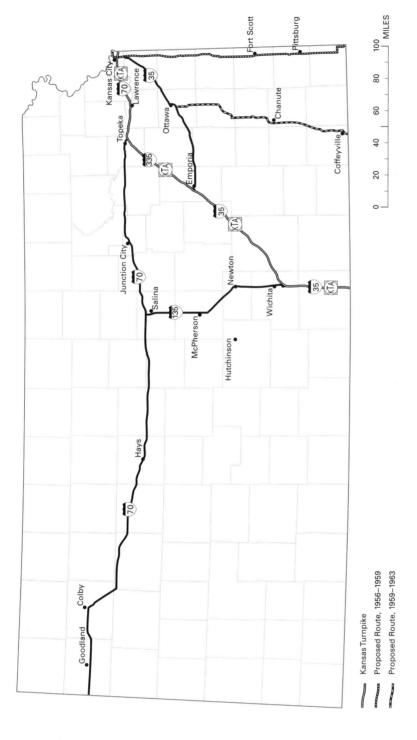

24. The Kansas Turnpike and Interstate Highway System. Information from Sherry L. Schirmer and Theodore A. Wilson, *Milestones: A History of the Kansas Highway Commission and the Department of Transportation* (Topeka: Kansas Department of Transportation, 1986), pp. 4:22, 4:29, 4:36.

25. Commercial Television Markets, 1982. Modified from a map in Susan Yerkes, "Low-Power TV: A New Broadcast Market Opens for Small Investors," *Kansas Business News* 3 (March 1982): 17.

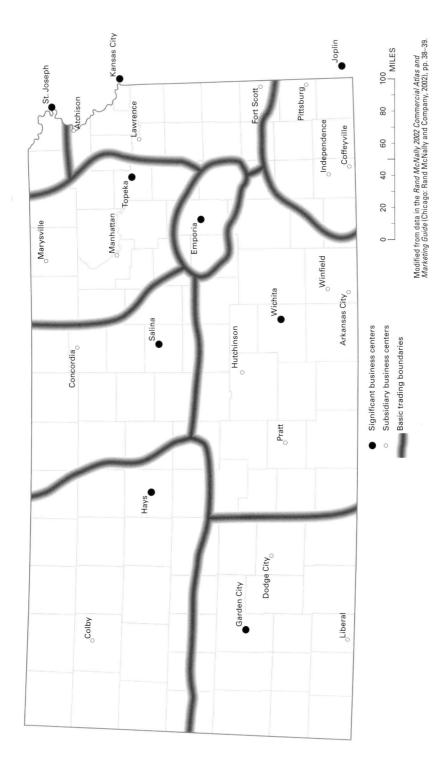

26. Basic Trading Areas Projected for 2010. Modified from data in the *Rand McNally 2002 Commercial Atlas and Marketing Guide* (Chicago: Rand McNally, 2002), pp. 38–39.

2

The River Towns to 1860

Before the prospects for travel over iron rails entered the realm of possibility, every would-be founder of an inland city knew the advantages of river location. The costs of shipping goods by water were so much lower than by land that, on large streams, observers reported town sites at nearly every break in the bluffs. Still, some general positions held more potential than others, and these relative likelihoods for success were carefully weighed on each successive frontier. When the vast Louisiana country came into American ownership in 1803, entrepreneurs knew that sites along the Missouri River, its obvious transportation artery, would boom. St. Louis, near the river's mouth, occupied the best position, but a general section of river some 250 miles upstream was held in nearly equal regard.

Old French voyageurs called this upstream section on what is now the Kansas-Missouri border *le grand detour*. The name reveals its significance. Between the present-day sites of Kansas City and St. Joseph, the river changes its general orientation from east and west to north and south. Americans replaced the French label for this place with a more prosaic term, the "elbow" region, but the importance remained. For any person wishing to travel west or southwest from St. Louis, the logical route would be by water to some site along this elbow and then overland from that point beyond. A city at this location would have tremendous future possibilities for trade.[1]

Development at the elbow remained no more than a dream for several decades after 1803. Euro-American settlement extended upriver only to the vicinity of Jefferson City when Missouri entered the union in 1821. It did not reach the western border of the new state for nearly another decade. Moreover, before 1837, the entire elbow except for its extreme southern end was part of Indian country. As a result, when local traders made their first successful venture to Santa Fe in 1821, they departed the river at Franklin, in central Missouri, the edge of American settlement at that time.

The first Euro-American outpost in the elbow region came in 1827. This was an army cantonment that had been urged by Missouri's most powerful citizen, Senator Thomas Hart Benton, as a means to protect the growing number of merchants who vied for the now lucrative trade overland to Santa Fe. Senator Benton pushed initially for a fort directly on the trail, somewhere along the Arkansas River in what is now central Kansas. Army officials deemed such a position too difficult to supply, however, and so substituted one on the easily accessible Missouri River. They identified the elbow region as the best general location and authorized Colonel Henry Leavenworth to select the specific site. The orders said it was to be on the north bank, somewhere within twenty miles of the mouth of the Little Platte River. As it happened, the Missouri River at that time and place flowed closer to the bluffs on the south side of its floodplain than on the north. The only potential sites on the north bank (where Parkville and Weston, Missouri, eventually would be founded) lay just outside his twenty-mile limit. Accordingly, the colonel elected to examine the other shore. There he found success, a site some 150 feet above the river level and about eighteen miles upstream from the Platte (Map 1; see p. 13).[2]

Colonel Leavenworth defended his choice to the War Department on the basis of geography and health: "In addition to the advantage of being on the same side of the Missouri as the road to Santa Fe, this position (the one I have selected) possesses the very material one of having a dry rolling country on the south and southwest of it. This will greatly contribute to the healthiness of the position." When the adjutant general approved the decision on September 19, 1827, and sent 188 men there the next month, the future Kansas has its first permanent Euro-American community.[3]

Missouri Cities in the Elbow Region, 1827–1850

Fort Leavenworth remained the lone settlement in the elbow region for ten years. During this period Santa Fe traders had moved their point of departure westward from Franklin to Independence, Missouri, about nine miles from the state line. No good landing tempted the drovers closer to the border than that, however, and they used the intervening upland prairies as pasture for their oxen, horses, and mules.[4] West of the state line settlement was prohibited. That the future Kansas was conceived as a permanent Indian territory from the 1820s to the 1850s is well known. Less remembered is the fact that the original northwestern boundary of Missouri was not the Missouri River. Instead, it was a continuation of the meridional line that separated Missouri from Kansas between Arkansas and the mouth of the Kansas River. This northward extension created a wedge of territory between the line and the Missouri River, a region known as the Platte Country.

Ownership of the Platte Country passed from Indian hands to the federal government in 1830. Between eight and nine hundred Iowas continued to live there, however, along with five hundred Sacs and Foxes. As Missourians began to realize the economic promise of this area, they feared that the president might decide to add it to the permanent reserve for Native Americans across the river in Kansas. They therefore lobbied to close the region to Indian resettlement. Success came in 1833, about the same time that several hundred Delawares and a similar number of Kickapoos were relocated to reservations just west and northwest of Fort Leavenworth (Map 2; see p. 14). The politicians then argued that the Indian and white races could be separated far more easily by an imposing natural barrier than with a mere line scratched on the ground. Conveniently, of course, such a barrier was at hand. With endorsement from Colonel Henry Dodge, head of the dragoons at Fort Leavenworth, their argument carried the day. New lands for the Sacs, Foxes, and Iowas were found north of the Kickapoos in Kansas, and in March 1837, President Van Buren declared that Missouri's northwestern boundary would now extend to the Missouri River.[5]

The opening of the Platte Country created a land rush. Many people sought farms there on rich loess soils, but others saw cities as equally profitable ventures. Serving the market needs of the new farmers was one obvious goal. Then, because town sites were illegal in Indian Territory, these entrepreneurs hoped to supply recreational services such as saloons and brothels to Fort Leavenworth and consumer goods to the newly relocated Indian groups. People with bigger visions saw two additional roles. They imagined a potentially huge military trade in foodstuffs, clothing, and freight services. Not only would the fort itself require such products, they reasoned, but also the other government outposts that inevitably would have to be constructed across the plains and maintained from the Leavenworth depot. Finally, they saw good possibilities for extending trade up the Missouri River itself. Just as Boonville, Lexington, and Independence had succeeded one another as gateway centers with settler advancement up the lower river, why would the same process not continue in the new northwestern Missouri and beyond?

Weston was the first town of consequence in the Platte Country. This success surprised nobody, for the settlement occupied the prime location with regard to the fort. Military officers and many other people knew from their local excursions that the closest upland town site along the Missouri side of the river lay six miles upstream. Joseph Moore, a former soldier at Leavenworth, platted the lots there in 1837, and the new community grew quickly to a population of three or four hundred. Steamboat captains, for example, made Weston the head of regular navigation on the river and, as such, the shipping point for tobacco, hemp, and other agricultural goods. In 1845, two local men boosted the economy further by erecting a substantial packinghouse to process cattle and hogs. Additional growth, however, was problematical because of the limited market provided by the garrison across the river, which rarely numbered more than two hundred men.[6]

Conditions at Fort Leavenworth and in the larger trans-Missouri West began to change rapidly after 1845, resulting in major growth for Weston as well as the creation of rival communities. First came the Mexican War of 1846–1848 in which the Leavenworth depot, because of its location near the Santa Fe Trail, served as a principal outfitting point. When the Army of the West under General Stephen Kearny marched from Kansas to New Mexico, for example, it took along five hundred pack animals and 1,550 wagons, most of which came from local suppliers. Troop numbers at Fort Leavenworth itself increased markedly as well and held at these higher levels for several decades. Business at Weston rose accordingly. Whereas only seven companies had operated in the town in 1842, thirty-four did so in 1848, and the local packinghouse changed hands for the princely sum of $10,000. Knowledgeable people surveyed its crowded wharf and agreed that the city had become the leading business center west of St. Louis.[7] The ebullient mood as it remained in the early 1850s was captured by traveler John McNamara:

> It was the immediate resort of the free and easy soldier after pay day at the garrison. Officers' families did most of their trading and marketing at this place. The Quartermaster threw millions into the coffers of the traders there. Much of the overland trade took its rise from there. I never was in a town of 2,800 inhabitants, in any state, where money, in gold and silver, was so plenty, and where wealth was so general in the possession of the traders.[8]

Profits on the scale generated by the merchants of Weston in the middle and late 1840s attracted additional businessmen by the score. Most of these took up residence in Weston itself, but some entrepreneurs, seeking lower initial investment costs and eyeing new trends in the development of the Far West, established a rival community at Blacksnake Hills, the next good bluff site up the river. This location, about forty miles from Weston, had been known about for decades. In fact, Joseph Roubidoux, one of the great St. Louis traders of the previous generation, had maintained a fur post there since 1803.[9]

Roubidoux counted some two hundred people living near his once-isolated store by 1842. Farmers had begun to occupy the northern section of the Platte Country, and they needed a marketplace on the river. Roubidoux thus platted a formal town site in 1843, naming it St. Joseph. His timing was fortuitous. The year before, Missourians and others had begun to travel overland to Oregon homesteads in significant numbers. Their favorable reports prompted some eight hundred to make the long trip in 1843 and then several thousand annually for the rest of that decade. The accepted method of travel was modeled on the earlier Santa Fe experience: a gathering of individual wagons in the spring of the year at certain spots and then a collective hiring of guides. Outfitting these expeditions clearly could make money, and the new merchants of St. Joseph enthusiastically advertised their location as an ideal point for departure.[10]

St. Joseph, the businessmen extolled, was now the practical head of navigation on the Missouri River. Its location, twenty miles farther west than Independence, was reason enough to expect status as a principal assembly point for western travel. Even better, migrants to Oregon from this upstream landing could avoid a difficult fording of the Kansas River. The message found its intended audience. During May 1845, a total of 954 people in 223 wagons left St. Joseph for the far Northwest. By 1847, after the United States had officially annexed Oregon Territory, this annual total had risen to 5,200. The local population increased accordingly. From 976 in late 1846, it grew to 1,542 in 1848 and continued to expand rapidly.[11]

Outfitting for the Oregon Trail was only the beginning of good fortune for the St. Joseph community. First, their rival merchants in Weston, busy with supplying the fort during its Mexican expeditions, did not contest the Oregon trade seriously. Then, in 1848, the potential for outfitting more than doubled with the discovery of major gold deposits in northern California. The following spring, because everybody realized that the best road to California was the same as that to Oregon, 1,508 wagons passed through town bound for mining glory.[12] A person would ascend the Missouri River to either Independence or St. Joseph, set out northwest from there to Fort Kearny on the Platte River in Nebraska Territory, and then follow the Platte Valley to the break in the Rocky Mountains known as South Pass (Map 1).

St. Joseph was the leading outfitting point for California in 1849, a feat partly attributable to superior location and advertising, and partly to an outbreak of cholera that year in the rival community of Independence. The city's population boomed past that of Weston, and steamboat companies made the St. Joseph wharf their new head of regular navigation. In 1850 the census reported 3,460 residents, two sawmills, and seven warehouses. Merchants were so prosperous, in fact, that they felt able to shun a less desirable, third group of migrants. These were Joseph Smith's followers, the Mormons, who had lived in northwestern Missouri during the 1830s but had angered fellow residents with their New England mannerisms and other perceived haughty ways. Since 1847 increasing numbers of these people had begun to follow a new leader, Brigham Young, to a western refuge in the valley of the Great Salt Lake. Once again, the most obvious route passed through the elbow region. Without encouragement from St. Joseph, many of these religious emigrants selected Weston as their gathering point.[13]

As the volume of American travelers to California, Oregon, and the Salt Lake Valley increased, so did the number of military installations built for their protection. In turn, this meant a proportional growth in the demand for supplies, especially from soldiers and miners. Merchants in the elbow region were optimally positioned to fulfill this need. Without really planning things, St. Joseph entrepreneurs found themselves thrust into the wholesale trade nearly as much as were the fort-oriented Weston merchants. Livestock was the first local specialty. By late 1848, St. Joseph boasted two new packinghouses that sold pork to the army and to private dealers alike. Beef cattle were important, too, for animals worth ten dollars in Missouri could be sold for fifteen times that amount in the mining camps of California. Buyers from St. Joseph ranged as far as Iowa and Arkansas in search of animals to sell during the spring migrations.[14]

Because so much of their business focused on outfitting expeditions and supplying distant outposts, the leaders of Fort Leavenworth, St. Joseph, and Weston focused much effort on issues of transportation. Their first priority, a good overland route to the Platte Valley, was officially marked out in 1850 and named the Fort Leavenworth–Fort Kearny Military Road (Map 1). This date and name are not particularly important, however, for the three-hundred-mile route was an obvious one that had been in use for many years. Heavily loaded wagons demanded a trail that was as high and dry as possible. From Fort Leavenworth this meant first traveling westward a few miles to reach the stream divide between numerous small tributaries of the Missouri River and those of Stranger Creek. This latter stream, a substantial one that flows nearly due south into the Kansas River, has headwaters about twelve miles west of where Atchison, Kansas, would be established in 1854. The military road then turned north along this divide and nearly paralleled the Missouri River from Fort Leavenworth to the Atchison area before angling to the northwest. Wagons then followed the high ground between the Delaware and Wolf Rivers south of present-day Sabetha, moved westward to a ford on the Blue River near what is now Marysville, and then pressed on toward the northwest and Fort Kearny.[15]

Because the military road ran northward from Fort Leavenworth before turning to the west, entrepreneurs from St. Joseph could intercept it by building a connector only about twenty-six miles long. This route, too, followed stream divides, an east-west pathway between branches of the Wolf River to the north and those of Rock and Independence Creeks to the south. The junction with the main trail was near the present-day intersection of Atchison, Brown, and Doniphan Counties.[16]

The business community of the elbow region was in the midst of a boom at the time of the 1850 census. The various communities were united in their dependency on long-distance trade but differed sharply on the destination of their goods and people. St. Joseph, of course, had garnered the majority of the business from emigrants to California, and Weston merchants controlled commerce with Fort Leavenworth and the Mormons. People whose destinations lay in Santa Fe or elsewhere in the Southwest, however, departed the river at other landings. Their concern, just like freighters anywhere, was a solid upland route. The key to this goal locally was avoidance of a Kansas River crossing. If a person transferred goods from a Missouri River steamboat to a southbank wharf below the mouth of this stream, he or she could avoid considerable effort and possible long delays. The ideal situation, realized by Santa Fe traders since the 1820s, was to dock a short distance one way or the other from the mouth of the Blue River, a tributary about six miles downstream from the state line. From such a position it was easy to follow a continuous upland that extended southwest through today's aptly named Overland Park, Kansas, and then along the high divide that separates the Kansas and Osage Rivers.

The issue that remained for Santa Fe traders in the 1830s was the selection of the best outfitting point within this general location. The initial town in the area was Independence, laid out by a group of commissioners in 1827 to be the seat of a newly formed Jackson County. By law these government offices were to be centrally located, but the officials compromised their charge to a degree. They placed Independence on a broad ridge of land just east of the Blue River and about two and one-half miles south of the Missouri. It was a disease-free site, they argued, still reasonably central within the county, and with decent access to transportation. Two landings soon evolved to serve the town, one six miles to the northeast and the other directly north.[17]

Santa Fe traders first used the Independence area as a rendezvous in 1829. With the establishment of Fort Leavenworth nearby and the last of the indigenous Osage peoples removed in 1825, the traveling merchants now considered the region safe. Independence also was a convenient site to meet two hundred troopers from the fort who, as an experiment, were to accompany them as far as the Arkansas River. When this venture brought back beaver pelts, gold, mules, and silver valued at $240,000 and earned its backers a profit of 100 per cent, the process was repeated the next season. By 1832, observers agreed that Independence had become the premier outfitting point for this trade. Warehouse construction was one obvious sign; so was the presence on the streets of mountain men, traders of both Anglo and Hispanic ancestry, and even a few Indians. One visitor in 1832–1833 called the town "full of promise."[18]

Independence merchants maintained control of the Santa Fe trade through the mid-1840s. The town's permanent population in 1844 was reported at 742, but this number swelled every spring in anticipation of the trading season.[19] Local businesspeople that year also had two special reasons to celebrate. First, by virtue of their upland location, they had escaped the June ravages of the worst flood ever seen by Europeans on the Missouri River. Second, they were seeing some diversification in their outfitting activities. Early travelers to Oregon had found the town nicely suited to their needs. Supplies for the long trek were plentiful, and their wagons could follow the well-established Santa Fe road for several days before turning north to cross the Kansas River and then to angle northwest toward the Platte Valley. The prospects for business expansion at Independence looked so favorable, in fact, that citizens began to call their hometown the "Queen City of the Trails."

Because economic success always creates rivals, the growth of Independence by the early 1840s made it inevitable that people would renew their search for potential south-bank landing sites farther upstream on the Missouri. No ideal spot existed, but a high bluff practically at the mouth of the Kansas River intrigued many of the explorers. This elevated position would be healthy, but the river flowed so close to the escarpment that any wharf would have to be narrow. Moreover, the only access between landing and bluff was a narrow, steep ravine. Most of the people who contemplated development at this locale lived in the village of Westport. This community lay four miles due south from the bluff and, as its name suggests, nearly on the state border. It had been platted by John C. McCoy in 1835 on high ground west of the Blue River and along one branch of the road to Santa Fe. The town had grown to a population of four hundred by 1840, with its merchants catering principally to the Indian trade in Kansas. McCoy and his fellow townsmen saw a good possibility for additional growth, however, if they could establish their own wharf and thereby supply the Santa Fe traders directly.[20]

Formal development of the levee began in November 1838 with the creation of the Kansas Town Company. Although its land title was unclear, the group's members went ahead and constructed two small warehouses in 1840. Three years later another Westport merchant, William M. Chick, moved his business to the wharf and erected a third, and more substantial, warehouse. Chick's building was the only one of the three to survive the great flood of 1844 and thus to be in a position the next summer to receive the first big shipment of goods from the trail. Following this success, Hiram M. Northrup (still another Westport man) opened a large store on the landing. He and Chick made improvements to the ravine road, and their trade increased accordingly. By beginning or ending the trail experience at Westport and its landing, people could save at least a day of overland travel as well as a crossing of the small but sometimes treacherous Blue River between Independence and Westport. Drovers also valued the abundance of grass available just south of Westport because this resource was decreasing farther east as rural Jackson County attracted more farmers.[21]

Westport and the Town of Kansas (the latter known more familiarly at the time as Westport Landing) might have developed faster than they did were it not for two natural disasters in 1844. The big flood in June plus a tornado in October required most of the available capital for rebuilding. Still, by the late 1840s observers reported that the southwestern traders were splitting their business about equally between the Westport area and Independence. To the superior grass and shorter haul of the newer towns, the better-established merchants in Independence could counter with lower prices and a greater range of services. Leaders in Westport also pursued the Oregon and California traffics as they developed, again following the lead of Independence. They made some progress, especially with an argument that grass greened up two weeks earlier in their vicinity than it did inland from the ports farther north. They were frustrated, however, in the peak California season of 1849. An epidemic of cholera that year broke out nearly simultaneously in Independence, Westport, and the Town of Kansas. It did not last long, but the timing was terrible. The combination of the actual scare and rumors about it drew almost all the western travelers upstream to St. Joseph.[22]

When officials from the census bureau surveyed the elbow region in 1850, the roles of each major city seemed to be well established. The logistics for assembling groups of people and supplies had become an orderly process, and travelers similarly knew which routes were best for particular destinations. As the greater Southwest and Northwest developed further, all the merchants in the elbow region expected to prosper as well. Their position, as they saw it, was one of geographic inevitability that had been finely tuned by two decades of hard work and improvements. With the Missouri River as the West's only major conduit for water transportation, and a permanent Indian territory adjacent to their position, the outfitting and freighting businesses should continue indefinitely.

With the advantages of hindsight, it is easy to find flaws in the merchant logic of the time. The most interesting aspect of their thinking, however, is how their insights into the future varied place to place. In particular, serious discussion about railroads, the force destined to become the most disruptive to the status quo of 1850, was limited to a single area city. This was St. Joseph.

Reporters at the *Gazette,* then as now the primary newspaper voice of the St. Joseph community, had joined their counterparts in neighboring towns throughout the 1840s in praising the importance of the steamboat industry. They were not entirely uncritical of this transportation medium, however, and complained about freight rates and the difficulties of navigation on the swift-flowing Missouri River. "Our country is destined to suffer much" from these conditions, they reasoned.[23] As a remedy, the writers suggested two familiar actions—additional boats on the river to lower shipping costs through competition, and dredging efforts to improve the river channel—plus a more visionary third path. They began by voicing openly what everybody knew: no amount of improvement would make navigation on the Missouri safe and reliable. Ice would always close the channel in the winter, and low water would create suspect conditions late in the summer. Its fast current would always carry dangerous debris and make upriver trips expensive and slow. "During the last few years," they concluded, "the river has been navigable for only four months, the risk great and the freight high. . . . The farmer is tied up; he is forced to lessen his production. The merchant is forced to exact cash from the farmer."[24]

Gazette writers thought that only a complete change in the transportation system would solve their problems. They expressed this revolutionary challenge in a calm manner, however. On November 6, 1846, they published the following prophetic words:

> We suggest the propriety of a railroad from St. Joseph to some point on the Mississippi, either St. Louis, Hannibal, or Quincy. For ourselves, we like the idea of a railroad at one of the latter places suggested, for this course would place us nearer to the Eastern Cities and make our road thither a through one; we like this road too because it did [*sic*] so much relieve the intermediate country which is now staggering and must always suffer so much for transporting facilities in the absence of such an enterprise.[25]

Given that considerable local money was invested in warehouses along the river landing as well as in other enterprises directly related to steamboats, one might have expected opposition to the railroad idea. Similarly, heavy bond issues necessary to construct so many miles of track should logically have produced additional nay-saying. Some resistance of this type did occur, especially from livestock dealers, but never in an organized manner. In the absence of opposing voices, therefore, a small group of entrepreneurs completed plans for the Hannibal and St. Joseph Railroad on January 8, 1847. Voters approved the necessary local financial arrangements shortly thereafter, and the state and federal governments added generous grants. Tracklaying began in 1851 and, after several delays, was completed eight years later.[26]

That the attitude toward railroads by St. Joseph business leaders in the late 1840s was unusual for the area is borne out by surveys of newspapers elsewhere. The editor of the *Frontier-Journal* in Weston, for example, although mentioning the St. Joseph plan many times, championed nothing similar for his community or for anyplace else in surrounding Platte County. A similar pattern existed in Independence and in Westport, behaviors that would have long-term consequences for the growth and prosperity of these young cities.[27]

The Early 1850s and the Opening of Kansas

During the spring and summer of 1850, thirty-five or so steamboats made 226 stops at the Weston dock. Up the river, St. Joseph reported 160 landings. Despite sandbars, shifting channels, and the necessity of special "snag boats" to clear the waters, the tonnage of waterborne commerce in the elbow region had increased regularly over the years. Government officials at Fort Leavenworth, for example, spent $853 for such freightage in 1844, $2,095 in 1845, and $17,648 in 1847. These and similar indicator numbers would continue an ir-

The Steamboat *Mary McDonald* at Wyandotte, 1867. Although already antiquated in an increasingly railroad world, this sidewheeler continued to ply the Missouri River for six years after this photograph was taken. Fire destroyed it near Waverly, Missouri, on June 12, 1873. Steamboats brought west many of the iron rails used on the Union Pacific, Eastern Division. (Courtesy Kansas State Historical Society)

regular rise to a peak year in 1858 when a hundred different steamboats plied the river. The next year saw the completion of the Hannibal and St. Joseph Railroad, however, and a reconstruction of the transportation picture.[28]

Although steamboats were the most obvious part of the regional economy affected by the first trains, the changes were many and sometimes complex. The catalyst these tracks provided for expansionist thinking by businessmen and politicians in St. Louis was especially important from the perspective of city growth. Because even before completion the Hannibal line looked as though it might be successful, St. Louis entrepreneurs began to consider financing a railroad of their own. Their ultimate goal became a track all the way to the rapidly filling territories of Oregon and California.[29]

Any dream of a railroad route from St. Louis to the Pacific Coast immediately encountered a host of obstacles. Some were financial and some physical, but the immediate challenge was a barrier formed by the series of Indian reservations that had been established along Missouri's western border (Map 2). More than ten thousand Native Americans lived there by the middle 1840s. These peoples, many of whom were reasonably acculturated, had been promised that the western prairie would be a permanent homeland; no more relocations would occur. Yet their properties stood in the path of any direct route for a railroad.

Ten thousand people were no small issue, but the fortunes that might be made from a transcontinental rail linkage prompted many schemes to reacquire the Indian lands. The leader of this movement was Thomas Hart Benton, the same dedicated expansionist from St. Louis who had promoted the Santa Fe Trail while serving in the U.S. Senate from 1821 until 1851. By 1853 he had a strategy in place. The most obvious ploy was to grant a portion of the anticipated financial gains to a variety of decision makers in exchange for their cooperation. This group included government officials, of course, but also military men and some of the acculturated Indians themselves.[30]

The second ploy was unique—a misleading map of the reservation lands. Historian James Malin has argued that Benton worked with Abelard Guthrie, a mixed-blood Shawnee-Wyandot, to produce a map that showed only the lands where the emigrant Indians actually were, and not their complete legal holdings. The document thus gave the impression that vast acreages in eastern Kansas remained unoccupied and therefore potentially open to white settlement. Benton and his associates circulated this map widely in the summer of 1853. Simultaneously, they made the legal argument that, because no limitation on settlement had been mentioned in the law that had attached this Indian country to Missouri for judicial purposes, any unassigned lands there should now be opened to all comers.[31]

The Benton logic was legally clever and effective with the general public. When George Manypenny, the fair-minded commissioner of Indian affairs, objected, people overwhelmingly branded him an obstructionist. Anyone from a Northern state who defended Indian sovereignty also was tarred. The argument here was based on the famous Missouri Compromise legislation of 1836 that prohibited slavery in any new territory north of that state's southern boundary. With slavery thereby impossible in the Kansas River country, Benton charged that people who insisted on complete Indian dominion there were actually trying to prevent the formation of another free-soil state.[32]

As the Benton plan gathered momentum, Commissioner Manypenny and his few supporters were reduced to a compromise. They began negotiations to reduce the size of the major reservations in the hope that Indian and white settlers might be allowed to coexist in the imminent new territory and state. The model treaty, ratified on July 11, 1854, with the Delaware who lived adjacent to Fort Leavenworth, not only reclaimed the larger part of that tribe's extensive holdings but also guaranteed the right of any railroad company to build through their lands. When all the other tribal negotiations came to include a nearly identical clause, the end of Indian Kansas was all but accomplished.[33]

The date of the Delaware treaty is significant, for it came not in anticipation of the official opening of the new Kansas Territory to white settlement (May

30) but nearly a month and a half after the fact. Technically, the land rush that June was thereby illegal. Such a hurried treaty process was, at once, a fundamental affront to justice and a testament to the effectiveness of Benton's campaign. He and his close associates would now stand to profit, of course, but not just they alone. The coalition was large and diverse. Land seekers, obviously, formed a huge contingent, actual farmers and speculators alike. Those who would plat town sites, my primary concern here, were numerous as well. And congressional maneuvering even managed to get around the political and social antipathy the opening obviously would have for many Missourians and other Southerners. The legislation that created Kansas and Nebraska Territories specifically repealed the slavery restriction of the old Missouri Compromise. Popular sovereignty for each new state was now the law.

Town foundings in Kansas began almost instantaneously after the opening on May 30. Most of the earliest plattings were along the Missouri River, for the advantages of a location on its great elbow were as logical from a Kansas perspective as they had been to Missourians. It also came as no surprise that Missouri furnished virtually all of the initial site planners and merchants. These next-door neighbors, after all, were the ones who had pushed the enabling legislation and who had the best knowledge of local potentials.[34]

Residents in all the river towns of northwestern Missouri realized that a new territory to their west would affect local economies. They also agreed that this change would be large and overwhelmingly positive. Opinion varied from town to town, however, on which strategies were best for which places. To better understand the geography of the new cities laid out in Kansas, this range of Missouri thinking needs to be appreciated.

At St. Joseph, by far the largest of the elbow cities at the 1850 census and counting perhaps five thousand citizens in 1854, the prevailing attitude was a confidence that their exalted status would continue. A week after the territorial opening, for example, a *Gazette* writer stated simply that, with the community's merchants and mechanics busy and its hotels full, "St. Joseph is destined to be the great city on the borders of [Kansas]."[35] The significance of this statement lies in what is not said. Local residents were not clamoring across the state line to lay out new town sites. Instead, they felt certain that most of the new businesses and businessmen would come to them.

The expectations in St. Joseph were logical because people there had known nothing except steady growth throughout the city's short history. After their initial success in attracting California emigrants in 1849, they had housed and equipped even more travelers the following two years. Early in 1850, for example, residents had "large and extensive outfitting stores" and "commodious hotels" ready for the crowds. They were not disappointed. One estimate had ten thousand wayfarers in town at one time that April, including five thousand campers just across the river awaiting departure. Another person put the season's total at fifteen thousand emigrants, while everybody agreed that the numbers were easily four or five times higher than those reported from rival Independence.[36]

Emigrant numbers fell off at St. Joseph during 1852. Part of this could be attributed to the decline of gold fever in California and part to the emergence of Council Bluffs, Iowa, a new competitor port farther up the Missouri River. These falling numbers did not worry merchants, however, because they believed their new railroad, already under construction, would reconfirm the advantages of trade at this site. The vision seemed foolproof. The railroad would bring manufactured goods from eastern cities into St. Joseph less expensively and more reliably than the river ever could have done. The same would be true for livestock and grain from across rural Missouri and Illinois. People in St. Joseph, with a regional monopoly on this new means of transportation, would reap all the benefits. They simply would put the arriving goods into warehouses and the animals into packinghouses or trail herds and get rich from the businesses of shipping and wholesaling. Some of this trade would go to the Far West, but even a permanent decline in the California market would be no real concern. Newspapers in the summer of 1853 already were full of stories about the imminent opening of Kansas and Nebraska. This would produce an immigrant stream lasting for decades.[37]

St. Joseph businessmen speculated in only three early Kansas communities. One of these was begun too late and perhaps too halfheartedly to have a chance for significant growth. The other two hardly counted because they were little more than extensions of St. Joseph itself. Because the floodplain just across the Missouri River from the city already served as a popular assembly point for western emigrants, it seemed logical to establish a few stores there to provide groceries and other "last-chance" supplies without the inconvenience of a ferry ride back to the main town. In the fall of 1854 two such clusters of stores became the towns of Bellemont and Roseport. Bellemont, seven miles upstream from St. Joseph, enjoyed a good physical site at the edge of a bluff but was too close to develop its own trade area and thus short-lived (Map 1). Roseport, even closer to the city, was prone to flooding. People there changed the name to Elwood in 1857, and it survives today as a small suburb.[38]

The nearest site to St. Joseph for a good port town lay eighteen miles downriver. A rock shelf there was ideal for a steamboat landing, and the location was on a particular river bend that happened to extend westward farther than any other in the elbow region. St. Joseph people recognized this considerable potential but evaluated it only in terms of serving a local area. When they assembled a town company called Doniphan (after a Missouri hero in the Mexican War), they did so slowly and deliberately. They did not organize until November 1854 and undertook no actual development until the following spring. This pacing all but fulfilled the prophecy of limited potential. It put the town at a competitive disadvantage with a more active rival community called Atchison, six miles farther downstream on the same river bend.[39]

If the business community in St. Joseph was only lukewarm about founding towns in Kansas during 1854 and 1855, people in the Westport area were absolutely cold to the idea. Their minds were too busy with other concerns. Although they had largely surpassed Independence by this time as the beginning point for trade with Santa Fe and the rest of the Southwest, they still had much to do to accommodate the demands of this activity. About six hundred wagons had departed down the trail from the Town of Kansas in 1850, and more the next season. Upon their return the volume of furs, hides, buffalo bones and meat, gold, and silver exceeded the local capacity for storage, and so some of these goods had to be secured temporarily in homes and even on sidewalks. With revenue of a million dollars or more coming into a pair of communities that counted no more than a few hundred residents apiece, local business facilities simply had to be improved. Town leaders, including Westport founder John McCoy and merchants Hiram Northrup and Washington Chick, promoted cooperation between the two places to coordinate development. They also obtained a charter in 1853 that incorporated the landing community as the City of Kansas.[40] Simultaneously, these men also financed new construction and advertised widely for the personnel needed to handle their pyramiding volume of business. One newspaper solicitation in 1851 read as follows: "It is true there are many good Mechanics, Merchants, Hotel keepers, etc. here. But still there is not half enough to transact the present business; and as the trade is so rapidly increasing, five times the amount could be done, had we the capital, the Mechanics, the Merchants, etc. We are not selfish, there is room for many more, and we desire to see it speedily filled."[41]

With businesspeople in the City of Kansas and in St. Joseph absorbed almost totally with their own concerns, the opportunities for town creation in the new territory fell to others. The most active entrepreneurs by far resided in Weston. These men, in fact, were amazing in their zeal. They laid their plans early, picked good sites, and (surprisingly to some observers) even moved to Kansas themselves. Atchison and Leavenworth, the two most successful cities during this period, were both creations of Weston people.

Why would the attitude in Weston have been so much in favor of Kansas development? Business there in the early 1850s, although not quite as successful as that in St. Joseph, was still quite good. The town had 1,915 residents in 1850 and 2,097 in 1853, numbers far larger than in Westport or Independence. Efforts to attract migrants bound for California and Oregon had largely failed, but many Mormons elected to purchase trail supplies there. More significant, Weston merchants beginning in 1851 had largely cornered the lucrative market in western military freighting. Following a series of problems with hauling their own supplies during the Mexican War, army officials decided to contract this business to private firms. Benjamin Holladay, a local saloonkeeper, obtained the contract to supply Forts Kearny and Laramie from the depot at Fort Leavenworth. The firm of Perry and Young did the same for Fort Mackay on the Arkansas River and the Santa Fe Trail. These two developments added to an already existing trade with the Mormon settlements near the Great Salt Lake, and to the local supply of goods and services for Fort Leavenworth as well. All in all, Weston could be fairly described in early 1854 as bigger and better than ever before in its history.[42]

Still, despite all the signs of prosperity, the interest of Weston people in Kansas settlement did not surprise contemporary observers. Some residents undoubtedly had been following the construction progress of the Hannibal and St. Joseph Railroad and realized that its completion would lessen many of Weston's trade advantages. More obvious were the facts that the Weston economy had been joined lockstep with that of Fort Leavenworth from the beginning and that this tie had been based entirely on proximity. Now, with the opening of Kansas Territory, everybody knew that a city would emerge on the bluffs adjacent to the fort on the south. Just as clearly, this new place would become the army's center for supplies and recreation. If Weston people did not build it, somebody else most surely would.

The physical changes that occurred in Weston during 1854 and 1855 are the stuff of legend. As late as June 10 that first year, a newspaper correspondent had called the town "a flourishing place" of about three thousand inhabitants. Just five days later, however, the scene was strikingly different: "At Weston emigration is immense to all parts of the [Kansas] territory. At the ferry, in one half day to noon, upwards of 500 people crossed."[43] As these townspeople went about their innocent and not-so-innocent businesses of fashioning new communities, fate was about to deal their former residence two blows that nearly ended its existence. First, in November 1854, the Missouri River shifted its channel. The wharf at Weston now stood half a mile from water. Then, four months later, a fire destroyed more than twenty buildings in the town's business district, including nine dry-goods stores and a bank. Although some rebuilding did occur, Weston's role would never again be more that a small marketing town. More than half of its earlier residents were now Kansans.[44]

Elbow Cities in Kansas

Every new town laid out in nineteenth-century America had its boosters. Usually the local newspaperman headed this group and wrote exaggerated claims that sounded nearly identical to those made in neighboring communities. Leavenworth, the city that Weston people established next to its namesake

Cathedral of the Immaculate Conception in Leavenworth, 1867. Bishop Jean Baptiste Miege's 1855 choice for the see of the new Kansas diocese suggests that he believed Leavenworth would remain the metropolitan center of the territory. Construction at 711–715 North Fifth Street began in 1864 and finished in 1868. This magnificent church lost its cathedral status in 1947 and burned in 1961. (Courtesy Kansas State Historical Society)

fort, was not like this. Its first newspaper did not bother to promote the town as a future seat of county government or center of regional trade. These things already were assumed even before the first building lots had been offered for sale. Instead, editor Lucian Eastin wrote as follows: "That this beautiful location is destined to be the Capitol and Metropolis of the rich and fertile soil of Kansas, no one who knows anything of its geographical position or of the country around it can doubt."[45]

The self-confidence evident in Mr. Eastin's words was not simple bravado. People as far away as New York and as near as the City of Kansas agreed with him. Instead of just putting in survey stakes among the weeds and brush, the town trustees invested $2,400 to prepare their site for prospective buyers. Instead of deriding a potential rival, the editor of the *St. Joseph Gazette* almost offhandedly admitted that "Leavenworth it is thought by many will be the great city of Kansas." Facts supported the assertions. The first territorial governor, Andrew H. Reeder, attended the initial sale of lots there and decided to purchase property himself. Jean Baptiste Miege, the Catholic bishop over much of the American West, toured the town in 1854 and promptly selected it as his cathedral site. Four years later he invited the Sisters of Charity from Nashville, Tennessee, to join him in the work by setting up a teaching center. Leavenworth definitely was the place to be.[46]

Although the presence of the governor and the bishop reassured the twenty-eight Weston men and three officers from Fort Leavenworth who had created the Leavenworth town company, these people all knew that the key to their economic venture was military freighting. Between 1848 and 1853 the army had contracted this large and growing business out piecemeal, either by the trip or through seasonal arrangements to particular locations. This meant frequent bids and short notices for assembling wagons and men. To gain more efficiency, officials decided to consolidate things in 1854. One firm would be awarded a two-year contract to supply all the posts in the West. The winner of this prize, a new company created through a merger of three individual operations, was called Russell, Majors, and Waddell. When they, logically enough, elected to make Leavenworth their headquarters early in 1855, the city's success seemed guaranteed.[47]

It is difficult to overstate the size and importance of freighting to Leavenworth in the 1850s. Beginning with $60,000 in initial capital stock, Russell, Majors, and Waddell owned 7,500 oxen and 500 wagons by the end of their first year. They spent $15,000 in town that season to construct a headquarters and other necessary buildings, and they employed 1,777 men. By 1858, however, these numbers looked small. Now the inventory was worth $2 million (including 45,000 oxen), and the teamsters numbered 6,000. Their government contract was $4 million for 1858 alone, while another $3 million came in during 1859. The best contemporary description of this enormity comes from Horace Greeley, the famous New York newspaper editor who toured the West in 1859: "Such acres of wagons! such pyramids of extra axletrees! such herds of oxen! such regiments of drivers and other employees! No one who does not seen [sic] can realize how vast a business this is, nor how immense are its outlays as well as its income."[48]

Leavenworth instantly became the largest city in Kansas and increased its lead over potential rivals with every passing month. The community had five hundred residents in the spring of 1855 but an estimated two thousand by the following autumn. A slightly more modest pace of growth characterized most of 1856, but things boomed again in 1857 and throughout much of 1858. The town's eight thousand people at that point made it the largest settlement between St. Louis and San Francisco.[49]

Any explanation for the extraordinary growth of Leavenworth during the 1850s begins with military contracts but involves several other developments as well. One of these is politics, or rather a careful avoidance of controversial issues in order to further economic development. Although most of the town's founders were Missourians from one of that state's larger slave-owning counties, they had a long history of partnerships with army officers, most of whom were from Northern states. By keeping the free-soil/slavery debate to a minimum locally, these Leavenworth pioneers kept good working relationships with all potential investors. In the words of one early historian, they "were of an entirely different stripe" from the people who established rival cities.[50]

Besides its tranquil political scene, Leavenworth also offered potential investors the best transportation system in the territory. This network, closely related to the presence of the fort, consisted of a short connector route southwest to join the main Santa Fe Trail plus three military roads (Map 1). Of these, the one in heaviest use ran northwest to Fort Kearny, for this was the best pathway to Utah, California, and Oregon. A second road, constructed in 1838–1839, extended due south along the Missouri border to Fort Gibson on the Arkansas River. This route held only modest attraction for local entrepreneurs, however, because they believed settler expansion and the city's potential for trade focused to the west. So thinking, townspeople were elated about the newest military road, built in 1853. This one ran right where they most wanted it—westward along the valley of the Kansas River.

The destination of the third road was Fort Riley, a post created that same year near the spot where the Republican and Smoky Hill Rivers join to form the Kansas. Army officials hoped that this site could be reached at least seasonally by steamboat. If so, the new fort would be able to take over some of the depot functions from Fort Leavenworth and thereby cut the cost for supplying frontier outposts.[51] From the army's original perspective, a road up the Kansas Valley was only to supplement the river. Private businessmen saw it in different terms. For them, this rough trail provided a means to open up Leavenworth's natural hinterland. Its immediate potential would be in wagon traffic to the developing Kansas countryside, both freight and emigrant. This is what editor Eastin of the *Herald* was referring to with his claim early in 1855 that "Leavenworth will be a considerable wholesaling place for merchants and businessmen in almost every branch of businesses."[52] Entrepreneurs also visualized an extension of the Fort Riley road farther west to the Rocky Mountains and perhaps beyond. Although more arid than the established pathway to the west along the Platte River, such a Kansas road would be the most direct way to the Pike's Peak region and to the entire central West. It also would focus this long-distance trade more precisely on Leavenworth and the elbow region. A related dream involved the entire spiderweb of trails out from the town. The business leaders, realizing that railroads would be coming to the area within a few years, logically argued that the most likely paths for these tracks would be alongside the older trails and to the same destinations.

Transportation issues beyond those of military freighting, although much discussed, were not acted on during the first four years of Leavenworth's life. It was not that these things were seen as unimportant. Rather, they simply were overshadowed by something even more alluring. Between 1856 and 1858 the city found itself the focus of a spree in land speculation. It was a true boom, not the small-scale maneuvering common on every frontier, and one made extralarge because it had been anticipated for several years. The cycle of events began in 1854 and 1855 when most of the highly desirable Kansas acreage remained in Indian hands and thus unavailable for sale (Map 3; see p. 15). Interest focused on the Delaware trust lands, tribal property being held temporarily by the federal government, with the sale to benefit the peoples who recently had occupied it. When word came that 209,148 acres of this trust would be auctioned off in November 1856, land that included Leavenworth itself as well as prime territory westward for some forty miles, big money began to flow into the city from across the nation. With as much as $5 million available for investment, sales and resales continued at a near-breathless pace for a year and a half.[53]

The most desirable business lots in Leavenworth sold for $1,500 to $2,000 at the outset of the bidding frenzy. Four months later people were paying $10,000 for the same properties. Residential holdings and rural acreages rose at comparable rates. So did wages. Studies have shown that most of the capital involved came from eastern cities. Bankers there lent money at 10 percent interest to a handful of local investors, particularly to the freighter king, William H. Russell (of Russell, Majors, and Waddell), and to the brothers Hugh and Thomas Ewing. These men bought heavily themselves (Russell owned 5,440 acres, the Ewings 5,400) and also lent money to others at rates as high as 50 percent. It was a giddy time, and naturally the population of Leavenworth soared. The end came suddenly in the summer of 1858, a result partly of a national financial panic the previous year and partly of a decision by President James Buchanan in 1858 to sell some forty-six million acres of public lands nationwide in order to balance the national budget. The glut of land on the market depressed prices for nearly a decade, and many property holders in the Leavenworth vicinity could not pay their mortgages.[54]

Leavenworth in late 1858, although a large city by frontier standards and the envy of other Kansas communities, was far from a healthy place economically. Its people had been so involved in land speculation that they had neglected more mundane matters of business. Now they were capital poor as well. William T. Sherman, the future Union hero during the Civil War who had lived in town until 1859 and had married into the prominent Ewing family, saw the situation clearly. Distress was likely, he wrote to his wife on September 25, 1858, because "thus far Kansas has been settled by lawyers and politicians instead of farmers and mechanics."[55] The only solid things, it seemed, were the military payroll and the gigantic firm of Russell, Majors, and Waddell.

Beyond the adventure in land speculation, two other events in the late 1850s affected Leavenworth in important but complex ways. Both occurred far from the city, first the so-called Mormon War of 1857, and then a gold rush to the farthest reaches of Kansas Territory near Pike's Peak in the Rockies. The Mormon issue began in May when President Buchanan perceived a threat to national sovereignty from Brigham Young and his growing numbers of followers in Utah Territory. Buchanan decided to replace Young as governor and, in order to help the new appointee sustain authority in a possibly hostile environment, also sent along twenty-five hundred men as a military accompaniment. This sudden news frightened not only the Mormons but also Russell, Majors, and Waddell. The freighters were told to ship two and a half million pounds of supplies to this new Army of Utah, but they had no wagons available. Their entire inventory was already out on regular seasonal routes. With assurances from the local quartermaster that Congress would pay any extra expenses, the company borrowed $300,000 to obtain 350 additional wagons and teamsters. This debt was doubled later that same year when angry Mormons burned 26 of the wagons and captured 1,760 oxen. The impact of these events, although not felt immediately, was profound. When government officials first delayed their payments to the firm and then refused to reimburse at all for the losses, Russell, Majors, and Waddell were plunged into a debt cycle from which they could not recover. Leavenworth's biggest employer declared bankruptcy in 1860.[56]

The gold rush to Pike's Peak was another mixed blessing for Leavenworth. News of the first strike circulated in September 1858 and initially provided an emotional lift to a community that had just seen its speculative bubble burst. Some people left for the mountains immediately, and a huge emigration was predicted for the following spring. To prepare, Leavenworth merchants restocked warehouses that had been neglected during the boom. Four local men (in cooperation with some New York capitalists) also decided the time was right to open a direct overland route to these mines. They created the Leavenworth and Pike's Peak Express Company, intending to carry passengers and freight on a daily schedule along the military road to Fort Riley and then westward across the unsettled plains. It was a risky venture when it opened in April 1859, and it could not survive the collapse of the mining fever the next year.[57]

In retrospect, the stage company was not as complete a failure as it seemed at the time. It served an important psychological function for the city, giving local entrepreneurs confidence that they could dominate a large area commercially much as they had been doing for the military. Denver City, the largest new community near the mines, had obtained a good portion of its groceries and similar supplies from Leavenworth wholesalers, for example, and local bankers deposited at least $59,000 of Pike's Peak gold in their vaults. Most important, the city now had an east-west transportation link across Kansas Territory that could compete with the older Santa Fe and Fort Kearny roads. In 1860, although the local population had dropped from its peak, Leavenworth's 7,429 residents were still very much positioned for great accomplishments. As one of their leaders, Daniel Anthony, had expressed things to his brother a year earlier: "My business thus far has proved successful beyond my most sanguine expectations.... Our town can't help, what seems to be its destiny, becoming the Metropolis of Kansas and the West."[58]

With the tremendous early success at Leavenworth, it is easy to forget that this was not the only town site on the Kansas side of the Missouri River to be touted for greatness during the 1850s. In fact, Leavenworth was merely one of three serious ventures plotted simultaneously by Weston businessmen during the early months of 1854. Each proposal had considerable merit. As I have already discussed, the acreage that Leavenworth was to occupy was not legally available in 1854. Would it thus not make more sense to establish a town still near the fort but at a place where clear title could be obtained? Other people noted that, although the Leavenworth site was ideal for trade in a political sense (i.e., adjacent to military decision makers), another location on the river might offer physical advantages that could allow freighters to cut their costs. National politics also were involved. The neutral stance of Leavenworth boosters toward issues of state's rights was logical in many ways but anathema to many Missourians. Perhaps a community avowedly Southern in tone would have a better chance to attract immigrants.

Town companies for the two Weston rivals to Leavenworth formed in July 1854. One of these created Kickapoo City. Their site, four miles upstream from Weston and two miles northwest of the military reservation, had been occupied previously by a Catholic mission to the Kickapoo Indians. It lay just north of the former boundary between the Kickapoo and Delaware reservations (Maps 2, 3). This old border was important to the town company. The Kickapoos had ceded this part of their land to the government earlier than had the Delawares, and officials opened it for general settlement on July 22, 1854. Land titles thus were secure in Kickapoo City at a time when Leavenworth pioneers had to rely on an unofficial claims association that might or might not hold up in a future court.[59]

Kickapoo City was admittedly "a premeditated rival of Leavenworth." Its initial sale of lots in November went well, with some properties bringing as much as $110. A newspaper, the *Kansas Pioneer,* began that same month and took a stance that was, at once, proslavery and anti-Leavenworth. The formula worked, too, at least for a while. Officials placed a coveted federal land office there, and by the end of 1855 perhaps a thousand people called the town home. The decision by Russell, Majors, and Waddell to locate in Leavenworth, however, soon made that slightly older community too large for anyone to doubt its continued existence. Kickapoo City faded rapidly after 1856.[60]

Although a good case could have been made in 1854 for Kickapoo City becoming a successful venture, Leavenworth's biggest rival in early Kansas turned out to be Atchison. This circumstance surprised casual observers of the time because Atchison was nowhere near the fort. In fact, with a location about halfway between Leavenworth and St. Joseph, one might have supposed that aspirations there were no more, perhaps, than to capture farmer trade in the area between the two larger communities. One look at the town's pedigree should have revealed the naïveté of such thinking. Missouri's David R. Atchison, a Platte County resident who had served in the United States Senate since 1843 and was its longtime speaker pro tempore, would never have lent his name and influence to a town company that sought only local trade. In truth, Mr. Atchison had his sights set on the entire West and saw in this new town a combination of natural advantages that could make it a large and prosperous gateway city. Transportation would be the key to capturing such a far-flung empire, of course, and transportation, in the words of a local historian, became the new community's "reason for existence."[61]

David Atchison's enthusiasm for the town site was based on two elements of physical geography. Of these, the most obvious was position on the river. The local site itself was beautiful, with White Clay Creek providing good access to a wharf and adjacent high bluffs available for scenic residential lots. Townspeople noted these advantages, of course, but stressed that the town also occupied the westernmost point within the entire elbow region. Travelers bound for the prairies or beyond could thus presumably save time by docking there. Local promoters clearly hoped so, because this fact served as the lead sentence for nearly every article or advertisement written about the town. Sutherland's business directory for 1860–1861 was typical: "This, the most important city in Kansas, is situated on the Great Western Bend of the Missouri River . . . exactly in the centre of the river border, and farther west into the Territory, by fifteen miles, than any other important point on that border."[62]

A savings of fifteen miles of overland travel was important to emigrants, but if this were all that Atchison could offer, it is doubtful that many freighters would have selected the town as a headquarters. The political connections available to such businessmen at Fort Leavenworth were even more valuable. In this game of city rivalry, however, Atchison promoters also held a second transportation card. Because of the location of local stream divides, the busy military road that ran from Fort Leavenworth to the Platte River and then on to the West Coast passed within five miles of Atchison (Map 1). This fact was not written about so frequently or so passionately as the river bend, but it was of considerably more practical value. Wagon masters who were three or four days on the road out of Leavenworth found themselves camped nearly at Atchison's doorstep. It took no genius to realize that an upriver relocation of their base of operations would mean major financial savings.

The founders of Atchison knew about the freighting potential of their town long before they ever hired a surveyor. French trappers and American mountain men had used the site as a western gateway in earlier years, while more recently, Mormon emigrants had gathered there every spring before setting off to Utah. Just west of town, a campground used by these pilgrims had already become known as Mormon Grove. David Atchison had great confidence, then, when he initiated the first sale of town lots with a rousing speech on September 21, 1854. The properties went fast, selling for prices nearly double those in Kickapoo City. The town company raised $400 from its members for a newspaper, $500 to create a proper wharf, and $2,500 for a log structure called the National Hotel. Significantly, they also pitched in to purchase the section of land at Mormon Grove so as to encourage and retain that important trade.[63]

Late 1854 and all of 1855 was a charmed time for Atchison. Although its promoters voiced strong proslavery views and refused to do business with certain immigrants from New England as a matter of principle, trade developed fast. About four thousand Mormons passed through town in 1855. More important, the two biggest private freight companies in the Utah trade—Livingston, Kinkead and Company; and Hooper and Williams—selected Atchison as their headquarters. So did Tutt and Dougherty, which specialized in cartage to Fort Laramie.[64]

Historians regard 1856 and 1857 as years of lost opportunity for the young settlement. Perhaps placing too much faith in natural advantages, townspeople spent much of this time in political activism throughout the surrounding counties. During these years, in fact, Atchison became known as "the most violent pro-slavery town in the territory."[65] Such a stance might have proved beneficial had Kansas culture evolved in a Southern mold. By 1857, however, it had become obvious that free-soil immigrants would dominate the future state. All at once, Atchison residents faced a political crisis that threatened to metamorphose quickly into an economic one.

When confronted with a choice between political ideals and economic well-being, the citizens of early Atchison reacted the same as most people: they gave priority to their wallets and purses. Realizing that fellow Kansans might not believe statements of contrition, members of the original town company went so far as to sell controlling interest in the community to a pair of outsiders who held solid free-state credentials. One of these, attorney Robert McBratney, was an agent for a Cincinnati emigration society. The other, Samuel C. Pomeroy, held a similar position with the influential New England Emigrant

Aid Company that had founded Lawrence and other area communities. This striking move brought about the desired infusion of money and energy. Pomeroy was elected mayor early in 1858. Nearly simultaneously, the local newspaper changed its masthead from *Squatter Sovereign* to *Freedom's Champion,* and the citizenry passed a $100,000 bond issue that would extend the tracks of the Hannibal and St. Joseph Railroad twenty miles farther to a point just across the Missouri River from Atchison.[66]

McBratney and Pomeroy made their investment at a perfect time. Trade with Utah boomed almost immediately after President Buchanan's decision to send troops there in the summer of 1857. This was followed a year later by the even bigger news of gold at Pike's Peak. Atchison people, despite having declined in numbers to only about four hundred by this time, were now united. They advertised their considerable assets, improved the connector road between the wharf and the main trail, and soon saw the wagons and emigrants begin to roll in and out in record numbers.

Local people realized that official military freight probably would continue to emanate from Leavenworth, and so they concentrated on attracting army sutlers and other private contractors. Newspaper editor John A. Martin forecast success in this gambit as early as February 1858, claiming that "the larger portion of the Salt Lake and California trade" and even "the chief portion of the supplies for the Utah army" would pass through Atchison. When these claims proved true, merchants could argue convincingly that their town's advantages truly were superior to those of other Kansas communities. For the record, they wrote that the "shortest, safest, and best road in all respects" begins at Atchison, and that only "the fool-hardy and insane" would choose to travel across Leavenworth's arid Smoky Hill route. Numbers supported the rhetoric, too. Some 775 wagons carrying 3,730,905 pounds of freight went west from the Atchison wharf in 1858, and these numbers increased substantially in 1859, to 954 wagons and 4,020,000 pounds.[67]

Under Pomeroy's leadership Atchison was a progressive place. Instead of relying only on their reputation for freighting, townspeople worked at diversification. They courted the Benedictine monks, for example, who had come to Kansas in 1855 with plans to establish a college at Doniphan. By 1859 the fathers had relocated to Atchison, and the town had a new St. Benedict's College to crown its northern bluff. Potentially more important than the college, town leaders also clearly saw the transforming potential of railroads. This was not so much vision as simple practicality, because the benefits nearby St. Joseph had experienced since becoming the terminus of the first line across Missouri in 1859 were obvious. Gold seekers from Ohio and Illinois now traveled by rail to the elbow region. So did many Mormon families, grocery and hardware goods, and the letters of the United States postal service. Soon after the first engine pulled into St. Joseph, Atchison men began attempts to capture these businesses. Their first coup was the mail contract for Utah, secured through the expedient of shipping the bags a few miles downriver from the railroad station and then loading them onto a Kansas coach. It was obvious that a rail extension southwest to the Missouri River at a spot already being called East Atchison would bring additional commercial success.[68]

The Atchison and St. Joseph Railroad opened in February 1860, just in time for the spring freighting season. Its impact was profound. Atchison was now the nation's westernmost point to have a rail connection with eastern cities. In part because the contract to carry the mail stipulated a route via the Platte River, the owners of the Leavenworth stage line abandoned their Smoky Hill road. Renaming themselves the Central Overland, California and Pike's Peak Express Company, they gradually made Atchison their terminus instead of Leavenworth. The railroad gave the Central Overland staying power, and it survived until after the Civil War. The biggest impact of the new iron tracks, however, was on freighting. At the grand opening of the Atchison and St. Joseph, one editor crowed that "Atchison must inevitably be the mart—the great entre-pot for the freight from the East and the exhaustless commerce of Kansas, New Mexico and Utah for all time." His puffery turned out to be true, at least for 1860. The volume that year was nearly twice that of any previous season, with 1,773 wagons carrying some eight million pounds of goods. Two-thirds of these supplies went to the gold region, while the thirty thousand or so residents of Utah formed another important market. In addition, a third of this volume was official army cartage. The prospect of cheaper goods via the railroad had persuaded military officials to ship from multiple points, no longer from Fort Leavenworth alone.[69]

Atchison's population at the 1860 census was 2,616, which was several thousand less than Leavenworth's total but considerably more than any other settlement in Kansas Territory. This number was even more impressive when compared with the city's status only two years earlier. Instead of a mere four hundred people and an economic situation that bordered on the desperate, four big commission houses now flanked White Clay Creek: A. S. Parker and Company; Home and Chouteau; D. W. Adams; and Irwin, Jackman and Company. Each of these employed more than five hundred men in season, and together they provided a definite cosmopolitan air to the young community. Atchison already had surpassed Leavenworth as a freighting center. With plans under way for a new railway west from town, prospects for the upcoming decade looked even brighter.[70]

Because of a scarcity of data, it is impossible to detail the economic structure of Atchison and Leavenworth in 1860. From all indirect measures, however, both communities, as well as their rivals across the river, were classic examples of the entrepôts that appear in the Meyer-Wyckoff model of frontier development introduced in chapter 1. Talk about freighting and the supply of emigrants dominates nearly all contemporary accounts about these places. Atchison and Kansas City, with their Utah-Colorado and Santa Fe trades, respectively, were about as pure in this regard as one could imagine, whereas the larger Leavenworth and St. Joseph had begun to add modest amounts of manufacturing and other activities to the older mainstay (Map 4; see p. 16).

Manufacturing employment, one of the few county-level business statistics collected in 1860, provides a means of comparison. After removing the typically rural-based sawmills from this category, the numbers are quite modest, with the differences reflecting age of community more than anything else. Jackson County (Kansas City, Independence, and Westport) recorded 488 manufacturing employees within a total urban population of 8,777. The corresponding numbers for Buchanan County (St. Joseph) were 317 in a population of 8,932, while Leavenworth posted 112 in a population of 7,429 and Atchison 6 in a population of 2,616. A flour mill was Atchison's lone reported industry, whereas the other places added a few tens of people each in the making of boots and shoes, bricks, metalware, saddles and harness, and wagons.[71]

If one narrows the focus to the two Kansas communities and examines goods and services available, the numbers generally are consistent with population. Thus the ratio between Leavenworth and Atchison for bakers in 1860 was 9 to 4, for bankers 5 to 3, for grocers 54 to 16, for hotels 9 to 6, and for physicians 27 to 14. Exceptions existed, however, all supporting the position that Atchison was the purer entrepôt community. Eight wholesale grocers lined Atchison's Commercial Street, compared with six in Leavenworth. The smaller community also possessed five wagon factories, compared with Leavenworth's three (these five apparently were missed in the census), and four freight companies to Leavenworth's one. As for the four big commission houses listed earlier, the larger city could counter with only three. So far as the elbow cities were concerned, Kansas in 1860 was developing in textbook fashion.[72]

3

Lawrence, Topeka, and the Placement of Public Institutions, 1854–1866

The average American businessman in 1854, whether from western Missouri, South Carolina, or New York, did not see the Kansas-Nebraska Act of that year as a divisive political statement. Instead, it was a practical response to an opportunity for development. Many places in the Far West were filling rapidly with new citizens, and Kansas was a conduit to them. Everybody realized that this territory soon would play an important role as granary, transfer zone, and supply point. Occasional mentions of slavery as an issue all were brushed aside. A plantation economy, these pragmatists argued, could never be important in a place too cold for cotton. Moreover, from a political perspective, the act had neatly resolved this issue by pairing Kansas with Nebraska. Although settlers in each place would determine the question for themselves, the plan obviously was for slavery to become legal in Kansas and illegal in Nebraska. In this way the balance of power would remain equal between North and South in the United States Senate, and people could get on with the things that mattered: land acquisition, city platting, freight contracts, and railroad charters.

If the potential for profit in Kansas Territory had not been as great as it was, it is likely that the political history of the region would have followed the script intended by the federal legislators of 1854. Instead, jealousy arose within the nation's business community. Why should Missourians reap all the benefits, thought many investors from the East. Was there a way for others also to compete for the economic prizes? Most of this envy was short-lived. Disaffected entrepreneurs soon found investment opportunities in other regions or decided to place their monies into Kansas through intermediary operators in St. Joseph, Leavenworth, or Atchison. A small minority, however, particularly those in New England, selected an alternate strategy. They took the issue of settler sovereignty in Kansas as a challenge rather than a foregone conclusion and decided to sponsor colonies of free-soil settlers.

At least six emigration companies were established in Pennsylvania, New York, and New England in 1854 and 1855. None of these hid their business agendas, but they did add to them topical political and social issues whenever possible so as to maximize investor dollars and settler numbers. Spokesmen rallied their crowds, for example, by asking how Congress could now allow slaves into territory that had long been promised to free laborers alone. To do so, they said, was breaking the "sacred pledge" made by the Missouri Compromise. Why were Southerners and not the spiritual sons of John Adams dictating such an important national policy? Talking in this manner, they played the card of regional pride and hoped to attract a certain number of patriots and missionaries to their ventures. Through it all, however, the core appeal remained solidly economic. Expectations of business opportunity and promises for reduced transportation fares, experienced guides, and free community sawmills attracted the colonists; the allure of solid returns on their capital investments drew in the Brahmin financiers.[1]

This chapter explores the major towns created by these colonizing companies. Because the prime sites along the Missouri River already were appropriated before these newer settlers could arrive, they faced challenging locational decisions if they were to be competitive economically. A second part of this chapter moves forward in time to examine the allocation of public institutions among the existing cities, especially in the early years of statehood. Such an allocation, obviously important to the potentials for future growth, is inherently political. In Kansas, it was doubly so. Much of the debate, whether on the surface or just below, pitted the slightly older river towns founded by Missourians against the communities created by or with the support of the Northern colonizers. Given the population size advantages of Atchison and Leavenworth, the outcome might seem obvious. But with the free-soil cause triumphant in the new state and bloody raids across the Missouri border an important part of everyday life, the politics of selecting sites for colleges and prisons were anything but simple.

The New England Emigrant Aid Company and the Foundings of Lawrence and Topeka

If David Atchison may be said to symbolize the aggressive colonization of Kansas by Southerners, Eli Thayer is his New England equivalent. Like Atchison, Thayer was a complicated man with mixed motivations for his interest in Kansas. He may have wanted to check the expansion of slavery but definitely saw involvement as a way to increase both his political profile and his bank account. A graduate of Brown University, he was living in Worcester, Massachusetts, in 1854, where he ran a school for women. He also had just been elected to the state house of representatives. A chance for expanded glory came as he listened to Senator William H. Seward of New York and others tout the settlement of Kansas as a noble challenge for good Yankees. This cause, he thought, required only a leader to become a reality, and so he volunteered himself.

Thayer chartered the Massachusetts Emigrant Aid Company in April 1854. Unlike similar companies that formed elsewhere at about the same time, this one proved successful because its founder worked hard to assemble an impressive cadre of financial and political backers. Boston businessmen John M. S. Williams and Amos A. Lawrence provided much of the considerable capital required, while the Reverend Edward Everett Hale stimulated support within New England's extensive network of Congregational and Unitarian churches and wrote an important promotional book for the group. Horace Greeley helped as well by endorsing the plan in his influential newspaper, the *New York Tribune,* as early as May 29.[2]

The name of Thayer's organization changed twice within the first year, ending up as the New England Emigrant Aid Company, but all else went smoothly. The influential John Carter Brown, Samuel Cabot Jr., and John Lowell joined the board of directors, and collectively they broadcast a statement of purpose. This was not to be an abolitionist organization in any way. It had no intentions to fight Southerners or to recruit settlers with cash payments. Instead, the idea was to make people aware of the opportunities in Kansas, to simplify the process of migration, and to provide tools for economic viability. Profits would come not from land speculation directly but from the sale of business lots and other properties donated by the individual town companies.[3]

While company officers in Massachusetts raised money and negotiated favorable fares for the railroad trip to St. Louis and the steamboat one to Kansas City, two agents spent the month of June 1854 in Kansas. Their job, of course, was to assess possible town sites in the territory. This important task turned out to be fairly simple. A recently abandoned army post, Fort Scott, located a hundred miles south of Leavenworth on the military road to Fort Gibson, was one possibility. Land titles were uncertain there, however, and the local streams unnavigable. Sites along the Missouri River or on the busy road to Fort Kearny offered secure titles and better transportation, but any town there obviously would be subservient to Atchison and Leavenworth. A better choice, agents Charles Branscomb and Charles Robinson reasoned, would be positions along the territory's second navigable river. Since steamboats were scheduled to ascend the Kansas River regularly in service to the new Fort Riley, a series of communities along this stream might well have bright economic futures. The river valley also held obvious promise as a railroad corridor to the west, and its upper reaches approximated what logically might become the population center of the future state.

The locational question for the new colonies now became simplified. Where were the best sites along the river? This process, too, proved surprisingly easy. Branscomb and Robinson saw that north-bank placements would be most desirable because the military road to Fort Riley was on this side. So, too, would be any future railroad that extended westward from Leavenworth. The Wyandots, Delawares, and Potawatomis controlled the frontage on this shore, however, all the way from the Missouri border upstream some ninety miles to the vicinity of present-day Belvue. Indians also owned parts of the south bank, but between the western boundary of the Shawnee Reservation and the eastern limit of the Potawatomi holdings, about twenty-five miles of river access lay open (Map 3; see p. 14). The two agents honed in on this region and identified a position near its eastern end as most favorable for development. It was, simply, "the first desirable site on the Kansas River to which the Indians had ceded their rights."[4]

With their destination known, an initial party of twenty-nine colonists left Boston on July 17. They arrived at their new homesite fifteen days later, having walked the last miles from Kansas City. Actual town organization began in September when a second settler group of more than eighty people arrived along with a hundred or so other individuals who decided that they, too, liked the prospects of the area.[5]

Lawrence, as the new settlement was named in early October to honor the company's most generous benefactor, had two transportation assets: the Kansas River, of course, and a position on one of the two main roads to California and Oregon (Map 1; see p. 13). Early accounts of the community barely mention either trafficway, however. Instead of the endless discussions about trade that dominated life in early Leavenworth and Atchison, these new people focused almost immediately on issues of politics and culture. Because of this, other Kansans judged them to be arrogant and eccentric. One outsider, who was somewhat sympathetic to the town, characterized the mind-set as follows: "They are advised, from headquarters, to avoid the use of all Western vulgarisms, and cherish their New England habits and customs. . . . Unless a man agrees with them in all their peculiar notions about building up a model State, he is charged as a 'Missourian'—as this is the worst epithet, in their opinion, they can apply to anyone they dislike."[6]

By February 1855 four additional parties of emigrants had arrived in Lawrence, and the first territorial census reported the local population to be 400. Of these, 149 were born in either Massachusetts or New York. A steam sawmill, purchased by the company in Kansas City, was hard at work preparing lumber for housing. Two months later, people began construction on the Free-State Hotel, a large, three-story, stone structure that would cost the company about $20,000. This hotel immediately became a symbolic structure. Admirers and detractors alike saw it as a harbinger of even greater Yankee immigration, but assessments of this prospect varied. If twenty thousand more New Englanders were indeed coming to Kansas (as rumor had it), Missouri control of the territory would be threatened. Southerners wondered if the hotel perhaps was secretly intended as a fortress.[7]

The Free-State Hotel, completed in the spring of 1856, was an obvious demonstration that Lawrence would be the center for future New England Emigrant Aid operations in Kansas. An earlier document, however, had provided a more precise blueprint for local development. When A. D. Searl was hired to make the initial town survey in October 1854, he was instructed to include two especially reserved tracts among the usual mix of streets and parks. Atop an escarpment known then as Oread Mount, his map clearly shows one section labeled "College Grounds," and another "Capitol Hill."[8]

The aspirations of Lawrence residents to claim the future capital of Kansas were too pretentious to voice regularly during the tumultuous first two years of territorial life. Those for a college, however, were known to all. Amos Lawrence had talked about endowing such an institution for his namesake town as early as 1855 and had assigned agent Charles Robinson to the project. Robinson began to raise funds the next year and petitioned Congress for 250,000 acres of public lands. When this grant failed to materialize, Mr. Lawrence donated generously himself: $1,000 in cash plus a hundred shares of stock in the Emigrant Aid Company and a pair of interest-bearing promissory notes for $5,000 apiece. Political turmoil in the region stymied further progress on the idea until 1859. At that time leaders of the Presbyterian Church in the United States of America promised to give $2,000 toward the erection of a building to be known as Lawrence University. Workers laid a foundation on the Oread hill based on this promise, but the money never materialized, and so construction ceased.[9]

By 1860, the citizens of Lawrence had seen their numbers increase steadily so that the city had become the fourth-largest community in Kansas. With 1,645 residents, they approached Atchison in size, if not in perceived potential for expansion. Even should the plans for a college be realized, nobody expected such an institution to employ many people. Also, the immigration from New England had virtually ceased once the free-soil cause had become ascendant in 1857, and the grand total of company-sponsored people in the territory never exceeded three thousand. On the positive side, Lawrence businesses did profit from the comings and goings of four steamboats that began to ply the Kansas River in 1857 once the political tensions across the region had lessened somewhat.[10]

If the same measures of commerce and industry examined for Atchison and Leavenworth are applied to Lawrence in 1860, the results are ones that would not have been encouraging to the city's financial backers in Boston. Beyond sawmills, the city had only one manufacturer of note, the Kimball Foundry, run by the brothers Edward, Franklin, Frederic, and Samuel. It also claimed only single entries under the important headings of forwarding merchants and commercial agents (A. K. Allen and Company), and of wholesalers (R. W. Ludington liquors and tobacco). The brightest statistic, perhaps, was the presence of four bankers, one more than in Atchison and only one fewer than in mighty Leavenworth. Boosters hoped that money resting in these vaults would soon be used for other city-building projects.[11]

In making their plans for growth, Lawrence residents had competition not only from the Missouri River towns but also from other New England settlements. This additional activity had begun late in 1854. With their initial efforts in recruitment and fund-raising going well, officials at the Emigrant Aid Company suggested the possibility for a second colony to their agents in Lawrence and Kansas City. This idea fell on eager ears because several of the men already had undertaken informal scouting trips. On one such venture, in October, Charles Robinson and a young Pennsylvania man named Cyrus K. Holliday had traveled twenty-five miles west from Lawrence along the Kansas River to the eastern boundary of the Potawatomi Reservation (Map 3). They had theorized beforehand that this general location would be ideal for a new city because it would combine good river transportation with a distance sufficiently far from Lawrence to allow both communities to develop adequate hinterlands. When the men located a low bluff near the mouth of Shunganunga Creek that would make a good wharf, they judged the site "especially desirable."[12]

Robinson and Holliday conveyed their enthusiasm to members of the most recent colonizing party from New England when they arrived at Lawrence in early December. These people, too, were impressed. Representatives quickly visited the site, and within a week, seven of them had joined with Holliday and Robinson to form an official town company. In exchange for a sixth of the town lots, the Emigrant Aid Company agreed to provide a sawmill for the new community and to construct both a school and a building that could temporarily house new residents. The naming of the town apparently generated debate, but the group eventually agreed on Topeka, a word of uncertain origin that appealed because it was euphonious, easy to spell, and unique.[13]

The Topeka community attracted about forty-five settlers during its first winter and considerably more the following spring when a steam engine arrived for its sawmill. The group garnered a post office in March and a newspaper, the *Kansas Freeman,* in July. Just as the case in Lawrence, however, early accounts of local life contain little discussion about trade and industry. Topekans, too, aspired more toward political and cultural glories. Like all freestate people, they had been energized during the summer of 1855 when the first territorial legislature had ousted their representatives. To organize an opposition, they met that September with other Northern sympathizers in the small settlement of Big Springs, midway between Topeka and Lawrence. There they agreed to assemble again the following month to elect representatives and to write a free-state constitution. The site for this assembly, selected apparently with little dissension, was tiny Topeka.[14]

The reasons that Topeka came to host the free-state convention of October 1855 are unclear, but they almost certainly had to do with the community's centrality within the New England colonies as they existed in Kansas at that time. In addition to Lawrence on the east, two small settlements had formed during the summer of 1855 directly south of Topeka: one at Council City (now Burlingame) and the other at Hampden (near present-day Burlington) in the Neosho Valley. A third, Manhattan, lay fifty miles to the west. Another factor in Topeka's favor may have been the greater proximity of Lawrence (the other logical choice for a meeting because of its size) to the dangerous Missouri border, as well as to most of the potentially hostile proslavery settlements within Kansas. For free-state Kansans, Topeka thus occupied a location that was central in one sense and yet remote in another. In this somewhat paradoxical manner, the city fit their needs well in 1855.[15]

While at the Topeka convention, delegates established their own version of a territorial legislature and began to label the original, proslavery one as bogus. These new free-state representatives would require a place to meet the following spring, of course, and once again, it was decided that Topeka should be the site. Local residents had lobbied for this privilege, to be sure, but had to face arguments that they lacked sufficient accommodations for such a meeting. "Good cheer and good fare" provided by the Chase and the Garvey boardinghouses, however, overcame the hesitations. Topekans also promised to erect a proper hotel before the spring session.[16]

Although the initial political successes of Topeka residents had come about almost by chance, once accomplished they became part of the community's self-identity. They provided local leaders with confidence that they could compete successfully for the state capital when that prize came to be awarded, as well as for other cultural institutions. By early 1857 the town had grown to 450 people, including several who had recently moved upstream a few miles from the rival, proslavery, and now declining community of Tecumseh. The change in statewide politics apparent by that date meant an easy local decision for voters the next year to transfer the seat of county government from Tecumseh to Topeka. One then can see a pyramid effect. With the county seat secured and the state capital a possibility, the population in Topeka rose to 512 in 1858. Town officials then set their caps on obtaining the colleges that officers from the Congregational and the Episcopalian Churches were planning to establish in the territory. They offered the Congregational ministers 160 acres as a site plus a two-story, brick or stone building and an additional 840 acres as an endowment. To the Episcopalians, who wanted to open a girls' seminary, they made a more modest proposal: a 20-acre site plus an endowment of thirty-three town lots. After some delays, both of these bids were accepted in 1860. The deciding factor may not have been the actual proposals, however. The previous October voters from throughout the territory had approved a new constitution under which Kansas would enter the union. The framers of this constitution, by a vote of 26 to 14 over Lawrence, had named Topeka to be the temporary seat of the new state's government.[17]

When the census takers made their rounds in 1860, the number of residents recorded for Topeka did not accurately reflect the town's status and potential. Although the total, 759, demonstrated a solid gain over earlier numbers, it did not yet incorporate the impact of the two colleges then under organization and the much larger surge that would occur if the temporary designation of capital could somehow be converted into a permanent label. All was tenuous, of course, because the pyramid principle could also work in reverse. If the capital were not to come, Topekans would be hard-pressed to keep the colleges. Then, without the colleges and with no industrialization on the horizon, ambitious townspeople likely would move on to more promising cities.

Selection of the Capital

Citizens in Lawrence and Topeka lobbied openly to have their communities named capital of Kansas Territory and then of the state to follow. An early example came in October 1854, when Samuel Pomeroy and other agents of the Emigrant Aid Company cornered the first territorial governor, Andrew Reeder, during his initial tour of the region. While in Lawrence, he was quickly escorted to the newly surveyed "Capitol Hill" and told that this choice spot would be "cheerfully yielded up" as soon as he might give the official word. As an added inducement, town officials also offered Reeder some choice building lots in the community at a reduced price.[18]

A blatant offer to sell property to a governor in exchange for future political favors would provoke a scandal in modern Kansas. In the 1850s, however, and for several decades afterward, such action was an accepted part of business and political life. Although he was regarded as an honest and capable official, Reeder therefore did not hesitate to purchase the lots offered to him in Lawrence. He did the same in several other towns.[19] To understand the process of selecting a capital for Kansas, this close relationship between politicians and land speculation must be kept in mind. Profit and power were the paramount issues, not fair play or centrality to the regional population.

Andrew Reeder was a Pennsylvanian by birth. He was a lawyer, a Democrat, and a defender of the principle of popular sovereignty in Kansas. When he arrived in the territory on October 4, 1854, the status of town development was such that the only practicable place to set up offices was at Fort Leavenworth, where he had docked. The post thus functioned as the de facto capital of the territory for about fifty days. Conditions at the post were crowded, however, and so a search began almost immediately for better quarters. This process revealed only one credible option: the Methodist Episcopal mission to the Shawnee Indians that was located just across the border from Westport,

Missouri. This mission, in operation since the 1830s, had three large and roomy buildings. Although its director, the Reverend Thomas Johnson, was reluctant to cede space to the governor and his staff, he finally did so, realizing it was "the most obvious necessity." Reeder moved into the mission on November 24, and his government functioned from there until the following July.[20]

The site selection for the permanent capital almost surely was made around the fireplace at the Shawnee mission during the winter of 1854–1855. Although the governor was not authorized to name its location directly, he did control where the legislature was to meet, which was essentially the same thing. Throughout the fall, and especially on the same October trip across the territory that had brought him to Lawrence, he had been assessing the possibilities. According to his private secretary, remembering back nearly fifty years after the event, Reeder warmly informed the people of "sundry" towns that "this would be a magnificent site for a capitol building." These places included Council Grove, Juniata (where the military road to Fort Riley crossed the Blue River), Lawrence, Leavenworth, and Tecumseh.[21]

Back at the mission, Reeder presumably evaluated what he had seen and received visitors from a series of existing and planned communities. Many of these peoples no doubt bore offers of additional financial incentives for his consideration. He may also have talked over the pros and cons of different places and proposals with at least some of the six other government officials who lived there with him. Although these men were strangers to one another initially, all of them shortly were to become key investors in the game of capital-making. Besides his Lawrence properties, Reeder himself had made investments in Leavenworth, Tecumseh, and Topeka. His biggest involvement, however, was with a 320-acre town site called Pawnee that military officials had marked off within the Fort Riley reservation in September. The governor had seen this place firsthand on his October tour and judged it to have unusually good prospects for success. With Fort Riley expected to replace Fort Leavenworth as the supplier of western military posts, this area naturally would become a transportation hub. A settlement there could thus expect a role as entrepôt comparable to that anticipated for Leavenworth. Reeder also liked the idea of having the territorial capital at some distance from Missouri. His concern here was danger, not the future distribution of population. The Shawnee mission, subject as it was to incursion by border ruffians at any time, was hardly "a proper or safe place." Finally, and probably not least important, the governor realized that the selection of Pawnee almost certainly would yield him greater personal fortune than would that of Lawrence, Tecumseh, or Topeka. Although he owned a few lots in those already established communities, with the new Pawnee he was an influential member of the founding company itself.[22]

Reeder is thought to have selected Pawnee as capital as early as December 1854. He told this first to his business partners so that they would have time

Advertisement for the Town of Pawnee, 1855. This announcement circulated in March 1855, several weeks before Governor Andrew Reeder declared Pawnee to be the territorial capital on April 16. Many insiders knew what would happen, of course, and so sales were brisk on April 10. The head of the Pawnee Association, William R. Montgomery, also served as commanding officer at Fort Riley. (Courtesy Kansas State Historical Society)

to erect a legislative hall by the following summer. Also, in confidence, he advised his private secretary to purchase shares in the company and he expanded the board membership to include friends from his home state of Pennsylvania plus four men from his official staff. Significantly, however, he excluded two other of his government associates. Both were Southerners who enjoyed popular support among the Missouri settlers: Samuel Lecompte (of Maryland), the chief justice of the Kansas court; and Daniel Woodson (of Virginia), the secretary of the territory.[23]

The governor officially announced his choice for the new capital on April 16, 1855. Despite the several theoretical arguments that could be made in Pawnee's favor, this declaration pleased almost nobody. Northern settlers were disappointed that Reeder had rejected the bids from Lawrence and Topeka. Missouri natives were even more upset. Assembling the legislature so far away from their strongholds was simply inexcusable. They also saw this decision as Reeder's second major sin within a month, for the governor had just set aside the results of a March election in several districts because Missouri residents had voted in them illegally. The Southerners vowed revenge. Using their connections in Washington, D.C., with Secretary of War Jefferson Davis and President Franklin Pierce, they not only removed Reeder as governor but also destroyed Pawnee as a place.

The process was swift. Officials declared the Pawnee lands illegally taken from a previous settler and Reeder guilty of abusing his influence as a public official. The real reasons were political, of course, with the governor and the town both judged as too biased toward the free-state cause in a territory that was supposed to become a slave state. The legislature did assemble at Pawnee as scheduled on July 2 but adjourned four days later on a motion to reconvene at the Shawnee mission. President Pierce dismissed Reeder on July 28, and a month later, all settlers were ordered off the Pawnee site.[24]

Pawnee would have made a good location for the Kansas capital, and had a chance to succeed as well. If the water levels in the Kansas River had been as high in 1855 as they were the year before, legislators might have arrived at Pawnee in steamboat luxury instead of in sour moods after three or four days of rough overland travel and primitive campsites. More important, had Andrew Reeder and his business partners included even a few influential Missourians in their schemes, the allure of profits all around might have overcome sectional prejudice. As it happened, however, the suddenness of Reeder's departure put power into the hands of an acting governor. Ironically, this was Daniel Woodson, the Virginia-born secretary of the territory whom Reeder had kept out of the Pawnee scheme.

Woodson had always been a consistent advocate for slavery in Kansas and thereby had the trust of the first territorial legislature as well as the similar-minded leadership in Washington. As early as April, when Pawnee was named as capital, Woodson seems to have realized the possibility of its failure. Along with several of the territorial officials, he acted on this impulse to create a paper town of his own, one to have in reserve, so to speak, should an opportunity arise. This was Lecompton. Located on the south bank of the Kansas River about midway between Lawrence and Topeka, the Lecompton town site was ideal from the perspective of its organizers. It had the same good potential for transportation and clear title to land enjoyed by its flanking New England settlements, and it was close enough to established communities to allow easy access for most legislators. Most important, because it was a new entity, its development promised to fill the financial pockets of its founders.

Lecompton's name comes from Samuel Lecompte, the chief justice of the territory and president of the town association. Daniel Woodson served as treasurer. The political powers that joined them can still be seen in the street names in the community. Woodson is the main east-west thoroughfare. Those intersecting it include Halderman (Reeder's secretary), Elmore (an associate justice serving with Lecompte), Isacks (the United States attorney), and Whitfield (the territorial delegate to Congress). The group met to organize in Westport, Missouri, and had the land surveyed in May. With six hundred acres included in the original plat, the "design and expectation" was at a metropolitan scale. On August 8, by a vote of the legislature at the Shawnee mission, this Lecompton dream became a reality.[25]

Compared with the debacle at Pawnee, people accepted Lecompton as capital without immediate incident. New England settlers were unhappy, of course, but took solace in having a capital near their own bastions of strength. The most disappointed people, perhaps, were the residents of Tecumseh. Rush Elmore, an associate justice of the territorial court, lived there, and joining him in the town company had been Governor Reeder, Representative John W. Whitfield, and Associate Justice Saunders N. Johnson. With a group of this stature, local people logically hoped for and may even have been promised the prize. Tecumseh likely had been Governor Reeder's second choice for the capital. The dream might even have been possible after the Pawnee collapse had Reeder remained in office. With the governor gone, however, Lecompte and Woodson held most of the territory's political clout. After engineering a victory for Lecompton over Tecumseh by a single vote in the legislature, the two men appeased Elmore with a share in the Lecompton company. Then, as a consolation prize to Tecumseh, they worked with the legislature to have that town named the seat for the Shawnee County government. This gesture did not compare to what had been lost, of course. The first taste of Lecompton's new exalted status came shortly thereafter when officials received $50,000 from the United States Congress to construct the capitol and other appropriate public buildings. By April 1856, the government there was fully in place.[26]

With Southern sympathizers in control of Kansas politics in 1855 and 1856, Lecompton prospered. Four churches and five boardinghouses soon appeared along its streets, including a landmark, three-and-a-half-story structure called the Rowena Hotel. With walnut woodwork and twenty-four comfortable rooms, the Rowena, in fact, was considered by many well-traveled visitors to be the finest facility of its type throughout the territory and perhaps even the entire West. The town's glory was short-lived, however. The spring of 1857 brought new immigrants to Kansas in numbers large enough to overwhelm the earlier residents in almost every county. Nearly all of these came from states north of the Ohio River. Although the legislature continued to convene in Lecompton throughout the remainder of the territorial period, an order was issued that year to stop construction on the stone capitol building. It soon became obvious to everybody that these new free-state people would join with older New Englanders in an effort to move the seat of government somewhere else. Lecompton, this combined group said, brought to mind painful associations that needed to be forgotten. Left unspoken were the concurrent desires to have government largesse for their own hometowns.[27]

The first territorial legislature to have a free-state majority met in January and February 1858. On the capital issue, its members faced a familiar dilemma. They could select an existing city and thereby offend the supporters of rival communities, or they could create a new paper town of their own. The latter option also would anger people, of course, but if key leaders were given shares in the town company, such a storm probably would pass quickly. The legislators, with shares themselves, would prosper as well. Not surprisingly, the group took this latter course of action. They agreed on a place called Minneola, a community projected in the northern part of Franklin County, a mile east of Centropolis. To further bolster the potential for growth at this site, the lawmakers simultaneously chartered several railroads whose lines centered at that identical spot on the map. Minneola, like Pawnee and Lecompton before it, was not by any means an illogical choice for the capital. Only one brief meeting was ever held there, however, before the attorney general of the United States ruled that any such movement of the capital violated the territory's organic act. A new location could be selected only if such an option were made available in the constitution that Kansans would write as part of their transition toward statehood.[28]

With the collapse of the Minneola scheme, thinking reverted to existing cities. Here Lawrence and Topeka were the front-runners. Both of these communities offered impeccable free-state credentials, of course, and each was reasonably central to the territorial population of the time. More important, leaders in both places long had coveted this honor and had been working diligently toward that end. If a bookmaker had assessed the situation in 1858, he or she would have had difficulty in determining a favorite.

Topekans began their quest for political glory early in 1855 when they followed the lead of Lawrence by reserving four large blocks (twenty acres) on their plat as capitol grounds. Their next successes, as noted earlier in this chapter, were to host the constitutional convention put together by free-state dissidents in October 1855, and then, in March 1856, the initial meeting of the same group's extralegal, shadow government. The legislative sessions for this government continued to convene at Topeka throughout their existence: once again in 1856, twice in 1857, and a last time in 1858. On two of these occasions, the territorial governor in Lecompton sent police officers into the meetings with orders for dispersal and, in one instance, the arrest of leaders. Each disruption, of course, heightened Topeka's status as a free-state center.[29]

Lawrence could match every political achievement at Topeka with a comparable one of its own. Charles Robinson, James Lane, and other local leaders never let anyone forget that Lawrence was the fountainhead for the territory's free-state movement and its largest community. In addition, the city had hosted two pivotal early meetings for the cause in 1855, one when the group's delegates to the first territorial legislature had just been expelled from the Pawnee assembly, and the other, in August, to organize the Big Springs convention. Lawrence also could assert symbolic leadership. This city was the site, in December 1855, of the first tense (although nearly bloodless) confrontation between free-state and proslavery forces known as the Wakarusa War. The county sheriff had burned the community's magnificent Free-State Hotel in May 1856 after a proslavery grand jury had labeled the building a "nuisance."[30]

Although Lawrence's credentials as a free-state center were impressive, leaders had an even greater reason to expect that the city eventually would be named the capital, namely, that during the three years between 1858 and 1861, Lawrence had been serving in this capacity in everything except official title. When members of the third territorial legislature (the first to have a free-state majority) met for their regular session on January 4, 1858, they assembled, as the law required, at Lecompton. The next day, however, they accepted a motion for adjournment and made plans to reconvene in Lawrence on January 8. Lawrence then hosted the group through the close of their business on February 13. When this pattern was repeated by the fourth, fifth, and sixth legislatures, it seemed evident to most Kansans that Lawrence had become "now, practically, the capital of the territory."[31]

The political rivalry between Topeka and Lawrence reached its peak in 1859. That year the territorial legislature called for a constitutional convention to assemble at the town of Wyandotte. Everyone knew that the document produced by this group would, among many other particulars, name a capital. Lobbyists from the two cities, unsure about the predilections of the various delegates, decided to hedge their bets and united behind the concept of having the convention identify only a temporary capital. A final determination would then be made later in a general statewide election. Still, everybody knew that voters likely would uphold the Wyandotte decision. The stakes were huge.

Analyses of the Wyandotte convention suggest that, although the delegates were hardworking and, by and large, representative of Kansas at that time, the decision about the capital was inexorably linked to two other issues: whether or not to annex that part of Nebraska Territory south of the Platte River, and which of the several incipient railroad companies would receive preferential treatment from the new state. Political deals apparently were made even before the meetings began. Nobody objected, for example, when the actual apportionment of delegates was different from that in the original legislation. Leavenworth and Doniphan Counties received a total of four more representatives than they deserved, whereas Atchison, Jefferson, and Wyandotte Counties suffered a comparable loss. Raymond Gaeddert, the expert on this matter, has argued that Cyrus K. Holliday of Topeka engineered the shift in delegates. The increased numbers in Doniphan County would help with a bid from that region to annex the Nebraska lands, while the larger Leavenworth total would promote the chances for assistance to the Leavenworth, Pawnee and Western Railroad. Holliday and his fellow Topekans likely would see their own railroad plans suffer from this arrangement, but as compensation, the Doniphan and Leavenworth people presumably promised their votes for the capital.[32]

Things worked out much as Holliday had planned. The motion to annex the Platte lands failed at the convention, but the Leavenworth railroad scheme received favorable attention and Topeka won the election for capital. Rumors circulated during and after the meetings that representatives from both Lawrence and Topeka offered money in exchange for votes, with one source reporting that Lawrence citizens somehow had raised $50,000 for the effort. Still, the votes suggest that the preconvention deal held true to form. With twenty-five votes needed for selection, Topeka received a total of twenty-nine, compared with fourteen for Lawrence and six for Atchison. All nine members from the Leavenworth delegation cast their ballots in favor of Topeka, as did all five from the Doniphan group. Such numbers made the outcome suspect indeed from the perspective of Lawrence stalwarts, but they never challenged the results.[33]

The final chapter in the complex lobbying for capital selection in Kansas occurred between April and November 1861. With statehood achieved in January of that year under the Wyandotte Constitution, the first state legislature set November 5 as the date for the required public vote. Although, in theory, the contest was open to any city, newspaper accounts suggest that people from the Lawrence and Topeka areas were the only active campaigners. Supporters of Leavenworth, an obvious potential choice because of its size, stayed out of the fray. Their silence on this issue almost surely was premeditated, part of still another political bargain by which the legislature that April had quietly awarded their town the site for the state prison. Atchison, the only other sizable place, was deemed too peripheral within the state to serve as its seat of government.[34]

Money apparently flowed freely throughout that summer and early fall. Official records show an appropriation of $19,500 by the Topeka city council for this purpose, an amazingly large amount for a community of perhaps eight

The State Capitol in Topeka, 1880. Occupying a twenty-acre site, the Kansas capitol was built in three main stages between 1866 and 1898. This photograph, looking northeast, shows the completed east wing in the background and the west wing under construction. Although the two sections have similar designs, the west wing was built faster and more cheaply than its counterpart. (Courtesy Kansas State Historical Society)

hundred residents. An audit revealed that most of this total was paid out in small-denomination bills, presumably to go into many individual accounts. Parallel documentation for Lawrence is not available, but the editor of the *Leavenworth Times* is known to have changed his endorsement from Topeka to Lawrence after Douglas County representatives offered him more money. One of the extreme (although undocumented) stories is that Charles Robinson of Lawrence, while sitting as the state's first governor, may have offered to give up Lawrence's bid for the state university in favor of rival Manhattan if Riley County voters would deliver the capital to his hometown. When all the maneuvering was over, the results closely mirrored those at the Wyandotte convention. Topeka, with 7,859 supporters, received the majority of the ballots cast. Lawrence was the choice of 5,334 people. When these numbers are broken down by county, they show that Leavenworth residents were evenly divided on the issue. The biggest anomalies were large majority votes for Topeka in four counties closer geographically to Lawrence: Atchison, Doniphan, Franklin, and Jefferson. Bribery certainly could account for these cases. Since three of these counties were bastions of Missouri settlement, however, old prejudices may have been involved. Lawrence, perhaps, still symbolized Yankee arrogance.[35]

The Prison and the University

It is easy to understand why cities would spend large sums of money to compete for the state capital. Prestige, jobs, and the political connections to influence future developments all would accrue to the winning community. For Kansas, the interesting question is not why the people of Lawrence and Topeka made that effort almost continuously from 1854 through 1861, but why those in Leavenworth, the new state's largest city by far, did not. Equally intriguing is that citizens there, after not contesting the capital issue, instructed their legislative representatives in 1861 to pursue the state's prison rather than its university.

An explanation for the attitudes and actions in Leavenworth must begin with the strong business mentality that dominated decision making there. Leavenworth people partook of sectional politics to a much lesser degree than did the residents of Atchison, Lawrence, or Topeka, and they displayed fewer aspirations for cultural trappings as well. Instead, they concentrated on currying favor with military officials, developing wholesaling and freighting operations, investing in real estate during the boom of 1857, and lobbying extensively for a variety of railroad projects. When the real estate market collapsed in 1858, it threatened the financial underpinning for this entire constellation of entrepreneurial activities. Boosters were reluctant to admit that a crisis existed for fear that such a confession would make matters worse, but when a drought hit Kansas in late 1859 and 1860, desperation became apparent and the local population totals began to fall. As historian David Taylor has described the situation with only slight exaggeration, "Leavenworth was a big town with nothing to support its inhabitants."[36]

Had the Leavenworth economy in 1861 been better than it was, it is quite possible that local politicians would have sought the state capital or, more likely, the slightly less coveted state university. They viewed success with the capital as improbable, however, given their location on the Missouri border and the inertia of existing movements in Lawrence and Topeka. The university, although an institution they probably could have secured with a concentrated effort, was seen to have two drawbacks. One was the reality that several other communities also were competing for this prize, so that a fight would be inevitable and the chances of success not guaranteed. The other was that a university in the 1860s was only a moderately good business investment. It would employ only a handful of professors and would require no more than a single building.[37]

Compared with the uncertainty of a pursuit for the capital and the limited economic potential of the university, the acquisition of the state's penitentiary offered considerable appeal. First, because no rivals had emerged, the Leavenworth people knew that this facility could be theirs for the asking. They only had to act quickly. This rationale was almost certainly why they insisted that the legislators assign the prison location before calling the vote on the capital or the university. Lawrence residents, for example, would not be thinking about the prison while they still had a chance at the capital. Profitability was another major attraction of this institution. The country in the 1860s had far more prisoners than college students, and therefore a bigger physical plant could be expected for a prison, together with more jobs in construction, maintenance, and the like. Beyond these sources of revenue, however, lay the even bigger bonanza of inmate labor. It was standard practice at the time for state authorities to invite private contractors to set up manufacturing plants inside prison walls. With inmates doing the many unskilled tasks associated with, say, brick making or wagon assembly, large profits were assured.[38]

Thinking in purely practical terms, then, Leavenworth merchants regarded the penitentiary as "a very desirable institution," and they united behind the effort to secure it. The path to do so was reasonably straightforward, although Governor Robinson apparently provided an obstacle. The bill awarding the facility to Leavenworth passed the legislature in April 1861. Later, after the November election that gave the capital to Topeka, Robinson decided that the prison, rather than the university, would be a preferable institution for his town of Lawrence. He therefore delayed implementation of the earlier act while reportedly working behind the scenes to alter the outcome. His moves proved unsuccessful, of course, and a board of commissioners purchased forty acres for the new facility five miles south of Leavenworth.[39]

By accepting the prison, Leavenworth residents effectively removed themselves from competition for the state university. This eased the decision-making process for the latter institution only a little, however, because a large immigration in the late territorial years had made the Kansas of 1861 much larger and more complex than it had been in 1855 or 1856. More people meant more town companies, each craving an advantage to help them survive and prosper. Once Leavenworth had claimed the prison in April, everybody realized that the university would be the last major public institution available for acquisition once Lawrence and Topeka had finished their campaigns for the capital.

Somewhat surprisingly, plans for a public university in Kansas actually go back to the days of Indian removal in 1855 when the original proslavery legislature approved a charter for the University of Territorial Kansas. This was to be located in the town of Douglas (a small community on the Kansas River between Lawrence and Lecompton) and funded by a sale of public lands. When the federal government failed to provide such acreage, however, the plan collapsed. A similar experience occurred in 1857, with the site changed to Kickapoo City, near Leavenworth. The writers of the Wyandotte Constitution of 1859 mandated a university for the new state, but serious debate on the issue

Administration Building of the Kansas State Penitentiary in Lansing, circa 1910. Designers hoped to make this prison imposing yet graceful. Using red sandstone and prisoner labor, they created a series of turrets and false battlements reminiscent of a medieval castle. Cell houses A and B, each with 344 compartments measuring 7 by 4 by 7 feet, flank the central building. Construction began in 1864 and was completed in 1882. (Courtesy Kansas State Historical Society)

did not begin until 1861, when eager city lobbyists forced the initial state legislators to confront their constitutional charge.[40]

Whereas the bill to establish the penitentiary at Leavenworth passed into law almost without incident, parallel action for the university became nearly as heated as that for the capital. Debate and maneuvers actually stretched out until 1863, and the issue was not resolved until the meeting of the third legislature. At first, everything appeared to go smoothly. A motion to locate the facility in Manhattan passed both the house and the senate by large margins. Presumably the thinking was that the college should go to one of the New England communities whose residents had long promoted the cause of higher education. With Topeka already having secured the Congregationalists' denominational school (Lincoln College, renamed Washburn in 1869) and Lawrence quite possibly poised to receive the capital, Manhattan was a fairly obvious choice. Even though this settlement in Riley County near the head of the Kansas River was even smaller than Topeka, with only a few hundred residents, Isaac Goodnow and other of its leaders had dedicated themselves to education from the beginning. In fact, they already had raised $14,000 from benefactors in the East to create a college called Blue Mont Central. Blue Mont opened in 1860, but the next year, seeking greater financial stability, its trustees offered their building, library, and 120 acres (the total valued at $20,000) as a nucleus for the state university. Financially strapped legislators realized they had a generous offer.[41]

Charles Robinson, the perennial champion of Lawrence, stopped the initiative for Manhattan almost single-handedly. Realizing perhaps better than most that Lawrence might lose the upcoming election for the capital, he saw that, without prompt action, the city could easily end up with no public institutions at all. The governor therefore vetoed the measure. Legislators attempted to override this decision, and nearly did so with a vote in the house of 38 to 20.[42]

Robinson's maneuver provided Lawrence residents with a year's reprieve. In the fall of 1861, with the capital now lost, they began to assemble a package of incentives that might match that of Manhattan. This year of preparatory time was not all to Lawrence's advantage, however, because it also enabled people from other communities to whet their appetites for political battle. When the legislators met again, they found three contenders for their approval. Manhattan people offered the same package as in 1861, while the Lawrence counter consisted of $15,000 in cash, $10,000 in real estate, and a twenty-acre site. A third bid—forty acres—came from Emporia, a tiny community about fifty miles southwest of Topeka that had been established by entrepreneurs from Lawrence in 1857. On paper, the Lawrence proposal was superior to those of its rivals, but the delegates in the house of representatives preferred the solid, already-completed campus in Manhattan. They therefore considered the Manhattan bill first and adopted it easily 45 to 16. The senators, however, saw matters differently and rejected the proposal twice: 13 to 10 and, after a vote to reconsider, 12 to 11.[43]

The action by the senate meant another year's delay for the university decision. During this hiatus, however, activity outside the state eased the anxieties of the contending Kansas factions. The United States Congress passed the Morrill Act in July 1862, whereby the federal government agreed to endow each state with ninety thousand acres or more of public lands if local officials would establish a college that emphasized instruction in the agricultural and mechanical arts. Kansans, sensing a solution to their problem, were the first to accept this offer. They promptly designated Manhattan as the future home for this "land-grant" school. Part of the assumption, of course (at least in some quarters), was that the state's liberal arts college would soon be approved for Lawrence.[44]

The legislative session of 1863 eventually fulfilled the hopes of Lawrence lobbyists, but the action was anything but straightforward. With the Manhattan people now satisfied, the contending delegations came from Emporia, which presented an increased offer of eighty acres, and Lawrence, which tendered a reduced one of $15,000 plus a twenty-acre site. The Lawrence proposal again looked superior, but like before, the original bill on the floor named its rival town as the choice. Clifford Griffin, a specialist in the university's history, has attributed this action to a belief by many legislators that the Lawrence group did not actually have the $15,000. Once more the city's lobbyists swung into action. People swore that the money was there, they developed an alternate educational plum to offer to the Emporia contingent, and reportedly, they bribed several voters for five dollars apiece. In the end, the word "Lawrence" was substituted for "Emporia" in the bill. When the house produced a tie vote, Edward Russell, a representative from Doniphan County, broke the impasse in favor of Lawrence. Russell, the stories say, did so after receiving assurances that Lawrence people would support railroad legislation that he favored. The senate easily concurred, 18 to 5, giving Lawrence a victory that, although tainted in nearly every way, was a victory nonetheless.[45]

Like Topeka in the campaign for the capital and Manhattan in the pursuit of the university, Emporia ostensibly was an unlikely place to contend seriously for a major public institution. All three of these communities were extremely small, with Topeka's 759 residents in 1860 making it the relative metropolis of the group. Strong free-state credentials were certainly factors in the Topeka and Manhattan successes, but an ability to back these things up with political acumen and solid financial initiatives was even more important. Early Emporia leaders possessed similar skills and inducements. Their community was a speculative venture put together by five influential men from Lawrence, including one who later became a general in the Civil War (George W. Deitzler), and another a United States senator (Preston Plumb). The site they selected, in the Flint Hills, did not look promising until one considered the possibilities for railroads. Emporia occupied the junction of two rivers (Map 5; see p. 17). It stood where the Cottonwood, one of only two good valley routes through the Flint Hills, fed into the larger Neosho. Since the Neosho lowland was renowned for its breadth and fertility, everybody expected that a railroad soon would follow its path from some city along the Kansas River south toward the Gulf of Mexico. A branch from it westward up the Cottonwood Valley and eastward into Missouri could make Emporia the transportation hub of east-central Kansas.[46]

The expectation of transportation developments meant that Emporia attracted skilled entrepreneurs and capital assets far out of proportion to its population. These people acted through Charles Eskridge, the agent for the town company and a state legislator since 1859. They also were probably the ones who realized that Lawrence (their former home) was vulnerable on the issue of the $15,000 pledge. The Emporia contingent were rewarded for their

The University of Kansas in Lawrence, 1867. Constructed intermittently over a period of two years, the university opened for classes on September 12, 1866, with three professors and fifty-five students. The photographer's view looks north toward the Kansas River and emphasizes the small size of this institution compared with the state's capitol and prison. (Courtesy Kansas State Historical Society)

political skills later in the 1863 legislative session when Eskridge's proposal to create a third public university passed the assembly without opposition. Just as agricultural education could be partitioned from the liberal arts for political purposes, so might teacher training. In this way, Emporia became home to the state normal school, endowed by the solons with a grant of 45,680 acres. The facility opened for classes in February 1865.[47]

Other Early Public Institutions

Although the placement decisions for the capital, penitentiary, and university took much legislative time and effort, these were not the only facilities on the public agenda during the 1860s. Article 7 of the Kansas constitution also required the state to foster and support "institutions for the benefit of the insane, [the] blind, and [the] deaf and dumb."[48] Civic leaders did not consider any of these latter enterprises glamorous in their own right, of course, and none of the three promised to employ large numbers of workers. Lobbyists from the larger cities, therefore, did not bother to enter the political competitions for their location. People in many medium-sized communities, however, such as seats of counties away from major rivers, took a different perspective. Although entrepreneurs in these places spent most of their energies on the pursuit of various railroad companies then organizing across the state, the ultimate pattern of this new transportation system was anything but certain. The acquisition of a state institution, even one that housed the insane or the blind, was a good way to differentiate one's town from the growing mass of county seats. It was simple gamesmanship. Not only would the facility generate some jobs and bring in state appropriations for construction, its presence would also increase the likelihood for better rail connections and, in turn, the establishment of a wide variety of manufacturing plants.

The allocation process for the three small institutions began in the legislative session of 1863, just after the decisions to award universities to Emporia, Lawrence, and Manhattan. The debate that year concerned the insane asylum, with those for the two special-education schools delayed until 1864. A person could almost predict the contending cities. The largest concentrations of Kansans so far excluded from their share of public funds were in the Missouri border counties that extended southward from Wyandotte to Bourbon. These places had not been positioned ideally for political favors from a free-state legislature because they contained large numbers of Missouri-born Democrats and had been subjected to raids from both factions during the troubled times of 1855–1856. Similar fighting continued during the Civil War years even as the legislators were meeting. Still, the population totals in these counties were larger than any others save Atchison, Doniphan, Douglas, and Leavenworth. A delegate from Seneca, the seat of Nemaha County on the Nebraska border, placed his town into consideration for the asylum for the deaf and dumb. Seneca was too far from the state's core population to merit serious attention, however. The same was true for Iola and LeRoy in their brief quests to obtain the same prize. With these extreme northern and southern locations excluded, the choices became much clearer. Bids from communities in Johnson, Miami, and Wyandotte Counties all would be well received.[49]

The initial bill debated for the insane asylum proposed a location at Wyandotte.[50] This city, the seat of Wyandotte County, certainly was a logical choice in many ways. In fact, considering that it overlooked the strategic junction of the Kansas and Missouri Rivers and had a population of 1,920 in 1860 that made it the third-largest urban center in the new state, a person might wonder why even bigger prizes were not coming its way. All is not straightforward in politics, however. Wyandotte was a relatively new creation. It also possessed a cultural heritage and several locational particularities that diminished its standing with the legislators of the time.

Wyandotte, the core of what a quarter century later would become known as Kansas City, Kansas, played little part in border troubles during the middle

1850s. This anomaly can be explained by a relatively late origin: the city was not platted until May 1857. Before that date, the council house, homes, and farms of seven hundred or so members of the Wyandot nation occupied its site atop a high bluff at the mouth of the Kansas River. These people, acknowledged by everybody of the time as "much more advanced in civilization" than any of the other Indian groups who had immigrated to the region in earlier decades, had adapted well to life in the shadow of Westport and Kansas City, Missouri. They operated a ferry across the Kansas River, created productive farms and orchards, and built substantial houses. They also counted a lawyer among their numbers (John M. Armstrong), as well as several white businessmen who had married into the group. A large majority of the community spoke English.[51]

The Wyandots had no desire to leave Kansas in 1854 when Euro-American settlement began. Instead, they hoped to become recognized as American citizens and then to participate in the state-building process as equal partners. The possibility for creating a city around their council house came in a treaty signed January 31, 1855. This gave citizenship to any Wyandot who requested it, including the right for the first time to sell property on the open market. Nothing major came from this change in status for two years. The triangle of land between the two big rivers was beautiful, but entrepreneurs perceived better business opportunities in Atchison and Leavenworth, as well as in nearby Kansas City and Westport. Whereas these other cities lay on established trade routes, the Wyandotte site had none. The way to the southwest was blocked by the Kansas River, of course, while any road westward or northwestward would have to pass through the reduced, but still sizable, Delaware Reservation. This was not a temporary matter either, for the Delawares, like the Wyandots, showed no signs throughout the 1850s and early 1860s of abandoning their well-watered lands that stretched westward to near Lecompton and northward from the Kansas River to the northern limit of Wyandotte County (Map 3).[52]

Plans for the city began in December 1856 when two men representing a group of investors from Kansas City, Missouri, and the East Coast met with Silas Armstrong, a principal leader of the Wyandots. From this came an unusual, seven-person town company. It consisted of three free-state Kansans, one Kentucky-born Missouri entrepreneur (Thomas H. Swope), and three Wyandot leaders, including Mr. Armstrong. After laying out the streets, they put four hundred shares of the company on the market the next spring for $500 per share. Their timing was excellent. It coincided with a large immigration into the state that year, and so the city grew. At the end of 1857 Wyandotte had four hundred residents, and a share in the company was worth $1,000.[53]

Growth at Wyandotte continued at a brisk pace for the rest of the decade, with the population doubling and then doubling again before the census in

1860. The reasons for this success were not obvious on the landscape but seem to reflect a belief that the community was backed by substantial money from beyond the state line. Kansas City and Westport people were most obvious in this regard, particularly Swope and the banker Hiram M. Northrup. Northrup, who had grown rich in the Santa Fe trade, actually moved to Wyandotte from Westport with his Indian wife. In retrospect, it is easy to surmise that much of the capital that flowed into early Wyandotte came in anticipation of the role this river junction might play in the railroad plans then under discussion in several dozen executive boardrooms. From the perspective of the state legislators in Topeka, however, all this was slightly suspect. These politicians were all for development, but they wanted to direct it themselves. Wyandotte, in their eyes, symbolized outside control and so was resented. For at least one longtime resident, this prejudice still festered some forty years after the fact: "In one thing only were the cities of the state united, and that was with tongue and pen to ridicule and denounce the absurd pretensions of Wyandotte. In former years every effort made by our legislative representatives for even the moral support of the state was met with contumely and contempt. We were denounced as a Missouri bantling, and were not entitled to aid or support from Kansas, even though politically we were forced to be recognized as a component part of the state."[54]

As members of the Kansas legislature debated the placement of the insane asylum in 1863, lobbyists for two additional communities joined Wyandotte's boosters. By chance, these towns were neighbors in Miami County: Paola, the county seat, and a smaller and more isolated place called Osawatomie. Of these two, centrally located Paola might seem to have been the better choice, but sentiment among the lawmakers immediately went to Osawatomie. In fact, the support for this seemingly modest candidate not only overwhelmed that for Paola but also did the same to Wyandotte. A revised bill passed the house easily with a margin of 56 to 6. The senate concurred 20 to 1.[55]

Something about Osawatomie clearly resonated with the legislators. Part of the answer may have been the influence of Dr. William W. Updegraff, an Osawatomie man who had been speaker of the house during the first meeting of the state legislature. A bigger factor, however, was the outsized role that the several hundred residents of this community had played during the free-state crusade of the 1850s. As the only settlement of the New England Emigrant Aid Company in the border tier of counties, Osawatomie had twice been plundered by Missouri raiders in 1856. It also had been home to the charismatic leader John Brown and had hosted the 1859 meeting that established the Republican Party in the territory. Famed journalist Horace Greeley, who had attended that 1859 session, later wrote that the town's "honorable eminence in the struggle" had caused it to be revered throughout the state.[56]

Osawatomie State Hospital, circa 1917. The towered, central structure was built in 1868 and could house 155 patients. Two large wings, added a decade later, raised this capacity to 1,400. Legislators in 1901 attempted to lessen the institution's stigma by changing its name from insane asylum to hospital. (Courtesy Kansas State Historical Society)

Osawatomie people felt honored to receive the asylum because their community had been losing population since 1857. They donated a site of 160 acres, prospered modestly from jobs provided by annual state appropriations of several thousand dollars, and in 1868 enjoyed a construction boom when the noted architect John G. Haskell came to town to oversee the erection of a $20,000 permanent hospital. Wyandotte residents, meanwhile, were not feeling sorry for themselves. When the state legislature met again, in 1864, it was common knowledge that one of the two remaining public institutions would go to them. The city's representatives selected the school for the blind, and after a donation of ten acres for the site, the state dutifully appropriated another $20,000 for its permanent building.[57]

The school for the deaf and dumb, the last public institution to be awarded, had two principal suitors. They were Baldwin City, a small town in southern Douglas County, and Olathe, the seat of Johnson County. Here the selection process pitted history against political clout. History was a factor because public support for the education of the deaf actually predated the debate of 1864. What became the state school had begun as a private venture in 1861 when a family with three deaf children happened to live near Philip Emery. Emery, who was partially deaf himself and an experienced teacher, agreed to help the children and so rented a building in nearby Baldwin City. Legislators looked favorably on this effort and in 1863 appropriated a state subsidy of four dollars per week for each student. Had Mr. Emery remained in Baldwin City, it is likely that this community would have been selected for the permanent state school in 1864. Following some bad advice, however, Emery moved his classes to Topeka that year in hopes of obtaining a larger subsidy. The lawmakers, loath to grant Topeka a second public institution, voted instead to accept an offer from Olathe of a 20-acre site and another 160 acres for an endowment.[58]

The decision of 1864 did not end the issue. In the months before the next year's legislative session, Baldwin people planned a strategy to reverse the earlier judgment. As a community with a strong free-state heritage themselves, they hoped to tap into this sympathy just as Osawatomie had done. They also resolved to attack Olathe directly, playing to a combination of real and imagined fears by claiming that "the border, and especially Johnson County, was too much exposed to the raids of the bushwhackers and rebels to allow the erection of temporary buildings." The ploy probably would have worked had not the new speaker of the house of representatives been an Olathe attorney, John T. Burris. Burris delayed the decision in 1865 and the next year (with the war over) engineered a reaffirmation of the Olathe site.[59]

The victory of Olathe supporters in 1866 marked the end of the public-facility allocation in early Kansas. Although many more such institutions would be built, beginning with a normal school at Leavenworth in 1870 to serve the northern counties and a second insane asylum at Topeka in 1875, these newer placements generally went to communities that were already prosperous through other economic activities. With some exceptions, especially in western Kansas, the time when an institution could almost literally elevate a town from obscurity into prominence was over after 1866.

In closing this examination of the institutional assignments, it is appropriate to assess how the various decisions worked out once the buildings were in place and the employees hired. From the perspective of the early twenty-first century, of course, the good fortune of Topeka, Lawrence, and Manhattan is apparent. A look at state appropriations provides a more accurate reading for the early years, however, because these numbers would have been primary in people's minds at that time. Comparable figures are available for all the facilities except the capital.

Of the three asylums, the first awarded fared the best. In the eight years from 1867 through 1874, legislators sent an average of $28,028 annually to Osawatomie. The figure at Olathe's school for the deaf was $15,841, while Wyandotte's institute for the blind received $10,179. Taking into consideration the population sizes and other economic activities in these communities, the importance of the asylum to Osawatomie becomes even greater. Whereas county offices employed many people in Olathe and Wyandotte, and Wyandotte had become a major railroad center by 1874, Osawatomie at this time had little besides its institution.

The relative state appropriations for the three universities are somewhat misleading in that the agricultural college at Manhattan derived additional income from its large land grant. When an allowance for this is made, however, the average numbers for 1867–1874 are roughly equal: $12,905 for Manhattan, $17,466 for Emporia, and $22,736 for Lawrence. Having a college in one's community may have been more prestigious than hosting an asylum for the blind or the deaf and dumb, but it was only marginally more lucrative in hard dollars. Other than Topeka with the capital, the clear winner in the game of institutions was Leavenworth. Its average appropriation, $87,533, dwarfed the others. This large number was just what businessmen there had expected. It also represented only a portion of the facility's total value because it excluded money generated by lucrative, prison-based manufacturing plants.[60]

4

Railroad Promotion and a Reconceptualization of Urban Kansas, 1863–1880

The associations between city growth and regional transportation systems are numerous and intimate. The two ideas are so interconnected, in fact, that they are difficult to separate. The developmental models discussed in chapter 1 assume that the primary relationship is causal. Transportation drives growth, they say, and lies at the heart of key mercantile terms such as entrepôt and the redistribution of goods. Judging from these models, a predictable system of cities will evolve so long as the principal means of conveying goods and people remain constant. What might happen, however, if the existing transportation system were disrupted and then replaced with something entirely new? This scenario, looking almost as if designed as a laboratory experiment, is precisely what happened in early Kansas.

In 1854, when the territory was opened, steamboats and freight wagons carried most heavy goods. Locations on navigable rivers were therefore ideal for aspiring cities. Less than seven years later, however, as the territory became a state, river traffic had declined drastically, and freight companies were scaling back on orders for new wagons. Railroads, the new kingmaker, had come to the elbow region and were about to spill across the entire state. In cases where entrepreneurs planned to push tracks ahead of the settlement frontier, the power of these new companies to create the initial urban system was obvious to all. How would they respond to such an opportunity? A more interesting question, perhaps, was what these railroads might mean to Leavenworth, Atchison, Lawrence, and the other sizable cities already existent at the time. Would business leaders in these places seek railroads aggressively, or would they assume that the tracks would come to them without active encouragement? In 1872, once the period of initial construction was completed and businesses had a chance to adjust, would the urban hierarchy of northeastern Kansas look much the same as it did in the years of overland freighting or radically different? These questions form the core of this chapter.

Early Proposals, the Pacific Railroad Act of 1862, and Leavenworth

Scholars traditionally date the beginning of the railroad period in Kansas to 1863, when workers for the Union Pacific, Eastern Division, began to lay tracks westward from the Missouri border near the town of Wyandotte. This is a critical year, but its use masks an important planning component for the movement that goes back at least fifteen years earlier. As the first large groups of Americans settled in California, Oregon, and Utah during the 1840s, politicians dreamed about grand belts of steel that would bind the continent together, ensure manifest destiny, and initiate lucrative trade relationships with China and Japan.[1] Such thinking underlay the very creation of Kansas Territory, of course, because before the big railroad plans could proceed, this most logical, central routeway needed to be wrested from Indian control. Local people, however, temporarily forgot the original transportation motivation for the Kansas-Nebraska Act once the territory lay open in 1854. More immediate concerns of city founding, land speculation, and freight contracts, not to mention free-soil politics, naturally beckoned.

Although reduced to secondary importance during 1855 and 1856, railroad talk was never absent from territorial Kansas. People were aware, for example, that in 1853 tracks had been laid to link Chicago with the urban centers of the East Coast, and that this city, in turn, was sending out lines toward the northwest, west, and south. They also saw the many advantages that the eastern railroads offered to merchants, including dependable shipping schedules in all seasons and enlarged areas of trading dominance. The Chicago example had more than theoretical interest as well, because the Hannibal and St. Joseph Railroad, then under construction to the very doorstep of Kansas, was directly linked on its eastern end to the emerging Illinois metropolis.[2]

The territorial legislature of Kansas exercised its power to grant railroad charters as early as 1855, but serious thinking about construction did not begin until 1857. This was the time of the Leavenworth land boom, of course, and the year in which a large Northern immigration quietly resolved the issue of slavery. It also was the third year in a row of water levels in the Kansas River too low to allow regular steamboat traffic. Since chartering railroads was easy and inexpensive, the first response of city leaders was to establish as many as possible. Several hundred such paper entities were created in the territory before 1860, and even smaller towns projected several intersecting lines for themselves. Although such dreaming was good for local egos, it ignored the practical issue of how people on the frontier could raise the millions of dollars needed to construct even one such line across their territory.

By 1860, leaders from each of the larger communities in the region had begun to focus their fund-raising talents on one or two railroads. To further this process, they also sought alliances with business colleagues in other communities. A railroad westward from Leavenworth, for example, might well be routed through Manhattan. Since both towns would benefit from such construction, it would make sense for them to share in its promotion. Such pairings of cities, although helpful, still involved too few people to raise the quantities of money required. Promoters soon realized that their collective railroad future could be ensured only through a substantial enlargement of the cooperative pool. If Kansans somehow could agree on a territory-wide plan for construction, they would have a better chance not only of financing it themselves but also of interesting outside capitalists and federal officials in their investment potential.

A convention to create a united railroad front met at Topeka in October 1860. As one would expect with so much at stake, it was a tension-filled event and not entirely successful. A proposal to allot each county a single vote was adopted, but only after vigorous opposition from delegates who represented Leavenworth and other relatively large communities. The Leavenworth people then withdrew from the meeting along with most of those from Douglas County. These losses notwithstanding, the remaining delegates agreed upon a plan for five trunk railroads. If all were to receive federal land grants, as was hoped, a reasonably complete transportation system might be in place within a decade.[3]

The Topeka plan of 1860 was biased against Leavenworth, as might be expected with the absence of that town's delegates, but otherwise was designed to reinforce the dominance of existing cities (Map 6; see p. 18). Two lines were to begin in Atchison, where they would join the extension of the Hannibal and St. Joseph Railroad that had opened earlier that year to a point just across the Missouri River. One of these would extend almost straight west from this ferry point; the other would run southwest to Topeka and then beyond, taking the general path of the Santa Fe Trail. A third railroad would originate in Lawrence. It would build south to Garnett and then on to Galveston or some other port on the Gulf of Mexico that could provide an outlet for agricultural products. The fourth line would start in Wyandotte and ascend the north side of the Kansas River valley, passing near Lawrence and Topeka, and through Manhattan and Fort Riley, on its way to the Pike's Peak gold country. Finally, the last projected line would serve the as yet sparsely settled counties of east-central Kansas. It would extend from the Osage River in Linn County to Emporia and then ascend the Neosho Valley northwestward to end at Fort Riley. Six places would become important rail junctions in this plan: Atchison, Council Grove, Fort Riley, Garnett, Lawrence, and Topeka. It probably made sense to most people in the region, but certainly not to those in Leavenworth and several border cities of Missouri.[4]

To better understand railroad development in early Kansas, one must pair the results of the Topeka Railroad Convention with attitudes and strategies found in St. Joseph and Leavenworth. As the largest urban places in the region by a considerable margin, these cities also possessed the most political influence and the best access to capital. Of the two, St. Joseph theoretically occupied the stronger position because it was larger and enjoyed the only railroad connection to the east. Its citizens, however, were divided over how much to push for an extension into Kansas. The newly completed Hannibal and St. Joseph line had cost more than $9 million, very expensive from their perspective. Most residents did not want to incur additional debt. Merchants also had just begun to enjoy the many advantages that accrued from having a railroad terminate in their town. The new tracks were opening markets to the east for area producers of beef and pork, while at the same time bringing in manufactured goods for redistribution at lower prices than ever before. Who could blame the local wholesale distributor, wagon maker, or pork packer for not wanting to disrupt the status quo?[5]

The people of St. Joseph were shaken from their uncertainty when Atchison promoters decided to extend the Hannibal and St. Joseph Railroad southwest to the vicinity of their Kansas town. If the Missourians did not act soon, they could lose competitive advantage. Thinking in this way, St. Joseph leaders joined with colleagues from Elwood, their daughter community just across the river, to create the Elwood and Marysville Railroad. The plan was to push the St. Joseph line ninety miles farther west to a community that had grown up where the Oregon Trail crossed the Blue River. Workers surveyed the entire route in 1859 and laid the first few miles of track in Kansas history during March and April 1860. The organization was badly underfinanced, however, and its hopes to raise additional monies were quashed later that year when Missouri River waters washed away several blocks of Elwood. Nearly simultaneously, laborers completed work on the twenty-mile track of the Atchison and St. Joseph Railroad. With these developments, St. Joseph leaders effectively granted Samuel Pomeroy and his Atchison associates an equal opportunity to construct the first steel connector between eastern manufacturing centers and the Kansas frontier.[6]

The business maneuvers in St. Joseph and Atchison were closely watched in Leavenworth. There a group of the most politically astute people in the state, convinced that the future for railroads was bright, worked hard to overcome the financial problems that plagued their rival to the north. Thomas and Hugh Ewing, sons of a former secretary of the interior who now worked as an attorney in Washington, D.C., were leaders of this effort, along with Andrew

J. Isaacs, James C. Stone, and James H. McDowell. Their plan was simple: to revive and expand a company called the Leavenworth, Pawnee and Western that, as its name implies, had obtained an early charter to build tracks westward along the existing military road to and beyond the original capital of the territory.[7]

The Leavenworth men conceived an innovative, elaborate plan to raise capital. Although borderline illegal in several instances, it also was highly successful. The first step, in 1860, involved a treaty with the Delaware Indians whereby the railroad company obtained authorization to buy 223,966 acres of tribal land for not less than $1.25 per acre. The idea, of course, was to generate capital through the resale or mortgage of this property. When not enough money was available to conclude this initial purchase, the wily entrepreneurs solved the problem with a supplemental treaty in 1861. Now the railroad was permitted to pay for the lands with bonds that were secured by a mortgage on part of this same acreage. It was a suspect procedure, to be sure, and one that required liberal bribery to pass the United States Senate. Later that same year the group made a similar arrangement to obtain 340,000 acres of the Potawatomi Reservation, some of which the railroad would need for its route between Topeka and Manhattan (Map 3; see p. 15).[8]

With the two land coups, the railroad directors expected their company to become an attractive investment for eastern capitalists. It probably would have, too, had not the Civil War intervened. Undaunted, the Ewings and their associates smoothly pushed ahead to an alternate strategy. They hired lobbyists and began to pitch their railroad to Congress as the perfect vehicle to form the eastern stem for the much discussed transcontinental connection to California. Although the war years might not seem a promising occasion for a legislative push to construct a western railroad, they actually were in a way. The concept had been in the air for more than a decade. While the political union had been intact, however, Southern congressmen had opposed construction along the best natural routeway, claiming that the Wyoming basin and the Great Salt Lake oasis lay too far north to serve their region. Now the project became politically feasible.

The railroad promoters all agreed that a single line would be appropriate from San Francisco eastward through the Rockies and onto the Great Plains. Where and how, though, would this line join with the emerging rail system in the midwestern states? By being first on the scene, the "Leavenworth ring" made a strong effort to locate this eastern end exactly in their hometown. Had they been successful, it would have been a coup among coups, an event that almost surely would have catapulted the city into metropolitan status. It came very close to happening.

The original bill before Congress in 1862 would have authorized the Leavenworth, Pawnee and Western company to construct the trunk line for this first transcontinental railroad all the way to the western boundary of Kansas. At that point a newly created Union Pacific corporation would take over and extend the tracks into the mountains. Two branch lines, added to satisfy regional interests, would also join this trunk somewhere in Kansas. One of these would be an extension of the Hannibal and St. Joseph. The other would emanate farther north, from an unspecified point in Iowa. As the debate proceeded, however, powerful people with interests in New York, Buffalo, Cleveland, and Chicago began to realize that a main line through Nebraska and Iowa might serve their business purposes better than would this Kansas route. A political battle thus developed along lines of latitude. The New York and Chicago crowd pitted themselves against not only the Leavenworth people but also representatives from a substantial group of cities that would benefit from this more southerly Kansas proposal; these included St. Louis, Cincinnati, Pittsburgh, Philadelphia, and Baltimore.[9]

The final version of the bill, passed as the Pacific Railroad Act of 1862, was a compromise that took away many of the Leavenworth advantages. The Union Pacific was now authorized to build the branch into Iowa, as well as the main track across Wyoming and the High Plains. This maneuver greatly increased the chances for the Nebraska-Iowa route becoming dominant over the others. Another blow was that Samuel Pomeroy, now a United States senator, had inserted a clause to have the St. Joseph branch extend west via his hometown of Atchison. This put a rival terminal only thirty miles from Leavenworth, too close for either city to maximize profits. As for the Leavenworth, Pawnee and Western itself, the company still was awarded a subsidy for construction, but it saw its role diminished from primary status. It was now to build west only to Fort Riley before turning northwest to form a junction with the Union Pacific track at the hundredth meridian in Nebraska (Map 7; see p. 19). The most important change, however (at least from the perspective of Leavenworth), was that the politicians modified the eastern terminus. At the insistence of St. Louis interests, the railroad now would begin not at Leavenworth but at a point on the Missouri border just south of the Kansas River. Although this new location made sense in a number of ways, including an easier connection with the Pacific Railroad of Missouri then under construction from St. Louis toward Kansas City along the south side of the Missouri River, it was potentially devastating for Leavenworth.[10]

Somewhat surprisingly, Leavenworth townspeople did not voice great concern over the loss of the eastern terminal. This was because local men still controlled the Leavenworth, Pawnee and Western. Surely, everyone thought, this management team could mitigate the damage. All that would be needed was for the main line to deviate north to their city from the new starting point before turning back west toward Fort Riley.[11] Although this particular routing was not specified in the legislative act (it said only that at least a branch line must enter the city), many people believed that Ewing, Isaacs, McDowell, and Stone would work their magic once again. All in all, residents did not know quite how to react to this complicated news of 1862. Certainly the investors in the railroad itself had to feel good because they now enjoyed a federal guarantee of financial support. Other citizens could take their choice of scenarios. Optimists proclaimed that getting a terminus of any type was a victory. They felt that Leavenworth's existing size would continue to attract most of the area's manufacturers and distributors. When a connection was built to the Hannibal and St. Joseph line, the city would easily retain is premier position within the state. Pessimists, of course, saw another future. Wyandotte, a town already associated with Missouri businessmen, would surely boom with the new railroad. In conjunction with Kansas City across the state line, this hitherto innocuous Indian community might now pose a serious challenge for the attraction of new entrepreneurs.

Armed with a new land grant and access to other federal financing, Leavenworth, Pawnee and Western leaders hired a construction firm in September 1862. They broke ground in December, first at Wyandotte as required by the law, but with most of their efforts concentrated at Leavenworth. Things almost surely would have continued in this locally biased manner had company officials not run out of money the following June. They were eligible for loans of $16,000 per mile on government bonds but could begin to collect these only after the completion of a forty-mile section of track. When an appeal to city leaders for additional funds was rejected, the directors had no choice but to sell controlling interest in their company to a group of outside investors headed by a New York banker named Samuel Hallett and the western explorer John C. Fremont. It was a move that would haunt Leavenworth residents for decades.[12]

Samuel Hallett, who assumed control of the railroad, was a practical businessman. He came to Kansas in August 1863; changed the company's name to the Union Pacific, Eastern Division; and shrewdly played the people of Leavenworth and Wyandotte against one another. He promised each group that he would build first from their particular city. He also requested financial aid from both communities: $150,000 in bonds from Wyandotte, and $100,000 from Leavenworth. Both requests were granted, but the Leavenworth mayor said that he would withhold his town's share until twelve miles of track had been completed. John Cruise, a witness to these events, recalled later that this decision was indicative of a haughty attitude that permeated Kansas's largest community: "The little city believed itself indispensable to Hallett, and the only possible terminus for the road he was building. Therefore it demanded exorbitant prices for land and for all supplies." Whether for these reasons or, as Hallett himself wrote the following March, because construction costs would be substantially lower and progress faster if he built straight west from

Wyandotte up the valley of the Kansas River, the company abruptly stopped work in Leavenworth that September and concentrated all efforts at Wyandotte. Leavenworth eventually received its branch (a connector to the main line at Lawrence), but work on it did not begin for nearly two years.[13]

Cities and the Union Pacific, 1863–1867

Once tracklaying began near Wyandotte in September 1863, the progress of the Union Pacific was more or less constant across the eastern part of Kansas. There were delays, to be sure, caused by wartime shortages of labor and supplies, and by the financial chaos that accompanied Samuel Hallett's murder by a disgruntled employee on July 27, 1864. Still, the railroad was opened to Lawrence in November 1864, to Topeka in January 1866, to Manhattan in August 1866, and to Fort Riley in October 1866.[14]

Because almost the entire territory east of Fort Riley had been settled before railroad construction began, and because the tracks paralleled the older transportation corridor of the river, the Union Pacific had limited effects on the urban structure within the Kansas Valley. There was an obvious boost to Wyandotte, of course, but otherwise company officials concentrated on laying tracks as efficiently as possible in order to secure their government funding. Instead of promoting north-bank towns that might compete directly with Lawrence and Topeka, they restricted their creations to small stations built as needed where the line passed through the relatively unpopulated Delaware and Potawatomi lands. Two examples that grew to modest size carry the names of company officials: Edwardsville in Wyandotte County, and Perry in Jefferson County.

The north-bank location of the Union Pacific within the Kansas Valley was a legacy of the railroad's original 1855 charter to connect Leavenworth and Fort Riley. Samuel Hallett and his successor, John D. Perry, would have much preferred to build on the river's south side. This would have increased the popularity of their company in Lawrence and Topeka and likely produced bigger bond issues from these places. It also would have eliminated the expense of constructing a bridge over the Kansas River just southeast of Wyandotte to connect the company's mandated starting point with their similarly required north-bank main path. Both of these issues carried geographic implications for city growth.

The bridge issue was resolved temporarily with a pontoon structure. A permanent design, however, would have to contain an expensive center section that could be raised periodically to allow the passage of boats up and down the stream. When Union Pacific officials complained about this cost, state legislators had their first opportunity to demonstrate support for the railroad industry. Their response, in early 1864, was immediate and without controversy. The lawmakers simply declared the Kansas River and four of its major tributaries unnavigable. By so doing, they thereby permitted construction of cheaper railroad bridges with much less water clearance. Union Pacific engineers promptly took advantage of the situation.[15]

The Kansas River bridge, in both its temporary and permanent forms, breathed life into Wyandotte. The railroad company built a short spur into the town proper, and in November, Hallett announced that this community would become the company's headquarters as well as the base for its machine shops and engine houses. With area newspapers advertising that a thousand laborers were needed at wages of $1.50 per day, Wyandotte suddenly became as important as its founders had hoped. A period of doldrums during the early Civil War years, when "loose clapboards, broken windows, and faded paint, indicate[d] a place where early growth surpassed its subsequent importance," now was quickly forgotten.[16]

The potentially awkward relationship between the railroad and all southbank communities along the Kansas River was exacerbated in the cases of Lawrence and Topeka because these particular town sites occupy southerly bends on that stream. In order for the tracks to reach a position directly across the river from Lawrence, for example, they would have to deviate two and a half miles away from a direct path up the valley. Hallett, possibly as a gambit, therefore proposed to bypass both cities in the interest of efficiency. He argued that his biggest concern was to beat the other (Iowa and Nebraska) Union Pacific company to the hundredth meridian because the federal legislation had stated that the first company to reach that point would receive the contract to build on toward the west. This race certainly was on his mind, but Hallett also hoped to exact more money from the towns if he were to construct the extra mileage. The announcement produced outrage from James Lane, the United States senator from Lawrence, who threatened to withhold the railroad's federal subsidy. The company finally conceded defeat and authorized the added distances.[17]

Once it was certain that the Union Pacific tracks would closely approach Lawrence and Topeka, leaders in the two cities planned how best to capitalize on the development. The biggest immediate issue was political. Neither city had legal jurisdiction north of the Kansas River where the depots would be. Because Shawnee County happened to span the river, incorporation of the station there into the Topeka city limit was possible. The situation was more complicated downstream, where the river formed the boundary between Jefferson and Douglas Counties. When the railroad arrived and local people created a community called Jefferson at the site, the Lawrence lobbyists moved into action. At the next legislative session, in early 1865, they managed to carve a township out of southern Jefferson County and have it added to Douglas. The

Headquarters of the Union Pacific Railway Company, Eastern Division, at Armstrong, 1867. Samuel Hallett's decision to move his railroad's operations from Leavenworth to the mouth of the Kansas River in November 1863 invigorated the economy of Wyandotte and adjacent communities. The office and nearby yards employed eight hundred people by 1870. (Courtesy Kansas State Historical Society)

town of Jefferson promptly rechristened itself North Lawrence, and five years later the two communities agreed to consolidate. Topeka and North Topeka had done the same three years earlier.[18]

Of all the preexisting communities in Kansas, the one with the highest expectation for the Union Pacific's arrival was surely Junction City. This town, founded in 1857 where the Smoky Hill and Republican Rivers unite to form the Kansas, was in many ways the successor to the stillborn capital of Pawnee (Map 5; see p. 17). It was three miles from Fort Riley and supplied this institution in many ways. As the Civil War ended, Junction City also was fast becoming a gateway community for land seekers in the fertile Republican, Smoky Hill, and Solomon Valleys that lay just to the west and northwest. To this end, government officials had placed a public land office there in 1860.

Junction City's population increased from a few hundred people in the early 1860s to 3,002 in 1865. Although some of this growth was a product of the fort and the land office, most of it was in anticipation of the Union Pacific. What made the expectations so much higher here than elsewhere? The answer lay in an analogy with the Missouri River elbow region of the previous generation. At Junction City the law mandated that Union Pacific contractors change direction. Instead of building on to the west, they were to turn northwest and follow the Republican River into Nebraska and a junction with the other branch of the transcontinental system. This pathway had been known since 1862, when the commissioner of the General Land Office had withdrawn acreage from public entry along the Republican as the first part of the railroad's land grant. The implication was obvious: Junction City would become the new Westport. Whether the destination be the gold fields in Colorado, the long-established trade at Santa Fe, or the new colonies in southern California, this would be the place for overland freighting to begin, for buyers of western goods to assemble, and for distributors to build their warehouses.[19]

Junction City served as the railroad's temporary terminus during the winter of 1866–1867, and community merchants sampled the glory they had been anticipating. It was to be a bittersweet moment, however, because residents knew it was not to last. A series of business decisions made by railroad executives over the previous two years had abruptly changed the emerging roles and growth potentials for this and many other Kansas communities. Construction delays in 1864 following the death of Samuel Hallett initiated the process. This disruption meant that the Eastern Division now had little hope of reaching the hundredth meridian before its Iowa-Nebraska rival, and therefore of obtaining the construction contract to build the main road on to the west. This reality, plus the growing influence of Chicago during the war years, made it apparent to Hallett's successors that the Kansas track, when completed, would be only a feeder to the one in Nebraska. This was a galling realization not only to state pride but also to the railroad's backers in St. Louis and the Ohio Valley, whose hopes of dominating western trade had been so high only a few years earlier.[20]

One way to cope with a losing proposition is to change perspectives and try to envision an alternative goal. Another is to alter the rules of the contest. John D. Perry (the railroad's new president) and his political allies adopted both strategies. The first hint of a transformation occurred in 1863 when Senator James A. McDougall of California proposed an amendment to the Pacific Railroad Act of the previous year. He wanted to add a branch line to Denver. This idea was supported in Iowa but opposed by Kansans, who feared that this plan might interfere with their hopes to control the main line west. The amendment thereby was defeated. The next year, however, as the population in Denver continued to grow, the possibility was revisited in an indirect manner. As part of legislation to increase funding for the Pacific railroad, a clause was inserted to allow any of the branch companies to depart from their routes previously selected and to join the main stem west of the hundredth meridian instead of to its east. Petitions would have to be approved, of course, and no additional federal monies or land grants would be available beyond those already approved.[21]

The somewhat contradictory terms of this modification to the railroad legislation—allowing route changes but not funding them—reflected mixed opinions in Congress. The Kansas legislature sponsored an official resolution to endorse the change, however, presumably with some preknowledge of the possibilities for a connection to Denver. The Eastern Division's board of directors probably made its decision to replace the route up the Republican Valley with one directly west to Colorado early in 1865. The directors waited several months to announce the decision publicly, doing so first in a letter to the secretary of the interior on November 27. By the following January, reports about the prospective new corridor were circulating among potential financial backers. The essence of the argument was that the Kansas road would now be a direct competitor with the Nebraska line rather than just a complement to it. The riches of the Colorado mines were the most beckoning attraction, of course, something that a route through Wyoming could not equal. The announcement also described plans to cross the Continental Divide and to lay track at least as far as the Salt Lake oasis, where a junction would finally occur with the transcontinental line. Opening up western Kansas with a route that followed the valley of the Smoky Hill River would be a part of the plan, of course, but hardly its central feature. The proposal was debated at considerable length in Congress during June 1866 and was signed into law that July.[22]

The implications of the Union Pacific's new route were profound. Most of the discussion in the national press focused on how the railroad would benefit Colorado, how the financing of the extra mileage could be arranged, and whether or not a branch should be built from it southwest to New Mexico Territory. Kansans, however, saw issues at closer hand. Junction City people were the most bitter, of course. They expected that Fort Riley would survive only a few more years, following the brief life cycle of most frontier posts, and had counted on the railroad bend to create a western shipping station that could replace those jobs. Settlers who had recently acquired lands in the Republican Valley felt cheated as well. Because of the terms of the Union Pacific grant, preemptors on government land there had paid twice the normal rate per acre ($2.50 versus $1.25) and had been limited to tracts half the normal size (80 acres versus 160). The voices on this frontier and in Junction City were small in number, however, and their positions isolated. They did not arouse general sympathy. Moreover, since most of these people had been resident only a short time, it was relatively easy for them to move on to more promising locations.[23]

Among the established communities in Kansas, Atchison was the one affected most negatively by the reorientation of the Union Pacific. This was because, under the terms of the Pacific Railroad Act of 1862, a feeder line was to run a hundred miles straight west from this city to a junction with the Eastern Division somewhere near present-day Clifton on its Republican Valley segment (Map 7). With the Eastern Division's abandonment of its Republican route, of course, the Atchison branch faced the possibility of becoming a dead-end railroad. A situation of this magnitude would have panicked people in a smaller or less secure community. Atchison, however, had a substantial population of 3,318 in 1865 and a United States senator, Samuel Pomeroy, in residence. Townspeople were disappointed but nevertheless were confident that this problem could be overcome.

Atchison entrepreneurs had endured a period of business inactivity during the Civil War. They had completed a rail connection to St. Joseph in 1860 but saw the western trade fall off to nearly nothing once the fighting began. In their lobbying efforts for the Pacific Railroad Act, they profited from two situations. One was having in their corner Senator Pomeroy, a man renowned both for political skill and for corrupt practice. The other was in having only half-hearted opposition from the St. Joseph business community and the owners of the Hannibal and St. Joseph Railroad. I discussed the attitudes in St. Joseph earlier in this chapter. The lack of insistence from eastern financial backers to extend the Hannibal and St. Joseph seems more inexplicable until one realizes that a conflict of interest was involved. The men in control of this enterprise, led by John M. Forbes of Boston, also managed the Burlington and Missouri River Railroad Company in Iowa. When the opportunity arose for the Burlington company to construct the main Iowa connector to the Union Pacific at Omaha, these men logically decided to concentrate their political efforts on this project. To achieve success there (which they did), they were willing to sacrifice the St. Joseph position.[24]

The vehicle chosen for the Atchison push to the west, the Atchison and Pike's Peak Railroad Company, was chartered in 1859. Its route was suitably vague so that the builders could bargain for local support, and Senator Pomeroy had overseen a favorable treaty in 1863 with the Kickapoo Indians, whose lands the track would have to penetrate. With 123,832 of the Indians' acres promised to the company for $1.25 per acre and permission to sell the land before making any payments, the future looked bright. In fact, this coup, in conjunction with the federal land grant and an additional subsidy of $16,000 per mile included in the Pacific Railroad Act of 1862, made things as rosy as could be imagined. Construction began in 1865, and the next year the directors changed the company name to a more appropriate Central Branch of the Union Pacific.[25]

In retrospect, it seems misguided that the Atchison group incorporated the Union Pacific name into their railroad venture just after John Perry had announced that the main Kansas line would head to Colorado instead of Nebraska. The name change was made with confidence, however. Congressional members all knew what was at stake here, and most people expected that approval of the Colorado destination would be accompanied by additional subsidies for the Atchison line. It thereby could be extended at least as far west as the hundredth meridian, where it could make its own connection with the larger Union Pacific system either near Fort Kearny, Nebraska, or somewhere on the Kansas High Plains. The year 1866 was a bad one to be seeking money, however. Dollars were needed for reconstruction in the war-torn South, and several representatives noted how generous the federal government already had been to Kansas railroads. The Atchison provisions thus failed to make it into the final bill.[26]

Kansas people fully expected that the funding oversight of 1866 would be rectified by future Congresses. They pushed ahead with their Central Branch, deciding on a nearly direct line to the west rather than a slight deviation to the north or south that would have aligned the path of the tracks with a row of county-seat towns. Construction, mostly along stream divides, went smoothly, with the hundred-mile limit written into the Pacific Railroad Act being reached in 1867. This spot, in southwestern Marshall County and about thirty miles short of the Republican River, became the temporary terminus. It was called Waterville to honor the New York hometown of the construction engineer. Meanwhile, funding efforts continued in Washington, D.C. Two of them came extremely close to fruition: a one-vote loss in the Senate, and later, a last-minute reversal of an executive order by President Grant in 1873.[27]

A reader of Atchison newspapers in the late 1860s would find little evidence of disappointment in the fate of the Central Branch. Like true boosters, the people there were still sure that the funding would materialize. They praised the tributary area that the enterprise already had opened up to local merchants and told anybody who would listen that the Eastern Division's tracks in the Kansas River valley were prone to flooding. They also saw the Central Branch as only one of many railroad lines that would soon enter their community. Like their rivals in Leavenworth, Lawrence, Wyandotte, and elsewhere, they envisioned themselves as the "Great Railroad Centre of Kansas." A line from Chicago via the Hannibal and St. Joseph already was in place. St. Louis people surely would be connecting soon, and plans were afoot for a partnership with Topekans to build a line southwest toward Santa Fe. What turned out to be the reality of the Central Branch (and in some ways for Atchison railroads in general) seems to have been expressed early on only by a writer from Lawrence: "Nobody knows where it is going. Its present terminus is Centralia. Its ultimate terminus is one of the problems of the future which the man in the moon could solve as readily as the managers of the road. . . . [It] is now hunting around on the prairies for a place to stop."[28]

Nodal Cities in Northeastern Kansas, 1866–1870

By the mid-1860s, with the paths of the Union Pacific's Eastern Division and Central Branch now known, people in the towns along both lines realized that they enjoyed greatly increased potentials for growth. The former railroad was tied directly to St. Louis, the latter to Chicago, and both seemed like sure bets for extensions farther to the west. The task for planners in these towns was now to capitalize on the initial good fortune by transforming their communities into transportation nodes where a second railroad would intersect the existing route. Given the east-west orientation of the two Union Pacific tracks, these entrepreneurs therefore concentrated primarily on the development of north-south connections. Southeastern Kansas, a well-watered land that was beginning to attract large numbers of Civil War veterans, was one general inducement. Another was a potential to develop a cattle trade with Texans and the residents of Indian Territory. People also talked about the creation of ports on the Gulf of Mexico where Kansans could perhaps export agricultural goods more cheaply than at the older and more distant cities on the Atlantic.

The first two nodal plans to emerge involved the four largest cities in the state working together in uneasy partnerships. Each arrangement took advantage of strategies and charters that had been in existence since territorial times, and both had been accorded tacit statewide approval in 1860 by the Topeka Railroad Convention (cf. Map 6 and Map 8 [see p. 20]). One of these would link Leavenworth with Lawrence and then extend south toward the Gulf of Mexico at Galveston, Texas. The other would connect Atchison and Topeka before heading southwest in the direction of New Mexico. The plans united the cities of Kansas against their Missouri rivals of St. Joseph and Kansas City. They also gave each Kansas senator, James Lane of Lawrence and Samuel Pomeroy of Atchison, a personal political plum.

Any railroad that hoped to be successful in the cash-starved Kansas frontier of the 1860s needed a federal land grant. Senators Lane and Pomeroy thus played critical roles. Lane initiated the process in December 1862. His bill was typically ambitious. It proposed support for a two-branched line that would run south from Lawrence. One set of rails would go straight to the border of Indian Territory; the other would angle southwest to Emporia and then extend on toward Santa Fe. Lane, of course, was attempting to combine the two older plans into one and, in this way, elevate the status of Lawrence. Pomeroy saw the future in different terms. When Lane's bill came out of committee in February 1863, Pomeroy offered an amendment that carried through both houses and into law with little discussion. One change extended the Lawrence railroad north to Leavenworth. The other made the Santa Fe line an independent entity. Both pathways across the state were left vague, with the former com-

Bridge of the Leavenworth, Lawrence and Galveston Railroad in Lawrence, 1867. Promoters of railroads running south from Lawrence, Topeka, and Junction City during the late 1860s could import supplies on the existing Union Pacific tracks but then had to negotiate the Kansas River. The photographer of this image was looking south across the company's temporary pontoon bridge. (Courtesy Kansas State Historical Society)

pany required only to reach Indian Territory and the latter to exit Kansas somewhere on the state's western border.[29]

The Civil War delayed implementation of the new grants, but fund-raising began in earnest in 1866. with hopes for construction the next year. Lawrence clearly occupied the favored position at this point, largely because that city had already become a railroad junction. Union Pacific officials, required by law to make Leavenworth a terminal for the Eastern Division, had decided to do this with a spur northeast from the main line at Lawrence. Work on this segment had begun in January 1866 and was completed five months later. The Leavenworth, Lawrence and Galveston, as the company officials had named it in 1865, thus found itself with a gift of sorts. Its northern segment was complete, leaving workmen to concentrate only on construction through southern Kansas. With Senator Lane as president of the company, it appeared as if no other railroad could compete and that the urban tandem of Lawrence and Leavenworth would emerge even more powerful than before.[30]

The Leavenworth, Lawrence and Galveston was indeed the first north-south railroad in Kansas to begin construction, and it opened twenty-six miles of service to Ottawa on January 1, 1868. The tracks did not advance farther for another eighteen months, however. Senator Lane had committed suicide in July 1866, and the company soon fell victim to mismanagement and lack of funds. Although this stagnation was good for Ottawa, the seat of Franklin County, it was bad for Lawrence's metropolitan dreams. Rival cities now had a second chance to compete.[31]

Cyrus Holliday, the same politically astute entrepreneur who had cofounded the city of Topeka, conceived the primary challenging company. Through an alliance of businessmen, he was able to create the Atchison, Topeka and Santa Fe Railroad, Kansas's largest and most famous line. He also made sure that Topekans were in charge from the outset. Holliday, in fact, had earlier promoted the influential Topeka Railroad Convention in 1860, doing so primarily to advance his personal business agenda. To gain influence, he partnered with Luther C. Challis, an Atchison banker he had met when they both served in the state legislature. Challis brought in Pomeroy, and things progressed smoothly from then on, at least from the perspective of Holliday.[32]

Construction began in November 1868 on a segment immediately south of Topeka. Both the direction and the starting point were controversial. Company officials justified them, in part, by the known presence of coal only sixteen miles away in Osage County. This mineral could provide fuel for the engines and a product to generate freight. In addition, a track in this direction could cut off the western trade anticipated by the rival Leavenworth, Lawrence and Galveston company and open up a large area of the state unserved by any line. Finally, the presence of the Union Pacific across central Kansas created a practical need to go south a considerable distance before turning west to meet the terms of the land grant. Atchison residents were not entirely convinced by this series of explanations, especially since they were voiced by the Topekan Cyrus Holliday, who had become president of the company. The grant required that the line be built to their city, however, and so they bided their time. Service on the first segment of track, from Topeka to the coal town of Carbondale in northern Osage County, began in June 1869. The finances remained solid, and so officials were already planning a rapid extension to Emporia and beyond.[33]

A third successful scheme to create a north-south railroad tied into the Union Pacific system occurred along the western fringe of settlement. The plan originated in Emporia, a community that had been founded with railroads in mind and that had just received an economic boost from being awarded the state's teacher-training college. In 1865, as rumors spread that the Osage Indians were about to give up their lands in southeastern Kansas, several Emporia men realized that the wide valley of the Neosho River, which ran south from their city into the Osage country, would soon fill with war veterans and their families. To tap this trade, as well as the cattle business in Indian Territory, Robert S. Stevens and Preston Plumb sought allies in Junction City and elsewhere and then petitioned for a federal land grant. They received it easily in February 1866, using the impressive company title of the Union Pacific, Southern Branch.[34]

The Emporia people acted decisively and with a common front in their railroad pursuits. This was partly a result of having watched developments in Council Grove, their neighboring community to the northwest. Because it had been the premier stopping point on the Santa Fe Trail, most observers predicted that Council Grove would become a major railroad center as well. It was one of six nodal cities projected by the Topeka Railroad Convention in 1860, for example, the spot where a Neosho Valley line was to intersect one running from Topeka to Santa Fe. The result of such construction, a local newspaper once had predicted, would be "the largest city in Kansas, south of the Kaw River." A sense of inevitability had led to less than vigorous lobbying efforts by townspeople, however, and residents were caught short when officials of the Santa Fe company decided to detour slightly from a straight-line route in order to tap the coal in Osage County. This detour took the railroad away from Council Grove and into Emporia (cf. Maps 6 and 8).[35]

Officials of the Southern Branch found money relatively easy to raise. Junction City people, having anticipated and then lost the advantages of being a major junction through the manipulations of Eastern Division officials, were anxious to participate. So were those in the other counties of the Neosho Valley. Robert Stevens, one of the railroad's Emporia founders, also proved to be an effective fund-raiser in the East. In late 1868, his work led to a major involvement by a New York consortium headed by Judge Levi Parsons, the president of the Land Grant Railway and Trust Company. Construction began in Junction City the next summer, and the road was completed to Council Grove by that October.[36]

The final community to promote a successful railroad south from the Union Pacific line was Kansas City. There a group of real estate developers led by Kersey Coates chartered the Kansas and Neosho Valley Railroad in 1865 with plans to build through the eastern tier of Kansas counties to Fort Scott and then on to Galveston. Kansas City residents had become aggressive with railroad promotion that year because, after more than a decade of anticipation, the Pacific Railroad of Missouri had finally been completed to their town. They quickly connected that line with the Eastern Division near Wyandotte and then plotted a series of new branches in several directions. In 1866, for their first move, workers laid several miles of track into Kansas along the Missouri River to connect the St. Louis railroad with the Leavenworth terminal of the Union Pacific. Then, simultaneously, they began to promote the line to Fort Scott (familiarly known as the border-tier road because of its location), as well as one to the northeast that would connect with the Hannibal and St. Joseph tracks.[37]

When examined just by itself, the border-tier enterprise (it was officially renamed the Missouri River, Fort Scott and Gulf in 1868) does not appear impressive. Its land grant was issued four years later than those of its rivals in Lawrence and Topeka, and one year after the award to the Southern Branch. Its officials also were not particularly aggressive with construction in the early years. The railroad's historian, Craig Miner, has noted that naming Olathe (the seat of Johnson County) as company headquarters merited little attention in local newspapers because so little money had been raised. Things changed in 1867 when the land grant materialized, and even more so in 1868 after the company purchased of a large piece of Cherokee Indian land along its projected route through southeastern Kansas. Trackage opened to Olathe in November 1868 and to Paola, the next county seat south, in May 1869.[38]

The transition from a transportation system based on steamboats and wagons to one dominated by railroads occurred so fast in the elbow region and the rest of northeastern Kansas that even the most progressive city leaders had trouble grasping all the implications of how the new economy would function and the steps necessary to make their particular towns competitive. Kansas went from virtually no railroads in 1862 to the beginnings of nodal communities in 1866. Before any of these successful places could pause for reflection or self-congratulation, however, entrepreneurs were already beginning to forge a more complicated series of railroad connections. Historian Charles Glaab has made a convincing argument that, by 1866, the decisions already had been made that would accelerate Kansas City into metropolitan dominance in the elbow area. Rival communities, of course, did not realize this state of affairs immediately and did not give up hope of changing the situation for at least another decade.[39]

What proved to be the critical issue for rapid urban growth in northeastern Kansas and northwestern Missouri during the 1870s was a nodality more complex than had been apparent in the mid-1860s. Following a logic developed in the earlier days of railroads, people had assumed that big rivers would always form the boundaries between major rail systems. In a westward-moving nation, this meant that Chicago businessmen would concentrate on trackage between Lake Michigan and the Mississippi River, and Leavenworth people on lines west from the Missouri River. Linkages between each of these newer cities and the metropolitan areas farther to the east were expected to develop, of course, but presumably without the need for great inducements from the western places. This last assumption proved to be shortsighted for any community that wanted a voice in its own future.

Kansas railroad history in the 1860s followed the accepted script for development. Residents pursued western, intercontinental lines first and then ones that stretched south toward new ports and another agricultural frontier. Only a few leaders grasped the critical importance of interlinking this emerging system tightly with the eastern centers. This is not to say that they did not want connections in that direction, but money was scarce. Kansans reasoned that if they developed good connections west and south, other people would pay for tracks to the east. The most common strategy was to play St. Louis interests and those in Chicago against one another. Leavenworth leaders, for example, tried to force St. Louis businessmen to build tracks to their city by threatening to vote bond issues for a line coming in from Chicago. Hardly anybody saw beyond this modus operandi and tried to develop a more proactive attitude toward either St. Louis or Chicago. People who could foresee the advantages of courting both of these cities actively were rarer still. As it happened, most of these unusual people resided in Kansas City.

The railroad program that proved successful for Kansas City and that was imitated later and in lesser quality by other area communities can be simply diagrammed (Map 9; see p. 21). In each case, the tracks resemble the letter X centered on the city in question. One eastern arm would extend to St. Louis. This would be either the Pacific Railroad of Missouri (later known as the Missouri Pacific) or a branch from it. A second arm would stretch toward the northeast and Chicago. The immediate goal would be the Hannibal and St. Joseph line, again either directly or with a branch. West and south from the cities would be the transcontinental and the Gulf of Mexico lines as already discussed.

Such a four-pronged strategy of nodality could have been achieved by Lawrence, by Topeka, or by any of the major Missouri River cities in the area. St. Joseph, as I argued before, enjoyed an early advantage but lost it when owners of the Hannibal and St. Joseph decided not to push for an early extension westward. Atchison people saw their star rise when the one for St. Joseph fell, and they made solid attempts to expand both to the west (the Central Branch) and to the south (the Atchison, Topeka and Santa Fe). Metropolitan hopes of residents there effectively ended with the reorientation of the Union Pacific, Eastern Division, in 1866, which left the Central Branch as a spur. This St. Joseph–Atchison version of the dream then passed to Topeka. Cyrus Holliday and his associates were located on the best line to Colorado and to St. Louis. With the Atchison, Topeka and Santa Fe they hoped to develop not only the southwestern trade but also, via Atchison and St. Joseph, a solid connection to Chicago.

Of the remaining three cities in the region, Kansas City and Lawrence shared the same transcontinental connection to St. Louis and the West enjoyed by Topekans. With their separate roads to the south under development in the mid-1860s, that left only a connection to Chicago via the Hannibal and St. Joseph. Kansas City entrepreneurs could make this link directly. Lawrence leaders, however, needed the cooperation of Leavenworth to bridge the Missouri River at that point. Finally, at Leavenworth, one finds the most complex feelings of hope and frustration. Having the largest population in the region gave people high expectations, but this size also fostered possibilities for factionalism and a sense of entitlement. All looked bright during the early years when the Ewing brothers were maneuvering the Leavenworth, Pawnee and Western, but hopes faded rapidly as St. Louis politicians forced that railroad to a new point of origin at Wyandotte. Having been burned once by out-of-state railroad powers, many residents were reluctant to invest their money in such schemes again. The possibility of factionalism thereby became a reality.[40]

The ascendancy of Kansas City in the late 1860s is usually attributed either to the failure of rival communities to maximize their railroad connections or to a single-minded, united effort on the part of Kansas City people to build these same lines. Elements of both scenarios were certainly present, but a tremendously important third factor also existed: the wishes of various eastern politicians and investors. Although newspapermen in Kansas City liked to brag to their colleagues in Leavenworth about their success in obtaining the transcontinental railroad, for example, it was obvious to anybody who examined the situation closely that the real power behind that decision came from St. Louis. The achievement of Kansas City people lay more in recognizing the advantage that this east-west link had given them and working diligently to develop the advantage farther.

Charles Glaab, an expert on railroad history in Kansas City, has stressed that leaders there never became obsessed with the retention of their river and overland-freight businesses. People with such conservative tendencies were concentrated instead in the older communities of Independence and St. Joseph. Banking and real estate development were more central to the Kansas City entrepreneurs, and after losing trade and growth potential to Leavenworth during the politically troubled, early years of Kansas settlement, these men had decided to work together for mutual profit. Their biggest insight, perhaps, was in noticing how traffic on the Hannibal and St. Joseph Railroad increased markedly during the early years of the Civil War. They realized that this was a sign that Chicago might well surpass St. Louis as the commercial marketing and supply center for the country's interior. Immediately they revived an earlier plan for a branch line and bridge that would link their city with the Chicago-bound tracks at Cameron, Missouri, fifty-four miles to the northeast (Map 9).[41]

Kersey Coates, John Reid, and other Kansas City men tried to interest eastern bankers in their Cameron project during 1864 and 1865, but they met with little success. Late in 1865 or early in 1866, however, they changed their tactics slightly to focus on James F. Joy, a rising executive within the Boston investment group that controlled the Burlington railroad system out of Chicago. Joy, who was president of the Chicago, Burlington and Quincy line and manager of the Hannibal and St. Joseph, had a taste for power and a weakness for real estate. The Missourians offered him some choice lots from their West Kansas

City Land Company, as well as financial control of the Cameron enterprise. Joy, in turn, agreed to build this line and to lobby for a land grant to support Kansas City's other big railroad project, the border-tier road toward Texas. The final component of the deal was congressional authorization for two bridges: one over the Mississippi at Quincy, Illinois, that would connect the Hannibal and St. Joseph Railroad directly with Chicago, and the other for the Cameron line over the Missouri at Kansas City. Robert Van Horn, the editor of Kansas City's *Western Journal of Commerce* and the area's representative in Congress, accomplished this job by tacking them on to an omnibus bill in June 1866.[42]

James Joy was true to his word and, in fact, became more involved in Kansas City affairs than some of his early supporters may have liked. He assumed control of the border-tier railroad in 1868, oversaw completion of the Missouri River bridge in 1869, and that same year, extended his (and Kansas City's) business reach farther to the south and west by taking over the management of a rival line to Texas, the Leavenworth, Lawrence and Galveston. Joy's plan for this Lawrence railroad, troubled financially ever since Senator Lane's death three years earlier, was simple. By building a twenty-mile section of track from Ottawa northwest to Olathe in 1870, he neatly redirected trade from a whole tier of counties away from Lawrence and into Kansas City. The contest for urban supremacy in the elbow region was over just that simply and quickly. Although rival cities dutifully erected their own impressive bridges over the Missouri a few years later (Leavenworth in 1872, St. Joseph in 1873, and Atchison in 1875), the pace of railroad interconnectivity at the once sleepy landing site for Westport merchants was growing too fast for anybody else to overcome. The rerouting of the Leavenworth, Lawrence and Galveston to the new Kansas City nexus was only one in a series of similar changes.[43]

Railroad Cities in Southeastern Kansas, 1868–1880

From the perspective of railroad history, Kansas can be divided into two regions. In the state's northeastern quadrant, the only area settled prior to 1865, preexisting towns maneuvered to create railroads. Elsewhere, throughout a full three-quarters of the state, this equation was reversed. Iron tracks either preceded the pioneer land seekers into these vast regions or trailed them by only a year or two. In such circumstances, railroad officials enjoyed almost dictatorial power. With investment money now beginning to flow, and charters almost always vague on the exact routes to be taken, these people created land development companies as subsidiaries to their transportation operations. They then hired publicists, recruited colonists from as far away as Russia, and platted town site after town site. With their inside knowledge about the locations of key junctions, division headquarters, and repair shops, they effectively assumed the mantle of regional planners for most of Kansas and the entire Great Plains.[44] The following pages lay out this railroad vision for the state's urban development as it was established between about 1870 and 1890. I do so region by region beginning with the southeast.

Three railroads set the initial urban character in southeastern Kansas, the same ones that the citizens of Kansas City, Lawrence, and Emporia had hoped would make their connections to the Gulf of Mexico. Two of these, the border tier from Kansas City and the Leavenworth, Lawrence and Galveston (LLG) from Lawrence, took routes almost straight south from their origins (Map 10; see p. 23). The Missouri, Kansas, and Texas (the name assumed by the Union Pacific, Southern Branch, in 1870), however, elected to follow the Neosho River valley and therefore angled from northwest to southeast (cf. Map 5). Such geometry forecast an intersection between this line and the Leavenworth, Lawrence and Galveston somewhere in Neosho or Labette County. That spot obviously would be a favorable site for a new city. Since town building by outsiders was prohibited in Indian Territory, people also had high expectations for development at the border locations where each company's tracks would leave Kansas. Such towns should have large trade areas to the south, a region already known for its cattle production.

Anybody familiar with the tactics of early railroad builders would expect that the three Kansas companies, in planning their routes south, would have played coy financial games with the little towns that lay along the first seventy-five or so miles of their corridors. Such dealings were rare in this region, however, because the owners had an even bigger prize in mind. To reach the Gulf Coast, they would have to obtain permission to build through Indian Territory. All three companies had negotiated this right separately with the Cherokee Nation in 1866. While tracks were being laid across the state in early 1870, however, the Cherokees reconsidered their position and decided to allow only a single line to pass through. The first company to reach their border would assume this right.[45]

Tracklaying in southeastern Kansas went at a breakneck pace after the Cherokee decision. The border-tier line, which was already as far south as Fort Scott in June 1869, looked like the probable winner, but its progress was slowed by a series of conflicts with squatters who lived south of that town on former Indian land that the railroad company had purchased. Workmen for the LLG laid track from near the Franklin-Anderson County border in January 1870 to southern Neosho County by December, but the fastest builder proved to be the Missouri, Kansas and Texas (MKT) company, which reached the state border near Chetopa, in Labette County, on June 8, 1870. This company, already becoming known as the Katy Railroad, thus won the right to cross Indian Territory into Texas, which it did during 1871 and 1872. The other two lines were temporarily truncated at the border.[46]

The first two towns to grow substantially from the new railroad expansion were Ottawa and Olathe, the seats of Franklin and Johnson Counties, respectively. Ottawa was formed in 1864, relatively late for east-central Kansas, when the Indians by that name ceded their territory. It grew modestly when the tribal-endowed Ottawa University opened in 1866. The arrival of the Leavenworth, Lawrence and Galveston in 1868 helped more, but the most important event was James Joy's decision to link this railroad with his border-tier tracks via a connector line from Ottawa to Olathe. This route, completed in 1870, transformed both communities into junctions and thereby made them better locations for business. Olathe grew modestly on this development, from 1,817 people in 1870 to 2,285 in 1880, but Ottawa did even better. The reason was a community initiative, first to secure the headquarters of the railroad and then, in 1871, to award a bond issue of $60,000 plus a free site if LLG officials would establish repair shops in town. When these facilities opened in 1872, their employment of about two hundred men helped the overall city population to rise from 2,941 in 1870 to 4,032 in 1880.[47]

The second important town to develop along the Leavenworth, Lawrence and Galveston corridor was Chanute, in northern Neosho County. In this region in the summer of 1870, the MKT tracks already lay along the west side of the Neosho River. LLG workers were progressing at a slower pace along the east side of the same stream. As these latter people passed through the last prerailroad settlements of Iola and Humboldt, it was obvious to everybody that, ten or so miles south of Humboldt, the railroad would have to bridge the Neosho and then cross the MKT tracks on its way farther south. One might have expected that LLG officials would have planned the exact location of this junction in private so as to plat their own town and thereby maximize profits. After all, the MKT line was already far ahead of them in the drive toward Indian Territory, and LLG investors would have to take whatever other opportunities were available. The problem, however, was that the general manager of the LLG, James F. Joy, was also in charge of the border tier, which still was racing with the MKT. Town building, therefore, was not a priority. This left area speculators to guess at the railroad's plans for the Neosho bridge and junction.[48]

Separate groups of local people established two towns at or near the junction site that summer: New Chicago, between the two lines just south of the crossing; and Tioga, just to the northwest and adjacent to the LLG tracks. MKT officials established a depot at New Chicago in September, but their colleagues with the LLG selected Tioga for their station. Similar fighting went on throughout 1871 for school sites and other facilities, but then common sense prevailed. In order to launch a united bid for the county seat, voters agreed to consolidate in 1872. They named the new community after Octavius Chanute, the original construction engineer for the LLG who had risen to the rank of company superintendent. For this flattery, Mr. Chanute agreed to move his depot from Tioga closer to the new urban center.[49]

Chanute had a population of about 800 at its birth in early 1873, along with a busy stockyard that catered to ranchers from Wilson and Woodson Counties. A national business panic that year and statewide plagues of locusts the next two seasons stymied the town's progress, however, so that the 1880 census showed nearly the same total as seven years earlier. Chanute also was handicapped in its first years by a failure to wrest the county seat away from Erie, its more centrally located rival. Railroad power gradually asserted itself in the 1880s, however, with Chanute rising to a population of 2,826 in 1890 while Erie stagnated below a thousand. Neosho County thus became an early example of a common Kansas phenomenon: a place where political and commercial powers were compartmentalized into separate communities.[50]

After Chanute, the next railroad towns in southeastern Kansas to attract attention were those on the state's southern border: Baxter Springs, Chetopa, and Coffeyville (Map 10). Of these, Baxter Springs had the earliest start and the biggest initial expectations. This was Kansas's first cattle town. Founded in 1866 along the old military road from Fort Leavenworth to Fort Gibson, Arkansas, it was only six miles from the extreme southeastern corner of the state. Several land speculators had noticed a gradual increase in the numbers of Texas cattle being driven north through the area to markets in Missouri, especially to Sedalia, the temporary end of the first railroad west from St. Louis. Fears that these cattle would introduce a fever to local herds had led Missourians to levy fines on drovers who brought diseased animals into that state. Kansas legislators, choosing immediate profits over long-term risks, responded to this restriction by opening their borders in 1866 to all such transient cattle. Baxter Springs people, in a perfect position to profit from this decision, were elated. Various trails across Indian Territory could funnel onto the old military road at this point before a last push north to the railroads at Kansas City. Residents also were confident that the imminent completion of the border-tier tracks would allow local people to usurp this stockyards function for themselves.[51]

Baxter Springs grew as the railroad neared, and its several hundred citizens cheerfully passed a huge $150,000 bond issue to aid in the construction process. The railroad's owner, the ubiquitous James Joy, faced a decision as the tracklayers reached Columbus in April 1870, however. On the one hand, he wanted to push on another twelve miles southeast to Baxter Springs because of the town's cattle trade and to qualify for its $150,000 donation. Doing so, however, would slow his progress toward the larger goal of reaching the border of the Cherokee Nation before the MKT. This was because the land directly south of Baxter Springs belonged to the Quapaw Nation, not the Cherokee. Further, this spot lay several miles east of the Neosho River valley, another specification for the railroad location included in the Indian agreement. Joy thought about constructing two lines but for unknown reasons decided to build only into Baxter Springs. When this was accomplished on May 14, workers for the Katy were still thirty miles from Indian Territory, giving Joy one last chance to press rapidly west-southwest another fifteen miles to the Neosho. Instead, however, he elected to file a legal appeal with the U.S. Department of the Interior. When this failed, the always-busy Joy simply turned to other business ventures and left the people of Baxter Springs to fend for themselves.[52]

Residents initially liked being the terminus of the border-tier railroad because the cattle trade came as expected. The Baxter Springs boom was short-lived, however, because newer Kansas legislation on the Texas cattle fever forced drovers to enter the state farther west, and the MKT tracks into Texas allowed direct shipment north by rail. Town residents also effectively killed their own prosperity. When the community was growing in the early 1870s and contained as many as four thousand people, they voted lavish expenditures of public money for street improvements, new schools, a jail, and a courthouse. Afterward, high taxes vied with the loss of the cattle market to drive residents away. The population fell abruptly to only eight hundred in 1876, and property abandonment became common.[53]

Perhaps because of their less-than-satisfying experience in Baxter Springs, the owners of the Leavenworth, Lawrence and Galveston Railroad decided to create their own border town instead of using one established by local speculators. This led to intrigue and accusations of bad faith, of course, a pattern of behavior that was to become increasingly common across the state. Actually, this particular conflict contained two episodes, both won by the railroad people. Activity began in June 1870 when residents of thinly settled Montgomery County learned that the LLG, having been beaten to the Neosho Valley by its rival line, would now head southwest through their territory. The plan, of course, was to establish another border town, but one far enough west of the Neosho to command its own trade area. Local people were elated and so voted to issue $200,000 in bonds to help construction through their county. Most urban residents at that time lived in Independence, the centrally located county seat. They assumed the LLG would pass their way. Railroad officials, clearly with a different scenario in mind, nodded politely to this view but never actually committed themselves to any one particular route. Once the money was theirs, however, they revealed a survey that missed Independence by seven miles. The objective, instead, was to intersect the valley of the Verdigris River (a rich lowland similar to that of the Neosho) and then to follow that valley south to the state border (Map 5).[54]

The decision to bypass Independence left the people there "fairly raging in indignation" but simultaneously set up a second round of speculation. Because the railroad was coming from the northeast, the most obvious site for the terminus town now would be where the Verdigris flowed into Indian Territory, and along its eastern bank. Knowing the preference of railroad builders for valley locations, several speculators already had established a town company at that spot as early as 1869. Known as Parker, after its founder, this place boomed in 1870 once the snub of Independence had become common knowledge. People abandoned several other paper towns, started fifty businesses, and immigrated in such numbers that the population grew to more than a thousand.[55]

Leavenworth, Lawrence and Galveston officials, seeing the activity in Parker, took a different tactic. Quietly, they purchased 160 acres on the west side of the Verdigris and then announced that they would bridge that stream three miles above Parker. The railroad's chief engineer, Octavius Chanute, surveyed and platted the new community. Perhaps to avoid the image of a company town, they named it Coffeyville, honoring Colonel James A. Coffey, who had laid out an earlier, failed town site nearby. The result was an urban experience similar to that of Baxter Springs, only without the lavish expenditures. Coffeyville became another classic cattle town, complete with magical growth and transient population. An early historian wrote that "everything was in a constant 'hurrah,' men were wild with excitement, and society was a chaos." It was an ephemeral glory, of course, but the richness of the Verdigris Valley helped to prevent a complete decline. From a low of 753 in 1880, the population climbed back to about 1,500 in 1883 and to 2,282 in 1890.[56]

Baxter Springs and Coffeyville, although expected to have successful futures, lost their chance for true metropolitan status when their life-giving railroads were unable to build on to Texas. In contrast, the winning Missouri, Kansas and Texas company, under the direction of its president, Levi Parsons of New York, and general manager Robert S. Stevens of Emporia, now possessed this power. An MKT town in southern Kansas would occupy a tremendous gateway position, receiving cattle and other frontier goods from the Southern Plains through its exclusive transportation funnel and sending out eastern manufactured products in like manner. The only potential blemish for this alluring picture might be if all these goods were to pass straight through the new MKT city and be redistributed instead from some point farther north and east, most obviously at the emerging railroad hub of Kansas City. Levi Parsons anticipated this possibility, however, and took a bold step to eliminate it even before his railroad tracks had reached the southern Kansas border.

On January 10, 1870, Parsons announced to his board of directors that, immediately after reaching Indian Territory, he would initiate a second railroad line. "In order to handle effectively the potentially great traffic between that area and the markets of the North and East," he said, the MKT must build a major connector route from southern Kansas to the main line of the Pacific Railroad of Missouri at Sedalia. This would bypass Kansas City, of course, and simultaneously create tremendous growth possibilities for whatever Katy town might be selected as the junction point for the Sedalia tracks. Visions of a true "New Chicago" did not seem far-fetched as people pondered the future riches of the Southern Plains. It is likely that glimmerings of this junction were in the heads of the Neosho County men who had applied the Chicago name to the town site where the MKT met the LLG. This location (i.e., Chanute) would have worked as a starting point for the Sedalia road, but because MKT officials did not control the town company there, it had no real chance of being selected. Most observers in early 1870 thought the choice would be Chetopa, the Katy border town in the Neosho Valley.[57]

Whether Chetopa ever had a realistic opportunity to become the principal city of the MKT line is open to debate. Townspeople certainly thought it did, but this belief may have been encouraged by railroad officials as a cover to allow their own land acquisitions elsewhere to go more smoothly. The Chetopa site was ideal in several ways: midway between Baxter Springs and Coffeyville, and in the fertile Neosho Valley. Its residents also had pledged $50,000 in bonds to aid the railroad's construction. The only drawback, really, was that the town had been created in 1868, without railroad involvement, and these pioneer settlers would now profit from the company's efforts. This issue had been negotiated, however, and by 1870 MKT officials owned "large interests" in the community and advertised it heavily.[58]

Perhaps no city in Kansas has ever experienced so rapid a rise and fall as Chetopa. In the summer of 1870 its prospects soared. Here was to be the headquarters for one of the nation's fastest-growing railroads and the focus of its great triangle of lines stretching northwest to Junction City, south to Denison, Texas (another company town), and northeast to Sedalia. Big repair shops would be here, as well as major switching yards and crew changeover operations. Such a railroad nexus also would mean almost unlimited business opportunity, of course, and large wholesale houses for groceries, liquor, and other goods quickly sprang into existence. The town's population reached three thousand within a month.[59]

If a person were looking for a weakness in the Chetopa vision of the future, it would have been in the observation that MKT officials scheduled the first work on their Sedalia line in Missouri rather than Kansas. This eastern portion of the track, which entered Kansas near Fort Scott, was completed in October 1870, just four months after the main line had reached Indian Territory. Then came the moment of decision. On October 24, Parsons announced not only his routeway from Fort Scott to the existing line but also the chartering of a new 2,560-acre town site in northern Labette County, about midway between Chanute and Chetopa. The location, completely owned by the railroad, just happened to be where the new transportation junction would occur. As a final, not-so-subtle hint about the future, he also volunteered that colleagues had

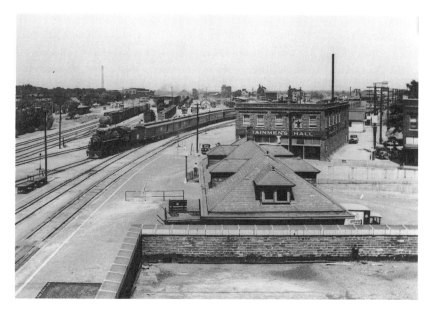

Depot Scene at Parsons, 1935. Parsons was created in 1871 to be the focus of the Missouri, Kansas and Texas Railroad. Until the company's reorganization in 1957, more than fifteen hundred people worked locally in its plush general office building, big repair shops (visible in the background), employee hospital, and similar facilities. No wonder that a "trainmen's hall" stood near the tracks. (Courtesy Kansas State Historical Society)

persuaded him to name this community Parsons. For Chetopans, of course, the news could not have been worse. Theirs was the experience of Baxter Springs, only magnified. Improvements to property became valueless literally overnight, and soon the valuation of the town was exceeded by its indebtedness. By 1880, conditions were still bad, but the population had rebounded slightly, up to 1,305 residents.[60]

Robert Stevens, the Katy's general manager, tightly controlled the development of Parsons. He began the sale of lots on March 8, 1871, a month after the railroad junction had been completed, and decreed that no speculators would be allowed. This brand of professionalism attracted more people than it repelled, however, and in a matter of days the town of Parsons had acquired two thousand residents and an appropriate nickname: "The Infant Wonder." Work on a first-class set of railroad facilities began just after the initial land sales. A huge headquarters building for the company opened in January 1872, followed by the roundhouse and machine shops later that year. The shops, ten massive stone buildings, cost more than $250,000. More than two hundred men worked there: carpenters, painters, blacksmiths, and whitesmiths, as well as machinists and mechanics. With annual wages totaling $180,000, these facilities were the core of Parsons and provided "a permanency to the prosperity of the city."[61]

Besides his official railroad duties, Robert Stevens also worked to diversify the Parsons economy. He created the first bank in town (the First National) and became its president. He did the same with the National Mill and Elevator Company. Using earlier business contacts, he persuaded the Chicago lumber firm of Melville, Pluto and Company to open a yard in town and Carney, Fenton and Company to establish a large wholesale grocery operation. Both enterprises received choice sites near the MKT tracks. Finally, Stevens set up his cousin E. B. Stevens with the Belmont House hotel. These operations all developed smoothly. Local railroad employment rose steadily to nearly nine hundred by 1890, and the overall population did the same. With 4,199 residents in 1880 and 6,736 in 1890, Parsons had become, if not yet a national center, at least true to its boosters' claim: "the metropolis of southeastern Kansas."[62]

Only a cynic would sneer at the steady growth of Parsons in the late nineteenth century. Those few who pondered why the city's population was only six thousand instead of sixty thousand, however, could identify at least three causes. All related to competition. Two of these I will develop elsewhere: the Railroad Act of 1886, which ended the MKT's monopoly of north-south routes through Indian Territory, and the exponential growth of Kansas City. A third, however, was the rise of a rival railroad city within southeastern Kansas. This was Fort Scott, in Bourbon County, and the rivalry was, to some degree, a creation of the same MKT officials who had founded Parsons.

Fort Scott was an unusual railroad city. Neither an older community that had created a major railroad company for itself, such as Lawrence, nor a new-style, railroad-created city, such as Coffeyville, it was instead a rare example of a reasonably successful, older town that managed to attract two prominent railroad lines through self-promotion. Fort Scott thereby had an opportunity for additional growth thrust upon itself in a way few other places in eastern Kansas could match. The city's origins were military, of course, with the fort itself having been established in 1842 along the road south from Fort Leavenworth. The army abandoned the post in 1854, just as Kansas Territory opened for settlement, and a town company formed in 1857 as soon as legal title could be obtained. By 1860, a total of 262 people lived there, many in old barracks.[63]

The city began to grow after the Civil War, in part because the fort had reopened during the fighting and several thousand soldiers had seen the prospects of the region at first hand. Local pioneers included George A. Crawford and Charles W. Goodlander, a pair of Pennsylvanians who between them built a series of successful flour mills, woolen mills, newspapers, and banks. Crawford also founded the Fort Scott Foundry and Machine Shops in 1869, which took advantage of abundant local coal supplies to employ a hundred men making boilers and engines. More important, this business also gave townspeople hopes of becoming a major manufacturing center.[64]

Fort Scott was important enough by 1865 that the builders of the Kansas and Neosho Valley Railroad, coming south from Kansas City, sought out town officials and offered to change the company name to the Missouri River, Fort Scott and Gulf in exchange for $150,000 in local bonds. These tracks, although delayed by other financial concerns, arrived in December 1869. Then, within the next month and while the initial railroad euphoria was still high, came the announcement by Levi Parsons that the MKT Railway would make Fort Scott a major stop on its new route to Sedalia. With these basic tools of economic development coming to town so effortlessly, it was no wonder that people began to expect urban greatness to be their birthright. Even before the Katy news, for example, a local newspaper had predicted "a city of 40,000 inhabitants [within a decade]—an Indianapolis as to railroads—a Pittsburg [sic] as to manufacture."[65]

As in many other communities whose initial gains came without difficulty, people in Fort Scott grew complacent. They placed too much importance on their coal resource, for example, forgetting that the veins of this same field became thicker farther to the south. They also expected both of the railroads to make their town a division point and the site for big-employment repair shops. These latter expectations were logical, they figured, since Fort Scott was approximately a hundred miles from both Kansas City and Sedalia, the standard length for a division at the time. What the local people forgot, of course, is that railroad officials were strictly businessmen and that neither company had a big investment in Fort Scott's future. The community lost out first to the MKT when, according to the memoirs of Charles Goodlander, a refusal in 1870 to permit the depot where Katy officials wanted it caused the company to look elsewhere for a shops site. You "can blame the imbecile mayor and council who were in authority at the time," wrote Goodlander, "that Fort Scott is not a city of 30,000." A similar incident in 1873, a city refusal to appropriate an additional $45,000, probably led managers of the border-tier line to keep their shops in Kansas City.[66]

Despite the failure to obtain the repair facilities of either railroad, Fort Scott was still a self-confident community in 1870. The population that year was an impressive 4,174, and plans were afoot for a new rail link to the west and another to Springfield and Memphis. People also argued that their town was far enough south not only to escape domination by Kansas City but also to be in a position to overtake that rising urban center at its own game. Kansas City did not have Fort Scott's coal, they noted, and its manufacturing base was only marginally larger. As a local poet, Eugene F. Ware, expressed the vision in 1871:

> And of those cities
> Gleaming on the prairies,
> Beautifully tall,
> Fort Scott towers up,
> The proudest of them all.[67]

The hopes of Fort Scott people, like those nationwide, were stymied by the panic of 1873. Town historian Edward F. Keuchel, however, has argued that local leaders lost more of their confidence from this experience than was the case in Kansas City and elsewhere. Instead of continuing to float new initiatives, they worried about losing what they had. The town sustained modest growth, reaching 5,372 in 1880, but never wrested regional railroad leadership from Parsons, let alone Kansas City.[68]

Railroad Cities in North Central and Northwestern Kansas, 1867–1880

Whereas three competing railroads vied for the right to develop southeastern Kansas, a single company controlled this power in the north-central and northwestern parts of the state for the two decades following the Civil War. This was the Union Pacific Railroad, of course, or as it was known between 1869 and 1880, the Kansas Pacific. This enterprise, which had been characterized by uncertain financing and slow construction east of Topeka, changed character completely about 1865. Under the direction of chief engineer Robert M. Shoemaker, crews efficiently and regularly advanced the tracks west to the hundredth meridian in present-day Logan County, where work had to be suspended for about a year while private sponsorship was arranged for the final segment into Denver. The tracks opened to Manhattan in August 1866 and to Junction City that October. Then, after a pause for the winter, they came to Salina in April 1867, to Hays in October, and nearly to the western border in the following August.

In order to understand the town-founding policies of the Union Pacific, one must realize that central and western Kansas were not very much on the minds of company officials during the period of construction. The St. Louis, Cincinnati, Pittsburgh, and Philadelphia men who dominated the board of directors thought in bigger terms. Empowered with by far the largest land grant in Kansas history (3,925,791 acres, or five times the amount received by the three railroads of southeastern Kansas) and emboldened by the old dreams of western riches and Pacific trade, they saw their chance to create a second transcontinental railroad that had even greater potential for profit than the original. Gold, timber, and coal in Colorado were the immediate goal. Simultaneously, however, these men were planning extensions and/or branches to Salt Lake City, New Mexico, and the coast beyond. They were efficient rail builders in Kansas only because they needed to get across.[69]

Knowledge that Union Pacific officials viewed western Kansas more as an obstacle to be overcome than as an opportunity for profit in its own right explains several of their actions. They spent less than two months on surveys for possible routes, for example, and adopted a course that deviated only slightly from a straight-line shot to Denver and the gold fields. Their corridor was essentially that of the old Smoky Hill Trail, which had been laid out from Leavenworth in 1859 during the initial mining rush to Colorado (cf. Maps 5 and 10). This roadway, originally touted as being a hundred miles shorter than alternates up the Platte or Arkansas Valleys, had been allowed to deteriorate during the war years of 1860 to 1865. It was revived when David A. Butterfield of Atchison proposed its use as a stage and freight line. Butterfield requested military assistance for this endeavor, since the Cheyenne, Arapaho, and other native peoples in this region had become increasingly uneasy about intrusions into their realm. This aid, easily procured because of the importance of Colorado, included the establishment of three major military posts at regular intervals: Forts Harker, Hays, and Wallace.[70]

The Butterfield Overland Despatch was never profitable and changed hands three times before 1868. The forts remained, however, as did the Indian danger. This combination provided good incentive for the railroad men to essentially duplicate the Smoky Hill Trail as their track bed. Another plus, equally advantageous for steam engines and horses, was in having the route follow the general course of the Smoky Hill River and yet keep mainly to the upland surface along that stream's northern flank. Stream crossings thereby were frequent enough to furnish water, while at the same time, the main path along the divide avoided the long stretches of sand that would plague any competing valley route.[71]

As Robert Shoemaker and his track crew pushed across the Kansas landscape in 1867 and 1868, they laid out stations at intervals but did little to develop these points into communities. Unlike the case with every other railroad of note in the state, a person can search in vain to find examples where Union Pacific officials have memorialized themselves in the names of major settlements. Similarly, no subsidiary organization existed that was dedicated exclusively to the platting and promotion of new towns. These duties, when required, were assumed by another subgroup, the National Land Company. But the National was headquartered in New York City and concentrated its efforts on selling farming and ranching properties in large packages and on promoting group settlement through a series of colonies. When all was said and done, Union Pacific people founded only four cities along the entire three-hundred-mile stretch from Junction City to the Colorado state line: Bosland, Brookville,

Ellis, and Wallace. All the other communities along the route, including Russell and Hays, were creations of either their colonists or other speculators, and many occurred several years after the tracks were in place.[72]

The first destination for the Union Pacific crews was Fort Harker in Ellsworth County. This post had been established in 1864, a year before Butterfield's trail. It controlled a slightly older route from Fort Riley southwest to Fort Zarah (near present-day Great Bend) on the Arkansas River and the Santa Fe Trail. As the Union Pacific built up the Kansas Valley, this military road had become popular for travelers to New Mexico, replacing the older passageway via Burlingame and Council Grove. Several small settlements were in place by 1866, including two county-seat towns: Abilene, in Dickinson County, and Salina, in Saline County. Both occupied good agricultural areas along the Smoky Hill River, but contemporary observers did not see Abilene as having potential for real growth. It was too close to bustling Junction City, for one thing. Moreover, according to a correspondent from Leavenworth, "Much of the land in the immediate vicinity is held by non-residents, as is the whole area of Dickinson County; hence, population is sparse."[73]

A third reason why people lacked confidence in Abilene's future was because its neighbor on the west, Salina, was being touted so highly. Closely analogous to Emporia, the site of Salina lies at the eastern edge of a dissected belt of hills and controls access routes to lands farther west (Map 5). With the reddish sandstone bluffs of the Smoky Hills in the background, three major lowlands converge at this point. Trending west-northwest are the Saline and Solomon Rivers, both north-bank tributaries to the Smoky Hill, whose valleys provided ideal track beds for railroads. The third lowland, the Smoky Hill itself, is even wider but deviates abruptly to the south at that point for more than twenty miles before angling westward once again. A direct east-west path through the upland is provided by a much smaller tributary. This stream, called Spring Creek, was the path selected for the military road from Fort Riley to Fort Zarah.

The advantages for transportation and trade inherent at this river nexus near the Smoky Hills became known soon after the opening of Fort Riley in 1853. In 1856, the same Preston Plumb who would cofound Emporia a year later scouted the region with a military officer. Despite being fifty miles beyond the frontier, they were impressed and made preliminary plans for a town site. Isolation proved great enough to deter Plumb's idea, but not a similar one two years later. A Scotsman named William A. Phillips, who also had traveled over much of Kansas in search of investment opportunities, judged this spot to have the greatest potential for growth and so laid out the Salina town site.[74]

Phillips and the small group of people who joined him called their main street Santa Fe to acknowledge their business with travelers bound for New Mexico. The gold rush to Colorado in 1859 also bolstered the economy to a degree, allowing the population to grow to perhaps two hundred before stagnating during the war years. Union Pacific officials, needing to ascend Spring Creek, necessarily built through their community, but there is no record of any special financial inducements or promised major facilities. Junction City, forty-seven miles to the east, already had been named the first division point on the line, which left only modest immediate possibilities. One sign that the railroad men saw the promise inherent in this location, however, is that they helped to finance a large, $20,000 hotel in 1867. Called by one observer "the most complete, perfectly arranged and handsomest hotel on the line," it served meals to passengers and provided rooms for those departing overland to points southwest and northwest.[75]

Once past Salina, Union Pacific crews were essentially building ahead of settlement. Six miles west of town they left the broad lowland and observed sandstone escarpments of the Smoky Hills gradually narrowing the valley of Spring Creek. The next town site of consequence lay near the far side of this upland where Fort Harker stood and where the railroad route would rejoin the Smoky Hill River after that stream's large meander to the south. With railroad officials again taking no initiative, a small group of private investors surveyed a property they called Ellsworth in May 1867, two months before the tracks arrived. They located it just west of the border of the military reservation, five miles from the post itself. Because people speculated that the railroad men might make the fort their western terminus for some time, buildings at the new city "began to sprout like mushrooms." This enthusiasm proved short-lived. A major cholera epidemic hit the post in July, killing three hundred people there and another fifty in town. The railroad crews did not pause as expected either, although they did build a blacksmith shop and a four-stall engine house to offer continuing employment. Many residents departed out of health fears, others to speculate at the next railhead. They left Ellsworth the seat of a newly organized county, and at least the temporary home of the railroad's second division, but with few hopes for sustainable growth. The town was only eighty-five miles from Junction City, the site of the previous division point, not far enough away for an efficient company operation. Fort Harker's life span was also thought to be short, and it indeed did close in 1873.[76]

If Union Pacific officials participated only minimally in the early affairs of Abilene, Salina, and Ellsworth, they did so even less in places farther to the west. About nine miles beyond Ellsworth, near Black Wolf Station, the tracks leave the valley proper of the Smoky Hill and ascend to the upland divide between it and the Saline River (Map 5). This move is easy to understand from the perspective of tracklaying because the valley is more incised from this point upstream, whereas the divide is flat and unobstructed. The change in terrain and associated vegetation did not help colonization, however. The railroad was now definitely west of the settlement frontier, and potential immigrants were unsure whether their traditional crops would be successful on these true plains. People also voiced concerns about how they themselves would cope with the wind and the general open country. Agents of the National Land Company eventually faced these challenges, but not immediately. Tracklaying went on at a rapid pace, but even local speculators for new cities now became rare.[77]

In 1867, as work on the Union Pacific progressed, Kansas legislators acknowledged the company's importance by juggling the geometry of the new counties they were creating. The tier that included Ellsworth was positioned slightly farther south than the one for Saline, for example, so that the city of Ellsworth would be central within its county. Similarly, the tiers that contained Russell, Ellis, and Trego Counties were moved northward so that the tracks would also bisect these units. This effort undoubtedly helped settlement in the long run, but it did little to overcome a wave of contemporary negative comments about the plains landscape that increased in volume as travelers sampled the region on the new tracks. Francis Richardson, who rode the line all the way to Denver in 1872, was typical in saying that, although the Smoky Hill Valley appeared to be good land, the rest of "the whole route is a barren waste."[78]

The most obvious drawback to town founding along the fifty-eight-mile stretch between Black Wolf Station and Fort Hays was lack of surface water. None at all existed along the route in Russell County, which accounts for that area having no more than five residents as recently as 1871, three years after railroad construction. Later that same year, when a colony from Wisconsin finally attempted a settlement at the county's center, its wells proved dry. Luckily for the settlers and the railroad land agent, similar digging was successful five miles to the west. There, in the headwaters of Fossil Creek, the group founded the community of Russell. A grateful railroad company sold them the site for a dollar.[79]

The water situation improved farther west in Ellis County. The tracks crossed the North Fork of Big Creek at a place eventually called Victoria and then did the same over Big Creek proper at two separate locations. Big Creek is by far the largest tributary to the Smoky Hill River west of Salina (Map 5). Its value in the seasonally dry plains environment had been recognized at least as far back as 1859, the first year of the Smoky Hill Trail, and its presence determined the location of Fort Hays and its predecessor, Fort Fletcher. The original post had been established at the North Fork crossing of the trail. Soldiers had ensconced themselves in the valley proper, a sheltered location, but also one prone to flooding. When the railroad survey was completed, military officials decided in May 1867 to relocate fifteen miles to the northwest. This placed the new fort, now called Hays, near the eastern crossing of Big Creek proper and close to the center of newly formed Ellis County. The relocation also created easily the best site for a city on the entire 220-mile railroad route from Ellsworth to the Colorado line.[80]

Three St. Louis businessmen led by William E. Webb had purchased three sections of land on speculation at the first Big Creek crossing in late 1866. When Webb learned that the new fort was to be in this same vicinity, he returned to the site and platted a town he called Hays City. Both it and the fort were in operation by June 1867, three and a half months before the first train arrived. Union Pacific officials again elected to stay clear of the planning venture.[81]

The immediate economic prospects for Hays City were quite different from those of Junction City, Salina, or even Ellsworth. No important military roads passed by. No plow agriculture of any consequence was expected either, although many people felt that the native grasses held good potential for cattle and sheep ranching. Even this grazing lay in the future, however. Probably because the railroad tracks began to penetrate the heart of buffalo country in this vicinity, Indian raids had become frequent enough to dominate everybody's attention. Fort Hays troops fought a combined Arapaho-Cheyenne force of four to five hundred men in a six-hour battle in the nearby Saline Valley that first summer, for example. Raids on railroad workers themselves occurred on August 1 and August 8, producing six deaths, and crewmen were forced to retreat to the fort for safety for many days at a time. A camp mentality thus evolved at the town, with elements of army and railroad life combining to produce a sizable population but coarse living conditions. Hays City claimed more than a thousand residents by its first anniversary, but it had no permanent church building until 1877. Its cemetery had been nicknamed Boot Hill before a still younger frontier outpost to the south, Dodge City, borrowed the term and made it even more famous. With liquor being sold at most of its hundred or so business establishments, a visitor in 1868 reported, "It is certainly a paradise for local editors, and [a] poor place for lawyers, the men generally preferring to do their own lawing, and in their own way."[82]

Hays City retained a rough image until about 1875, when the Indians reduced the frequency of their raids and the military the size of their garrison. That same year another event transformed the agricultural potential of the area. Adam Roedelheimer, a land agent for the Kansas Pacific, sold acreage north and east of town to a large group of ethnic Germans who had been living on the steppes of Russia for several generations. When their efforts to grow small grains proved successful and some 1,200 of their kinsmen immigrated to join them, Hays found itself a market town by the early 1880s. Federal officials opened a land office, and the local population, which had fallen to a low of 370 after the railhead had moved west, climbed back to 850.[83]

More than 150 miles of Kansas remained on the Union Pacific route west from Hays City. None of it excited the imaginations of railroad men or anybody else except for those interested in the buffalo hunt. The state legislature had

Building the Union Pacific Railroad in Ellis County, 1867. Crews laid an average of two miles of track daily in this region. The photograph shows work at Milepost 300, two miles east of present-day Ellis. White oak ties are being placed at a rate of twenty-five hundred per mile. (Courtesy Kansas State Historical Society)

created a series of counties along the route in 1868, but these dissolved themselves after a few futile years and did not begin to reappear as functioning entities until 1879. The same traveler who had commented on the liquor in Hays in 1868 also provided a typical description of the train-side view in this region: "For the entire distance . . . not a tree or a shrub relieved the monotonous landscape of level, parched prairie. . . . The remains of sod houses, which have sheltered railroad hands, speculators, and their goods, . . . are scattered along the route."[84]

Except for a place called Sheridan that existed for a year at a waterless site where tracklaying stopped in 1868–1869 while awaiting new funding, the only Union Pacific attempt at urban speculation in far-western Kansas was near the last of the state's military posts: Fort Wallace. This site, harsh but functional, was where a tributary called Pond Creek flows into an only slightly larger Smoky Hill River to produce the first regular flow in that latter channel. Significant also was its distance of 130 miles from Fort Hays. Railroad officials realized that they needed a division point in this vicinity and that private speculators would be unlikely to invest money in a place where "not a spear of grass four inches high, dare show its head for fear of the wild winds." Consequently, they decided to construct their first company town.[85]

Wallace, the obvious name for the new community, was platted in 1870. It was never larger than a few hundred people and retained its status as a division point only a few years until the fort closed in 1882.[86] Its establishment is important, however, because it initiated a ripple effect of town foundings by the railroad men. Once Denver had been reached, time was available for railroad officials to improve efficiency along the now-completed line. This included the optimal spacing of division points, each about a hundred miles apart and with access to good water. The planners realized that water was the critical element in their choice and that their calculations needed to focus around the small High Plains oasis at Wallace. Working their way eastward from that point, a near-perfect combination of distance and water presented itself in western Ellis County, where the tracks crossed Big Creek thirteen miles west of Hays City. This, they decided in 1870, would become a new community called Ellis.

Farther east in Kansas, the water issue lessened, but the railroad men, now operating as the Kansas Pacific, decided to keep their distances per division as regular as possible. A hundred and two miles east of Ellis, in the narrow confines of Spring Creek Valley, they created another new town: Brookville. From this location to Kansas City the mileage was exactly two hundred, meaning a need for one additional division break. This time the men opted for an existing station that was close to the ideal location. The citizens of Wamego thereby found themselves pleasantly endowed with railroad jobs that they had not solicited. All of this happened during the month of April 1870.

The practice of a railroad company creating its own towns for division points is nothing unusual, of course. To do so after the line was already in operation, however, and to ignore nearby, larger communities that willingly would have bid for the opportunity to capture new jobs, marks the Kansas Pacific as unique on the Kansas scene. The people in Hays City were upset at the cavalier moves, of course, as were those in Salina, only fifteen miles from Brookville. Angrier still were the residents in Ellsworth and Junction City, who saw their existing facilities transferred to Ellis and Wamego, respectively. These decisions all were justified in the name of efficiency and of the lower cost of land outside of existing communities. They clearly stymied the hopes of many entrepreneurs, however, who dreamed of accumulating a given area's railroad, government, and manufacturing employment all at a single site and thereby creating a major city. Junction City people were able to overcome the problem when the Kansas Pacific sold its existing shops and other railroad buildings to the newer Union Pacific, Southern Branch company, which had its northern terminus there. Ellsworth was not so fortunate; its equipment was simply loaded onto flatcars and shipped to Ellis. The recipient towns all grew modestly, but each of them had larger rival communities nearby, and both Brookville and Ellis lay in rugged terrain not suitable for traditional agricul-

Loading Texas Cattle onto the Kansas Pacific Railroad at Abilene, 1871. This drawing by Henry Worrall originally appeared in *Frank Leslie's Illustrated Newspaper* on August 19, 1871. Buyers in Abilene handled an estimated forty thousand animals that year, enough to fill 2,034 rail cars. Prices were low, however, in part because stormy summer weather made the cattle nervous and therefore thin. (Courtesy Kansas State Historical Society)

tural settlement. Wamego had about a thousand residents in 1880, Brookville and Ellis six hundred apiece.[87]

The unusual behavior of Kansas Pacific officials in creating Brookville and Ellis is almost certainly related to cattle drives, another important regional event of the period from 1867 to 1871. This phenomenon, still the source of much imagery and symbolism for the entire Great Plains, was a huge financial bonanza for Kansas Pacific stockholders because they had their tracks in place before any of the rival railroad companies in southeastern Kansas or the Santa Fe line in the southwest. These cattle drives also directly modified the incipient urban structure along the Kansas Pacific in one important instance.

The creation of the Kansas cow town is usually attributed to Joseph G. McCoy, an Illinois native who had made considerable money from livestock trading in the Springfield area during the Civil War. He watched the construction progress of the Union Pacific in central Kansas in 1866 and came to the state early the next year in hopes of contracting with company officials to deliver cattle at a set rate per carload. He found the way already prepared in some ways, because other people, including former governor Thomas Carney, had been lobbying the state legislature to allow such cattle importation despite the objections of many who feared trampled fields and the splenic fever these animals carried. The original bill would have permitted Texas cattle only in the western, unsettled third of the state. This was beyond the immediate reach of the railroad, however, and so a substitute motion proposed the opening of the southwestern quarter of the state instead. This would allow animals as far east as present-day Wichita and as far north as present-day McPherson and Great Bend. Another provision would allow bonded companies to establish trails from points within this zone north to railheads on the Union Pacific, so long as these trails were all west of Saline County. This substitute bill became law in the spring of 1867, at the time when the railroad's tracks had just reached Abilene.[88]

McCoy has written that Union Pacific officials were skeptical of his plans to bring cattle north in quantity but agreed to pay five dollars for each carload he could deliver to a station yet to be identified. He then set out for Junction City to examine the possibilities between there and the most recent railhead at Salina. All of these sites were technically illegal under the recent legislation, but he had informal assurances from Governor Samuel J. Crawford that this would be no problem. McCoy liked Junction City and made an offer to purchase property there for a large stockyard. The asking price, however, was "exorbitant," and so he scouted further. He found suitable sites at both Solomon City (on the border between Dickinson and Saline Counties) and Salina, "but after one or two conferences with some of the leading citizens, it became evident that they regarded such a thing as the cattle trade with stupid horror."[89]

Abilene, "a very small, dead place" in the words of McCoy, became the choice of last resort. It rapidly evolved into the entrepreneur's private fiefdom. McCoy bought 250 acres, built the Great Western Stockyard to hold eight hundred animals, and purchased a ten-ton set of scales. He then erected an "elegant," three-story hotel called the Drovers' Cottage, stables, and a bank. Although the effort had begun too late in the season to permit exhaustive advertising, still eighteen to twenty thousand cattle passed through his facilities by September. The next year, 1868, the number jumped to fifty-three thousand head, and McCoy was soon elected mayor.[90]

Abilene's cattle boom lasted through the 1871 season. Although not totally a positive experience because it ushered in "a crowd of cutthroats, black legs, thugs, gamblers, and prostitutes" along with money and business, it most certainly elevated the community's status within the urban hierarchy. From a "dead place" in 1867, it came to house 525 permanent residents in 1870 and 2,360 in 1880. This success provoked envy, of course, and therefore rivals. The most formidable of these appeared several counties to the south where the Santa Fe railroad in 1871 was in the process of intercepting the state's principal cattle trail. Kansas Pacific officials were not about to concede the lucrative trade without a fight, however. As more farmers entered Dickinson County and Abilene townspeople themselves began to tire of the fray, they made two moves, both of which proved futile.[91]

First, as a temporary measure, the railroad men allowed a St. Louis company to erect a stockyard at Ellsworth and reported to their board of directors that they expected this town to become a major shipping point for 1871. Then, in an effort to control the trade entirely by themselves, they decided to create a new town devoted exclusively to cattle. Their choice was a heretofore neglected station in extreme northwestern Ellsworth County. No settlers were in the region to object, and they hoped that the climate and terrain might dictate a permanent grassland economy. The National Land Company revealed its intentions at the platting in September 1871. Utilizing the genus name of cattle, they called the place Bosland. As it happened, the greater distance drovers had to travel to the Kansas Pacific line once the Santa Fe was operational in 1871 cut Ellsworth's glory to a single season and never allowed Bosland's to begin. Settlers there quietly changed the name to Wilson.[92]

Railroad Cities in South-Central and Southwestern Kansas, 1869–1880

Like the case of northwestern Kansas with the Union Pacific, a single railroad company heavily influenced the cities in south-central and southwestern Kansas. This enterprise, the Atchison, Topeka and Santa Fe, had a quite different and more benevolent corporate philosophy than did its slightly older rival, however, and ended up dominating much of the economic life across the entire southwestern quarter of the nation. As I recounted earlier in this chapter, this railroad originated in Topeka during the territorial years with a vision of continuing the trade legacy of the Santa Fe Trail. A federal land grant, obtained in March 1863, made the project feasible and provided almost complete freedom on the selection of route. The document specified only that the tracks must cross the western border of Kansas within ten years or the unsold grant acreage would revert to the public domain.[93] Nobody thought much about these restrictions in 1863, but the date and the destination both proved frustrating to company officials and critically important to the pattern of urban evolution throughout much of the state.

By the time the Civil War was over and sufficient money had been raised to begin construction in October 1868, the alternatives for locating the first sixty miles of track had been carefully studied. From Topeka, the route was to go south into Osage County, where coal seams of minable thickness were known to exist, and then turn southwest to intersect the planned Union Pacific, Southern Branch, Railroad at Emporia (Map 10). Farmers and small towns already occupied these first miles, but Thomas J. Peter, the general superintendent of the railroad, envisioned coal mining on a large enough scale to supply the demands of his own steam engines plus the heating requirements of Topeka and other communities along his route. To better control this valuable resource and to house the hundreds of miners he anticipated needing, Peter decided to establish two company towns. The first, sixteen miles south of Topeka, he called Carbondale. Because of its proximity to the capital city, this was never intended to be a large settlement. Railroad officials leased several thousand acres of coal land, however, and created the Carbon Coal Company to undertake the mining proper. This company also was charged with erecting houses for the miners and a store for general merchandise; it did so in the spring of 1869. Several months later, when a contract was announced to supply coal for the Kansas Pacific Railroad as well as the Santa Fe, the future looked assured.[94]

Nine miles southwest of Carbondale lay Burlingame, the seat of Osage County at the time and the oldest settlement in the area. Peter dutifully ran his tracks through this town, but because no good coal seams had been located in the immediate area, the company made no investments.[95] Instead, he cast his eyes on another tract of land nine miles farther south where surveys had revealed the best coal deposits in the region. Coincidentally, this site was midway between Topeka and Emporia, and so a city there potentially could command a good-sized trade area.

Osage City, as the new community was christened, probably had its origins in 1868 when Peters and other Santa Fe officials learned that a coal seam thirty-six inches thick had been discovered in the vicinity. Peters himself bought the land parcel, but the actual development was assigned to a new subsidiary organization of the railroad, the Arkansas Valley Town Company. John F. Dodds, the Santa Fe's chief land agent, made the original survey in late 1869, several months before the tracks were completed, and he stayed on to ensure a quality development. All went as planned. The mines opened in the summer of 1870, and a decade later, twelve hundred men were employed in twenty-eight shafts spread over thirty thousand acres of coal leases. About half of total production was directly controlled by the Santa Fe, which had shifted its mining unit, the Carbon Coal Company, to the new town from Carbondale.[96]

Osage City, although never envisioned as a metropolis, was planned to be much more than a rough mining camp. Consequently, the town company laid out wide streets, installed lamps on every corner, and encouraged the use of brick and stone in construction. A visitor in 1883 was impressed, noting not only a general neatness of appearance but also educational and religious institutions "of a very superior order."[97] The community never was able to garner the county-seat designation, but it claimed 2,089 residents in 1880 and 3,469 a decade later, both figures the largest in Osage County.

Santa Fe Depot in Newton, 1878. Newton was seven years old at the time of this photograph and home to two thousand residents. Its days as a cow town were past, but a position at the junction of the Santa Fe's two main lines generated considerable traffic. Four years in the future this business would justify a new depot built as part of the Arcade Hotel, a grand, 120-room building. (Courtesy Kansas State Historical Society)

Following the decision to create Osage City, no great strategy sessions were necessary for the next section of track. Peters laid nearly a straight-line survey southwest to Emporia and built this line during the summer of 1870. The choice of Emporia was easy, even though the Santa Fe had no financial interests in this preexisting community. People there had voted for a $200,000 bond issue. This had been preceded, however, by several years of more fundamental, private discussion between the town's leading citizen, Preston Plumb, and the Santa Fe's founder, Cyrus Holliday. Both men obviously saw mutual benefits in creating an intersection between Holliday's line and Emporia's own creation, the Union Pacific, Southern Branch Railroad (i.e., the MKT), which extended along the Neosho Valley. A final, and probably even more fundamental, reason why the Santa Fe people selected Emporia was the same one identified by that town's founders in 1857. The Cottonwood River, which joins the Neosho at this point, carves an excellent route west through the Flint Hills (Map 5). Savings in construction costs would be considerable by utilizing this lowland, and this was the way tracklayers went during the winter of 1870–1871.[98]

Once through the Flint Hills, the railroad entered unsettled territory, and its owners, who by now were dominated by a group of Boston investors, confronted a major dilemma. In order to meet the terms of the land grant, it seemed prudent to head straight west to an intersection with the Arkansas River (perhaps near present-day Sterling) and then to follow this stream to the western border of Kansas. This course of action would yield no immediate profits, however. An attractive alternative was to participate as fully as possible in the cattle trade from Texas, which had been growing steadily in volume since 1865. This, however, would dictate construction to the south. To give themselves time to consider the predicament, a compromise order was issued to Superintendent Peter. He was to take a southwesterly course to a new town site that had been recommended the previous summer by David L. Lakin, the head of the railroad's land development office. Lakin's choice was odd in some ways, still fifteen miles northeast of the Arkansas River, and with only a small supply of surface water from a stream called Sand Creek (Map 5). The site's redeeming virtue, however, was livestock. This was where the railroad's path toward the Arkansas intersected the biggest current cattle trail in Kansas.[99]

A name for the Sand Creek settlement, Newton, was announced in February 1871, five months before the arrival of the first train. Officials said this place would serve as the terminus of the line for that season, and that a stockyard sufficient to hold six thousand cattle would be in place by summer. All this forecasting was to alert Texas drovers to a superior shipping opportunity. No longer would animals have to follow the established Chisholm Trail north all the way to Abilene on the Kansas Pacific Railroad, as they had done for the previous four seasons. By utilizing the new facilities of the Santa Fe, they could save seventy-five to eighty miles of time and animal weight loss. It was still another plan that turned out successfully for the company, at least in the short run. Herds arrived in town beginning in May, and forty thousand head were shipped out that summer.[100]

Nobody on the Santa Fe directorate disagreed with the establishment of Newton as a cattle town. The decision meant a needed infusion of capital for the company and the first substantial return on investment for the Boston men.[101] A major difference of opinion existed about the city's future, however, especially as compared with other possibilities for urban development in the immediate vicinity. These officials knew that their tracks now stretched out some 136 miles from Topeka, about as far as was practicable for a single division to run. A permanent site was therefore needed for repair shops and a division headquarters. Should Newton, with its suspect water, be this place or not? The future of the cattle market also was tricky. Just as the Santa Fe people were in the process of intercepting business from the Kansas Pacific, some other railroad could reach farther south in Kansas and repeat the process. Why not build the Santa Fe in this direction now to head off such a future challenge? Several officials favored this latter idea, including Holliday and Peter. They suggested what they perceived as an intermediate position. Extend the main line south from Newton another twenty-seven miles to a small settlement called Wichita, established the year before on the Arkansas River, and then proceed west from there.[102]

The proponents of the Wichita strategy had several items in their favor. Reliable water would make this a better division point than Newton, and a spur on south to the boundary of Indian Territory would corner the cattle market for the foreseeable future. In addition, a straight-line route west from Wichita to the Colorado border arguably could make more sense than the anticipated path from Newton. The key here was a large northward bend of the Arkansas River in central Kansas (Map 5). Following this bend northwest from Newton and then back to the south would require many more miles than would the overland Wichita alternative. Moreover, after bypassing the bend, this latter, direct-line route would meet the Arkansas again just east of Fort Dodge for the passage across the High Plains.

Opponents of the Wichita proposal, predominantly the Boston contingent of the directorate, acknowledged its attractiveness but worried about several things. Because the town already existed, the railroad could not control development there. There also was a technical matter arising from Wichita's position not on public lands per se but rather within the former reservation of the Osage Indians. Although this territory had been ceded to the United States, it was supposed to be sold to benefit the tribe and not given to a railroad as part of a grant. Could the old treaty be modified? The answer probably was yes, but such a resolution would have taken time that the railroad officials did not have to spare. They realized that any significant delays in moving west would seriously compromise the company's chances to qualify for its full federal land grant in a more fundamental way. In January 1872, only fourteen months remained for track completion to Colorado out of the ten years that Congress had allotted. In this time, approximately 285 miles had to be built west from Newton. So did 50 hitherto neglected miles from Topeka to Atchison. The company therefore could perhaps not afford either luxury: going to court or constructing an additional 27 miles to the south without assurances of a grant in that region. As for the proposed savings by avoiding the big bend of the Arkansas, these dollars probably would be more than counterbalanced by those required to bridge that large river at Wichita.

Despite much debate, the Santa Fe officials were unable to resolve their thorny issues of priority during the critical year of 1871. The next season, while attempting to pursue all of their agendas simultaneously, they inadvertently helped to fabricate a far more complex urban system for south-central Kansas than otherwise might have prevailed. Ostensibly, the initial developments seemed to favor Newton. The town was proclaimed the first division point southwest from Topeka and the site for sizable repair shops. The directors also elected to construct their main line west from town instead of first digressing south, and to utilize the Arkansas River lowland as its path to Colorado. A third decision of 1871, however, initiated a process than eventually cost Newton its chance to dominate urban life in the region. Ironically, it was an action endorsed by the local residents themselves.

The fateful third decision was the passage in August of a large, local bond issue to finance a new railroad called the Wichita and Southwestern. Essentially a spur for the Santa Fe, these tracks would run only from Newton to Wichita, but in the process, they would guarantee a new cattle center. Wichita people obviously were in favor of such construction, and so was Thomas Peter and several other members of the Santa Fe leadership. Peter, in fact, can be said to have initiated the bond issue. When one of Wichita's founders, James R. Mead, learned that the main Santa Fe line likely would go west from Newton rather than south, he had asked Peter directly what it would take to obtain a track for his community of six hundred people. The superintendent's telegram in reply, now famous in Wichita circles, read as follows: "In answer to yours of 2d [i.e., June 2, 1871] will say, if your people will organize a local company and vote $200,000 of county bonds, I will build a railroad to Wichita within six months."[103]

To understand the bond issue, one must first realize that Newton and Wichita both lay within Sedgwick County, which at that time was a twin to Butler County in size and shape. Wichita needed Newton's votes and obtained enough of them to carry the bonds by pushing two arguments. First was a supposed gain in sophistication that would accrue to Newton residents if they were to abandon the vulgarity of cow town life now that numerous, high-quality railroad jobs were coming in as replacements. The other was a promise to support a plan to carve out a new county from the existing structure that would be centered on Newton. This latter maneuver would allow both Wichita and Newton to become county seats and thereby gain additional desirable employment. With some grease from Santa Fe lobbyists, the state legislature passed a bill to this effect during its 1872 session. A new Harvey County thus emerged, containing ten townships taken from Sedgwick and three from McPherson (plus, the next year, two from Marion). By 1878 or so, the two thousand residents of Newton were well pleased with their fate. They enjoyed quality railroad and county jobs and were seeing the beginnings of a substantial milling industry that took advantage of the area's flat terrain, fertile soils, and good shipping connections. Finally, the spur line to Wichita had made Newton a junction point, and therefore a likely spot for the development of wholesale trade.[104]

The picture of contentment at Newton was to be fairly short-lived, at least from the perspective of its promoters who had hoped that Santa Fe favoritism would lead to metropolitan status. The first negative issue to reveal itself was water. Sand Creek supplies proved too small to provide for a growing community and too poor in quality for use in steam locomotives. Engineers in the shops tried various additives, but the problem grew severe. In 1879, railroad officials decided that the best recourse was relocation of all major facilities to a better-watered site. This judgment was wonderful for their new choice, Nickerson, forty-five miles farther west in the Arkansas Valley, but traumatic for Newton. As a local writer expressed the situation, the town "was left to languish on the prairie."[105]

Languish is actually much too negative a word to describe Newton accurately in the 1880s. Local agriculture, milling, and the town's still excellent railroad facilities produced a steady increase in population, from a total of 2,601 in 1880 to 5,605 a decade later. Still, the loss of Santa Fe employment hurt local egos and reduced the chances for additional rail connections and therefore for even faster growth. In particular, the 1880s was a time when several railroad companies were building east-west across southeastern Kansas and deciding where to join with the Santa Fe system. Had Newton kept the shops and the population these implied, this city would have been the logical choice. Without these facilities, however, the junction site became more problematic. People in the cow town of Wichita were thereby given a critical opportunity to diversify and expand that they otherwise would never have enjoyed. So were several other places farther to the south.[106]

The Wichita and Southwestern Railroad, built as promised during the spring of 1872, transported 70,600 head of cattle that summer. Its investors, who had leased the line to the Santa Fe, were happy. So, obviously, were the pioneer merchants at the new railhead, who could hardly believe they had rescued a considerable measure of economic success from a proposal that could have failed completely. The railroad, in fact, was only the most recent in a series of good and bad fortunes that had made life in the young community anything but dull. Wichita's physical location, for example, was a mixed blessing, depending on a person's needs and desires. The surrounding area was too sandy for many agricultural endeavors. But the specific site, at the junction of the Arkansas and Little Arkansas Rivers, was well watered and had been a traditional spot for the Osage peoples to exchange goods with various plains tribes. In the years just before the Civil War, the presence of these seasonal markets had attracted several Anglo traders, as well as a larger contingent of buffalo hunters who needed a base of operations. These settlements became permanent in 1864, when two thousand pro-Union Wichita Indians were relocated nearby for their safety during the war. Because these people were wards of the government, money was made available to supply their needs, and one of the area's Anglo traders, the same James R. Mead who later was to promote the railroad, saw a business opportunity. He moved his post to what became known as Wichita Town and commissioned a friend, Jesse Chisholm, to import cattle for food.[107]

When the Wichita Indians moved back south to their former homes in 1867, Mead decided to retain and promote his trading site. The trail that Chisholm had fashioned to bring livestock north from the Canadian River was in the process of being extended at both ends. Abilene, a city just a hundred miles north of Mead's outpost, had become the first big cattle town on the Union Pacific Railroad, and the herds thus passed right by his door. The possibilities for intercepting this trade in the near future seemed obvious, as did the opportunity to continue supplying the Wichitas and other peoples to the south. Within a year, Mead had several entrepreneurial partners, including the Kansas governor, Samuel J. Crawford, and together they formed a town company in April 1868.[108]

The hopes of 1868 were frustrated initially by problems with the land policy on the Osage Reservation similar to those later encountered by the Santa Fe's planners. That tribe's treaty of cession, made in 1867 and modified in 1869, contained provisions only for the sale of agricultural lands, not town sites. Clear title thus was not possible until the spring of 1870, by which time it was nearly too late to participate in the business of cattle drives. Farmers were coming into the region in large numbers and threatening to either ban the herds or impose tighter quarantine laws. Somehow, however, the enterprising Mead, together with William Greiffenstein and other local promoters, managed to work their will against all opposing forces. They simultaneously gathered support for the railroad bond issue, as I have already related, while paying off local farmers to ensure an unmolested path into the city and creating a contingency fund to handle any local claims for damages from the animals. They also successfully lobbied the trail bosses to the south for their business and the Kansas legislature to keep their town just outside of an expanded quarantine line that was about to shut down Abilene's business. The big herd numbers of 1872 made everything worthwhile, of course, and the process repeated itself in 1873, 1874, and 1875. Wichita, now with a population of 2,703, had arrived, and its leaders were already beginning plans for the next phase of their intended development.[109]

As Wichita people frantically scrambled with their newfound cattle business, members of the Santa Fe's board of directors, Superintendent Peter, and their track crews had no time to reflect on possible future developments for that site or at any other locations to the south. Their challenge lay in a different direction. With less than a year remaining for them to reach Colorado, the railroad men hurried back to Newton and put their concerted effort into laying track to the west. This work proved commendable. The new construction soon reminded many reporters of the push MKT workers had made down the Neosho Valley two years earlier. Crews from each enterprise accomplished the considerable feat of setting two miles of track per day on a regular basis. Occasionally, they laid an amazing three. Both companies also beat their deadlines. The Santa Fe's tracks reached Colorado on December 28, 1872, slightly less than eight months after workmen had left Newton. The land grant was now secure, but obviously there had been precious little time and money to spare on the creation of cities.[110]

Anybody looking west from Newton could see that the first town site with superior potential for growth would occur at the railroad's junction with the Arkansas River (Maps 5, 10). The distance (about thirty miles on that bearing) would be far enough from Newton to command a trade area of its own, and waterpower should be abundant for industry. The only mystery concerned where exactly this spot would be. Railroad people usually camouflaged their surveying operations, but time pressures did not allow for secrecy in this case. Consequently, an enterprising New York man named Clinton C. Hutchinson, who previously had helped to found Ottawa, simply followed the crew's line of stakes, projected their end point at the river, and negotiated a plan of action with railroad officials at their annual board meeting in Topeka. If the company would sell him the land for a town, he would develop it. Both parties would share equally in the sale of lots.[111]

Hutchinson, as the new settlement was christened, developed in a regular but not spectacular manner. This was much as one would expect with the competitors of Newton and Wichita both nearby. Mr. Hutchinson opened a real estate office and successfully promoted himself for election to the Kansas House of Representatives. There, with assistance from the railroad lobby, he engineered a shift in the boundaries of his Reno County similar to that done for Newton. Instead of being on the northern edge of the old unit, the town of Hutchinson was now nearer the center and thus able to secure the county seat. Two mills operated on the river by 1876, and four years later, Hutchinson had a population of 1,540. Its future looked to be that of a slightly larger than average county-seat town.[112]

One measure of the speed at which Santa Fe workers built across western Kansas is that in June 1872, as the first regular-service train pulled into Hutchinson, survey crews were already locating the right-of-way near Fort Dodge, more than 150 miles farther to the west. This rapid pace, plus the decision to follow the north bank of the Arkansas River, meant a route that was essentially fixed. Existing towns in the pathway would benefit, and the railroad company would insert a few extra stations as needed. On the long stretch toward Fort Dodge, the Santa Fe built through two already-existing, county-seat towns (Great Bend and Larned) and partnered with a Chicago group to create what they expected would be a third at Kinsley (originally called Petersburg after the general superintendent). The railroad's Arkansas Valley Town Company added four others in the gaps: Nickerson, Raymond, Ellinwood, and Pawnee Rock. None of the latter group were expected to grow. In fact, Superintendent Peter offhandedly contracted for depots at all four places

Railroad Yards at Dodge City, Early 1900s. The histories of Dodge City and the Santa Fe railroad are deeply intertwined. The railroad employed 406 people in 1935, many at the depot and division offices located adjacent to the tracks on the left (north). (Courtesy of Kansas State Historical Society)

on the same day in July. The three county-seat communities fared moderately better, mostly through local trade. Great Bend and Larned grew to just over a thousand residents apiece by 1880, while Kinsley, farther west, recorded 457. The only special railroad facilities in these places were a small stockyard at Great Bend and an eating house at Kinsley, "where the noonday meals are furnished eastern and western passengers."[113]

Upstream from Kinsley, the Arkansas River makes a U-shaped bend, angling southwest for about twenty-five miles before turning back toward the north (Map 5). By leaving the valley at this point and cutting overland instead, a traveler could save a dozen miles. One branch of the Santa Fe Trail had taken this option; so did the railroad surveyors. People who followed this project realized that the spot where the tracks would rejoin the river valley would be an obvious city location, especially since the elapsed distance from Newton was such as to require another division point and set of repair shops. As a result, a group of officers stationed at nearby Fort Dodge joined with the Santa Fe's own planning staff to create the Dodge City Town Company. They did so in August, a month ahead of the track crew.[114]

Dodge City had a favorable location from several perspectives. On a local scale, it lay just beyond a liquor-free zone that the army had imposed within five miles of Fort Dodge, its westernmost Kansas outpost on the Santa Fe Trail. Looking more broadly, everybody with an interest in the cattle business judged the potential for marketing there to be superb. To begin, Dodge City occupied the southernmost point on the main line. More important, most observers judged the climate in this part of the state sufficiently arid so as to discourage homesteading. If this proved to be the case, then Dodge City might not face the short life span characteristic of more eastern cow towns where coalitions of farmers, ministers, and others soon formed to force the boisterous, transient society away. Even if Wichita and other places were to maintain the cattle trade for several more years, the founders of Dodge City were not worried. This area in 1872 lay at the heart of buffalo country, where an estimated two thousand hunters were participating in what had become a very lucrative, if bloody, business.[115]

Buffalo did indeed prove to be the first mainstay in the city's economic growth. One prominent merchant, Robert M. Wright, bought and shipped more than two hundred thousand hides in the winter of 1872–1873, when he reported the streets of town to be "lined with wagons, bringing in hides and meat and getting supplies from early morning to late at night." These numbers declined rapidly over the next few years, but cattle replaced them almost animal for animal. At the annual Santa Fe board meeting in 1875, officials were told that the Wichita area was rapidly becoming too crowded for the drovers, and that new facilities needed to be developed elsewhere. The planners thought about Great Bend, but both its distance from the grazing lands of Indian Territory and the likelihood that the state quarantine line against Texas splenic fever would be moved at least that far west, argued against this site. They decided instead to focus on Dodge City.

The year 1876, when drovers brought 9,540 head up what came to be known as the Western Trail, was the first notable cattle season for the new community. Herdsmen found saloons in abundance and the Santa Fe ready with an impressive stockyard that stretched two miles along the tracks. By 1882, when the shipment level had jumped to sixty-five thousand head, the station operation had expanded to include nine switches, separate pens for sheep, two roundhouses with twenty-six stalls apiece, and a substantial machine shop. Despite having recorded a population of only 996 in the recent census, the town already had come to dominate a huge area. In addition to southwestern Kansas, this reach extended deep into the Texas panhandle with the Jones and Plummer freight trail to the small outpost of Mobeetie and a military road to Fort Supply and beyond. People thought of Dodge City as a cattle town, but it was very much a railroad one as well.[116]

Ford County, with Dodge City as its seat, was the last organized political unit traversed by the Santa Fe on its passage across Kansas. With speed still on the minds of construction engineers, town planning was minimal throughout the remaining 120 miles to the state line. The three largest stations, established at equal intervals along the route, were all named after company officials: Pierceville, Lakin, and Coolidge. Only Coolidge, near the Colorado line, was given any facilities beyond a depot, and it only a small machine shop and the temporary offices for the western division. In retrospect, the most glaring omission was a stop of any kind at what was to become Garden City. The first settlers at this site, a group from Sterling, in Rice County, were relatively late arrivals. They had to use the depot at Sherlock, seven miles to the west, for several months in 1879 until they were able to barter a selection of town lots to the railroad in exchange for a depot. Growth in Garden City did not begin until after 1880, when an irrigation ditch dug in desperation proved to be successful and an attraction for new settlers and investors.[117]

5

Later Railroads and Railroad Towns, 1877–1910

The rush to the Colorado line during 1872 severely strained the financial resources of the Santa Fe. Although the company's long-term prospects remained solid, it had little money available for construction on toward Pueblo or for improvements to the existing line. Other Kansas railroads faced similar troubles at this time. After overextending themselves in the years following the Civil War, they had to face a major national business panic in 1873. This was followed the next year by a locust invasion so massive that it was compared regularly to a biblical plague. The combination of events was enough, in fact, to throw the whole credibility of western development temporarily into question. Santa Fe engineers extended their main line only twelve miles into Colorado during 1873, and hardly any new construction occurred in Kansas until 1879.

The stagnation of 1873 to 1878 proved to be only a hiatus, not a permanent end to development. It provided time for entrepreneurs to reflect, and in this way perhaps even contributed to what proved to be a robust recovery. Over the course of the next decade, the older railroad companies added to their trunk lines and constructed an extensive series of branch routes. More impressive still, three new competitors entered the Kansas market and built hundreds of miles of main line. By 1889, these entrepreneurs had completed the basic structure of routes and intersections that was to underpin city growth in Kansas for the next half century (Map 11; see p. 24).[1]

Kansas City and Wichita

Two of the earliest moves in this second round of railroad activity were critical to the region's future premier cities: Kansas City and Wichita. Kansas City, as I discussed in the previous chapter, had begun to grow rapidly in the early 1870s because of the relative superiority of its rail connections to the east, west, and south. Meatpackers were especially big employers. The presence of these packers was a logical outgrowth of cattle shipments from Abilene and Ellsworth on the Kansas Pacific. By 1872 or 1873, however, an anomaly in this business arrangement had become obvious. The Santa Fe railroad was now the leading shipper of cattle, but the company lacked direct access to the Kansas City market center. The railroad's leaders saw that they had three options. They could continue to transfer cattle onto Kansas Pacific cars at Topeka (and thereby give their rival line a share of the profits), they could attempt to lure the packers from Kansas City to their terminus at Atchison, or they could develop their own rail entranceway into Kansas City. The first of these choices was distasteful and the second quite risky, since Atchison still lacked a bridge over the Missouri River. In contrast, the third option was both pragmatic (given the growth potential of Kansas City) and more affordable than expected. The railroad owners encouraged enthusiastic local companies to construct separate lines along the south bank of the Kansas River from Topeka to Lawrence and from De Soto to Kansas City. Then, by leasing both tracks plus an older line that connected the two segments, the Santa Fe company acquired access to the packers with no immediate capital investment. The new service became operational August 30, 1875.[2]

Santa Fe officials replaced the lease arrangement into Kansas City with outright purchase of the trackage in 1877. They then began to make major investments of their own in Wyandotte County, including switching yards and a division headquarters. This arrangement proved immensely beneficial to the railroad's investors and to the Kansas City business community. Traffic on this new segment of track rapidly surpassed that to Atchison, and the company began to consider ways to bolster its presence in Kansas City even further. One of the most important of these, with implications for other communities as well, was to realign its new portal into the city. Instead of having trains head almost due north from Emporia and then make a right-angle turn at Topeka, officials decided to build a new line directly northeast from Emporia that would pass through Ottawa and Olathe and then into their new yards at the community of Argentine. Actually, the company had to construct only the segments from Emporia to Ottawa and from Olathe to Argentine because it had acquired an existing line between Ottawa and Olathe through purchase of the Leavenworth, Lawrence and Galveston company in 1880. The new linkage was put in operation in 1884 and received main-line status shortly thereafter. Kansas City prospered even more, this time joined by Emporia, which became an important switching point between the old and new tracks. People in Lawrence and Topeka, in contrast, saw fewer trains pass their way.[3]

The decision by the Santa Fe officials to enter the Kansas City area made an already dominant city even more so. In contrast, the selection of Wichita as a principal destination for the St. Louis and San Francisco Railway Company, a new entrant to the state's railroad system, was arguably the most important factor in that community's ability to make a successful transition from former cow town to the major urban center for south-central Kansas. Owners of this railroad had no thoughts of building to or even near Wichita when they began

Box Cars at the Santa Fe Yards in Kansas City, circa 1949. Santa Fe officials have increased the size and quality of their facilities at Argentine almost continuously since 1875. The yards contained more than twenty-seven miles of track in 1890 and at the time of this photograph encompassed an area four miles long and a half mile wide. (Courtesy Kansas State Historical Society)

operations in 1866 as the Atlantic and Pacific Railroad Company. Their plan, as the name implies, was transcontinental in scope. It was to take an existing railroad that ran southwest from St. Louis toward Springfield, Missouri, and extend it farther in that direction to the thirty-fifth parallel of latitude near present-day Oklahoma City. From there they would go west to Los Angeles and then to San Francisco. The idea seemed solid, for the route would serve a new realm and avoid many snow problems and the western mountains. Construction foundered, however, when Native American groups refused to let the tracks pass through Indian Territory, and the company declared bankruptcy. Later, reorganized as the St. Louis and San Francisco Railway Company (i.e., the Frisco), its leaders decided to attempt an alternate route.[4]

The new plan of the Frisco officials amounted to a simple detour. They would buy an existing short-line railroad that ran from their main tracks northwest to Carthage, Missouri, and then on west to Columbus and Oswego in extreme southeastern Kansas. From Oswego they then could build farther west around the Indian obstruction before rejoining the original route at some undetermined point. This news, announced in 1878, bolstered the spirits of people in the two Kansas towns already on the route but brought them only modest population gains because the two north-south railroads through these places (the border tier, and the LLG) were, like the Frisco, still blocked from entering Indian Territory. The choice of which route to take west from Oswego proved easier than one might expect, because company officials placed a priority on linking with the Santa Fe system. This could be accomplished only at a few locales. Although no announcements were made at this time, money weighed on the minds of these men, and they hoped to lease the right to use the existing Santa Fe tracks across western Kansas before heading back southwest on their own.[5]

The voters of Sedgwick County passed a $230,000 bond issue in May 1879 as part of a plan to make Wichita the Frisco's junction point (Map 11). This money helped, of course, but the company probably would have built to Wichita in any case.[6] This destination simply made the most sense. Had the tracks been laid straight west from Oswego instead, they would have passed south of Wichita through the prosperous county-seat towns of Winfield and Wellington but would have made no junction with the Santa Fe. Plans were afoot to extend the Santa Fe rails to both of these communities, but nothing was yet a reality. In addition, a train continuing on west from either of these two places on Santa Fe tracks presumably would have to pass back through Wichita to that company's main line. The Frisco men's other choice for a junction was Newton, north of Wichita. This would have meant additional miles of construction, however. In addition, that city had just lost the Santa Fe shops because of water problems. The timing of the new line thus was impeccable from a Wichita perspective. Had the Santa Fe spur never been built to that city in 1872, Newton would have been the only practicable junction for the Frisco. Similarly, had the Frisco's decision to build come even a year later (after the Santa Fe had extended its tracks farther south), they might well have opted for a straight-line route into Winfield.

Shortly after the Frisco tracks arrived in Wichita in May 1880, its owners joined with Santa Fe men to announce a merger of sorts, whereby the two companies would jointly use the Santa Fe line to Pueblo and then, in the future, together build along the Frisco's thirty-fifth parallel through New Mexico and Arizona Territories. It was a profitable deal all around. The Frisco people obtained financial relief, the Santa Fe directors eliminated a competitor, and Wichita became a railroad nexus. The new connections to timber from the Ozarks and coal from southeast Kansas were especially useful and helped to inspire city merchants to think how they could become suppliers of these and other goods to emerging communities in the southwestern quadrant of the state.[7]

One additional community benefited substantially from the Frisco construction across southeastern Kansas. This was Neodesha, a small trading site in southern Wilson County at the junction of the Fall and Verdigris Rivers (Map 5; see p. 17). The Frisco engineers utilized Fall River and its tributary, Salt Creek, as their path into the Flint Hills upland, but Neodesha also happened to lie midway along the two-hundred-mile run between Wichita and the route's origin in southwestern Missouri. For a token payment of $5,000, the town became the headquarters for the railroad's Kansas Division. With a roundhouse and repair shops, Neodesha went from nearly nothing to 924 residents in 1880. It was, as two local historians have said, "a stroke of good fortune that could not be overvalued."[8]

Arkansas City, Wellington, and Winfield

The area directly south of Wichita, somewhat surprisingly in several ways, was another major focus for railroad-developed cities in the 1879 to 1886 period. The surprising part is that, by these years, this fertile region had been blanketed by farmers who were opposed to any more cattle drives or new shipping towns. Santa Fe officials had no need to contest the farmers' position either, because the company's newer stockyard facilities at Dodge City were working well. Nevertheless, the railroad men decided to open several important routes through the area and to place major company facilities in two communities. First, in early 1879, they built a single line south from Wichita to Mulvane at the southern edge of Sedgwick County. Then, strangely to some observers, they split the tracks. One branch continued south along the Arkansas River through Winfield to a small settlement called Arkansas City, about four miles from Indian Territory. The other branch went southwest from Mulvane to Wellington, the seat of Sumner County. This was its terminus for about a year during 1879–1880, but then the tracks were divided and extended once again. One spur went south to Caldwell, three miles from Indian Territory. The other ran west, first to Harper in Harper County via a track obtained as part of the LLG purchase, and then southwest to Kiowa on the state border in 1884 to 1886 (Map 11).[9]

Three motivations underlay the new construction. One was competition. The LLG Railroad, which had reorganized itself as the Kansas City, Lawrence and Southern, recently had built a line west from Montgomery County through Winfield, Wellington, and Harper, and therefore into territory that Santa Fe people considered rightfully theirs. The Santa Fe extensions were therefore partly a rebuttal. A second, interconnected factor was a variation on the older cattle trade. Although drives into this part of Kansas were illegal after 1875, animals still could come as far north as the border and be shipped from a suitably located Kansas facility. Such thinking clearly was the reason for the extension to Caldwell, which was on the Chisholm Trail. Many of the Texas cattlemen still preferred this route over the Western Trail to Dodge City because it usually featured better pasturage.[10]

The third motivation behind Santa Fe construction in south-central Kansas was longer term and much more important than either local competition or lingering cattle shipments. The company's officers, together with many other interested parties, had been lobbying for freer access into and across Indian Territory. The two branches from Mulvane were designed primarily in anticipation of such construction. The moment came via a congressional act of July 4, 1884, which, among other things, gave the company the right to two routes. One extended from the Winfield area south to the vicinity of Denison, Texas. The other ran southwest to and beyond the vicinity of Fort Supply, near the hundredth meridian. The development of these corridors was a key element in an aggressive strategy by William B. Strong, the general manager of the Santa Fe at the time, to convert the company from a regional into a national power. The Texas thrust would compete head-on with the MKT Railroad for the rapidly growing trade in that section of the country. The one to Fort Supply initially was to tap the new cattle country of the Texas panhandle and Pecos Valley. It then could connect with the existing trunk line in New Mexico to provide a passage to Los Angeles that was more direct and lower in gradient than the current routing west to Pueblo and then south over Raton Pass to Albuquerque. What looked like mere spurs when they were constructed to Arkansas City and to Kiowa were thus really intended as future main lines.[11]

As crews constructed the two new extensions in 1886 and 1887, the question arose of where to locate the division headquarters and machine shops for these important ventures. Such facilities normally would have been placed far south of the Mulvane junction, but white settlement was still prohibited in Indian Territory. This reality meant a bonanza for Arkansas City and for Wellington, which almost by default were named home bases of what became known as the Oklahoma and the Panhandle Divisions, respectively.

Of the two cities, the people in Wellington owed the Santa Fe men the most gratitude, because their situation otherwise would have been unremarkable. This community had been founded near the center of Sumner County with a quest for the county seat obviously in mind. The surrounding area was flat and fertile, but the possibilities for trade were fairly limited. No large river was nearby, and although railroad spurs ran south to Caldwell and Hunnewell on the state border, it seemed unlikely that either would be extended. Trade to the west via the Santa Fe line was possible but similarly improbable, since the city of Harper, one county to the west, commanded a better gateway position. An initial flurry of economic activity from railroad construction, the business of county government, and local milling had been enough to attract 2,694 residents by 1880. Santa Fe employment nearly doubled that number by the next census, when 4,391 people called Wellington home.[12]

The founders of Arkansas City had trade and industry in mind, not agriculture or local government. Their site, near the state border, held early potential for cattlemen who wanted to control nearby sections of the rich, unoccupied grassland in Indian Territory known as the Cherokee Outlet. Slightly later, after the Osage, Kansa, Ponca, and other tribes had been moved into this outlet region, trade replaced ranching. Arkansas City had more potential than other border towns, however, because of its rivers. Two large streams join here: the Walnut and Arkansas Rivers (Map 5). Both have wide, fertile floodplains, and together they held considerable potential for water power. Mills were among the earliest industries in town, and in 1881, local men formed the Arkansas City Water Power Company to attract manufacturing plants through a diversion of the Arkansas River into a canal having a twenty-two-foot fall. This industrialization, which included windmill, mattress, and chair factories, provided a substantial complement to Indian trade and railroad employment. Together, the amalgam pushed the local population from 1,012 in 1880 to a very impressive 8,347 by 1890, the ninth-largest city in the state.[13]

Finally, and unexpectedly, a third city of note also joined the competitive field in south-central Kansas during the period of railroad construction. This was Winfield, twelve miles north of Arkansas City in the valley of the Walnut River. Winfield, like Wellington, had obtained designation as a county seat and as an intersection of a Santa Fe line with the east-west tracks of the Kansas City, Lawrence and Southern. It was a rather inconsequential community from the time of its founding in 1870 until the railroads came almost a decade later, but then its leaders became active pursuers of an eclectic mixture of industries, public institutions, and private colleges. Following the example of Arkansas City, they dammed the Walnut River for power. Then, in 1884 and 1885, they lobbied the Kansas legislature to acquire the state's school for feeble-minded youth and outbid seven other communities to host a new college for the Methodists of the Southwest Kansas Conference (today's Southwestern College). The offer to the Methodists of $60,000 plus a forty-acre-site and free building stone was truly impressive for a town of only three thousand residents. Finally, and not so successfully, the people tried to promote themselves as a gateway center to southwest Kansas as a rival to Wichita. Even so, Winfield residents had much to congratulate themselves for in 1890, when the census recorded a population of 5,184.[14]

Railroad Cities in North-Central Kansas

While an energized Santa Fe company and its ally, the St. Louis and San Francisco, were briskly developing cities and new trade areas in the southern half of Kansas during the late 1870s and early 1880s, leaders of the Kansas Pacific were only moderately active in adding branch systems to their main line across the north. Many historians have noted this lack of initiative. They attribute it, in part, to the owners' single-minded obsession with Colorado and other larger-scale western issues, but even more to poor management and excessive capitalization. The company's singular lack of success with an early branch line may have been an equally important factor.[15]

In 1867–1868, while still operating as the Union Pacific, Eastern Division, company officials had commissioned a survey southwest from Ellsworth along the general course of the Santa Fe Trail. Their plan was to make this a main route to the Pacific Coast. Had these men been successful in securing additional funding to support such construction, they would have captured the market and territory that eventually became that of the Atchison, Topeka and Santa Fe Railroad, and their company probably would have become a dominant economic force nationally. The city of Ellsworth, obviously, would have boomed. Although Congress denied the request, the idea persisted in Kansas Pacific boardrooms through at least 1870, and officials actually followed through on a modified version in 1872. They authorized and built what they called the Arkansas Valley Branch across fifty-six miles of the Colorado High Plains from Kit Carson to a junction with the Santa Fe line at Las Animas. This construction, however, was too little, too late. Serving no real purpose, since both main lines went essentially to the same destination, it was a failure from the start and was abandoned six years later.[16]

Despite its monetary and other problems, the Kansas Pacific company did manage to construct four major feeder routes in the state during the period between 1872 and 1887. These obviously were important in generating traffic for the trunk line, and they also prompted significant growth for four cities that occupied nodal positions in the new system. Two of these, Junction City and Salina, were points along the main line. The others, Beloit and Concordia, controlled sites where branches intersected with another east-west railroad, Atchison's nearly forgotten Central Branch (Map 11). A generation later, in 1910, this process was repeated when the company installed a new trunk corridor northwest from Topeka and created a major junction point at Marysville.

Junction City was the first town to grow as a result of the new construction. This community, which had become a railroad node in 1869 when workers for the Union Pacific, Southern Branch, began to lay tracks to the south, was the most obvious location to begin a branch to the northwest as well. The Republican River runs in that direction, providing good lowland soils for prosperous agricultural communities (Map 5). In addition, a survey for such a route already existed in the company's offices because this had been the original path anticipated for the main Union Pacific trackage before 1865. Construction began for what was called the Junction City and Fort Kearny Branch in 1872. After halting one county to the north at Clay Center for several years, officials

Cavalry on the Parade Ground at Fort Riley, 1897. Once this post was named a permanent site for schools of cavalry and light artillery in 1887, Junction City's economic prospects blossomed. The developments meant at least five hundred civilian jobs plus opportunities to sell goods and services to a thousand soldiers. Each of the barracks in the background of the photograph housed 136 men (Courtesy Spencer Research Library, University of Kansas)

pushed it ahead another thirty miles to Concordia, the seat of Cloud County, in 1878. Junction City, whose voters put up $100,000 for the construction, obtained the shops for the line. These were small but much-needed boosts to the local collective ego because the community had recently suffered two other major railroad losses: the Kansas Pacific's divisional headquarters to Wamego (1870) and the MKT shops to Denison, Texas (1873). As a result of these changes, the city's population fell during the 1870s from 3,100 to 2,684.[17]

Things brightened considerably for Junction City during the 1880s. Because of mechanical advances that enabled longer distances between service stops, Union Pacific (the name again of the Kansas Pacific after 1880) officials announced in 1889 that they were eliminating one of their Kansas divisions. The Brookville facilities would be closed and those at Wamego moved back to Junction City where they originally had been housed before 1870. An influx of 150 brakemen, conductors, engineers, and firemen, plus seventy mechanics and laborers meant perhaps five hundred new people to the town. In addition, partly because of the city's efficient system of railroad tracks in all four directions, the army had announced in 1885 that nearby Fort Riley would escape the closing that was the fate of most frontier posts. Instead, a permanent garrison would be maintained there that could move by rail to crisis spots over a broad region. Junction City people rejoiced, and civilian construction crews helped boost the town's population to 4,502 by the 1890 census.[18]

The second trunk-line city to prosper because of branch railroads was Salina. In this case, however (as opposed to Junction City), the feeder routes were the primary reason for the change instead of merely a contributor. As I described in the previous chapter, Salina occupies a position in the Smoky Hill Valley where the stream turns south for twenty miles and flows through an extremely wide and fertile plain (Map 5). This lowland actually extends even farther south another forty miles or so, where it merges into the valley of the Arkansas River. Altogether it forms arguably the best farming country in the state, and a railroad spur in this direction was easy to justify. The Salina and Southwestern was built to McPherson, one county south, in 1879.[19]

One spur line would not have created a huge impact on Salina's economy, but when a second and longer one was added beginning in 1886, the town became the marketing and distribution center for a large area. This second branch ran to the west-northwest and followed the course of the Saline River, which joins the Smoky Hill at Salina (Maps 5, 11). These tracks opened initially to Lincoln, one county away, but the next year were extended west another 160 miles to a new town on the High Plains called Colby. This construction, called the Salina, Lincoln and Western, funneled the trade of seven additional counties directly to Salina merchants. Flour mills and wholesale groceries sprung up seemingly overnight, and the Methodists decided this was the right place to locate their college to serve northwestern Kansas (Kansas Wesleyan). The community became one of the fastest-growing in the state. From 918 residents in 1870, it ballooned to 3,111 a decade later, and to 6,149 in 1890.[20]

The counties of central Kansas north of the Union Pacific tracks, agriculturally prosperous but unserved by railroads, were a primary focus for development about 1880. The key was to juxtapose a good location in a river valley with multiple railroad connections. Because of the richness of the Republican Valley north and west of Junction City, it was obvious to most observers that a city of considerable regional importance would emerge there, somewhere along the new Kansas Pacific spur. Clay Center, the seat of Clay County, filled the role initially, but this community was too close to Junction City (only thirty miles) to command a large trade area. A superior location became obvious in 1876 when the long, quiet Central Branch railroad was acquired by Boston financier Oliver Oakes II and began to extend tracks westward from its terminus at Waterville in Marshall County. This company decided to build toward the Republican River as well, specifically to a site in southwestern Washington County where the river began a thirty-mile stretch where it flowed almost directly west to east (Map 5).[21]

The junction of the Central Branch with the Junction City and Fort Kearny line occurred in June 1878 at Clifton, a town that straddles the boundary be-

Nazareth Motherhouse of the Sisters of St. Joseph in Concordia, circa 1992. After coming to the city in 1884, the sisters established first a school, then a hospital, and then a home for the elderly. In this way they contributed greatly to local economic development. The motherhouse at Thirteenth and Washington Streets was built in 1902–1903 of red brick and limestone. Its central tower is 125 feet tall. Part of the building housed an academy until 1922, when the sisters opened Marymount College in Salina. (Courtesy Kansas State Historical Society)

tween Clay and Washington Counties. This occasion tripled the local population, much as one might have expected, but the new total was only about five hundred residents by the early 1880s. This was hardly the size many business observers had anticipated. The problem was not lack of traffic or poor business recruiting. Rather, it was competition. Because of the east-west path of the Republican in this area, the two railroads paralleled one another for a considerable distance before the Fort Kearny road turned north to Belleville and the Central Branch (now ambitiously called the Atchison, Colorado and Pacific) continued on to the west. This situation created plenty of room for rival communities that shared the same set of advantages as Clifton. Two of these—Clyde in northeastern Cloud County and Concordia in that same county's north-central section—became more successful than their slightly older rival. Clyde, described by one observer as having "a large number of substantial and ornamental business houses," recorded a population of 956 in 1880. This number was a good showing, according to that same writer, "considering the disadvantages she had had to contend with."[22]

Concordia proved to be by far the most successful of the three communities in the middle Republican Valley. It enjoyed two advantages over its competitors. First, because of a location reasonably near the center of its county, it was able to obtain the seat of local government. The town also was fortunate to be the home of two politically astute men: James M. Hagaman and Samuel D. Houston. Together, they were able to convince Senator Pomeroy to locate a federal land office in their community in 1870 at a time when it had no railroad connections and almost no population. Hagaman also ran successfully for a seat in the Kansas House of Representatives. From this position of power, he obtained state financing for a wagon road to Junction City. Next, in 1874, he used his influence with Governor Thomas A. Osborn to create a state-sponsored normal school to serve the growing population of north-central Kansas. Although this new college threatened the hegemony of the original teacher-training school in Emporia and led to political pressure that closed the branch institution after only two years, a tone of civic pride and business acumen had been established that boosted local immigration. The town had 1,853 residents in 1880 and grew again in 1886 when officials of the Catholic Church decided to subdivide Kansas into three dioceses and named Concordia (along with Wichita) as a new see. A large French-Canadian immigration into the area already had attracted an order of nuns called the Sisters of St. Joseph. A cathedral (Notre Dame du Bon Secours), a boarding school for girls (Nazareth Academy), and a hospital would follow. With a population of 3,184 in 1890, Concordia was a thriving and diversified community.[23]

As workers pushed the Central Branch Railroad west from Concordia in 1879, their next immediate destination was the Solomon River. This stream, like the Republican, flows almost directly west to east in this part of the state and is renowned for the width and fertility of its floodplain (Map 5). The point of intersection between railroad and stream was Beloit, a small settlement twenty-five miles from Concordia that had been established in 1870. Beloit was close enough to the center of Mitchell County to have become the county seat without difficulty. It also was far enough north from both Salina (forty-five miles) and Ellsworth (fifty miles) to escape their economic pulls, and it enjoyed a site for milling that was superior to Concordia. Whereas the banks of the Republican were low and the bottom quicksand, the Solomon at Beloit combined rapid flow with the moderate incision needed for good dam construction. The assets at Beloit became even greater in 1879 when Kansas Pacific officials decided to lengthen the last of their spur lines, the Solomon Valley Branch, into the city. This railroad, which had originally been constructed from Solomon City twenty miles north to Minneapolis in 1877, was eventually extended on west from Beloit in 1886 (Map 11). It thereby brought trade into the city from parts of six or more counties as far west as Norton and Rooks. The population, 1,835 in 1880, rose to 2,455 by 1890. A competition with Concordia for economic dominance in north-central Kansas also began, a rivalry that continues to the present day.[24]

The final railroad town on the Union Pacific system in Kansas emerged thirty years after the booms at Beloit and Concordia. It was a result of a company decision in 1909 to establish a major connecting route between its main lines across Kansas and Nebraska. Historians find irony in this realization, of course, because the strategy nearly duplicated the original plan for construction outlined in the Pacific Railroad Act of 1862. This time, instead of a road from Junction City, officials elected to build northwest from Topeka across the flank of the sparsely settled Flint Hills. Their immediate destination was a station near the community of Marysville, where the new line would link up with an east-west railroad known as the St. Joseph and Grand Island that had come under Union Pacific ownership.[25]

The St. Joseph and Grand Island, a comparatively minor player on the Kansas railroad stage, had a history similar in some ways to that of the Central Branch. It had been built "after numerous delays and disappointments" to bring railroad service to the northern tier of Kansas counties and to provide St. Joseph with a link to western resources. Its asset to the Union Pacific in 1909 was that, at Marysville, the line turned north to make a junction with Nebraska's Union Pacific system at Grand Island. Once the new road, known as the Topeka cutoff, was completed in 1910, Marysville began to grow. This was an old community, founded early in the territorial period where the military road between Forts Leavenworth and Kearny crossed the Blue River. It had become the seat of Marshall County as well as a milling center of note, taking advantage of the relatively reliable, north-to-south flow of the Blue. Charles F. Pusch, a Prussian-born industrialist, boosted the town's status further. His initiative first made Marysville one of the Midwest's leading manufacturers of cigars. Then came a push for more railroad facilities. From a position on the board of directors of the St. Joseph and Grand Island, he engineered the acquisition of that company's division headquarters in 1918. Company employment and facilities at Marysville subsequently grew steadily throughout the 1920s. By 1930, more than 450 men worked there for the railroad in various capacities, including the running of a company stockyard. The town's population, about 2,000 in 1900 and 1910, rose over the next two decades to 3,048 and then to 4,013.[26]

The Missouri Pacific Cities

With the establishment of Beloit and Concordia as regional centers for north-central Kansas in the early 1880s and their equivalents of Arkansas City, Wellington, and Winfield to the south, a person could argue that the period of notable railroad construction and town building might well have been completed for the state. This reasoning ignores larger-scale motivations on the part of national railroad companies, however, plus a period of general business prosperity during the 1880s and the long-standing role of the plains as a transit region. Following such logic, leaders of two major corporations decided to build additional steel trails across Kansas late in this decade. In the process, they added eight more pure railroad towns to the state's urban network and, through their intersections with other railroad systems, boosted the trade areas and status of several other established places.

New York financier Jay Gould personified the first of the two new entrepreneurial ventures. This ambitious man had taken advantage of the national business panic of 1873 to purchase several midwestern railroads at discount prices, including the Missouri Pacific (the old Pacific Railroad of Missouri) and the Wabash system, which connected St. Louis eastward to Cincinnati and Toledo. This pair of companies, his core holdings, stretched across nearly a third of the country and tempted Gould with thoughts of his own transcontinental network. Kansas, once again, provided the obvious passageway.[27]

Two options existed for crossing the state. The most logical, perhaps, was to start from Kansas City and utilize another of the Gould purchases, the Missouri River Railroad, for travel to Atchison. From Atchison, the company then could move west along the Central Branch (still another purchase) to Beloit. Only half the state then would remain for new construction toward Denver or some other city in Colorado. Gould obviously liked the possibilities for Atchison as a transportation hub. He extended his Missouri River line farther north to Omaha in 1882, acquired Atchison's old spur line into St. Joseph, and bought a hundred-acre tract on which to build a twenty-stall roundhouse and various machine shops. This all was to constitute a Missouri Pacific subsystem known as the Northern Kansas Division. Local newspapermen and merchants were ecstatic about this development, of course, as well as the prospects for more to come. Here was good news for Atchison after two decades of transportation disappointment. A series of publications broadcast a new sobriquet, "The Railroad Centre of Kansas," in hopes that the image might at last become the reality.[28]

Atchison did indeed grow as a result of its central role in the Missouri Pacific system, from 7,054 residents in 1870 to 15,105 at the zenith of the Gould excitement a decade later. At this point it was the third-largest city in the state and trailed both Leavenworth and Topeka by fewer than 1,500 people. Unfortunately, this total for Atchison never increased significantly in the years that followed. One of the reasons for the stagnation was that Jay Gould elected to build his major western line elsewhere rather than expanding the Central Branch into this role. The rationale for his action remains obscure, but a longtime student of Atchison history has speculated that two factors were most important. Gould wanted to put pressure on the managers of the Santa Fe railroad by building closer to their trunk line, and he was disappointed that Atchison leaders as late as 1881 continued to give preferential treatment to the Santa Fe over the Missouri Pacific in matters of local rights-of-way and similar issues.[29]

The new main line that the Missouri Pacific crews constructed across Kansas in 1886–1887 passed through almost the exact middle of the state and was designed to split the trade area of the Union Pacific and the Santa Fe systems (Map 11). Since these two established routes were as close together as thirty miles in the vicinity of Ellsworth and Great Bend, and never farther apart than about eighty-five miles, this task was more than a little foolhardy. In fact, the construction was a tribute to personal arrogance rather than good economics. The reasons for its location aside, however, this Missouri Pacific road had important positive impacts on three Kansas communities, and lesser influences on several others.

Gould designed the new pathway to depart from the existing trunk line at Holden, Missouri, and to proceed straight west into Miami County, Kansas. There, because he was about a hundred miles from the company's big machine shops at Sedalia, he decided to establish a division point and repair center. The residents of Paola, the county seat and largest town in the area, assumed this prize would be theirs. Their cross-county rivals in Osawatomie, however, were more enterprising. They passed a bond issue to purchase a site for these facilities and then presented the land to Gould personally. The owner, obviously impressed by this initiative, promptly accepted the offer and, in 1887, erected a large set of buildings for what was called the Central Kansas Division. The four hundred employees hired at Osawatomie rivaled the number in Atchison. They immediately transformed the community from a backwater place known only for its insane asylum into a classic railroad town. The local population, only 681 in 1880, ballooned to 2,662 in 1890 and to 4,191 in 1900. Proud Paola, in contrast, dropped to become the number two city in Miami County.[30]

West of Osawatomie, the Missouri Pacific surveyors laid out a nearly straight path to the west. Their junctions with, in turn, the lines of the former LLG, the Santa Fe, and the MKT boosted local economies at Ottawa, Osage City, and Council Grove, respectively. Farther west, this trajectory brought the tracks into southern Saline County, close to the Union Pacific trunk at Salina. Here arose a double dilemma: whether or not to bypass Salina and continue straight on west, and whether or not to accept a financial offer from that city to establish a second division point there. Gould himself visited the site in October 1886 to listen to local proposals and ultimately decided on a compromise. His main line passed Salina fourteen miles to the south as originally planned, but he also ordered construction of a short, half-circle branch loop with the city at its apex (Map 11). This seemingly minor decision proved immensely beneficial to local businesspeople. As the Missouri Pacific continued on west, its tracks brought a tier of eight new counties under the economic influence of Salina's merchants. Without intending to do so, Gould's line thus functioned exactly as if it had been designed as a spur for the Union Pacific. The city became a dream location for wholesalers of all types, and local boosters rightfully began to advertise their town as the "capital" not only of north-central Kansas but of the northwestern sector as well.[31]

The final Kansas city to emerge as a significant urban place because of the Missouri Pacific line was Hoisington in Barton County. This was the site selected for the company's second set of machine shops instead of Salina. Hoisington actually was created expressly for the railroad by a now familiar combination of local entrepreneurs and railroad officials. The originator and town namesake, Andrew J. Hoisington, was a newspaper editor in Great Bend, eleven miles to the south, as well as a large landowner. He also, obviously, enjoyed good political connections. The facilities at Hoisington started small but increased in 1910. The local population rose accordingly, from nothing to about 2,200 by 1912.[32]

The Rock Island Cities

The last major railroad corporation to build extensive trackage in Kansas was the Chicago, Rock Island and Pacific. Its timing was identical to that of the Missouri Pacific, and the two companies epitomized the old rivalry between the business titans of St. Louis and Chicago for economic control over the central plains and the West beyond. Rock Island officials saw two corridors through Kansas that they might develop. One was across the northern tier of counties. This path would lie midway between the Union Pacific's main line in Kansas and that of the Burlington Railroad across southern Nebraska. It also would give them access to the rich Colorado market. The second possibility was to bisect the territory between two or more of the major lines of the Santa Fe. Southwestern Kansas, for example, still lay open south of the Arkansas River and north of the Santa Fe route through Wellington and Kiowa. A line through this territory could then be extended to the Southwest and the Pacific. Another opportunity lay in a route toward central Texas that would run west of Arkansas City and east of the Santa Fe's line from Wellington to Kiowa.

The trunk road of the Rock Island, which had been developed in earlier decades, trended southwest from Chicago through its namesake city on the Mississippi River and then on to Altamont, Missouri, forty miles east of St. Joseph. The Kansas initiative began in 1885 when the company purchased the small St. Joseph and Iowa Railroad that connected Altamont directly to both St. Joseph and Atchison. One of these two communities obviously was going to provide the entranceway to the plains, and the place from whence the new construction plans could be implemented.[33]

All contemporary evidence suggests that the Rock Island leaders originally favored Atchison as their Kansas portal. That community was where Marcus A. Low, the head of the building company, had established his offices initially and where negotiations with town officials for access to their bridge over the Missouri River were most intense. These negotiations did not turn out as the Rock Island people had anticipated, however. Just as the Atchison business community had favored the Santa Fe railroad over the Missouri Pacific and thereby angered Jay Gould, they then decided to favor the Missouri Pacific over the Rock Island. Thinking that giving a preferential right-of-way over the bridge to Rock Island people might harm their chances for Gould to extend the Central Branch and to build a company hospital in town, they were lukewarm to Low's requests. As a result, the general manager turned to St. Joseph and decided to build his line from there. For Atchison, this last chance for true railroad glory was especially ironic. The trackage eventually constructed by the Rock Island to the southwest was almost exactly the route that townspeople had always envisioned. It also was the dream that had been denied them once before when a Santa Fe executive decision made Kansas City that company's eastern gateway instead of Atchison.[34]

Marcus Low originally planned his new line to run west some forty miles to Hiawatha, the seat of Brown County, where he would establish a junction for his branches to Colorado and to the Southwest. Hiawatha was already a railroad town of some status, having recently acquired the headquarters for the freight division on the Missouri Pacific line between Atchison and Omaha. Surprisingly, however, merchants there told Low that "they did not want another railroad on any conditions." They apparently feared competition from new communities that might grow up along the tracks.[35]

Low, a practical man who had been granted complete control of the construction process, did not try to argue with the Hiawatha position. Instead, he left his original right-of-way at Troy and built southwest to an open area near the common boundary of Atchison, Brown, and Jackson Counties (Map 11). There he purchased 620 acres and created Horton in the fall of 1886, a pure Rock Island company town. Horton was intended not only as a major switching point for the railroad but also as the site for large shop facilities. Like its kindred community of Parsons in southeastern Kansas, it grew so quickly that

Railroad Shops and Yards at Horton, 1936. Officials of the Rock Island railroad created Horton where their main line split into two western divisions. Repair shops built there in 1887 served twenty-five hundred miles of track and employed 960 people at their peak. As reliable diesel engines began to replace steam locomotives in the late 1930s, however, railroads needed fewer shops. The impressive Horton facilities closed in 1939. (Courtesy Kansas State Historical Society)

it acquired a nickname. "The Magic City" had four thousand residents within months of its founding and then settled into a steady population of about thirty-three hundred for the next several decades. Six to eight hundred men worked in the shops, which were designed to be the second largest on the entire railroad system. These employees formed the basis for a blue-collar, but quite prosperous, community.[36]

At Horton, Low divided his work crews. One group built south to Topeka, obtaining funds from Jackson County along the way by agreeing to make a slight detour through the county seat of Holton. Topeka was not a critical link in the Rock Island strategy to capture trade territory. Instead, going through the capital city was a simple political bow, since "help from the legislature might be necessary at any moment." The sensible Low even decided to move his company headquarters there soon after the tracks were complete in early 1887.[37]

The way west to Colorado from Horton was complicated by the existence of two older railroads, the Central Branch and the St. Joseph and Grand Island, both of which served the northern borderland of Kansas as far west as Marshall County. Low decided that his best option was to bypass this competition by looping northwest to the southern Nebraska town of Fairbury and from there back into Kansas. His reentry point to the state was important because the accumulated mileage from Horton demanded a division point in that vicinity. The place selected was Belleville, the seat of Republic County, and a community hungry for railroads.

Ten years before a position along the Rock Island had loomed as a possibility, Belleville residents had voted $130,000 in bonds to induce leaders of the Central Branch Railroad to extend their line from Waterville northwest toward their city instead of southwest to Clifton, Clyde, and Concordia. When that company decided on the latter route, Belleville had to settle for becoming the terminus for the Junction City and Fort Kearny Branch of the Union Pacific. These tracks entered town in 1884. Rock Island strategists saw this branch as a way to increase traffic on their new main line, and Belleville people were elated to gain their first major railroad. From having only 238 citizens in 1880, the community grew to 1,868 in 1890 and 2,224 by 1910. During that period, the Rock Island employed about a third of Belleville's adult males and thereby created still another almost-pure railroad town.[38]

West from Belleville, the Rock Island had northern Kansas nearly to itself except for a pair of spur lines from the Burlington system that reached south from Nebraska. Marcus Low therefore laid his line directly to the west, connecting the county-seat towns of Mankato, Smith Center, Phillipsburg, and Norton. At Norton, which he reached in January 1888, he encountered the first Burlington spur and so decided to angle to the southwest. By doing so, he not only avoided competition but also was able to create a junction with the branch line of the Union Pacific that terminated at Colby, the two-year-old seat of Thomas County. This area was still very much frontier at the time, but the railroad junction put Colby ahead of its peer communities with a population of 516 in 1890.

Low needed to establish two division points and another set of machine shops somewhere along his High Plains section of track. He selected the first simply and logically by identifying the county-seat town at the proper distance from Belleville. This was Phillipsburg, a community of only 309 people in 1880, but one with access to good surface water at nearby Deer Creek. The local population promptly tripled, although it still remained under a thousand because the superintendent elected to place no repair shops there. A site for the final division headquarters was more difficult. Norton, in the valley of Prairie Dog Creek, was the last community to have a dependable supply of surface water, but it was too close to Phillipsburg. Low opted instead for Goodland, the seat of Sherman County on the Colorado border.[39]

The Goodland site had several advantages. First, unlike the experience of the Union Pacific people thirty miles to the south in Wallace County, the area was known to possess groundwater supplies adequate for the needs of a parade of steam locomotives. Goodland also was a new community, having obtained its post office only in December 1887, six months before the first Rock

Island train arrived. Low therefore was able to create a partnership. The community would get major railroad facilities, including shops, division headquarters, and a $20,000, two-and-a-half-story building that would serve as both a depot and a hotel. Town leaders, in return, made Marcus Low a member of their county development board, the group that controlled all city planning. With the Rock Island's influence and contributions, the Goodland community seemed poised to dominate the economy of a large section of northwestern Kansas and adjacent Colorado. Its population of 1,027 in 1890 dwarfed that of neighboring county seats, and the same still held true in 1920 when its count had risen to 2,664.[40]

The Rock Island line across northern Kansas was constructed with speed and quality, a remarkable feat considering that, for most of this time, company crews were working simultaneously on two other projects in the southern half of the state. This other half of the story began in early 1887 with strategy sessions at the temporary terminus of Topeka. Whether the company intended to build to the south or to the southwest, it obviously would require permission to pass through Indian Territory, and so Congress had been petitioned to that purpose. Approval came on March 2 in the form of a double authorization: one route to the vicinity of Galveston and another through the Southwest toward El Paso. Marcus Low and his fellow officials, obviously backed by solid finances, decided to pursue both goals at the same time. They elected first to build a single line west from Topeka along the south bank of the Kansas River. This orientation would avoid direct competition with the Santa Fe, whose tracks ran straight south from that city. The plan also would allow them to utilize the valley of Mill Creek for passage through the rugged Flint Hills of Wabaunsee County (Maps 5, 11). From the far side of that topographical barrier, the idea was to proceed south to either Council Grove or another stop on the old Santa Fe Trail called Lost Springs in northeastern Marion County. There the company would establish a division point, shops, and a junction for the two new main lines.[41]

The plan for facilities development materialized almost exactly as envisioned, except neither Council Grove nor Lost Springs obtained the prize. Instead, the new construction and jobs went to the community of Herington. The details of how this arrangement came to pass have become the stuff of legend in Kansas. Monroe D. Herington, the namesake and founder of this town in southeastern Dickinson County, had come to the region in 1881 where he quickly became rich buying and selling land on a large scale. He platted Herington in 1885 and, the next year, lured the Missouri Pacific's main line through the new settlement with an offer of eighty town lots, $1,000 in cash, and a free right-of-way through his extensive ranch lands. Hearing in early 1887 that Rock Island people might build to Council Grove, he brought his own proposal to Low's Topeka hotel before a delegation of business leaders from that rival town could find the designated room. The offer—half interest in 160 acres of the town, a deed to 80 acres for shop grounds, free rights-of-way across Dickinson County and most of Morris County, a personal guarantee for a county bond issue, and free building stone—reportedly "amazed the railroad man," who authorized a new route for his tracks later that spring. In this way Herington became another company town, a fact its residents later proudly acknowledged by naming their high school sports teams the "Railroaders." The local population, even without the benefit of county-seat jobs, grew steadily: 1,353 in 1890, 3,273 in 1910, and 4,065 in 1920.[42]

From Herington, the two lines that were laid on to the south each had quite different impacts on regional development. The one toward Galveston essentially reinforced the status quo of the time. It passed through Marion (the seat of Marion County), crossed the Santa Fe line at Peabody, and then went on south through Wichita and Wellington before exiting the state at the old cow town of Caldwell. The *Wichita Beacon* editorialized in 1886 that passage of a bond issue to support this new railroad would mean a turning point for the city, with new people and manufacturing firms. Such gains were undoubtedly true, but Wichita was well on its way to impressive growth before the Rock Island, and would have continued to do so even without it. The same was true for Wellington on a smaller scale.[43]

Southwest from Herington, and especially beyond the crossing of the Arkansas River, the Rock Island railroad assumed the role of kingmaker in regional development. There the route passed through High Plains territory that was just being settled in the late 1880s, where county-seat battles still raged, and where transportation routes remained unfixed. The process of exactly where to locate the new tracks therefore could be fluid. As things turned out, this potential for mutability continued even after construction began in 1887. Bond issues or their absence led to three major changes in course before the tracklayers reached the state border. The only original planning decision that came to fruition, in fact, was for the first section of track, which ran southwest from Herington to the county-seat town of McPherson.

Rock Island officials intended to cross the Arkansas River at Sterling, in southern Rice County, and then build west to Dodge City. This plan was altered first by an aggressive campaign from boosters in Hutchinson. They offered an attractive bond issue plus a free right-of-way through their community. They also pointed out prospects for greater commercial growth locally than would be the case at Sterling, primarily because the directors of the Santa Fe railroad had recently (1886) completed a new section of line in this area. This new trackage, known as the Kinsley cutoff, ran from its namesake town straight east to Hutchinson (Map 11). It eliminated the extra mileage of going around the Arkansas River's big bend and thereby reduced traffic on that section of line through Larned, Great Bend, and Sterling.[44]

Hutchinson, because of the Santa Fe decisions of 1886, went from being just another station along the line to a switching center of some importance. The town also began to develop a larger trading hinterland, especially northwest toward Great Bend and west along the Kinsley cutoff. The presence of the Rock Island propelled this trading function into even greater importance, especially when the railroad men decided to locate their new line considerably southwest of the Kinsley route where it could open up new territory for Hutchinson merchants to exploit. Soon after the new company's locomotives began to appear in July 1887, the wholesale industry boomed. A survey in 1908 enumerated thirty-two such enterprises in town, together doing an annual business of $11,500,000 and employing more than four hundred traveling men. Their territory—southwestern Kansas plus parts of Colorado, New Mexico, Oklahoma, and Texas—was growing rapidly, and Hutchinson began to look a lot like Salina, its sister railroad and wholesaling center for the northwestern section of the state. The population, matching this remarkable business explosion, jumped from 1,540 in 1880 to 8,683 in 1890.[45]

Beyond Hutchinson, any route to the southwest would pass near the middle of Pratt County. Iuka, the seat of this political unit, was the planned destination, but because local people realized that this place was approximately 110 miles from Herington, and therefore a potential site for the next division point, competition arose. Iuka's rival was a newer settlement (1884) at the exact middle of the county, Pratt Center. According to a former Rock Island employee from this time, Iuka representatives inexplicably failed to appear at a county meeting to vote on the railroad bonds, leaving Pratt Center to claim the prize. Whether or not the Iuka absence was the result of Pratt Center interests having purchased liquor for an Iuka party (as local stories persist), the deed was done. With the railroad now coming to Pratt Center, and with it the division headquarters and maintenance shops, local leaders also were able to relocate the county courthouse to their community. The new town, soon known simply as Pratt, swelled to 1,418 residents in 1890. A gradual increase in local railroad workers to about three hundred pushed this total to 3,302 by 1910.[46]

Rock Island crews built west from Pratt to the next county-seat town of Greensburg still with the intention of continuing on in that direction to Dodge City. This routing soon was stymied, however, by a refusal of two key counties west from there to pass bond issues. Voters in both Gray and Haskell Counties took this negative stance because each was in the midst of a county-seat fight. Instead of issuing threats or bribes to get his way, however, the pragmatic Marcus Low simply reassessed his options and then asked the voters of Meade County (south of Gray) if they would like a railroad. When the answer came back yes in the form of a bond issue, the Rock Island men angled their tracks southwest out of Greensburg, bypassed Dodge City and the planned route beyond through Ulysses and Johnson City, and built instead to Meade Center.[47]

Rock Island Depot in Liberal, 1972. The Spanish design selected for this building in 1912 highlights the role that Liberal played as a conduit between midwestern and southwestern markets. The depot closed in 1961 when passenger service ended, but townspeople renovated it in 1997–1998 to provide office space for the local chamber of commerce. (Courtesy Kansas State Historical Society)

At Meade Center, Low presumably was out of options. If he were to reach his long-expressed goal of the Colorado coal mines at Trinidad, he would need to build straight west through the southernmost tier of Kansas counties, presumably via the county seats of Richland, Hugoton, and one of the two rivals for that position in Seward County (Fargo Springs or Springfield). Predictably, however, the voters in Seward County could not agree on a route and so defeated their bond issue. Low reassessed things once again. This time he changed his larger objective and decided to forgo the coal traffic of Trinidad in favor of the cattle business of the Neutral Strip (now the Oklahoma panhandle) and the Southwest beyond. He therefore built southwest from Meade to the state border and installed himself as president of a new town company there. This site became known as Liberal.[48]

Liberal, founded in 1888, served as the southwestern terminus of the Rock Island line for twelve years. This stoppage was unplanned, but the railroad company had expended most of its financial reserve during the recent frenzy of construction, and during the 1890s the most severe drought in Kansas history essentially depopulated the western portion of the state. A few cattle came to the railroad's new stockyard, but not many, and the census enumerator for 1900 could find only 426 people in town despite the presence of good groundwater supplies and another division headquarters for the Rock Island. Things brightened locally in 1901, however, when the rains began to return and the company announced plans to extend its tracks on southwest into New Mexico, where connections could be made to El Paso and Los Angeles. A second branch from Liberal, this time in 1928 and running south to Amarillo, made the city even more important. The community thereby rapidly moved from bust to boom, acquiring the county seat from abandoned Springfield and recording 1,716 residents in 1910, 3,613 in 1920, and 5,294 in 1930.[49]

The story of the gradual reorientation of the Rock Island tracks during their construction across western Kansas from one projected path to an entirely different, northeast-southwest alignment is one that can be read at several levels and applied to other companies, cities, and times across the entire plains region. First of all, it is testimony to the practicality of railroad and other national-level decision makers. It also provides a cautionary tale for city leaders, such as those in Gray, Haskell, and Seward Counties, who are wont to see expected developments as inevitable. Geographically, of course, the most interesting aspect of this particular cumulative decision has been its impact on future developments. Once the Rock Island "angle" was established, it essentially determined the paths of two additional railroads in the area. Santa Fe workers laid them both, the first southwest from Wichita to create a rival border cow town at Englewood in Clark County. Then, in 1912, they paralleled the Rock Island on the west, building southwest out of Dodge City through Hugoton and on to a final border town called Elkhart. As highways were constructed during the 1920s, they tended to follow the older railroad patterns. In this way the angled and parallel U.S. Routes 54 and 56 today serve the southwestern counties, and people there can still travel west to Trinidad only by an indirect path.[50] As with much else in human affairs, seemingly minor decisions can have important and long-lasting effects.

6

Mining, Irrigation, and Newer Institutional Towns, 1876–1950

Observers in eastern Kansas during the late 1870s and in western Kansas a decade later reported that the urban system for the state was much less fluid than it had been in earlier years. A definite pattern was emerging, one they felt would endure for the foreseeable future. Superior railroad connections had made Kansas City, Missouri, the regional metropolis, and the same was true on a smaller scale for Atchison, Hutchinson, Parsons, Salina, and Wichita. Although the fate of Leavenworth remained uncertain, as did that of the string of cities along the border of Indian Territory, nobody expected any significant new urban competitors to emerge on the scene. It was a railroad world, after all, and the major routes had already been decided.

Like every vision of the future, the one seen by Kansas planners during the 1870s was based on a limited perspective. People then thought primarily in terms of local agricultural productivity, the control of trade areas within the state and on the frontier beyond, and the creation of mercantile and manufacturing industries at entrepôt locations to serve these markets. In retrospect, their biggest oversight was the mineral resources of the state. Poor assessment in this area is odd in some ways, especially with regard to coal, because fears of fuel shortages worried not only settlers who ventured onto prairie homesteads but also urban businessmen who hoped to start factories. Perhaps people were misled after preliminary explorations in northeastern Kansas during the 1850s and 1860s had discovered only a few thin and discontinuous lenses of coal near the surface.

Still, signs of future developments were present at many locations, and early settlers had often indicated them with place-names. Greenwood County has a Tar Creek, for example, and Coal Creeks exist in eleven different eastern counties, plus an isolated occurrence farther west near Wilson. Evidence of salt deposits was even more abundant, especially in central Kansas. The names Salina, Saline County, and Saline River are the most obvious examples, but four counties there have Salt Creek Townships, and Mitchell County once had a community called Saltville. The potential inherent in the place-names first became a commercial reality in 1876. Developers of new lead and zinc mines near Joplin, Missouri, decided to market their products via nearby Kansas railroads and to promote local coalfields in the process. By 1918, the variety of minerals production in the state had increased to include lead and zinc, natural gas, petroleum, and salt, all on a large scale. Kansas assumed a rank among the top ten mineral-producing states in the nation that year and held that lofty status nearly continuously for half a century.[1] This production, needless to say, had major impacts on urban development. I will discuss this process historically, beginning with the coal, lead, and zinc mining of southeastern Kansas.

Pittsburg, Galena, and Other "Towns That Jack Built"

Southeastern Kansas first came to the awareness of the American public just after the Civil War. Union veterans hoped to settle this fertile area but were rebuffed by a land speculator. The U.S. secretary of the interior had sold essentially all of Cherokee and Crawford Counties to James Joy as an incentive for Joy to construct the Missouri River, Fort Scott and Gulf Railroad. Joy wanted to resell this land for two to seven dollars per acre, but many settlers claimed they had been present in the region before the sale and so deserved a cheaper price. This conflict, in conjunction with the railroad race south to obtain the right to build on through Indian Territory, blinded people to the fact that coal seams beneath this part of Kansas were much thicker, closer to the surface, and numerous than those found farther to the north.[2]

Residents of Fort Scott, the oldest community in the area, were the first to realize that mining might become an important part of the regional economy. They used a local coal seam called the Mulky as fuel for several foundries. Although continuous and covered by only ten feet or so of overburden, this deposit was (and is) less than ideal. It is only six to fourteen inches thick, and half of the overburden is solid limestone. Thirty miles to the south, however, the situation improves markedly. The Mulky and its surrounding strata have been eroded in this area, leaving older layers of rock near the surface. These include nine different coal seams, in particular, a twenty-two-inch one called the Mineral and another, known as the Lower Weir–Pittsburg, that ranges in thickness between thirty-two and forty-two inches. Access to these potential riches is restricted. West and north of Crawford and Cherokee Counties, the desirable seams are buried too deeply to exploit, whereas to the east and south they have been eroded (Map 12; see p. 25).[3]

Knowledge of the Lower Weir–Pittsburg seam goes back at least to the 1850s, when a few people began to mine its exposures in the valley of Drywood Creek in northeastern Crawford County. Its uses, however, were limited by the size of the local market and poor transportation. Things changed for the better in March 1870, when Joy's border-tier railroad line opened service to the county seat of Girard. Before entrepreneurs could initiate plans to ship coal north as a domestic fuel, however, another regional development initiated talk of a different and more lucrative use for the mineral.[4]

The coal-bearing formations of southeastern Kansas end abruptly on their southeastern flank. Beyond a line running northeast from just west of Baxter Springs, much older rocks lie at the surface. These happen to contain rich concentrations of lead and zinc (Map 12). Like the coal in Crawford and Cherokee Counties, the metallic ores had been known for about twenty years before 1870, but isolation again limited exploitation. Knowing that both the Missouri River, Fort Scott and Gulf and the Atlantic and Pacific (i.e., the Frisco) Railroads were building into the area from the north and the northeast, respectively, people in southwestern Missouri gradually began to look more carefully for major deposits. A big strike occurred in August 1870 in the valley of Joplin Creek, about six miles east and six miles north of the corner of Kansas. Five hundred miners worked this lode within a year of discovery, and a town called Joplin was platted to serve as a center for supplies and recreation. This community, as rough and coarse as any Kansas cow town, grew rapidly to 4,200 residents by 1873, but railroad companies held off building connections to it. They feared the boom would soon collapse. In 1875, however, E. R. Moffett and John B. Sergeant, the two men who made the original strike, took matters into their own hands. They organized the Joplin Railroad Company to construct thirty-nine miles of track north-northwest from their city to a junction with the Missouri River, Fort Scott and Gulf line at Girard.[5]

Moffett and Sergeant could have connected their Joplin track with the border-tier route at Baxter Springs instead of Girard and saved themselves more that half the cost and mileage. They selected Girard, however, because the route there would pass through the center of what was becoming known as the southeastern Kansas coalfield. Obviously, they had a plan. The two men purchased twenty-five thousand acres of these coal lands from the railroad and, with their own tracks complete, had their construction engineer lay out a new town site in the midst of this acreage. When he did so in the spring of 1876, they decided to call it New Pittsburg.[6]

New Pittsburg, or just Pittsburg as it was known after 1880, originally was conceived simply as a mining camp, and its coal as a supplement to the lead as a means of generating traffic for the railroad. This perspective changed quickly, however. The reason was another mineral, zinc sulfide or sphalerite, that existed in close association with the lead in the Joplin area. Zinc had been difficult and expensive to process before the early 1870s, and so found few uses. Then new techniques led to the popularization of zinc coatings on steel to create rustproof surfaces and to the employment of brass (an alloy of zinc and copper) on steam locomotives and in munitions. Profit now was possible if processing costs could be controlled, but the conversion of every ton of sphalerite into zinc metal still required about three and a half tons of coal. Instead

St. Louis and Pittsburg Zinc Company, 1909. Between 1882 and 1896, a combination of local coal and nearby ore made Pittsburg the nation's leading producer of metallic zinc. Six smelters employed more than a thousand men. Each of the eight big furnaces photographed here required eighteen tons of coal to convert three tons of ore into one ton of metal. Most of the city's smelters soon moved elsewhere in search of cheap natural gas, but the St. Louis and Pittsburg complex remained active until 1917. (Courtesy Kansas State Historical Society)

of smelting at the mining site, the solution was to ship ore to a coal source and do the processing there.[7]

Historians have summarized the history of early Pittsburg by saying that, although coal transformed farmlands there into a mining camp, it was zinc that converted the camp into a city. Using the popular nickname for sphalerite, this meant that Pittsburg, like Joplin, was a "town that jack built." The idea of smelters at this site was pioneered by Robert Lanyon, a man from the lead-mining area of southwestern Wisconsin who had become prosperous in the smelting business at La Salle, Illinois. Lanyon came to Pittsburg in 1877. He liked what he saw and built a smelter in 1878. This venture proved very profitable, and within four years, his family and a company from Granby, Missouri, had opened three more. All were mammoth operations for their time, and their appetites for coal even larger. Pittsburg, which had a population of only 624 in 1880, four years after its founding, jumped to 6,697 residents by 1890 and to 10,112 by 1900. Thousands more lived in several dozen smaller mining camps that had sprung up around the city.[8]

The industrial statistics for early Pittsburg are astonishing. A writer in 1905 reported that sixty-two shaft mines operated in the area at that time, and together they provided employment to seven thousand men. The four smelters produced some ninety thousand pounds of metallic zinc per day in 1889, and their success had prompted companies to plan two additional facilities as well. A survey in 1893 showed that, of the twenty-five smelters active in the entire United States, Kansas had nine and Pittsburg six. Fifteen years after its founding, it was clear that the little community had become the zinc-smelting center of the world.[9]

The success of the zinc smelters and the quality of local coal greatly affected every railroad in the region. Not only did each one build into Pittsburg, but they also developed their own coal mines and coal-mining communities. Frisco people initiated the action by purchasing the original Joplin Railroad Company in 1879. Next, in 1882, officials of the Missouri River, Fort Scott and Gulf built a branch south from Bourbon County and opened mines near Mulberry (northeastern Crawford County) and Minden (Barton County, Missouri, seven miles from Pittsburg). The Missouri Pacific people, who controlled the MKT system at the time, ran a line northeast from Chetopa in 1885 and developed mines at Cherokee in northern Cherokee County. Finally, in 1886, Santa Fe leaders authorized a track southeast from Chanute that terminated in their new mining town of Frontenac, adjacent to Pittsburg on the north. Each of these companies required large amounts of coal for their own locomotives and feared discriminatory practices if the Frisco maintained exclusive control. They also recognized a growing home-heating and industrial market for this product throughout their respective midwestern empires.[10]

With four highly capitalized railroad companies in the Pittsburg area, coal mining became a large and competitive business. The number of miners rose steadily throughout the 1880s and 1890s, peaking at more than ten thousand during the early years of World War I. With competition for local labor from the lead and zinc mines and the zinc smelters, it proved difficult, in fact, to find able-bodied workers. As the companies necessarily began to recruit abroad, Pittsburg not only grew but became an ethnic mosaic. Approximately half of the coal miners were European-born, the largest numbers coming from Austria, Belgium, France, Germany, Italy, Slovenia, and Wales. Heavy investment in coal mines also spurred the development of other industries that could make use of this fuel. Among the most successful were brick and tile makers, which took advantage of suitable area clays beginning around 1890. Three companies in this business—the Nesch and Moore Brick Company, the W. S. Dickey Clay Manufacturing Company, and the Pottery Plant—together employed 275 people in 1927. Thomas McNally began another important operation, the Pittsburg Boiler Works, in the late 1880s. This grew into McNally Pittsburg, Inc., a major manufacturer of coal preparation and building equipment.[11]

Strip Mining in Crawford County, circa 1925. Because stripping extracts all of a coal seam instead of only half with an underground system, it rapidly became the mining method of choice in southeastern Kansas during the late 1920s. The key element was a powerful, revolving steam shovel. This one from Marion, Ohio, cost $150,000 at a time when miners earned $7.50 per day. (Courtesy Kansas State Historical Society)

Two additional commercial assets of Pittsburg came about as indirect products of the coal business. People especially celebrated the acquisition of a major railroad and its repair shops. In the late 1880s, industrialist Arthur Stilwell was building his Kansas City Southern line through western Missouri, on the way to Joplin and then the Gulf of Mexico. Pittsburg entrepreneurs, especially proprietors of independent coal companies who felt shipping discrimination from the existing, mine-owning railroad companies, desperately wanted another transportation outlet. They also realized that their town was the correct distance from Kansas City to become a division point. Consequently, they courted Stilwell with an offer of choice real estate plus $40,000 in local bonds. The result, beginning in 1893, was a short detour of a Missouri railroad into Kansas territory, three hundred acres of shops and yards, and nine hundred new jobs. Not surprisingly, the railroad company also became involved in the local mines, and its owner erected a signature building downtown that still remains, the Stilwell Hotel.[12]

A second important economic legacy of the early coal development was Pittsburg State University, which was established in response to a need for skilled workmen in the mining industry. It began with a set of manual training courses in the local high school in the 1890s. In 1903, following an extensive lobbying effort, the Kansas legislature authorized the state normal college at Emporia to set up a two-year auxiliary program at Pittsburg that would emphasize industrial education. Bearing the awkward title of the Kansas State Manual Training Normal School, this program acquired independent status in 1913 and added courses in other fields. The name was changed to Kansas State Teachers College in 1923, and it gradually became an important part of the local economy. In 1928, for example, the school employed 161 faculty members and enrolled approximately a thousand students.[13]

During the 1890s, the presence of the brick and tile industry, the McNally corporation, and the Kansas City Southern shops all proved important to Pittsburg, as much symbolically as for their actual employment. This was a time of potential crisis because the city lost its smelters as rapidly as it had attained them. The reason was the development of natural gas fields a county or two to the west, which offered a new and practically cost-free fuel that could substitute for coal. Robert Lanyon himself initiated the move, and as the fiery furnaces shut down, Pittsburg lost a payroll of approximately $25,000 per month and some 2,500 people from its population base. People worried, but the coal companies realized they had many other markets, and the community had become firmly enough established in other ways that the economy rebounded quickly. The town's population of 14,755 in 1910 represented a solid gain over the previous decade, and that for 1920 (18,052) the same thing again. Pittsburg was now the largest city in southeastern Kansas.[14]

Although Pittsburg was never challenged as the coal capital of Kansas, it was only one of several urban centers that arose from the early symbiosis between this fuel and the mining of lead and zinc ores. The two metals actually created the bigger communities at the outset, but their glory was shorter-lived. With coal available for smelting, local lead and zinc ores immediately became much more valuable. As people searched for additional deposits, they found an ore region that extended mainly southwest from Joplin. The biggest of the second round of strikes occurred a mile across the Kansas border in the valley of Short Creek. This discovery, in March 1877, prompted rival mining companies to found twin cities: Galena (the name of one of the ores, lead sulfide) and Empire City.

Both of the new Kansas towns claimed between two and three thousand residents within three months of their births, but when the Galena company installed a lead smelter on its site in 1879, it became the dominant community. Both the Frisco (from Joplin) and the border-tier (from Baxter Springs) railroads extended their tracks to the town, and by 1900, it briefly looked as if the community might overtake even Joplin in population. With 250 mines in the immediate area, sixty-five ore crushers, and two smelters, Galena's population reached a pinnacle of 10,155 that year before newer discoveries south and west in Oklahoma initiated another boom and cheap natural gas lured most of the smelters away. The decline was not absolute, however, because 6,096 people still lived there in 1910, and more than 4,000 for the next four decades until the mines finally closed.[15]

Kansas eventually was home to seven lead and zinc mining camps, all in southern Cherokee County. Of these, Treece contained some of the richest deposits, but its ores were processed in nearby Picher, Oklahoma, and so it remained small in population. Baxter Springs, halfway between Galena and Treece, was better positioned for growth. This old community, bypassed by the cattle drives and deeply in debt in the mid-1870s, was nearly a ghost town at the time of the Galena strike. Its railroad connections then enabled it to serve a minor role as supply center to these mines. Much larger change came in 1903 when vast ore deposits were found just south of town. These were developed slowly, but between 1916 and 1921 they formed the most active part of what had become known as the Tri-State Mining District. During those years, Baxter Springs mines are said to have produced 65 percent of the national output of lead and zinc, and residents asserted that their town was the richest in the state on a per capita basis. With natural gas already on the scene before the Baxter Springs strike, however, no local smelters existed. This, plus the availability of electric trolley cars to bring miners from Joplin and Galena, kept the population smaller than local boosters anticipated. The town had 3,608 residents in 1920 and 4,541 in 1930.[16]

The final important urban elements within the Cherokee-Crawford mining district were satellite coal towns around Pittsburg. None of these ever became large, but three did exceed populations of 2,500 early in the twentieth century. The first of these was Weir City in northern Cherokee County. Weir City's initial advantage was a location near the southern limit of the best coal deposits and thus closer to the lead and zinc mines of Joplin than other Kansas sites. It therefore was the site chosen in 1872 by the Chicago Zinc and Mining Company for the first zinc smelter in the state. Based on this development, officials with the Missouri River, Fort Scott and Gulf Railroad ran a spur line three miles east into town from their main tracks, and the population crested at 2,977 in 1900, about the time that the smelter closed. The other two coal towns to achieve modest population peaks—Frontenac in 1910, with slightly more than 3,000 residents, and Mulberry in 1920, with 2,697—both lie just north of Pittsburg. Their zeniths correspond to the maximum employment of miners in the area, before switches to natural gas for the home-heating market and to laborsaving, strip-mining techniques for coal itself began to change a regional way of life.[17]

The Salt City

Following the opening of the coal district at Pittsburg in 1876 and the Galena lead and zinc strike in 1877, Kansans had a ten-year wait for their next big mining news. Both the location and the mineral were unexpected, but the deposit was so rich that it attracted national interest and major development almost immediately. The date was September 27, 1887, the product salt, and the place Hutchinson.

Salt marshes and salt creeks are common occurrences in Kansas along a north-south axis that extends from near Belleville, through Salina and Wichita, and then on to Wellington. In the 1870s, James G. Tuthill of Republic County made an attempt to develop one of these marshes into a commercial venture. When its brine proved to be weak, however, the idea was quickly abandoned. Then, a decade later, an exploratory well for oil or natural gas drilled by Ben Blanchard struck a twenty-foot vein of almost pure salt five hundred feet beneath the city of South Hutchinson. Blanchard, the founder of that community just across the Arkansas River from Hutchinson proper, was a promoter and so made no secret of his finding. When his drill encountered a second seam of salt, a local newspaper writer immediately penned the headline "Salt for the World" and added speculation that was true to the spirit of the times. By estimating the cost of production at seven cents per bushel, and noting the current market price of twenty-five cents and the annual national production total of thirty million bushels, he concluded that "a bonanza is at our doors."[18]

The reporter's assessment of 1887 did not miss the mark by far. Within a year, Hutchinson people saw $600,000 invested in the business and the construction of an amazing thirteen salt plants. Blanchard happened to have drilled at almost the ideal spot in the state. Beneath his feet, within the Wellington Shale Formation, lay a salt deposit four hundred feet thick—a resource so abundant that even professional geologists have termed it "practically inexhaustible." The Hutchinson salt layer (as it was soon named) is a remnant of a dried-up sea from the Permian age. Spatially, it extends north to the vicinity of Beloit, south into Oklahoma, and west nearly to Garden City. On the east, not surprisingly, it stops abruptly along the line marked by the salt marshes and creeks. Mining potential is determined in part by this geography but also by the deposit's variable depth. Like all rocks in the state, the salt deposits dip gently to the west. In a place such as Hodgeman County north of Dodge City, for example, they exist eighteen hundred feet beneath the ground. In contrast, only fifteen miles or so east of Hutchinson, the mineral lies close enough to the surface that groundwater penetrates to its level and so dissolves it. The best mining therefore occurs in a narrow band just west of the groundwater contact. This zone extends from Lincoln County south through Ellsworth, Rice, Reno, Kingman, and Harper Counties.[19]

It is interesting that the rush of entrepreneurs to exploit Kansas salt in the late 1880s and 1890s did not produce any new cities. Instead, and logically, these people realized that it made more sense to open mines in or near existing communities. This would assure labor supplies at low cost and access to the existing system of railroads. The railroads were the most important consideration, because with a bulky, low-cost product such as salt, transportation rates can easily make the difference between profit and loss. At the peak of the local salt boom, in 1890, twenty-three plants were in operation. North of Hutchinson these sites included Ellsworth, Kanopolis, Lyons, and Sterling. To the south, Kingman and Anthony had facilities (Map 12). Thirty years later, when competition had whittled the plant number down to twelve, Kanopolis and Lyons were the only competitors left for Hutchinson. Again, this distribution was largely a product of transportation. Although all the early salt towns had regular rail service, Anthony and Kingman were only on minor routes. The same was true for Sterling after the Santa Fe built its Kinsley cutoff. In contrast, Ellsworth and Kanopolis (a community that had grown up at the site of old Fort Harker) were on the main line of the Union Pacific, Lyons on an important Santa Fe route, and Hutchinson on the main lines of both the Rock Island and the Santa Fe. A related factor was that Jay Gould, owner of the Missouri Pacific, also was a major investor in Kansas salt. That his railroad built a branch to serve Kanopolis, Lyons, and Hutchinson is no coincidence.[20]

The hundred or so workers at the salt plants in Kanopolis essentially created a town where only a village had existed before. A similar number at Lyons helped a small county seat to triple in size from 509 in 1880 to 1,736 in 1900. The biggest impact by far was at Hutchinson, where three individuals—Edward Barton, Emerson Carey, and Joy Morton—established themselves as industry leaders. Barton, with his brothers Frank and William, was the first on the scene, in 1892, but the other two companies became larger. Morton, who had previously worked in Michigan's salt industry, bought an existing plant in Hutchinson in 1893. He was a good manager and so was able to build a new facility of his own in 1898 and then, in 1907, to open another plant said to be the largest and most modern in the world. Carey was similarly progressive. His special interest was in developing not only the salt industry itself but also factories that could make use of this product. He previously had owned one such business, a tanning operation, and in 1906 led a local movement to build a soda-ash plant. Soda ash (anhydrous sodium carbonate), made from salt and limestone, found use at the time in the manufacture of glass, soap, and explosives. The venture was a success for many years and employed five hundred workers at its peak of production during World War I. Several small meat-packing companies, also big users of salt, came to town as well.[21]

Although the soda-ash plant closed in 1921 and big meatpackers found it cheaper to transport salt to Kansas City and Chicago than to move their facilities to Hutchinson, salt itself remained a mainstay of the local economy. Community leaders touted the nickname of the "salt city" at least as early as 1908, local high school teams called themselves the "salt hawks," and six hundred or so people found work in the plants. With annual sales that reached $1 million in 1900, $2 million in 1917, and $3 million a year later, only milling rivaled salt extraction as the number one industry in town. With these two manufacturing mainstays, plus an equally important wholesale trade to southwestern Kansas, Hutchinson's population rose impressively decade after decade. From 1,540 residents in 1880, it jumped to 8,682 in 1890 and to 16,364 in 1910. By 1920, with still another solid gain to 23,298, the city had become the fourth largest in the state, behind only Kansas City, Topeka, and Wichita.[22]

The Natural Gas Boomtowns of Southeastern Kansas

As knowledge of the quality coal deposits in Crawford and Cherokee Counties began to spread across the rest of southeastern Kansas in the early 1870s, people in most communities conducted at least preliminary drillings in the hope that they, too, might be blessed with a rich fuel source. One such effort occurred at Iola, the seat of Allen County, in 1872. Instead of coal, however, this well hit a pressurized mixture of water and natural gas at a depth of 736 feet. The result was a memorable geyser. For many years, at internals between fifteen and forty-five seconds, it erupted, making what was reported to be "a remarkable and very beautiful site, particularly when the gas was set on fire and the spraying water looked like a fountain of liquid flame." The residents of Iola admired their new phenomenon but did not know what to do with it. When the waters were discovered to contain dissolved minerals, however, Roswell W. Acers and his son Nelson saw opportunity. By erecting several cottages and a hotel, planting trees, and using the gas to create warm mineral baths, they created a fashionable summer resort. This spa prospered until floodwaters from the Neosho River filled the well in 1885 and stopped the flow.[23]

The Acers mineral well was unique, but the experience of finding deposits of natural gas or oil and not knowing what to do with them was fairly common throughout southeastern Kansas between the 1860s and the 1890s. Oil springs attracted early interest in Miami County, for example, where several wells hit small pools at just over a hundred feet in depth. A decade later, many towns in the area funded itinerant drillers in hopes of finding enough gas to operate a set of local streetlights. Oil was used to make lubricants, of course, but demand

and prices were both low. Internal combustion engines were still in the future, and no one yet imagined the possibility of using natural gas directly as an industrial fuel.[24]

The first suggestion that Kansas oil might become a real industry occurred in 1892. Entrepreneur William M. Mills, having leased eighteen thousand acres in Wilson County, hit a deposit near the town of Neodesha that yielded three to four barrels per day. Because his contract said that any oil found must be refined locally and his funds to do so were limited, Mills then sold these properties to two better-capitalized men from Pennsylvania, James Guffey and John Galey. They optimistically leased another one million acres in the region and drilled fifteen wells. When thirteen of these hit productive pools, they built the mandated refinery at Neodesha, and an industry was born. This refinery, which opened in 1897 under the auspices of the Standard Oil Company (a group that had bought out the Pennsylvania men), was a modest success. It faced no competition closer than Chicago, and its site next to the Verdigris River was low enough in elevation that gravity-flow pipelines could easily bring in oil from surrounding counties (Maps 5 [see p. 17], 12). These advantages were countered, however, by low demand and the expense of transporting a bulky product to the larger markets of the country. The refinery's forty-five employees and $3,000 monthly payroll caused only a small boost to the Neodesha population.[25]

Interest in Kansas oil, stimulated by the Neodesha refinery, increased further at the turn of the century with the discovery of a huge new pool on the Texas coast near Beaumont. This extra oil on the market soon led to a glut, however, so that prices plummeted in 1904 and did not recover for ten years. The Neodesha facility held on, but just barely, and it looked as though a petroleum-based industrial boom in Kansas had been stillborn.[26]

Although local oil producers spent considerable time and effort in 1904 blaming their problems on either Standard Oil or newer producers in Texas and Indian Territory, most residents of southeastern Kansas were not overly concerned about the fate of this mineral. They focused instead on natural gas, another regional product that already had proved to be much more lucrative. From being seen as nearly worthless in the 1880s, the value of gas rapidly eclipsed that of oil in the 1890s. Drillers were finding it in quantities so large as to be almost unimaginable; simultaneously, industrialists all at once realized that they could pipe it directly to furnaces and thereby power a whole variety of manufacturing endeavors. The promise, in fact, was seen as truly revolutionary. The contemporary mood was perhaps best conveyed in 1903 by an observer from Iowa:

> The products are right there, in such immense quantities that the commercial world has been almost appalled. . . . There is no meter in these

Standard Oil Refinery in Neodesha, circa 1900. This long-lived facility was constructed in 1897 and symbolically initiated the oil industry in Kansas. The two tanks on the left are stills where crude oil was heated to begin the distilling process. An output of five hundred barrels per day was mostly kerosene in the years before automobiles. (Courtesy Kansas State Historical Society)

Kansas towns, the people paying by the month or year and consuming all the gas they want. In fact, some of the towns keep their street lights burning all day, finding it cheaper than to hire men to shut them off and relight them. It is so cheap that it "makes a man [sic] head swim" to look at the infinitesimal cost of operating, as compared with the old methods using hard or soft coal, coke, or wood. Natural gas has brought industry, men, and money to southeastern Kansas. It is destined to make this one of the greatest manufacturing centers of the United States. Nothing has been found to compete with gas as fuel.[27]

As the Iowa writer makes clear, the industrial boom that developed from the gas discoveries benefited several communities (Map 12). Iola, however, was its birthplace. This region, the home of the Acers geyser, had seen sporadic prospecting throughout the 1880s. On Christmas Day 1893, a well that penetrated deeper than the others hit a major pool at 850 feet. Its success attracted other leasers and additional wells until a large field was established that extended eight miles east of town and four miles north and south. Estimates of the productive capacity of this field, which rose steadily to an incredible one billion cubic feet per day, soon came to the attention of Robert Lanyon, the head of the zinc-smelting empire in nearby Pittsburg. He realized that, although his coal-fired smelters were efficient, ones that used the superabundant new fuel would be even more so. A botched experiment using the wrong kind of furnace slowed his enthusiasm temporarily, but Lanyon moved into a new plant just east of town in 1896.[28]

Once the Lanyon smelter proved to be successful and rumors began to spread about production costs there being 30 percent lower than at Pittsburg, a mass translocation of the industry occurred. The W. and J. Lanyon Company came in 1897, then the George E. Nicholson facility. By 1902, the Iola area was home to five new smelters, and all the older ones in Pittsburg had been decommissioned. It was a momentous change, in terms of both speed and of magnitude. Moreover, these smelters were only part of Iola's recent industrial acquisitions. Three plants to manufacture bricks and another to produce portland cement had joined the group by 1901. Altogether, the new factories created nearly four thousand jobs for a community that, in the census of 1890, had been able to muster only 1,706 residents.[29]

No records exist about the role played by Iola's two major railroads in the decision of Lanyon and other industrialists to relocate to the Allen County gas field, but they were almost certainly a key contributing factor. Zinc, bricks, and cement are all heavy products that require good rail transportation to reach markets efficiently. Iola initially had been served by the old Leavenworth, Lawrence and Galveston system, which later came under the control of the Santa Fe company. The town received an east-west line in 1883 when, as part of Jay Gould's plan to become a leader in regional transportation, he constructed a subsidiary of his Missouri Pacific Railroad from Fort Scott to Wichita. These Missouri Pacific tracks were especially well positioned for industrialists because they ran atop the gas for the field's entire eight-mile length. Factories along this line actually created two suburbs for Iola: Gas City, three miles to the east, and La Harpe, three miles beyond Gas City.[30]

By 1907, at the peak of local expansion, the three Iola area communities combined to host nine zinc smelters (60 percent of the American capacity), a zinc roller mill, a sulfuric acid plant, two cement factories, three brick plants, and a large iron foundry said to be one of the largest in the West. A total of 308 gas wells fueled this hive of industrial activity, and the thousands of jobs it created pushed the local population total to more than fourteen thousand. If a visitor took the time to look beyond the factories, he or she would see a small town transformed in other ways as well. A new $80,000 combined water works and electric plant was in place, a trolley system ran east to La Harpe, new schools and churches were under construction, and the owners of the Iola

State Bank had just completed a new Renaissance revival building (using local bricks) that seemed to guarantee permanence and prosperity. At the very least, an infrastructure was in place that could sustain the community for years to come.[31]

Nobody in 1896 or 1903 knew the life expectancy of the Iola gas field. It was a new phenomenon on the national scene. Although some industrialists were reluctant initially to uproot their factories from other cities and move to the new fuel source, the great financial success of their competitors who had done so made them look like fools. People in Iola were optimists, of course. Two local historians, writing in 1901, were typical in their appraisal: "At the rate at which it [the gas] is now being used, it is the opinion of experts that the field will not be exhausted during the life of this generation, and perhaps not for sixty or seventy years."[32]

The unprecedented transformation of Iola in the late 1890s naturally produced a drilling frenzy throughout the region, and several other modest-sized pools of gas were discovered. A field near Cherryvale, in Montgomery County, proved large enough to attract the S. C. Edgar Zinc Company from St. Louis, Missouri, in 1898, as well as several brick factories. Altoona, Fredonia, and Neodesha in Wilson County had similar success, as did Caney in southwestern Montgomery County. These communities all evolved in this way from small towns into slightly larger places (thirteen hundred in the case of Altoona; about four thousand for Caney, Fredonia, and Neodesha). Significantly larger gas fields, however, were found in northern Neosho County and in south-central and southeastern Montgomery County. Nearly simultaneously, these resources produced industrial booms at Chanute, Coffeyville, and Independence in 1901 and 1902 that rivaled the ongoing one at Iola.[33]

Chanute, with 2,826 people, was a considerably larger community in 1890 than was Iola. To that date, it had prospered principally through the largesse of railroads. The town had formed originally where the Leavenworth, Lawrence and Galveston line intersected that of the Missouri, Kansas and Texas when the two groups were racing south twenty years earlier. Prospects brightened in 1880 when the Santa Fe company purchased the poorly capitalized LLG system. The new owners added a profitable extension southeast from town to the Crawford County coal field in 1884 and two years later selected Chanute as the site to consolidate both the headquarters (from Lawrence) and the supporting machine shops (from Ottawa) for what they now termed the Southern Kansas Division. The package of railroad-related activity also included a few small manufacturing concerns attracted by the new availability of cheaper coal.[34]

The stability of a railroad-town existence did not prevail long in Chanute, however. Because the community lay midway between Neodesha's oil and Iola's gas, the countryside was filled with drilling rigs by the middle to late

1890s. The city itself actually sponsored the first successful wells in Neosho County in 1899. These were split about evenly between the production of oil and gas, a pattern that continued in the region in future decades. The boom hit in 1901, and, as one observer wrote, "The next two or three years read like a fairy tale. Farms were dotted with derricks, gas wells blazed, great volumes of oil spouted heavenward, . . . and agriculture became a secondary consideration with the farmer, who sat in the shade of the haystack and figured his royalties. Property prices and rentals became exorbitant, houses sprang up like mushrooms, everybody was busy, prosperous and happy."[35]

The primary difference between the industrial experiences of Chanute and Iola was in balance. Because most of the zinc smelters in the region already had relocated to Iola, Chanute promoters concentrated on other businesses. The Girard Smelting Company did come to town, but people were just as excited about the Chanute Refining Company and the Kansas Co-op Refining Company. Everybody expected these locally owned processors to restore prosperity to the petroleum business by paying better prices than the outsider-controlled Standard Oil plant in Neodesha. Similarly, while Chanute attracted its share of brick production with four plants, its leaders also wooed two of the earliest manufacturers of portland cement in the state, including the Ash Grove company, which became a mainstay of the local economy. Finally, the town's centrality in the new oil business and its good railroad connections made it the favorite location for petroleum production companies. About a third of the hundred or so enterprises in the state established their headquarters in Chanute, as did two manufacturers of well apparatus: the Star Drilling Machine Company and the Western Drilling Tool and Supply Company. This balance made Chanute's brand of prosperity a little calmer than the one at Iola but still, in the words of almost every commentator, "phenomenal." In 1910 the local population total, 9,272, matched that of its neighbor to the north almost exactly.[36]

Midway between Iola and Chanute, along the converging lines of the MKT and the LLG/Santa Fe, lay the town of Humboldt. This was still another boomtown, similar to Chanute in having economic development balanced between oil and gas, but much more modest in scope. From a population of 1,402 in 1900, the town acquired a small oil refinery in 1905, two brick companies, and a large plant of the Monarch Portland Cement Company in 1907 and soon had enough jobs to support a thousand additional residents.[37] These new factories all were appreciated, but boosters privately were disappointed that the growth was not larger. After all, they noted, their oil and gas pool overlapped so much with the one at Chanute that the two were essentially the same. The explanation, analogous to that for the salt industries in central Kansas, is directly related to transportation quality and population size at the onset of development. Chanute offered companies east-west railroad connections; Humboldt did not. Chanute also provided the amenities of a bigger library, better street maintenance, and the like. With pipelines available to move the oil and gas easily, companies often had the option to locate either in a small town adjacent to the field or in a larger, more dynamic community a little farther away. Most business leaders quite logically elected to live in the latter places.

The most direct challenger to Iola as the gas center of southeastern Kansas proved to be Coffeyville, a community sixty miles to its south, where officials of the LLG Railroad had created their terminus after losing the race to be first to the Indian Territory. Coffeyville had spent most of the 1870s as a small cattle-marketing center and also had developed some trade with the Cherokees across the state line. Its population in 1890, a total of 2,282, was expected to remain stable for the foreseeable future. Local lumberman William P. Brown had drilled a productive gas well that year, which was used to supply lighting for area homes and streets, but this was not a particularly remarkable occurrence for that time and region.[38]

The first explorations for natural gas in the Coffeyville area were hampered by an inflow of water into the wells. Whether because of this or some other reason, local drillings progressed at a somewhat leisurely pace during the 1890s, although enough success had occurred by 1894 to attract one large manufacturer, the Coffeyville Vitrified Brick and Tile Company. After that point, subsequent drilling suggested that the potential output from the field north of town was much larger than expected. This finding fostered a debate. Officials of the Coffeyville Gas Company, which controlled most of the supply, initially urged caution in development, saying that extracting the resource quickly and selling it cheaply might lead to rapid depletion. They recommended not to emulate Iola in this respect but to sell the product at a more "reasonable" rate of six cents per cubic foot.[39]

The counsel of the gas company was perhaps wise from a broad perspective, but this policy frustrated members of the local commercial club who were trying to attract industry, particularly glass companies from Indiana whose products required an abundant and low-cost fuel source. Expediency soon triumphed over conservation. The city bought out the gas company in 1902, voted to sell its product at cost (about three cents), and immediately found several dozen companies more than happy to move into what came to be advertised as "the very heart of the gas belt."[40]

Coffeyville in the decade beginning in 1902 transformed itself even more completely than had Iola or Chanute. Reporters noted "new life, new energy, [and] new enthusiasm" and by 1907 could list fifty-one factories either in town or in the process of building. Six hundred homes had been erected in the previous year, and everybody expected that number to be surpassed during the current building season. The source for all this enthusiasm, of course, was some 466 wells, which together brought three million cubic feet of natural gas to the surface each day, along with a smaller but still significant supply of petroleum. Coffeyville's industries included the same set of energy-intensive users that characterized the other boomtowns, but in a different mix because of the timing of the recruitment. Ten glass companies formed the core and most distinctive group. These were considered highly desirable acquisitions because they required a smaller investment than either cement companies or zinc smelters but employed a similar number of men. Their jobs also were highly skilled and therefore paid well. The larger of these facilities, including the Coffeyville Window Glass Company, Sunflower Glass Company, and Mason Fruit Jar Company, each hired well over a hundred workers, and the group as a whole sixteen hundred. Supporting the glassmakers were an even larger number of brick and tile companies, two oil refineries (National and Cudahy), one smelter (Ozark Smelting and Mining Company), a large foundry (Acme), and, starting in 1905, the shops and headquarters for the Southern Kansas Division of the Missouri Pacific Railroad. Most of these enterprises employed two hundred or more men, making the city the envy of its competitors. Its population increased faster than even that of Chanute or Iola, more than doubling itself to 4,953 between 1890 and 1900, and then doing it again to 12,687 by 1910.[41]

Coffeyville's neighbor in Montgomery County, Independence, took a somewhat different path to oil and gas growth, one that roughly parallels the experience in Chanute. A large gas field southwest of town opened in 1902 and brought in its wake three manufacturers of glass, two of brick, and two of cement. As elsewhere, these enterprises created jobs by the hundreds and were largely responsible for the transformation of a somewhat sleepy county-seat town into a diversified community of 10,480 by 1910. Independence, however, was not defined by these industries to the same extent as Coffeyville. This was partly an outgrowth of the city being a county seat, but more so because leaders of the region's main petroleum company, Prairie Oil and Gas, decided to make Independence their headquarters in 1904. This action occurred without any recruitment and set a new tone for the community. When Prairie's successor enterprise, Sinclair Oil, erected an elegant five-story office building downtown in 1916, the city added a substantial component of white-collar employees to its workforce, including many engineers. A later decision to establish offices for the Kansas Natural Gas Company there reinforced this image, as did the relative growth of the community's professional segment after 1920, when gas-related manufacturers began to fade.[42]

Urban Growth and the El Dorado Oil Field

Pipelines have been associated with the Kansas oil and gas business nearly from its outset. They connected producing areas to the Neodesha refinery, for example, and gas wells to zinc smelters. Before 1904, however, the scale of such movement always had been local. This mind-set began to change when a few entrepreneurs proposed bigger and longer tubes that could convey these energy sources to existing industrial sites tens or even hundreds of miles away. This arrangement, they argued, would be more economical than having companies move to the wellheads. The new concept was first put into operation regionally by the same Kansas Natural Gas Company that later made its headquarters in Independence. The company bought up several smaller producers in southeastern Kansas in 1903 and 1904 and made plans to construct a pipeline to Kansas City. Although this venture was ambitious and potentially very profitable, it found little support among local people. Quickly realizing its worst-case implications—no more factories in their small towns—they made banners saying "Kansas Gas for Kansans," raised some $15,000 to fight the matter in court, and even dynamited a section of the new construction.[43]

The protests of 1904 were all in vain, of course, but people were largely correct in their fears. The positive experiences of Iola, Coffeyville, and other gas boomtowns of southeastern Kansas were not to be repeated in the central and southwestern sections of the state, even though energy resources there were soon found to greatly eclipse the volume of earlier wells. The assemblages at Chanute and Independence had been products of a particular time and place. Technology and scales of development had been relatively limited, long-distance shipments were expensive enough to give regional manufacturers a captive trade area, and nobody was aware of the short and precarious life spans of individual oil and gas fields. As the century unfolded, it turned out that pipeline construction was just the first of many industrial changes that would deeply affect the relative potentials for growth among the cities of Kansas, as well as everyplace else.

Despite its heavy overtones for the future, the concern over pipeline construction in southeastern Kansas proved to be short-lived. The threat to local industry certainly was real, but this point was rendered moot by a decline in oil and natural gas output from the area. Oil production in Kansas peaked in 1906 and did not reach that level again until the development of new fields ten years later. Similarly, statewide gas production crested in 1908 and remained below that volume until the 1930s. The adjustments of Chanute and the other towns in the state's southeastern quadrant to these changes were complex, as might be expected, and drawn out over several decades. I will discuss them in the next chapter. Here I want to continue the story of new petroleum discoveries and their impacts on the existing city structure.[44]

With the production declines of the older wells in southeastern Kansas came a wave of new drilling. Originally, this was concentrated near existing fields and at similar depths, but a preponderance of dry holes gradually led to exploration farther west in the state. This period, around 1910, also saw the first widespread involvement of professional geologists in the search process and, with them, the development of the theory that oil and gas accumulate in subsurface arches or domes where sedimentary rock structures are folded. The initial instance where these activities came together to produce large-scale success was near an appropriately named, but otherwise rather obscure, county-seat town in the Flint Hills called El Dorado.[45]

El Dorado had been platted in 1868 where the east and west branches of the Walnut River join to create a wide valley (Maps 5, 12). In its early years it served primarily as an agricultural oasis within the Flint Hills pastureland, a locally important but not especially large community. The town grew to a population of 1,411 in 1880 after the Santa Fe railroad had built a spur south from Florence in 1877, and to 3,339 at the next census after this line had been extended south to Winfield and Missouri Pacific officials had constructed an east-west railroad through it on the way from Fort Scott to Wichita. After this, the number of citizens held constant for the next twenty-five years. A flour mill, a gristmill, and a woolen mill along the river furnished most of the industrial activity, together with the division shops of the Missouri Pacific during the years 1887 to 1900.[46]

The possibility of large oil and gas deposits in the El Dorado area became known to a few geologists in 1913 when Charles N. Gould mapped out a promising north-south trending anticlinal fold in the deep subsurface rocks while working for a company called Wichita Natural Gas. This company purchased a number of leases in the vicinity and drilled several wells near Augusta (ten miles southwest of El Dorado) in 1914. These proved Mr. Gould correct when they intercepted a substantial flow of gas and oil at a depth of 2,466 feet. The gas portion of this news was heartening for Western and its parent company, Cities Service, because their goal had been to supply this fuel to area communities for their municipal needs. The presence of oil, however, initially occasioned no particular celebration. Petroleum prices were still low in 1914, and the site was far from any existing refinery or pipeline.[47]

The well that changed people's attitudes about oil in Butler County was called Stapleton Number 1. This also was drilled by the Wichita Natural Gas Company, but the place was El Dorado instead of Augusta, the date more than a year later (December 1915), and the flow of oil in the well much greater. At the same depth as before, in what became known as Stapleton sand, the crew encountered a "steady pulse" of liquid that produced one hundred to four hundred barrels per day. The timing of the Stapleton discovery could not have been better. World War I was then well under way, and with it a big increase in the demand for petroleum. Prices, which had been only thirty to forty cents per barrel, rose dramatically to as high as $3.50, and the drilling crews around El Dorado and Augusta worked hard to maximize their share of the unexpected bonanza.[48]

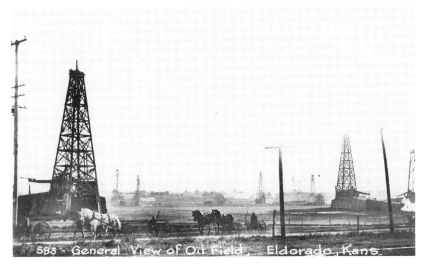

El Dorado Oil Field, circa 1917. Starting from a single well in late 1915, the El Dorado field became the most productive in the nation within two years. Drillers concentrated thousands of derricks on forty square miles just east of the city. Each one stood seventy to a hundred feet tall and tapered from a width of five feet at the base to two and a half at the top. (Courtesy Kansas State Historical Society)

As early as June 1916, two hundred oil wells were in production just west of El Dorado and another fifty near Augusta. This was only the beginning. El Dorado alone had six hundred by the end of that year, while in 1917 the Empire Gas and Fuel Company out of Wichita drilled a thousand new holes by itself. It was a boom on a scale that dwarfed even the gas discovery at Iola in 1893, and the crowds of people who pushed into El Dorado reminded more than a few longtime Kansans of the excitement of the cattle drives during the 1870s. The scale certainly was comparable, only instead of animals by the thousands, "the spectator can look for miles at an endless field of derricks set out in rows with all the regularity of a new apple orchard." A notable difference between the cattle towns and El Dorado, however, was order. Despite the hard work and the big money, the town never became known for saloons or prostitution. Some people attributed this demeanor to the Kansas work ethic, but certainly the knowledge that Butler County oil was critical for the war effort in Europe must have contributed to the calm as well.[49]

The scope of the El Dorado oil field can best be told with numbers. From an average annual production for Kansas of about three million barrels in the preceding years, the new wells pushed this number to nearly nine million in 1916, 3 percent of the United States total. In 1917 the statewide output, again almost all from this site, rose to more than thirty-six million barrels, or 11 percent of the American total. Then, when it was needed most overseas in 1918, the totals were forty-five million and 13 percent. It was no wonder that local people felt they had helped "float" the country to victory with their prodigious production.[50]

The impact of El Dorado oil money on area towns is harder to quantify than the number of wells drilled or barrels pumped. The main production area extended west from El Dorado about fifteen miles and stretched some thirty miles north and south along the course of the Whitewater River. The village of Potwin lay near its northern edge, Towanda at its western limit, and Augusta in its southern portion. Two of these communities became sites for large refineries: the White Eagle Oil Corporation at Augusta in 1917 and the Vickers Petroleum Company at Potwin in 1918. Both businesses remained active for decades and, with their employment of several hundred people apiece, became mainstays of the local economies.[51]

El Dorado became the focus for oil-related industry. The process began with the El Dorado Refining Company in 1916, the first operation of its kind in the region. This was financed in part by pioneer oilman A. L. Derby, who later created his own large Derby Oil Company. Then, the next year, came the Midland Refinery organized by local resident Fred A. Pielsticker with backing from another soon-to-be-famous man, William G. Skelly. This plant assumed the Skelly name in 1919 and expanded to become the largest in the state. By 1930 it had a capacity of twenty-five thousand barrels per day, employed more than twelve hundred people, and produced an annual payroll in excess of $1 million. El Dorado actually was home to six refineries in 1918 before a round of consolidations occurred, and the community also served as the headquarters for the large Empire Oil Producing Company and of several smaller drilling, production, and oil field service operations. Such an impressive array of industry obviously boosted the local population. The town moved quickly from its earlier steady-state existence of 3,000 residents to a total of 10,995 in 1920 and then remained at this new plateau for the next several decades. Such prosperity was a function both of the local oil field maintaining high production for half a century and of the output of smaller fields, especially east into Greenwood County and north into Marion County, being brought to the El Dorado refineries for processing. Pipelines and railroad cars took the refined products far from the city, however. As the protesters of 1904 had feared for their communities seventy-five miles to the east, no major manufacturing firms using oil as a fuel decided to move to Butler County.[52]

The direct influence of the El Dorado oil boom also was felt in three neighboring cities, all in desperate need of energy sources. Community leaders in Arkansas City, Wichita, and Winfield had spent much of their time in the 1890s and the following decade trying to capitalize on good railroad connections by attracting a variety of wholesalers and manufacturers. They generally were successful with the first part of this goal but found industrialists hard to interest. A lack of cheap fuel was the problem. City officials had searched diligently for coal in the 1880s, but without success, and then sponsored several largely futile drillings for natural gas and oil.[53]

When word about Butler County's riches began to spread in 1916, the unfulfilled needs of the people from Arkansas City, Wichita, and Winfield made them among the most enthusiastic of investors. All three communities also promoted pipelines from the field, a round of new drilling in their home counties, and the development of local refineries. Winfield's entry into the refinery business, the Kansas Gas and Gasoline Company, was modest in size and lasted only into the 1930s. Arkansas City was more successful. Two companies built refineries there, Shell Petroleum, whose operation was short-lived, and Kanotex Refining. Kanotex, a home-owned company, moved to Arkansas City in 1917 from Caney, in Montgomery County, after the oil fields of southeastern Kansas had begun to fade. Its owners expanded onto the old Shell site in 1930 and developed an extensive pipeline feeder system that covered all of Cowley County plus the Blackwell area of northern Oklahoma. This gave the company reliable supplies while, in turn, Kanotex furnished Arkansas City with more than two hundred permanent jobs. The arrangement endured some eighty years.[54]

Wichita's involvement in the El Dorado oil business was much more extensive than that of any other city. This community, which lay only twenty miles west of Towanda and Augusta, was physically closer than either Winfield or Arkansas City and also contained many more potential investors. Wichita in 1910 was already home to 52,450 people and had recently eclipsed Topeka to become the second-largest city in the state. This success, as local boosters always liked to remind people, had come with "no special advantages geographically over any other part of the state." Instead, it was a growth created by "wide awake" businesspeople who knew how to work hard and together to accomplish their goals. With four major railroads (the Frisco, Missouri Pacific, Rock Island, and Santa Fe), a company located there could easily reach markets in any part of the state. An impressive 138 jobbing and wholesale firms called Wichita home in that census year, as did five flour mills led by the Kansas and the Watson companies, both of which had daily production capacities of more than a thousand barrels. A stockyard industry had become big as well (five hundred employees) and supported two substantial meatpacking plants, one owned by Jacob Dold and the other by John Cudahy; these employed another two thousand men.[55]

As residents of a progressive, businessman's city, Wichita people were quick to realize the possibilities El Dorado oil meant for attracting new industry and generating money on its own account. After all, a local firm, the Wichita Natural Gas Company, had drilled the initial wells, and its successor organization,

Empire Oil, controlled the largest share of the field as a whole. As the El Dorado explorations moved north and west from the original centers toward Potwin and especially Towanda, Wichita investment increased. Then, as the oil continued to flow, the bulk of the money it produced found its way into Wichita bank accounts. A tour of the city of El Dorado reveals few expensive houses that date from the zenith of oil activity. Such structures certainly were constructed, but largely in Wichita. Oil money has been credited with boosting that city's assessed valuation by $5 million in 1917 alone. The total amount of money Wichita people received from the boom is more difficult to calculate, but a figure of $65 million has been suggested.[56]

The most obvious transfers of the oil economy to Wichita were three refineries. Two of these, Barnsdale and Golden Rule, were small and ephemeral, but one constructed by the Derby Oil Company in 1920 provided hundreds of jobs for the rest of the century. A second major, and continuing, impact came from Wichita people applying knowledge and money they obtained in Butler County to the development of newer oil fields throughout central and north-central Kansas in the 1920s and 1930s. Although various producing areas temporarily led the state in petroleum output, the real center of the business after 1917 has remained firmly ensconced at Wichita.[57]

A third direct economic tie between El Dorado oil and Wichita development involves the city's signature industry, airplane construction. Historian Craig Miner has made this connection most succinctly, saying straight-out that it was "Butler County oil money and Butler County oilmen's needs" that "started that city on its way to being the Air Capital of the World." The key linkage is the role played by local oilman Jacob M. "Jake" Moellendick and several oil-rich Wichita banks in providing seed money to bring pioneer aircraft manufacturer E. M. Laird to town in 1920. From Laird's early factory came Walter Beech, Clyde Cessna, Lloyd Stearman, and other men who later started their own extremely successful companies.[58]

Oil and Gas Cities of Central and Southwestern Kansas

Oil and gas production in Kansas has had three distinct historical and geographic phases. The earliest, in the southeast, attracted major manufacturers. The second, at El Dorado, raised the state into a position of national prominence and generated a large amount of revenue for a relatively small number of investors. Most of the actual oil products were shipped out of state, however, and only a few cities in Kansas benefited directly. The last phase, almost as if oil discoveries were reenacting the frontier experience, occurred still farther to the west. Very little of this oil and gas was used locally either, but unlike the case at El Dorado, the production was spread out over a much larger area. This dispersal had the effect of putting money into the pockets of a larger number of landowners and thereby lifting the overall standard of living in many places. Although no one town can be said to have had a boom comparable to those in Iola or Augusta, many communities saw their populations double and the quality of their public schools and roads increase in a similar manner.

Oil in the Kansas interior is concentrated near a major arch in the subsurface rocks known as the Central Kansas Uplift. Pools associated with this structure accumulate some five hundred feet deeper than those at the Butler County anticline and lie along an axis that extends from Norton County south-southeast to the vicinity of Hutchinson and Pratt (Map 12). No field that produces even half the amount generated at El Dorado has ever been found in this region, but the total output there over the years actually has exceeded that at the older center. The first successful well, Carrie Oswald Number 1, was drilled in November 1923 in northwestern Russell County. This was followed the next year by discoveries near Gorham (also in Russell County) and west of Lyons in Rice County. Similar exploratory success continued for more than twenty years, with sites ranging from near the Nebraska border almost to Oklahoma, and from Graham and Ness Counties eastward to Ellsworth and McPherson. By 1956, cumulative production totals had exceeded a hundred million barrels in six counties: Barton, Ellis, McPherson, Rice, Russell, and Stafford. These places, especially their county-seat communities of Great Bend, Hays, McPherson, Lyons, and Russell, were where the economic benefits of this wealth came to be concentrated. Collectively, they were known as the oil-patch towns.[59]

The transformation of central Kansas by oil first came to the attention of outsiders in the 1930s when that area's relative prosperity stood out during the Great Depression. A business periodical in 1934 noted, for example, that although the industry's development had been so rapid that "few realize its magnitude," it actually employed thirty thousand Kansans, or about one worker in twenty. The important matter of leases contributed even more money to the economy. Companies had some six million acres of the state under contract by 1935. From these arrangements, property owners received about a third of the total value of the oil product through royalties, bonuses, and rentals. For 1934, this meant a third of $47,600,000, a lot of cash for that time and place. Some writers even predicted that "Kansas—the wheat state, bids fair to becoming Kansas—the oil state, if the present trend . . . continues."[60]

As they had been in El Dorado, oil refineries were the most visible sign of the industry in this interior region. Six of the twenty-eight plants that existed in the state in 1936 were in central Kansas. Great Bend had one called the Falcon Refining Company, Hutchinson the United Oil Company, Natoma (in Osborne County) Krueger Oil, Russell the Russell Refinery, and McPherson the twosome of the Dickey Refining Company and Globe Oil and Refining. The most interesting aspects of these enterprises were their small size and short life spans in comparison with the much larger and more enduring output of their supporting wells. Only one of the six operations, Globe Oil in McPherson, survived as long as a decade, and even it had to change ownership to do so (the National Co-operative Refinery Association [NCRA] took over in 1943). Although the later addition of the Phillipsburg Cooperative Refinery meant a second operation for the region, it was clear that the oil industry in the 1930s had entered another stage of development. The first Russell County oil had been pumped into Union Pacific tank cars for shipment to larger and more efficient refineries that had been constructed in Kansas City. This practice of processing the product outside the region quickly became standard. The tactic obviously worked to the detriment of central Kansas, although it was hard for people there to miss refineries that had never been fully developed in the first place.[61]

Despite the absence of big refineries and other major industry, oil production bolstered the economy of interior Kansas in two important, although diffuse, ways. One was through lease money, as I suggested earlier. An equally critical component was perhaps best described by a Pratt County man who spoke with a reporter from the *New Yorker* in 1983:

> Well, I say oil is jobs. It's a whole bunch of jobs. It's a dozen different jobs before the oil even comes out of the ground. Somebody has to go out and write the lease for the land, and arrange the right-of-way. Then the surveyor comes on. Then the dirt contractor. He gets the site ready for the rig. Then the drilling company. Then the casing goes in. Then the valves, the tanks, the pipeline. Then there's transportation. Then pumpers. And then, way down there at the end of the road, there's the refinery. All those guys working, that really makes a community boom. . . . And the money they make! My God, a roustabout—even a roustabout—he makes eight or ten dollars an hour.[62]

The cumulative effect of lease money and oil field jobs spread itself over a host of small communities. It also contributed to the doubling of populations at Lyons (in Rice County) and at Hays (in Ellis County). Lyons and Hays are difficult to evaluate in this regard, however, because of the concurrent presence of the salt industry in the former town and a growing college in the latter. In contrast, local accounts from Great Bend and Russell, the most classic of the area's oil towns, leave no doubt that this was the business that made them substantial communities. Great Bend, which serves as the oil center for two counties (Barton plus Stafford to the south), saw its population nearly triple from 1920 to 1950 (4,460 to 12,665). Oil money was a key to financing both a successful industrial park on an abandoned World War II bomber base in the 1950s and an acceptable bid for a junior college a decade later.[63]

The developments at Russell, although more modest than those in Great Bend, are perhaps even more impressive. Despite a position on the main line of the Union Pacific, this town had been relegated to obscurity before its oil years. The neighboring communities of Hays to the west and Salina to the east dominated both regional trade and area services. Oil did not completely reverse these conditions, of course. No junior college or major industrial park materialized, but Russell definitely progressed in many ways, including following the lead of Great Bend in tripling its population between 1920 and 1950 (from 1,700 to 6,483). A local reporter in 1951 went so far as to call petroleum the town's "magic wand." It was the stimulus, he said, "for hundreds of new homes, public buildings, new streets, schools, churches, theatres, parks, new business firms, and an enduring industrial overtone."[64]

Of the two oil refinery cities in central Kansas, McPherson grew more during the years of big local production than did Phillipsburg, but this development had more to do with McPherson County's better soils and rail connections than anything else. The oil business employed several hundred people in McPherson during the 1940s, an obvious contribution to the community's population gain from 4,595 in 1920 to 8,689 in 1950, and the NCRA Refinery has steadily expanded its capacity and workforce since that time. In 1995, the company employed 511 people. Phillipsburg, a much smaller community in 1950 at only 2,589 people, is interesting in having been chosen as the site for two fuel-related companies despite being at the extreme northern edge of the oil region. Both selections were made precisely because of this peripheral location, however. Owners of the Kansas Nebraska Natural Gas Company and the Phillipsburg Cooperative Refinery each decided that, because their markets were in the northern plains (an area without oil or gas at the time), a position midway between supply and destination made sense. The two operations had become the biggest employers in town by the 1950s, appreciated additions to a local economy that had recently lost its division point on the Rock Island railroad because of a switch from steam locomotives to more reliable diesel engines.[65]

Kansas's final mining frontier, known as the Hugoton natural gas region, developed in the nine southwesternmost counties of the state. By some measures, this was the state's most important mineral deposit because it led the world in gas production for several decades during the middle of the twentieth century. This industry also produced Kansas's most invisible mining landscape, however, and hired the fewest workers. Its impact on local towns and cities was therefore mixed. Populations doubled but remained small. As in central Kansas, the primary benefit came from a widespread distribution of lease money across the area.

The first discovery of the southwestern gas deposits occurred nearly simultaneously with that for oil in central Kansas. An exploratory hole twenty-five hundred feet beneath the surface of Seward County in 1922 hit a pocket more than adequate to supply heat to the residences of nearby Liberal. This was the extent of ambition at that time, however. Remoteness from markets and an absence of pipelines prohibited dreams about more general wealth and development. Another five years passed before a second well was attempted, this one fifteen miles to the northwest to supply the needs of Hugoton, the seat of Stevens County. This second well was the one that inspired larger-scale operations. Two local men envisioned a series of gas contracts with area towns and then a pipeline from Hugoton to the regional trade center of Dodge City, eighty miles to the northeast. The completion of this project in 1930 created a demand for additional drilling, but the entire area still contained fewer than two hundred wells as late as 1938.[66]

Major pipelines built in the 1930s by two companies—Panhandle Eastern and Northern Natural Gas—finally initiated rapid development of the region. New wells drilled to keep these conduits full soon revealed a vast field. Extending over some two and a half million acres, it blanketed all of Grant and Stevens Counties, as well as portions of seven others (Map 12). The new pipelines, the former running northeast as far as Detroit, Michigan, and the latter to Nebraska and the northern plains, were voracious consumers of gas, and production climbed rapidly. As the numbers hit a hundred billion cubic feet per year in 1941, two hundred in 1947, and three hundred two years later, it became obvious that the Hugoton region contained a resource of epic proportions. Panhandle Eastern added a second big pipeline in 1949 and a third in 1955, bringing its services to thirteen states.[67]

Southwest Kansans, having observed the lack of industrialization at the oil fields near El Dorado, Great Bend, and Russell, did not expect that aluminum smelters or similar big users of power would come to their counties. Although politicians lobbied occasionally for such developments, residents were grateful just to have lease money. These dollars were critical, in fact, because they began to arrive in the midst of the Dust Bowl. By chance, devastation was worst almost exactly where gas was most abundant. Later, as annual production soared past the mark of four hundred billion cubic feet in 1951 and five hundred in 1956, people turned these dollars into new courthouses, renovated public schools, bigger combines, and luxury automobiles. The mood in the core cities of Hugoton and Ulysses, as well as in peripheral Garden City, Johnson, Lakin, Liberal, and Sublette, echoed exactly the ones I have described for Russell and Great Bend.[68]

From a more particular urban perspective, the most interesting locational decision was where the Panhandle Eastern and Northern companies would place their regional headquarters. Centrality to production would suggest Hugoton or Ulysses, and Northern followed this logic. This company's choice of Hugoton actually was based more on history than on position, however, because Northern grew out of the pioneering Argus pipeline enterprise of that city. After company officials announced their first big construction project to Omaha in 1930, Governor Clyde Reed of Kansas declared Hugoton the "Gas Capital of the Southwest." This name, modestly upgraded by local boosters to read "Gas Capital of the World," still adorns highway signs and is a reasonable reflection of the local economy. Northern's offices, plus those of a branch of Panhandle Eastern and of several independent companies that drill and/or service the wells, remain the main industries in a prosperous, but still small, town. Together, these 250 or so employees pushed the population from 1,368 in 1930 to 2,781 twenty years later.[69]

Officials at the Panhandle Eastern Pipeline Company, a larger operation than Northern, established their headquarters in Liberal on the eastern edge of the gas field. With this choice, they had to travel slightly farther to wellheads, but they gained amenities. Liberal was a much larger community at the outset of gas development than either Hugoton or Ulysses. Its location on a main line of the Rock Island railroad also made access to the outside world relatively easy. The company decision meant the addition of 150 jobs in the office itself, plus a larger number in subsidiary and support operations. Panhandle Eastern also established a plant in town to manufacture butane, propane, and similar products. This latter unit operated for some thirty years until it was phased out in 1963 in favor of a similar, but larger, joint-owned factory that produced helium. In these ways, the natural-gas industry provided Liberal with much needed diversification from its railroad base. The town's population, 5,294 in 1930, increased to 7,134 in 1950.[70]

The Garden City Phenomenon

Although natural gas and other mining economies contributed considerable variety to the somewhat rigid system of mercantile cities that the railroads had laid out across Kansas, these were not the only urban modifications of note for the period between 1880 and 1920. The placement of a second round of public institutions created specialized communities in several instances. The most important addition to the list of cities at this time, however, was singular in occurrence and nearly as unexpected as the gigantic gas reserve at Hugoton. The stimulus this time was irrigation, and the community the aptly named Garden City along the Arkansas River in southwestern Kansas. When sugar beets proved profitable there after 1906, that town became as "magical" in its way as had Horton or Iola in earlier decades.

The group of adventurers from Sterling, Kansas, who rode the Santa Fe tracks west to found Garden City in 1879 picked a bad time to resettle. That year was the first of two droughty growing seasons for western Kansas, and

Factory of the United States Sugar Company in Garden City, 1910. Garden City's community leaders believed that this new facility, built in 1906 by investors from Colorado Springs, would solidify their irrigation economy. The plant employed three hundred men by 1912 and processed nine thousand acres of beets. Professional colonizers such as H. C. Wiley's Grand Western Land Company (sign on train) helped to recruit laborers. (Courtesy Kansas State Historical Society)

nearly every person in the small community saw their crops fail. The exception was a man called Squire Worrell, who had previous experience with irrigation in California and Colorado. Worrell converted a millrace built by another colonist into a canal and then used it to water several acres. His success brought praise from two fronts: fellow farmers who wanted to share in the venture, and speculators who wanted to boom the general area. The former group helped to enlarge the original enterprise into the Garden City Irrigation Company, which served about a hundred acres the following season. The latter people found enough outside investors to charter ten additional canal companies by the end of 1881.[71]

The physical conditions for irrigation at Garden City are favorable in several ways. The Arkansas River changes gradient about twenty-five miles upstream near the hamlet of Hartland. A reduced rate of flow there causes the stream to meander across its floodplain and to deposit part of its suspended sediment. This alluvium, which exists mainly north of the river proper and expands downstream toward Garden City, consists of well-drained, silty soils that produce excellent crops. Lying close to the river, it also is easy to serve by ditches. Business speculators constructed five major canals across this surface by 1882, and because of their economic promise, the town of Garden City grew from nearly nobody to about a thousand people.[72]

Irrigation in western Kansas was not without its problems, of course. Water levels in the Arkansas varied greatly from month to month as well as from year to year. The region also experienced many seasons wet enough to produce crops without the need for and expense of ditch water. A series of these "good" years extended from 1882 through 1886, during which time the canal companies could find few buyers for their water and so either failed or let their facilities deteriorate. Such tension over whether or not to irrigate and how best to farm in Finney County persisted through the 1890s, and Garden City's population held steady at about fifteen hundred residents.[73]

A change that restored irrigation to a central role in the region came in the late 1890s when George W. Swink, a promoter from Rocky Ford, Colorado (farther upstream on the Arkansas), urged local producers to try sugar beets as a crop. Beets require precise amounts of water at their various stages of growth and therefore the use of irrigation. Their main appeals were subsidies provided by the federal government and guaranteed purchase contracts offered by sugar factories. Swink persuaded several local investors to renovate the old ditches and try his idea even though the closest processing plant was 170 miles away. Later, when he was able to interest a Colorado mining tycoon named Spencer Penrose in building such a sugar factory in Garden City, the pace of development quickened into a boom.[74]

Penrose headed up a syndicate called the United States Sugar and Land Company. This group took control of thirty thousand acres, administered two of the five local canals, and, in November 1906, opened a large processing plant. The city contributed both money and land to the project, but people felt more than adequately rewarded when the price for rural and city property promptly doubled and new residents poured in by the hundreds. For its part, the syndicate provided seeds, oversaw a pumping system that supplied all the area's ditches, and even imported migrant labor from Russia, Mexico, and elsewhere to do the bulk of the hot field work. Irrigation in Finney and adjacent Kearny County grew to some forty-seven thousand acres by the late 1920s, and the factory's annual payroll increased to more than $265,000. Garden City's population jumped accordingly: 3,171 in 1910; 6,121 in 1930; and then 10,905 in 1950. Residents of the railroad center of Dodge City, who had claimed all southwestern Kansas as their tributary area before Penrose came to the state, now found themselves with a formidable regional rival.[75]

A Second Round of Public Institutions

The same years of excellent agricultural weather in the mid-1880s that had prompted farmers in Garden City to temporarily abandon their irrigation ditches also encouraged other families to push into the last unsettled portions of the state. Farther east, a similar prosperity and spirit of accomplishment imbued people with the sense that Kansas was ready to develop its infrastructure and service system beyond a rudimentary stage. One demonstration of the mood was the establishment of additional public institutions, including asylums, colleges, and prisons.

As I have described in chapter 3, the allocation of the state's first major public facility, the capital, was accompanied by extreme political maneuvering during the 1850s and 1860s. The fate of many cities and would-be cities hung in the balance. The same could be said for the early designations of university and prison communities. The stakes were not nearly so high in the 1880s. The public institutions available at this stage were not as prestigious as earlier ones and did not promise as many jobs. Prosperous cities such as Salina, Topeka, and Kansas City therefore did not pursue them. On the other hand, small towns, for which such construction and employment truly could have been transforming, rarely had a realistic chance of obtaining one of these modest plums. They lacked the necessary political clout.

By a process of default and minor political initiatives, the new public institutions awarded between the 1880s and 1920 tended to go to moderate-sized communities. There, although certainly contributing to the local economies, they did not alter relative positions in the urban hierarchy. The first instance of this phenomenon occurred in 1884 when city officials from Winfield donated a tract of land to obtain the Asylum for Idiotic and Imbecile Youth. This institution, with its name changed to the State School for Feeble-Minded Youth in 1900 and then to the State Training School in 1919, employed about seventy-five people in 1910 who cared for four hundred patients. Similar examples include the Soldiers' Orphans' Home awarded to Atchison in 1885, the Industrial School for Girls to Beloit in 1889, and the Hospital for Epileptics to Parsons in 1899. The Atchison facility, which began accepting all orphans after 1899, employed about fifty people early in the twentieth century. Beloit's school—slightly smaller than the orphanage—grew out of a cooperative venture between local citizens and the Women's Christian Temperance Union. The hospital at Parsons was obtained through the efforts of state senator Charles H. Kimball. It was the largest institution of the three, with more than four hundred patients in 1910 and ninety-four employees.[76]

A fifth example of what might be called the Winfield model of institutional location was Hutchinson's acquisition of the State Industrial Reformatory. This city, with its salt- and trade-based economy, would seem at first to be too prosperous to seek out a reform school. The decision there to cede 640 acres worth $24,900 in exchange for the facility was made in 1885, however. This was several years before the salt discoveries, at a time when Hutchinson people saw their economic prospects fading in comparison to those of neighboring Newton and Wichita. The reformatory was designed for first-time male offenders between the ages of sixteen and twenty-five. It was essentially a trade school, offering instruction in farming, blacksmithing, carpentry, shoemaking, and the like. The first prisoners arrived in 1895, and 217 lived there by 1900, where they were looked after by forty teachers and other employees. Promotional literature on the city, as might be expected, chose to ignore the facility's presence.[77]

Three other public institutions from this period merit attention because they went to communities in western Kansas. This region's relatively short settlement history, in conjunction with its variable and somewhat forbidding climate, meant that the urban hierarchy there was still somewhat fluid in the early decades of the twentieth century. A new public facility, even one of modest size, therefore had considerable potential to elevate a city's status and size relative to those of its peers.

The largest of the three western institutions, what is now Fort Hays State University, came about through individual initiative more than political maneuvers. In 1889, when the time came for the seventy-six-hundred-acre Fort Hays to be deactivated as a military post, people debated the fate of its lands. A group of army officers wanted to purchase them for eventual resale to settlers. General Philip Sheridan suggested it might be more appropriate to give them back to the Indians, and others wanted the facility converted into a home for old soldiers. Hays residents generally opposed all these scenarios. They themselves were divided in their opinions, but many supported a plan suggested by their state senator, Martin Allen, to use the land as a government testing site for agriculture on the High Plains. Nothing was resolved for several years until Albert R. Taylor, the president of the state normal school in Emporia, linked the Allen proposal with an idea to use some of the acreage for a branch campus of his college. This was the substance of a petition made to Congress in 1900 and accepted by that body later in the same year. The state could have the entire tract if it would agree to establish an agricultural experiment station on part of it, a college on another section, and a park on a third.[78]

Kansas officials accepted the federal offer in 1901, and the Western Branch of the Kansas Normal School opened the next year with 57 students and 2 faculty members. The facility grew steadily. It cut its ties with Emporia in 1914 to become the Fort Hays Kansas Normal School, and by 1920 it enrolled 242 students and employed 40 faculty members. Twenty years later these numbers had increased to 1,085 and 110, respectively. Such statistics, together with jobs generated by the central Kansas oil field, formed the basis for Hays becoming one of the fastest-growing cities in the northwestern quadrant of the state. From 1,136 residents in 1900, the community expanded to 3,165 in 1920, and to 8,625 in 1950.[79]

One of the ironic twists in western Kansas history is that, although the location for the region's state college never was contested, that for its mental hospital most certainly was. The difference was not a matter of relative prestige, of course, but of which contest was open to public debate. By early in the twentieth century, a combination of crowded conditions in the original asylum at Osawatomie and long travel times from there to much of the state prompted the legislature to fund an additional facility. The size of its 1911 appropriation—$100,000—attracted many bidders, and the restriction that only communities west of the ninety-eighth meridian (the vicinity of Hutchinson) were eligible made even smaller places feel they had a chance. Fifteen towns expressed interest, with the primary contenders emerging as Dodge City, Garden City, Great Bend, and Larned.[80]

Larned, the smallest of the four rival cities, won the prize despite having a population of only 2,911. Partly the legislators were aware that the town had suffered economic stagnation since its famous namesake fort had closed in 1878. Partly they thought the small size might provide the most restful setting for mentally ill patients. The new facility proved to be beautiful and quite large. It was built in a dispersed "cottage plan" that resembled a college campus and was situated on a 953-acre farm in the valley of the Pawnee River, three miles from town. It also was designed to be nearly self-sufficient and therefore possessed its own power supply, post office, dairy, and the like. From forty patients and thirty employees at the start, these numbers grew rapidly to more than fourteen hundred and four hundred, respectively. This employment significantly boosted the city's population, which increased steadily to 4,447 in 1950.[81]

The final important public institution placed in western Kansas during the period of 1880 to 1919 occurred neither by chance (as with the college) nor by geographic edict (as with the mental hospital). Specialized treatment centers for tuberculosis were becoming common at this time, and accepted wisdom suggested that an ideal location should have abundant sunshine, cool nights, and low humidity. Western Kansas could not equal sites such as Colorado Springs, Colorado, or Santa Fe, New Mexico, which owed much of their growth in the early twentieth century to this business. The High Plains nevertheless fit the list of desired traits quite well. When the same legislature that had appropriated money for the Larned hospital in 1911 also set aside $50,000 for a sanatorium, it launched another competition.[82]

The legislature appointed a committee of physicians to find the site "best suited" for the new facility, but it attached a proviso that this community would have to donate 160 acres to the cause. Using these guidelines, the committee selected Norton, a county-seat community of 1,787 people near the

Nebraska border. This was a good choice. Norton's elevation was 2,260 feet, high enough to ensure good cooling on summer nights, and it lay adjacent to the sheltered and wooded valley of Prairie Dog Creek. The town also was served by one of the Rock Island's main lines and so had good access to the rest of the state. These features hardly distinguished the community from two dozen or so others on the High Plains, however. The deciding factor was probably the presence of Dr. Chauncey S. Kenney, who had been treating tubercular patients at Norton since 1910 and who was named the sanatorium's first superintendent when it opened in 1914. The facility employed just over a hundred people during the 1920s and 1930s, caring for between two and three hundred patients at any one time. These numbers account for most of Norton's growth during these decades. The city had 2,186 people in 1920 and 3,060 by 1950.[83]

Not that anybody noticed at the time, but the 1920s represented the end of an era for Kansas. It was the passing of a frontier of sorts, although not the conventional one of free agricultural lands that had faded more than a generation earlier. Rather, this was the end date for a set of cities to emerge that would have a chance to compete in the faster-moving world that was to be twentieth-century America. The major railroad lines had all been laid by this time, and cities not well positioned on their grid had few possibilities for expansion. Similarly, new cities or major expansion of older communities based upon significant minerals discoveries, new public institutions, and a few other assorted means also were accomplished facts. Hugoton, Larned, and Russell were the last of their kind. This is not to say that these three cities and their competitors across the state enjoyed guarantees for success in the decades to come. Instead, it was a matter of their having survived an initial round of economic struggle. The roster of places with a potential to flourish further had been trimmed drastically by the 1920s from what it had been during the halcyon days of town platting and local boosterism from the 1850s into the 1880s. Kansans had founded more than forty-two hundred communities during that earlier period. By 1930, fewer than seventy of these had found a means to maintain populations of twenty-five hundred or more and therefore to be viable in roles beyond that of small service centers (Appendix).

7

Urban Consolidation in a Railroad Mode, 1880–1950

The founders of cities in nineteenth-century or early twentieth-century Kansas usually had clear goals in mind. One site might have potential to garner a share of the Missouri River steamboat trade, for example, another to secure offices of a county government, and a third to provide homes for employees of a nearby coal mine. The series of historical forces was so specific, in fact, that it provides an obvious framework for the previous chapters in this book. Once cities have been brought into being, however, they assume lives of their own. The original deposit of ore may become exhausted, or railroad moguls may decide to relocate a division point, but the people of the affected communities do not necessarily have to move on. They may be able to find a new economic niche by attracting a manufacturing enterprise or taking advantage of an old fort or trail to develop a tourism industry. Some towns do die, of course, and others decline. Still others seem to lead charmed existences, either continuing to serve their original business roles or somehow being able to parlay one advantage into another to produce a history of steady or, rarely, even mercurial growth.

My goal in this chapter is to examine urban Kansas in the years between 1880 and 1950 as a system of cities. As with any such system, functional roles necessarily exist to supply the regional populace with a range of services, manufactured goods, retail outlets, and the like. Some places tend to specialize, others to provide more varied offerings, but the roles can and do change over time. The period I discuss here is unified by the dominance of railroad transportation. The corporations that owned these steel webs had founded or cofounded many of the larger towns in the state. They also had invested heavily in many of the coal mining, meatpacking, and other businesses that utilized their tracks. The pattern they and other power brokers had created by early in the twentieth century—one dominated by Kansas City, Topeka, and Wichita—was seen by nearly all residents to be the enduring form for the central plains. The past river glories of Atchison were forgotten, the future, highway-oriented ones of Overland Park unimagined.

Unequal Growth

One of the principal puzzles of Kansas urban history is a huge differential in city growth rates. Although most farming villages faced absolute population declines in the decades before 1950 (enough so that historian Daniel Fitzgerald had been able to compile two books on the state's "ghost" towns), the experience of larger early communities tended to range between stagnation and modest growth. Arkansas City and Fort Scott, for example, contained 8,347 and 11,946 people in 1890, respectively, and 12,903 and 10,335 in 1950. A few places, but only a few, were able to grow substantially faster than the state as a whole (Chart 2).[1]

If one begins with Kansas as it existed in 1870, the urban hierarchy is simple (Map 8; see p. 20). Leavenworth, with 17,873 residents, is clearly the dominant city in the state, having more than twice the population of either Lawrence or Atchison. Because of the machinations of railroad builders, this advantage proved to be short-lived. By 1880, a small drop in Leavenworth's population combined with rapid growth for Atchison and Topeka produced a near tie in overall size, and uncertainty over how the future urban system might look. A decade later this uncertainty had lessened in one sense, because Atchison dropped from its pinnacle. Newer cities added to the complexity, however. An amalgamation of Wyandotte with neighboring communities had produced a large new city called Kansas City, Kansas. Five urban places with populations of 7,500 or more also had emerged beyond the old core region of northeastern Kansas. These included Fort Scott in the southeast, Emporia in the east-central region, and Arkansas City, Hutchinson, and Wichita in south-central Kansas.

The pattern of 1890 held more or less constant through 1900, but by 1910, differential population growth had occurred on a scale that amazed nearly every observer. Kansas now had three cities that dwarfed all potential rivals. Kansas City led the way with 82,331 residents. Wichita, having doubled its population in a decade, recorded 52,450 citizens, while Topeka had 43,684. The great surge in growth rates for these places continued over the next forty years, producing two centers with well over 100,000 residents and an increasingly large gap in size between Topeka and the lesser urban centers of the state. Another interesting development during this period was the emergence of three intermediate-level cities: Hutchinson by 1920, then Salina, and finally Lawrence. Although populations in these places never approached the levels of the big three, their growth rates clearly exceeded those for smaller competitors.

Perhaps the best way to begin to understand the range and pattern of urban growth in Kansas in the years 1880 through 1950 is to look again at the state's railroad system. Instead of concentrating on initial construction and individual

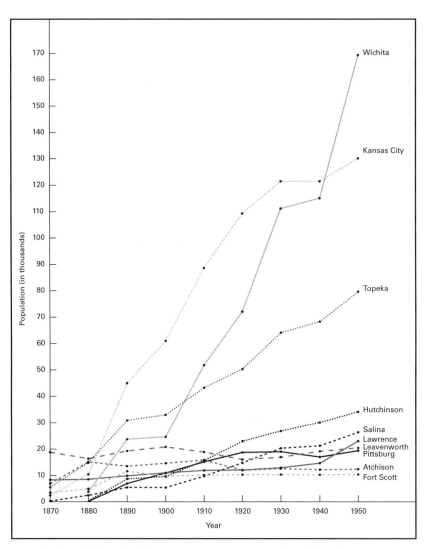

2. Populations of Selected Cities in Kansas, 1870–1950. Data from federal census records.

lines, however, this time the focus is on traffic. The flow of goods and people between places is an objective measure of connectivity. This flow also can serve as a useful surrogate for the overall strength of trade, manufacturing, and other components of modern urban economies. Railroad freight traffic would perhaps be the best single measure for these years, but reliable, comparative data are not available. The number of passenger trains serving each route is known annually since 1868, however, thanks to a comprehensive timetable publication known familiarly as the *Official Guide of the Railways*.[2]

Information for the year 1878 shows a Kansas system that simultaneously had begun to consolidate even as it continued expansion into unserved regions (Map 13; see p. 26). Spurs that dead-ended in Beloit, El Dorado, Independence, and Wichita soon would be extended, of course, but officials for the Missouri Pacific and the Rock Island systems had not yet begun to contemplate construction of their trunk lines. Predictably, the Kansas Pacific company was able to dispatch one more train across the state daily than could the Santa Fe. Denver, the Kansas Pacific destination, was considerably bigger than the Santa Fe's Pueblo. All of the cities I described in chapter 4 as becoming nodal points by the late 1860s have maintained that status: Atchison, Junction City, Kansas City–Wyandotte, Lawrence, Leavenworth, and Topeka. In addition, several new intersections of note have developed south of these relatively established places, in Chanute, Emporia, Fort Scott, and Parsons.

Traffic counts reveal that all was not equal among the railroad nodes of 1878, however. One obvious change from earlier in that decade was a decline in travel along the Missouri, Kansas and Texas tracks between Junction City and Parsons. As the Katy's president had predicted when he authorized a new line northeast from Parsons to Sedalia, Missouri, far more goods and people were flowing eastward into the nation's older urban centers than westward toward the frontier. Junction City and Emporia were thereby losing importance as shipping points and manufacturing sites at this time. A second important trend visible on Map 13 is increasing dominance by the well-interconnected Kansas City–Wyandotte center over any and all rivals (cf. Map 8). Officials for the Santa Fe railroad already were sending as many trains from Topeka east to Kansas City as they were northeast to their original terminus at Atchison. Greater Kansas City in 1878 had five major rail connections to the east, two to the north, and three to the west and south, all carrying two to four passenger trains per day. No other city could come close to this combination of trackage and traffic. Atchison and Leavenworth each boasted nearly as many connections (seven apiece), but they both lacked the Kansas City volume.[3]

A second delineation of railroad passenger traffic, this one for 1892, reveals a system nearing saturation throughout the state (Map 14; see. p. 27). Main lines for all the major companies are now in operation, and every town of any size has daily freight and passenger service. The map is complex and remains so even when (as here) routes carrying a single train per day are eliminated. So many additional tracks also have produced new points of intersection, and it is easy to identify nodes at Concordia, Garnett, Herington, Newton, Ottawa, Paola, Winfield, and other places. Two of the new centers, Hutchinson and Wichita, appear to hold special promise, Hutchinson for its volume and Wichita for the sheer number of lines that converge there.

Although the map suggests that Hutchinson and Wichita were well positioned in 1892 for future economic dominance over south-central Kansas and

The Fred Harvey Dining Room in Hutchinson, 1926. Fred Harvey operated an extensive system of railroad restaurants under an agreement with Santa Fe officials and maintained high standards of quality and efficiency. His establishment in the Hotel Bisonte was partitioned into a lunch counter ("pay as you go") and a more formal dining room ("six bits"). Note the quality woodwork and fine glassware. Servers, known as "Harvey girls," earned a starting wage of $17.50 per month. (Courtesy Kansas State Historical Society)

possibly the southwestern sector as well, no equivalent development had yet occurred in north-central and northwestern Kansas. Uncertainty in this region contrasts even more markedly with the situation in eastern Kansas, where all the hierarchical trends noted for 1878 have accelerated and a single system now appears firmly in place. Greater Kansas City, with more than 130,000 people on the Missouri side and another 43,000 in Kansas, has become the undisputed entrepôt for the entire state. Three major rail lines carry heavy traffic to the east, while the Santa Fe and Union Pacific companies, now joined by the Rock Island, run eighteen trains daily along the Kansas River valley. These tracks and their extensions north, south, and west make nearly the entire state tributary to this single urban center.[4]

The rail dominance of Kansas City by 1892 has meant greatly reduced status for that community's onetime and hoped-to-be rivals. In the southeast, the strategy of Levi Parsons to divert traffic to and from the southern plains through his namesake city and then on to St. Louis and Chicago is in the process of collapse. Demand by Katy customers for access to Kansas City forced the company to build such a connection north from Parsons in 1886–1887. The bigger city thereby grew even faster, while Parsons stagnated and the older, more western MKT towns of Emporia and Junction City suffered even more. Dreams in Fort Scott for an entrepôt role similar to that of Parsons came to a nearly identical end. As for the metropolitan expectations in other communities of northeastern Kansas, the one at Leavenworth is most clearly in jeopardy, while Lawrence and Topeka have been co-opted into the larger urban system and share the general flow of goods and people into and away from Kansas City. Atchison and St. Joseph, in contrast, remain largely apart from the growing regional economic hegemony. St. Joseph, about half the size of Kansas City with some fifty-two thousand residents, obviously possesses linkages to the east sufficient in number to match its muscular neighbor, but traffic on these lines is light and that on routes farther to the west lesser still.[5]

A third depiction of passenger rail traffic in Kansas, for 1948, fifty-six years after the second, reveals a much-simplified system of flows (Map 15; see p. 28). The somewhat amorphous trade connections apparent for the western half of the state in 1892 seem to have resolved themselves. Hutchinson, by maintaining its linkages to the west and southwest while Wichita allowed its corresponding lines to lapse, has assumed the role of entrepôt for a quarter of the state. This development, of course, carries with it a related economic question. If Wichita people have ceded dominance in regional trade to Hutchinson, what is left to explain the larger city's continued rates of high population growth during these years? Meanwhile, in northwestern Kansas, where no strong center was present in 1892, Salina has clearly moved into a position of dominance. Three railroads bring it the trade of twenty-four counties, including influence over a subsidiary distribution center for the far northwest at Colby.

Eastern Kansas in 1948 reflects a continuation of trends under way for three-quarters of a century. Trade in southeastern counties remains focused on Fort Scott and Parsons, but a greater number of trains connect this region with Kansas City than in 1892. This trend indicates that both local centers are losing ground to a larger competitor. Winfield, farther to the west, obviously has declined as a rail center, but Emporia, following the Santa Fe's development of a direct route from that point northeast into Kansas City, has grown in stature as a major switching center for that company. The railroad roles of Herington, Newton, Ottawa, Paola, and Topeka remain much the same as five decades before.

The focal point for Map 15, of course, is the Kansas City metropolitan area. An even greater percentage of the state's rail traffic converges on this point in 1948 than it did in 1892. Twenty-five passenger trains per day ran along the Kansas River in 1948, seven more than on the earlier map, and another ten entered from the south via the Frisco and MKT lines. These increases are not the only factors that produce the focus on Kansas City, however. Lawrence and Leavenworth have almost completely disappeared as railroad junctions in their own right, while Atchison and St. Joseph maintain only the barest shell of independence. At this date, Kansans from communities as widely spaced as Coffeyville, Goodland, Liberal, and Paola all fell under the control of this urban center for their rail passenger service and, presumably, for their business connections in general. With approximately half a million people on the Missouri side and 130,000 in Kansas, Kansas City alone had achieved the metropolitan status about which people in so many area communities had dreamed.

A Functional Classification

Beyond an examination of railroad interconnectivity, a second useful perspective on differential growth rates can be provided by the functional roles that individual cities have played over time. This approach seems straightforward. It necessarily involves the creation of a taxonomy, however, and with it, the fundamental challenge of any classification to reconcile theoretically ideal measures with ones for which comparable data are available. Time and size considerations pose special problems in this regard. A wider range of statistical information on economic activity exists for larger cities than smaller places, of course, and for recent years as opposed to the nineteenth century. The federal census office did not collect reliable numbers for the wholesaling industry, for example, until 1930.

Geographer Chauncy D. Harris created the oldest and most respected system for the classification of American cities in the 1940s. He considered both occupational and employment data but stressed the latter measure because it enabled him to better delineate the manufacturing and trade sectors within the economy. The most difficult aspect of the procedure was how to identify and control for the employment that was internal to the city (what other scientists came to call city-serving or nonbasic production) from "workers employed in the primary activities" of the place (city-forming or basic production). Harris made adjustments for this issue by assigning varying threshold percentages to the workforce in the different economic sectors based on his personal experience. Later scholars, testing these assignments by determining more objective, minimal "support" levels for each category, confirmed that the initial judgments were close to the mark. Given this approval, along with the reality that employment numbers plus city and county population totals constitute the only consistently available statistical information for Kansas towns from 1870 through 1950, I deemed a slightly modified version of the Harris method the most appropriate system to employ.[6]

The taxonomy as implemented recognizes seven urban functional categories plus a large residual grouping for less specialized places. To be listed as an army or college town, the number of local soldiers and support people (or students and faculty) must equal at least 25 percent of the total urban population. Hospital, mining, manufacturing, railroad, and wholesaling centers, in contrast, are generally identified by percentages of the local population working in such activities. The threshold number for the manufacturing sector is 8 percent. For hospitals and mines this figure drops to 4, and for both railroads and wholesalers it is 3.[7]

Specialization has characterized Kansas cities from the territorial period. As I discussed in chapter 2, both Atchison and Leavenworth owed their initial prosperity to the interrelated businesses of overland transportation and wholesale trade. The wholesaling component of this pair continued important in both communities long after steel rails had replaced military roads. Although details on employment are not everywhere available, state gazetteers list the number of such firms for each city. Atchison and Leavenworth, together with Lawrence, each recorded totals for 1870 and 1880 that exceeded twice the number a person would expect if such industry had been distributed strictly in proportion to the size of each city in the state. Atchison alone maintained this level of prominence through 1891, the last year such gazetteer data are available (Table 1, Maps 16, 17; see pp. 29–30).[8]

Beyond wholesaling and the army's obvious presence in Leavenworth, the only other specialized communities of more than 2,500 people in 1870 were the adjoining towns of Wyandotte and Armstrong. There, according to the census, the Kansas Pacific Railroad employed 770 people at its division headquarters and main shops. The number of places with concentrated activities grew rapidly as the state developed, of course. By 1880, two mining centers had emerged in southeastern Kansas along with focused railroad employment at Ottawa (the Leavenworth, Lawrence and Galveston), Parsons (the Missouri, Kansas and Texas), and Topeka (the Santa Fe). Manufacturing began to prosper at two other places: Leavenworth, in conjunction with available prison labor, and Wyandotte, where businessmen established meatpacking plants near the Armstrong rail yards. Finally, the army town of Junction City had grown enough to be officially recognized as an urban place.

The decade between 1880 and 1890 saw Kansas expand substantially in both total urban population and city specialization (cf. Maps 16, 17). Manhattan, reaching the population threshold of twenty-five hundred, qualified as the state's first college town during this period, while similar growth at Osage City and Pittsburg added two coal-mining communities to the system. The addition of a variety of industrial and service activities at Fort Scott, in contrast, made that community less defined by its mining economy. Four counties in Kansas contained two urban places apiece by 1890, the first time that

Table 1a. A Functional Classification of Cities in Kansas, 1870–1910.

	1870	1880	1890	1900	1910
Army	Leavenworth	Junction City Leavenworth	Junction City Leavenworth	Junction City Leavenworth	Junction City Leavenworth
Asylum/hospital			Osawatomie	Osawatomie	Osawatomie
Education			Manhattan	Manhattan	Emporia Manhattan
Manufacturing		Leavenworth Wyandotte	Argentine Atchison Hutchinson Kansas City Lawrence Leavenworth Pittsburg	Argentine Cherryvale Coffeyville Hutchinson Iola Kansas City Lawrence Leavenworth	Caney Chanute Cherryvale Coffeyville Humboldt Iola Kansas City Leavenworth Neodesha
Mining		Fort Scott Galena	Galena Leavenworth Osage City Pittsburg	Galena Osage City Pittsburg Weir	Frontenac Galena Pittsburg
Railroad	Wyandotte	Ottawa Parsons Topeka Wyandotte	Argentine Arkansas City Chanute El Dorado Horton Junction City Kansas City Osawatomie Parsons Topeka Wellington	Argentine Arkansas City Chanute Fort Scott Hiawatha Horton Junction City Kansas City Newton Osawatomie Parsons Pittsburg Topeka Wellington	Arkansas City Chanute Dodge City Fort Scott Herington Horton Junction City Kansas City Newton Osawatomie Parsons Pittsburg Pratt Topeka Wellington
Wholesaling	Atchison Lawrence Leavenworth	Atchison Lawrence Leavenworth	Atchison	No data	No data

Source for Tables 1a and 1b: Calculated by the author from data in various federal censuses, annual reports of the Kansas Bureau of Labor and Industry, state gazetteers, and local histories.

Table 1b. A Functional Classification of Cities in Kansas, 1920–1950.

	1920	1930	1940	1950
Army	Junction City	Junction City	Junction City	Junction City
	Leavenworth	Leavenworth	Leavenworth	Leavenworth
Asylum/hospital	Norton	Norton	Larned	Larned
	Osawatomie	Osawatomie	Osawatomie	Osawatomie
Education	Emporia	Emporia	Lawrence	Lawrence
	Lawrence	Lawrence	Manhattan	Manhattan
	Manhattan	Manhattan		
Manufacturing	Augusta	Augusta	Augusta	Atchison
	Cherryvale	Cherryvale	Cherryvale	Augusta
	Coffeyville	Coffeyville	El Dorado	Cherryvale
	El Dorado	El Dorado	Neodesha	Coffeyville
	Humboldt	Humboldt		El Dorado
	Hutchinson	Iola		Fort Scott
	Iola	Kansas City		Iola
	Kansas City	Lyons		Kansas City
	Leavenworth	Neodesha		Marysville
	Lyons			Neodesha
	Neodesha			Winfield
Mining	Baxter Springs	Baxter Springs	Baxter Springs	Baxter Springs
	Frontenac	Galena	Galena	Galena
	Galena	Pittsburg	Pittsburg	Pittsburg
	Mulberry			
	Pittsburg			
Railroad	Arkansas City	Arkansas City	Arkansas City	Arkansas City
	Chanute	Chanute	Belleville	Belleville
	Dodge City	Emporia	Chanute	Ellis
	Emporia	Fort Scott	Emporia	Emporia
	Fort Scott	Goodland	Fort Scott	Fort Scott
	Goodland	Herington	Goodland	Goodland
	Herington	Hoisington	Herington	Herington
	Horton	Horton	Hoisington	Hoisington
	Kansas City	Kansas City	Horton	Kansas City
	Marysville	Marysville	Kansas City	Marysville
	Newton	Newton	Marysville	Newton
	Osawatomie	Osawatomie	Newton	Osawatomie
	Parsons	Parsons	Osawatomie	Parsons
	Pittsburg	Pittsburg	Parsons	Pittsburg
	Pratt	Pratt	Pittsburg	Pratt
	Topeka	Topeka	Pratt	Topeka
	Wellington	Wellington	Topeka	Wellington
			Wellington	
Wholesaling	No data	Atchison	Wichita	Atchison
		Concordia		Marysville
		Hutchinson		Salina
		Salina		Wichita
		Topeka		
		Wichita		

phenomenon had occurred in the state. In each instance, predictably, the non-county-seat community had a specialized function. Coal was king at Pittsburg (Crawford County), of course, while railroad work was the emphasis at Arkansas City (Cowley County), Osawatomie (Miami County), and Parsons (Labette County). Railroad towns, in fact, nearly tripled their numbers between 1880 to 1890, with over half of them associated with the rapidly expanding Santa Fe system.

For the larger urban centers in the state, much of the growth can be associated with specializations. Atchison, Kansas City (plus adjacent Argentine), and Leavenworth were unique at this time in that each had two economic concentrations. Atchison had added a manufacturing component to its older emphasis on wholesaling, while the recently created Kansas City (a merger of Wyandotte and Armstrong with Armourdale, Kansas City, and Riverview) featured both railroad shops and manufacturing. Leavenworth's combination was different again. Entrepreneurs there opened a series of deep-shaft coal mines to provide local fuel for their ongoing manufacturing activities. Elsewhere in northeastern Kansas, the presence in Topeka of the main Santa Fe shops and offices, in combination with the varied service functions associated with the state capital, went far to explain that city's growth. Lawrence, where the business community financed a dam across the Kansas River to furnish waterpower for industry, added a significant manufacturing component to its older but smaller role as a university community. Of the larger cities in southern Kansas, Emporia, Fort Scott, and Wichita all had developed diverse economies. Hutchinson was unusual in having a manufacturing specialization (salt primarily) despite its rather isolated position at the periphery of the state's urban system.

Since many of the urban patterns established by 1890 held true throughout the entire period until 1950, a decade-by-decade tracing of the details of population change and economic specialization would be an exercise in tedium. Changes in emphasis and overall competitive success can be summarized more efficiently by increasing the map interval to twenty years. Doing so, and looking first at the scene in 1910, one sees the inception of a series of urban communities west of Hutchinson. None of these were large, but they symbolized a maturing of what had been almost totally a rural economy in that region (Map 18; see p. 31). Two of the eight towns in this group, Dodge City and Pratt, met the criterion to be called railroad towns. Slightly farther east, the enigma of Wichita's remarkable growth has continued. That community, which had increased its size advantage over rival Hutchinson from some 15,000 people to more than 36,000 during the previous two decades, still laid no claims to any of the economic specializations listed in Table 1.

Despite the appearance of a new railroad town at Horton to add diversity to northeastern Kansas, the overall size structure of cities in that region remained stable between 1890 and 1910. Functional roles did begin to change, however. Data on the wholesaling industry are unavailable to assess Atchison's continuation in that market, but both that town and Lawrence no longer qualified as manufacturing specialty centers by the end of this period. Diversification in both communities explains part of the loss, but some also was a result of many industrialists deciding to concentrate their efforts in fewer and bigger locations as railroad efficiency began to reduce transportation costs relative to other expenses. In the central plains region, that choice site often was Kansas City, which managed to increase its percentage of manufacturing employees from eleven to fourteen while at the same time growing overall from 38,316 residents to 82,331.

The events in Kansas City notwithstanding, manufacturing activity for the state as a whole actually tended toward dispersal, not concentration, at the turn of the century. A total of nine towns qualified for this specialization in 1910, three more than in 1890. Kansas City and Leavenworth maintained their positions from the earlier count, joined by seven new places clustered together in four adjacent counties of the southeast (Table 1). These seven represented the first oil and natural-gas boom, of course, and extended from Iola in Allen County south to Caney and Coffeyville near the Oklahoma border. Among other developments, the exploitation of these cheap and abundant fuels produced four industrial towns within the confines of previously rural Montgomery County.

Changes within the Kansas urban system between 1910 and 1930 were subtle, with nothing to match the reconfiguration of economic roles that had accompanied natural-gas discoveries during the previous period (cf. Map 18 and Map 19 [see p. 32]). With the notable exception of wholesaling, all specialty categories contain similar numbers of cities in both years. This wholesaling sector, for which no census data are available before 1930, is interesting from several perspectives. It reveals that Atchison people had been able to sustain major warehouse activity even seventy years after the demise of the town's original wagon routes. The numbers also lend credibility to the mercantile theories about city growth discussed in chapter 1. With the exception of Kansas City, the four largest cities in the state all reported significant jobber employments. Most notable, this specialty category is the first one in which Wichita appears. Ignoring the influence of St. Joseph and Kansas City, Missouri, for the moment, the map suggests that Wichita and Topeka may divide much of the wholesaling business of Kansas along an east-west axis, with Atchison, Concordia, Hutchinson, and Salina controlling smaller trade areas west from their respective locations.

The most notable functional changes in urban northeastern Kansas during the 1910–1930 period occurred in Leavenworth, where losses were apparent in both manufacturing and absolute population, and in Lawrence, which for the first time met the qualification to be called a college town. These longtime rival communities, where visions of preeminence in trade and manufacturing once seemed close to attainment, were in the process of becoming similar again in an unanticipated manner. Both were largely institutional towns by this date, Lawrence with its university enrollment rising and Leavenworth with sizable employments at the state prison, a federal penitentiary, Fort Leavenworth, and a Veterans Administration hospital beginning to overshadow other activities.

Elsewhere in the state, successful searches for petroleum produced new manufacturing centers based on oil refineries at Augusta and El Dorado in Butler County and contributed to the emergence of Great Bend, Hays, and Liberal farther west. The longtime salt-mining community of Lyons, also aided by oil field money, reached the urban threshold population of twenty-five hundred at this time as well. Finally, Dodge City tripled in size. Its growth, primarily in trade, was enough to lessen an earlier economic dependence on railroad employment and put the community in a good position to continue influence over a large hinterland.

Moving forward again another twenty years, to 1950, one finds the geography of urban Kansas continuing to hold steady (Map 20; see p. 33). Slow, relatively even growth is the rule, a pattern that actually stretches all the way back to 1900. Such quiescence led to a tone of inevitability in contemporary writings about the state's structural form and economic activities. The sizes of specific cities and their various industrial roles became things almost taken for granted.

A close examination of the state in 1950 reveals much evidence to support the common assumptions of the time but also several items to challenge them. Nine new communities had grown to at least twenty-five hundred residents during the previous two decades, and two others dropped below that figure. The additions were widespread across the state and include only one community (Russell) that grew to more than four thousand people at this time. The deletions from the map are more interesting because they portend general structural changes that had begun to work their way into the economy. One was Humboldt, in the old gas belt, which had managed to retain its signal industry, the Monarch Cement Company, but lost other manufacturing plants as the town's fuel supplies faltered. The second was Horton, once home to the second-largest repair shops on the entire Rock Island railroad. As reliable diesel locomotives began to replace steam engines in the late 1930s, however, shop employment there began to dwindle, and with it, the local economic base.

If observers of the time chose to ignore the messages of Humboldt and Horton either because of wishful thinking or for other reasons, they could take solace in the fact that the overall pattern of specialized cities in 1950 was remarkably similar to what had come before (Table 1). The railroad list showed two new towns (Belleville and Ellis) to balance the loss of two others. Strong manufacturing remained in both southeastern Kansas and Kansas City, while among the institutional communities, only Emporia temporarily lost its status because of a decrease in enrollment. Wholesaling data, in contrast, suggested that this business had become somewhat more concentrated over the years, with Concordia, Hutchinson, and Topeka dropping from the list. Marysville, the seat of Marshall County, garnered the unexpected distinction of being the only community in the state to qualify for three different economic specializations: manufacturing, railroads, and wholesaling. Wichita, with 168,279 people, had surpassed Kansas City to become the state's largest city sometime in the previous decade, but the rest of the urban hierarchy remained largely as it was.

Meatpacking, Milling, and the Rise of Kansas City, Kansas

The dynamics of the Kansas urban system are best understood when summary analysis is combined with close inspection of individual places and particular economies. The obvious spot to begin a geographic survey is Kansas City, the state's largest urban agglomeration throughout most of the 1870–1950 period, and home to its greatest concentration of three salient plains industries: railroads, meatpacking, and milling.

In the past, a person could infer much about the heritage of a particular city by looking at the names given to its professional sports franchises. Although this thinking does not apply to either of Kansas City's current major teams, the Chiefs and the Royals, it does to local entrants in the American Basketball Association during the 1970s and in the Northern Baseball League beginning in 2003. Those names, the Steers and the T-Bones, respectively, pay homage to a venerable business whose presence also helped to ensure that many other economic ventures could develop. Simply put, "nothing was so important to Kansas City's industrial growth before 1900 as meat packing."[9]

As I have discussed in earlier chapters, the roots of a large-scale packing industry in Kansas City are intertwined with those of railroads. The initiating event was the entrepreneurship of Joseph G. McCoy, who established a shipping point for Texas cattle along the Union Pacific tracks at Abilene. When McCoy sent his first carloads of cattle into Kansas City in 1868, railroad officials were caught unprepared. They lacked even the feeding pens required to hold animals temporarily before transshipment on east. Potential for profit quickly generated a proactive business plan, however. Why should live animals continue to be shipped to the East Coast for slaughtering? With Kansas City's growing network of railroads to the south and west, conditions would now seem ideal to make it the nation's primary destination for beef cattle. Animals slaughtered at this site, close to the pastures, would be heavier than those that had to endure longer train rides. Moreover, packers could save money by shipping only meat instead of hooves and horns as well.

The capital required to implement a cattle market and a packing industry on a national scale was beyond the capacities of the several small slaughterhouses then present in the city and surrounding region. Money came instead from two outside sources in what seems to have been a coordinated effort. First was a decision by Philip Armour, a prosperous young Chicago packer, to enter the Kansas City market. He did so in 1870 via a firm called Plankinton and Armour, headed by his brother Simeon and an older business partner, John Plankinton from Milwaukee. This company slaughtered thirteen thousand cattle and fifteen thousand hogs in its first year of operation. These numbers overwhelmed anything done previously in the city, and constituted 61 and 42 percent, respectively, of the total local output. The owners judged the results promising enough that they constructed a new and much larger plant during the summer of 1871.[10]

The new investment by the Armour family led directly to the creation of a second business—a stockyard—because packers on their scale obviously would require an organized market from which to purchase animals. James Joy was the primary financier here, together with fellow officials from the Hannibal and St. Joseph Railroad (which had recently made a connection into Kansas City) and its parent company, the Chicago, Burlington and Quincy. In 1870, L. V. Morse, the superintendent of the Hannibal line, purchased five acres of bottomland along the Kansas River between Wyandotte, Kansas, and Kansas City, Missouri. There, astride the state line and adjacent to a recently completed railroad bridge that connected the main lines of the Kansas Pacific and the Missouri Pacific, the new Kansas Stock Yards Company erected eleven stock pens and a brokers' exchange building. Success was immediate. By 1876, the company's acreage had been expanded to fifty-five, the presidency had passed to financier Charles Francis Adams Jr. of Boston, and the name had become the Kansas City Stock Yards Company. Seventeen years later the group's pens would sprawl across 207 acres and have a capacity of 70,000 cattle, 40,000 hogs, and 45,000 sheep.[11]

The presence of major outside investments in the Kansas City livestock industry created an atmosphere of dynamic growth. This mood was heightened further in 1875 when Santa Fe officials, who controlled the largest share of cattle shipments at the time, decided to extend their tracks from Topeka into the city. Still, the process of business expansion was not all straightforward during that first decade. Impacts from a national financial panic in 1873 lingered for several seasons, for example, as did those from a grasshopper plague in 1874. Most serious, because it was an ongoing concern, was a quirk in the buying habits of American consumers. Although these people readily accepted pork in various cured forms, they much preferred their beef to be fresh. In a time before refrigeration, this meant that most cattle had to be slaughtered near the nation's population centers. Kansas City operators could produce salt

Kansas City Stock Yards Company, 1934. The vastness of the holding pens in this photograph is overshadowed by the cattle's pitiful condition. As part of a drought-relief effort, federal buyers paid an average of $14.44 per head for these animals. (Courtesy of Kansas State Historical Society)

beef and canned beef at a price eastern packers could not match, but markets for such products were small, limited primarily to workers in isolated settings: sailors, lumbermen, and miners. Pork, rather than beef, therefore produced the greater share of profits for local packers in the 1870s.[12]

The successful introduction of refrigerated railroad cars early in the 1880s ended earlier frustrations. Shipping fresh beef long distances soon became a reality, and with this technological change, the packing industry truly boomed in Kansas City. Local cattle slaughters went from 31,000 in 1880 to 677,000 in 1892, an increase of more than 2000 percent, and a series of big packing companies came to town looking for sites near the Armour plant, the stockyard, and the railroads. By 1890 or so, the city had firmly established itself behind Chicago as the number two packing center in the world. Just as significant, packing was now the largest industry in the local metropolitan area, providing jobs to more than 4,600 workers and thereby the financial support for an estimated 35,000 residents.[13]

The march of the packers into Kansas City began in 1880 when the Fowler Brothers of Chicago relocated one of their plants from Atchison and Jacob Dold and Sons from Buffalo, New York, purchased and expanded an existing local facility. Three more firms came in 1884, including Morris, Butt, and Company of St. Louis; another in 1885; and then, in 1887, a second giant. This was

Armour Packing Company in Kansas City, circa 1939. When erected on James Street just north of the stockyard in 1892, this building and two companion structures constituted the largest beef processing plant in the world. The company ceased local operations in 1967 as part of a massive relocation of the packing industry. (Courtesy Kansas State Historical Society)

Swift and Company, a Chicago firm like Armour, which proceeded to erect the first local facility dedicated exclusively to the production of dressed beef. The Armour brothers responded to these challenges in 1892 with a new specialized plant of their own. At $1 million, it cost twice the amount of even the mammoth Swift enterprise and soon employed four thousand men and women. By 1900, after the Cudahy Company of Omaha had erected a large new plant and Schwarzchild and Sulzberger had purchased the old Morris, Butt operation (later reorganized as Wilson and Company), Kansas City joined Chicago as the only city to house the "Big Four" of the packing business: Armour, Cudahy, Swift, and Wilson.[14]

As the various packers came to Kansas City, they all built plants within sight of one another. Like the stockyard and the railroad people before them, they selected the bottomlands of the Kansas River. Moreover, with the exception of the Dold plant, they all ended up on the Kansas side of the state line, most within a new industrial community called Armourdale that a group of Boston investors had platted in 1880 directly south of the bluff towns of Wyandotte,

Armstrong, and Riverview. Proximity to railroads and the stockyard explain much of this locational thinking, but at least two other issues also were relevant. The first, freely admitted at the time, was waste disposal. As the writer of an early county history put it directly, "What drew the packing men here was the Kansas and Missouri Rivers, into whose immense bosoms they could unload their offal with impunity."[15]

The second geographic issue for the plants, rarely stated directly but obvious as a subtext, was a desire of the business community in Kansas City, Missouri, to separate itself physically from the noise and smell necessarily associated with this industrial process. Such an attitude, while justifiable on its own terms, carried with it more than a hint of social prejudice. This was because the unpleasant side of the industry was seen to permeate beyond the plants and stock pens to encompass workers who toiled there and who might bring the offensive smells home every night. Then, as demands for labor at the plants quickly exceeded local supplies and companies recruited new men from Germany, Ireland, and then Croatia, Italy, and Slovenia, ethnic biases came into play as well.

Although social and ethnic differences across the state border grew more obvious every day during the 1880s and 1890s, hardly anybody voiced a complaint. Entrepreneurs on the Missouri side were becoming wealthy enough that city leaders undertook construction of an extensive system of parks and parkways. They gratefully named one of the latter Armour Boulevard. Kansans, whether longtime residents or new immigrants, were equally elated at having more high-paying jobs than they could have imagined, even with the possibilities of employment with the Union Pacific, Santa Fe, and other major railroad companies. When the new Armour plant opened in 1892, both communities wholeheartedly agreed with the assessment of a news reporter who called this "probably the greatest event in the history of Kansas City."[16]

To better coordinate area growth, local Kansans decided in 1886 to merge Wyandotte with Armourdale and three other adjoining communities to create a large, new unit called Kansas City, Kansas. This would not have been the wisest choice of names had identity separate from their older Missouri counterpart been a primary concern, but again, the focus was on management of an existing boom rather than image manipulation. Local efforts reached a pinnacle in 1900 when the federal census reported that their blue-collar city now ranked as the seventeenth most important manufacturing center in the nation. Not surprisingly, this ranking quickly became a mainstay of promotional brochures, along with praise that such glory had been achieved by a population (51,418) considerably smaller than that of other competitors.[17]

Looking at the components of Kansas City's exalted industrial status for 1900, one finds a listing of 492 manufacturing establishments that employed a total of 10,544 men and women. Packing accounted for an amazing 7,664 of these jobs, or 73 percent of the total. Ten years later, when packing employment peaked locally at 10,556, an almost unbelievable 90 percent of the total value of products for the city could be attributed to these Armourdale plants. Equally amazing, they also constituted 50 percent of the value of the entire state's manufacturing.[18]

The presence of large-scale slaughterhouses at the turn of the century typically implied a constellation of related industries that would make use of animal by-products. Chief among these were companies manufacturing fertilizers, glues, leather goods, and various products from grease and tallow, including soap. Tanning and leather works never developed in Kansas City for some reason, but the others all found local conditions nearly ideal for their activities. The packing plants themselves controlled most of the production of fertilizer, glue, and grease, but the census of 1900 also reported two independent fertilizer operations in the city, one lard refinery, and two makers of oleomargarine. All of these companies were small. Soap production, in contrast, was both independent of the packers and national in scope. One small plant existed as early as 1878, and three were in operation by 1890, employing a total of twenty-eight people. One of these three, Peet Brothers, then began an expansion that attracted the attention of the established Cincinnati firm Procter and Gamble. Soon, in 1903, this latter company built a large Armourdale plant of its own, and Peet Brothers matched the new scale of operations with similar construction in 1910. Nine years later Peet Brothers (soon to be purchased by the Colgate-Palmolive Company) employed 561 workers and Procter and Gamble 680. These two rival plants, which together produced a quarter of the nation's soap in 1910, remain at their familiar sites to this day.[19]

The industrial juggernaut on the Kansas River bottoms that emerged in the 1880s had several components beyond packers and soap makers. The Union Pacific Railroad, which had initiated the process in 1870, employed nearly nine hundred men by 1920 at its repair shops just to the north of the Armourdale district. Similar, but smaller, facilities for the Missouri Pacific and Rock Island stood nearby. Across the Kansas River, two additional bottomlands provided good sites for similar activities. One of these, called Rosedale, lay due south from Armourdale in the valley of Turkey Creek and served as the pathway into the city for the tracks shared by the Frisco and the MKT lines. The second site, Argentine, lay adjacent to the Kaw itself upstream from Armourdale and played a similar role for the Santa Fe railroad. Beginning in 1881, this company also established a network of repair shops there more extensive than even the nearby Union Pacific facilities. More than thirteen hundred machinists and other people worked at these in 1920.[20]

Rosedale was platted by officials of the Missouri River, Fort Scott and Gulf Railroad in 1872, four years after that company had first built through the Turkey Creek Valley. Bluffs isolated this lowland site from other Kansas communities, but the valley led directly into Kansas City, Missouri. A bucolic atmosphere abruptly gave way to an industrial one in 1875 when A. B. Stone, president of a large rolling mill in Decatur, Illinois, decided to relocate closer to his market. Stone, who manufactured iron rails for various railroad companies, liked the centrality of Kansas City and hoped to find coal deposits nearby that could fire his eleven furnaces. When this coal proved to be a reality (a twenty-inch seam at 335 feet just across the tracks), Rosedale became a boomtown; five hundred employees worked at the Kansas Rolling Mills in 1881. Although Stone's factory fit in well with the overall pattern of industrialization in the area, it was short-lived. A change by railroads from iron to steel tracks forced both the mill and the mine to close in 1883, and the Rosedale community struggled for several years.[21]

Argentine, the second Wyandotte County town site south of the Kansas River, started slightly later than Rosedale but enjoyed a more stable economy. The area first acquired an urban identity in 1875 when Santa Fe officials purchased 128 acres on which to build terminal facilities for their new line into Kansas City. Its name and official platting came six years later, however, after William N. Ewing, a man who had worked for the railroad as superintendent of the company's coal mines at Osage City, decided to erect a large smelter to process silver and gold ores from Colorado and lead and zinc ones from Missouri and Kansas. Ewing liked the Kansas City area because of labor availability and its good rail access to mines, markets, and a coal supply in southeastern Kansas. With backing from Kersey Coates and other businessmen from Kansas City, Missouri, he opened his facility in 1880 and named the surrounding community after the Latin word for silver.[22]

In 1881, Ewing's company passed into the control of August R. Meyer, who developed it into a large and profitable enterprise called the Consolidated Kansas City Smelting and Refining Company. Three thousand people worked there during the 1890s (more than half of Argentine's total population), and most of them made good wages. At its peak year, in 1898, the company produced 8 percent of the nation's gold, 12 percent of its silver, and 20 percent of its lead. Lead processing also produced toxic fumes that killed vegetation downwind from the plant as well as numerous dogs and cats, but this hazard was an accepted aspect of Argentine life. When the smelter closed in 1901, it was because of rising shipping rates, not environmental risk.[23]

Unemployment affected Argentine less severely than Rosedale. Partly this was because Santa Fe officials progressively enlarged their local terminal facilities and repair shops. Five hundred men worked at these facilities by 1890. Argentine also benefited indirectly by the inability of Rosedale's ironworks to adapt to new technology and by having skilled workers available after the failure of the smelter. Realizing that steel beams were replacing iron ones in most American construction, two Minnesota men, Howard A. Fitch and Olaf C.

Smith, found in Argentine a combination of labor and shipping facilities that might support such a business. From fifteen men in 1907, when they first opened, their Kansas City Structural Steel Company grew into the largest metal fabricating plant west of Pittsburgh. More than five hundred people worked there in 1919, six hundred in 1929, and the list of projects gradually came to include almost all the large buildings and bridges in greater Kansas City plus many others as far away as California. This company, like the railroad yards, remains a strong, continuing presence in the Argentine bottoms.[24]

The concentration of blue-collar industry in Argentine, Kansas City, and Rosedale was eventually gathered together under a single governmental umbrella through the larger city's annexation of the Argentine community in 1909 and of Rosedale in 1922. Both Kansas City proper and Rosedale then became sites for the last phase of the area's early industrialization, the flour milling business. That these places would become a focus for the storage and processing of wheat is not surprising, of course, since the nation's largest growing area for this crop lay conveniently west and southwest of the urban area and within the webs of the Santa Fe, Union Pacific, and other city-oriented rail systems. What is mysterious is that large-scale milling did not begin in Kansas City until a full half century after this region had been opened for settlement in 1854.

A near absence of specialized wheat mills in early Kansas is simply explained: farmers originally did not concentrate on this crop. Spring-planted varieties did not survive well in the local summer heat, and fall-planted ones often died during the winter or became diseased. Minnesota, where spring wheat grew well, was the nation's milling center along with Buffalo, New York, close to the big northeastern markets. The beginnings of a change occurred with the introduction of flintier and more cold-resistant varieties of fall-planted wheat in the 1880s. These grew well throughout Kansas and the southern plains and also had higher protein contents than the relatively soft wheats of Minnesota. Millers did not like hard varieties, however, because they did not work well with traditional grinding stones. Housewives also were leery of the new flour even though the higher protein content ensured better breads. They had become used to the older products and accepted the advertising mantra that Minnesota flour was "best."[25]

The processing problem was resolved by the development of roller-mill technology later in the 1880s. Changes in public relations took longer, and for the remainder of the nineteenth century, sacks of southwestern, hard-wheat flours continued to be discounted seventy-five cents or more below northern varieties. Because of this price differential, Kansas City was not a profitable place for milling until about 1900. This delay was beneficial in one way, however, because it enabled local entrepreneurs to utilize a newer technology of concrete and steel and therefore to build their initial structures much larger than the older wooden models. The Bulte Mills Company erected the first of the new-generation facilities in Kansas City in 1904, one with a capacity of 1,800 barrels of flour per day. Four more were in place by 1917, including a 2,500-barrel plant by the Ismert-Hincke Company of St. Louis that employed 153 people and an even larger one by the Southwest Milling Company that required 304 workers.[26]

The state of Kansas achieved the number one ranking in flour production for the first time in 1932. Five years later, a reporter in a business magazine judged that this industry now deserved to be joined "with meat-packing and petroleum production to form one of Kansas' BIG THREE . . . industries whose income provides the largest amount of revenue." Kansas City itself (including mills on both sides of the state line) ranked as the second-largest milling center in the nation for most years during this period, followed closely by three other regional entries: Hutchinson, Salina, and Wichita. The volume was high and the employment substantial, though still far below that of packinghouses. The city's mills peaked at 1,099 workers in 1939 before increased consolidation and mechanization caused a slow decline.[27]

A promotional booklet published in the 1930s made claims for Kansas City that echoed almost exactly numbers from a generation before. This was still the number two meatpacking city in the country, the third-largest producer of soap and flour, and home to the largest structural steel plant west of the Mississippi River. Its residents also claimed three oil refineries, fifteen terminal grain elevators, and thirteen trunk-line railroads. Despite these facts, however, the outlook for the city was not entirely positive. Its industrial ranking, now number twenty-four in the nation, had fallen from its apogee and showed evidence of farther declines still. The various plants, which had epitomized modernity in 1880 or 1900, now were becoming outmoded. Moreover, although rail access remained exceptional, both Armourdale and Argentine lacked good connections to major highway systems for the truck traffic that was becoming increasingly important to business performance. Local packers, in particular, had begun to cut back operations.[28]

The economic future that faced the industrial community of Kansas City in the 1930s was potentially grim. Fortunately, however, sage planning and simple luck combined to transform the manufacturing base. The process was so seamless and successful, in fact, that the average blue-collar worker experienced only a small shift in job location and a different type of assembly line. The savior was a new industrial district called Fairfax, located on a large lowland tract adjacent to the Missouri River. This lowland, known originally as the North Bottoms, abutted the northeastern corner of the downtown business district but had remained undeveloped before the 1920s because no levees protected it. Once these structures were in place, however, a local man named Guy E. Stanley promoted the 1,280-acre site as a regional terminal for food wholesalers. He induced officials of the Union Pacific Railroad to spend a million dollars on development, including the construction of numerous railroad spurs and an industrial airport. Seemingly all was well, but people from Kansas City, Missouri, then blocked the project in court, claiming that a distribution center at that location would hurt their city market. Resolution of this legal problem took two years. By then, World War II was upon the country, and both the railroad and the city wisely reconsidered their potential clientele. Fairfax, as Stanley called the area, now was opened to manufacturing plants of all types. As such, its timing was perfect—one of the nation's first modern industrial parks built just ahead of the general realization that this was to be the model for future business expansion.[29]

Because Fairfax was developed with industry in mind, Stanley and his associates gave high priority to transportation needs. Access to barge docks and to railroad lines was easily accomplished, including trackage laid parallel to major roadways and directly to each manufacturing site. Construction of a business airport required somewhat more money, but highways proved to be the most expensive infrastructure component by far. Two main routes emerged: Fairfax Trafficway, a new construction that ran directly north from downtown through the center of Fairfax itself, and Seventh Street Trafficway, a parallel road some twelve blocks to its west. This latter project was extremely ambitious. It began in Rosedale, at the southern edge of the metropolitan district, where the route connected that community with the rest of the city via a bridge over the Kansas River and a series of viaducts above the maze of Armourdale industries and railroad tracks. Workers then widened an existing street through downtown Kansas City and extended construction north through Fairfax. The road then crossed the Missouri River to make connections with existing highways in Platte County, Missouri. The Seventh Street project took twelve years to complete, becoming fully operational in 1935 with the opening of the Fairfax Bridge over the Missouri.

Industrial leaders loved the concept of a district designed especially for them, a site isolated from residential and shopping areas and the accompanying calls for less noise or fewer trucks. Because Fairfax technically lay outside the city limits of Kansas City, savings in school and property taxes also were substantial. Wide streets, maple plantings, and a fairly uniform practice of using buff brick for construction gave a decided air of dignity to the development. With this came a renewed sense of pride for local residents who had begun to hear derisive comments from outsiders about the grime and outdated facilities associated with their older plants. The Chrysler Corporation installed a 210,000-square-foot parts depot in the new park to serve the entire Midwest, the Sunshine Biscuit Company decided to make Fairfax one of its four main bakery sites for the nation, and General Motors took over a hundred-acre site that had manufactured B-25 bombers during the war and converted it into an assembly line for Buick, Oldsmobile, and Pontiac automobiles. By 1948 the district had been expanded to two thousand acres. Forty major industries were present, and more than twenty thousand people worked in state-of-the-art facilities. An annual payroll of $60 million made Kansas City the subject of many complimentary news stories and restored local faith in the power of industry to sustain a modern economy. The local population, which had stagnated in the 1930s, grew again to a total of 129,553 in 1950.[30]

Life in the Shadows: Leavenworth and Atchison

The economic ascendancy of Kansas City over Leavenworth is often symbolized by the opening of the Hannibal Bridge across the Missouri River in 1869. Whether or not Leavenworth people were immediately aware of the tremendous commercial opportunities this structure implied for Kansas City remains uncertain. Two years later, however, one of Leavenworth's own enterprises provided clear evidence that the town's mercantile aspirations were no longer on a national scale. This was city sponsorship of the Kansas Central Railroad west from the community. A new railroad ordinarily would connote lofty ambitions, but the Kansas Central was different. By using narrow-gauge tracks, its owners demonstrated to all observers that they were more interested in saving money than in being able to make interchanges with the Union Pacific and other main lines.[31]

With a public admission through the railway that their claim to be the plains metropolis was no longer sustainable, Leavenworth businessmen began to plot strategy for a different kind of city. Their settlement remained the largest in Kansas, but according to two men who were part of the process, "we had our eyes painfully opened to the unwelcome fact that we had built [sic] without a foundation; . . . that we were a city without employment." Their solution, apparently easily made, was a turn to manufacturing. It seemed a perfect fit. The plains region was growing rapidly in the 1870s, yet had no factories of note and certainly no city that specialized in these endeavors. "Here," the Leavenworth men reasoned, "was a field of industry that was not occupied—a field that contained a mine of untold wealth."[32]

Manufacturing was not an entirely new idea for Leavenworth entrepreneurs. It had been overshadowed temporarily only by even greater allures in land speculation and maneuverings for transcontinental railroads. The first obvious move toward industrialization, in fact, took place as early as 1861 when city leaders instructed their legislators to acquire the state penitentiary. As I discussed in chapter 3, the accepted model for prison operation at that time was to have inmates work at various commercial endeavors. The institution in this way could be self-supporting at a minimum. Ideally, however, prison enterprises would make considerable profits for both the state and the companies that sponsored the selected factories.

Inmates at the prison site five miles south of Leavenworth were kept busy throughout the 1860s quarrying stone to construct a massive, castlelike building in which they would live. By 1871, however, with this structure nearly complete and 397 convicts on hand, the institution accepted the first in a series of sealed bids for the labor of about half of these men. Wages averaged forty-five to fifty cents per day, considerably lower than rates outside the walls, and the state agreed to furnish needed buildings, machinery, and utilities. One of the first successful bidders, not surprisingly, was the Kansas Wagon Manufacturing Company, which catered to both the military-freighting and farm trades. Its owner, Alexander Caldwell, had been elected to the United States Senate in 1871. With the advantage of prison labor, Caldwell soon became the leading industrialist in the city. From a start in 1873, he expanded to 233 employees by 1880 and turned out seven thousand wagons per year. This figure rose as high as ten thousand by 1884. Other early prison contractors included shoemakers Kellogg and Burr (fifty men) and furniture manufacturer John Sorrenson (fifty men).[33]

The decision to expand the city's factories beyond the prison walls was a good one for Leavenworth people in several respects. The local railroad system, although clearly inferior to that of Kansas City, nevertheless enjoyed easy national connections via the Missouri Pacific, Union Pacific, and several other lines and was therefore more than adequate for most industrial purposes (Maps 13, 14). In addition, the Missouri River still flowed at the foot of Delaware Street and could provide "the means of cheap carriage for iron, lumber, and other raw materials used in manufacturing." National conditions also favored industrialization at sites near the frontier such as Leavenworth. Concentration within the country's various manufacturing sectors was still rudimentary at this time, and hand labor was common. Small shops were therefore not at the disadvantage from economies of scale that they would be a few decades later. Freight rates were relatively high as well, which provided considerable protection for local and regional markets.[34]

As Leavenworth investors evaluated the scene about 1869, the only potential flaw they saw in their economic plan was an absence of fuel. This was a deficiency thought to be easily remedied, however, because small amounts of coal had been discovered just northwest of town in the Salt Creek Valley. Deeper shafts were sure to find abundant supplies. Such faith was rewarded in 1870 when a company headed by Lucien Scott, president of the First National Bank, hit a twenty-four-inch seam at 713 feet, another at about 1,000 feet, and an even thicker one at 11,000 feet. By 1880, Scott was employing three hundred miners at his site on the Fort Leavenworth military reservation. These men brought an average of 300 tons of the black mineral to the surface every work-

Abernathy Furniture Company in Leavenworth, circa 1949. In the city since 1856, Abernathy's was once the largest furniture maker in the West. Its facilities at 205 Miami Street employed more than four hundred people in 1939. (Courtesy Kansas State Historical Society)

ing day, and townspeople believed their future was ensured. Prison officials soon decided to sink a shaft of their own within their walls, while local businessmen opened a second mine in 1886 and then a third just south of town at the hamlet of Richardson. Besides increasing the fuel supply greatly, these new shafts also were important for advertising. Since they all encountered the same coal seams as had the original mine, they proved that the field extended over an area at least twelve miles square. Boosters calculated that this meant a minimum of 259 million tons of coal beneath their city, enough to keep a thousand miners busy for 576 years. In other words, all manufacturers should know that "the supply of coal in Leavenworth County is practically inexhaustible."[35]

Leavenworth's industrial strategy was to attract small factories and let them grow as the plains region developed. All went as scheduled throughout the 1870s as some fifty entrepreneurs came to make the city their base. Enough progress had occurred by 1880 that boosters organized themselves as the Leavenworth Board of Trade and began to change their tone from optimistic to extravagant. Two local men wrote, for example, that the city "occupies the same relation to the States west of the Mississippi that Pittsburgh occupies to the Middle States." Another claimed that to discuss Kansas industrialization without making Leavenworth the center of attention "would be like attempting the play of Hamlet with that important character left out."[36]

The reality of Leavenworth manufacturing, although impressive, was considerably less than the claims. Entrepreneurs continued with the wagon-making trade until the turn of the century, and the city became an early flour-milling center of note, but these two activities declined because of technological change and competition from Kansas City, respectively. Local people then shifted their praise to the furniture industry. With three major plants, the city was truly successful in this arena, and claims that it ranked "well up with Grand Rapids [Michigan] as a production center" were not exaggerated. The leader, the Abernathy Company, was held in especially high regard by the community, for it was homegrown and had expanded from an extremely modest stature in 1860 into "the most extensive furniture factory in the West." Its red-brick, multiple-story building housed four hundred workers and large supplies of local cherry, sycamore, and walnut woods. Almost as expansive was the Helmers Manufacturing Company, which emerged from Henry J. Helmers's barber supply business to become a major maker of chairs, tables, and cabinets. Third, and only a little smaller, was the Klemp Furniture Factory.[37]

Two of the remaining three jewels in the Leavenworth industrial economy had overlapping ownership: the Great Western Manufacturing Company, which started as a general foundry and gradually came to specialize in equipment for flour mills, and the Great Western Stove Company. The former operation began in 1858 under the leadership of an ambitious immigrant from New York, Edward P. Willson. He and his partners divided off the stove business in 1875, and each enterprise employed about 150 men in 1880. The third big employer was another foundry, the Missouri River Bridge Company. Like those who worked at its rival in Argentine, the Kansas City Structural Steel Company, the men here specialized in large projects. After expanding from bridges alone into stock tanks, farm bins, power-plant equipment, and river barges, the company changed its name to the Missouri Valley Steel Company and employed between 75 and 125 people depending on demand.[38]

All major Leavenworth industries prospered throughout the early twentieth century, and the total number of plants peaked at around sixty-nine in 1913. Data from 1920 show the three furniture makers with a total of 305 employees and the two Great Western facilities with another 299. Beyond this, a newer (1907) sash and door company called Goodjohn hired 60 workers and the largest mill (the J. C. Lysle Company) another 71. This was a decent assemblage, but Leavenworth's industrial dreams slowly faded as mechanization gradually lowered the number of employees required, transportation improved, and companies in better locations elsewhere grew at faster rates. Reformers forced contract work at the prison to stop in 1909, and a special edition of a newspaper in 1913 was one of the last to tout the city's manufacturing greatness. Then the factories and mines began to shut down. The last local coal sold in 1939, and workers assembled the last piece of furniture in the 1940s, leaving the town with no more industry than an average city of its size. Local residents, seeing this trend, began to seek opportunities elsewhere as well. From a population peak of 20,735 in 1900, the number fell to 15,912 over the next two decades before beginning a slow revival.[39]

With the loss of much of their manufacturing strength, Leavenworth people in the 1940s once again began to assess their role within the state. Gradually they recognized that the local economy had come to be based predominantly on a series of public and private institutions. A reliance on such payrolls or even a deliberate quest for the initial facilities had never been a special goal for city leaders, but several organizations had materialized as adjuncts to past metropolitan aspirations. Now, and especially with a realization during the 1930s that institutional finances were more apt than others to remain intact through bad economic times, these facilities came to be cherished and advertised.

The original institution at Leavenworth, of course, was the military fort. This post seemed likely to be closed in the early 1870s but was saved first by a chance action and then by a lucky personal connection. In 1874, the United States secretary of war decided to locate the nation's military prison on the reservation. Seven years later, General William T. Sherman, who had married into the prominent Ewing family of Leavenworth, established there the army's School of Application for Infantry and Cavalry. In this way, the fort acquired two roles that continued over the ensuing decades with only slight changes. In 1901, when primary instruction in military applications was moved to various other posts, Fort Leavenworth was made the site for more advanced training. The General Service and Staff College thus created endures to the present.[40]

Saint Mary College, a school for young women established by the Sisters of Charity, joined the fort as a Leavenworth institution in 1860. These nuns had come west from Nashville, Tennessee, at the request of Bishop Jean Miege after he had selected the city as headquarters for his new diocese in 1855. Although Saint Mary has never been large, its six to seven hundred students and thirty to forty instructors added significantly to the overall character of the city. The Kansas State Penitentiary, beginning in 1861, became the third institution. A fourth followed in 1885 when several local men led by former governor George T. Anthony lobbied successfully to obtain the western branch facility of the National Military Home, which had been established earlier in Dayton, Ohio. For a cash payment of $50,000 plus a section of land, the town acquired thirteen domiciliaries to house some four thousand old soldiers plus a hospital and other support buildings. These grounds came under the direction of the Veterans Administration in 1930, and this agency added a large, five-hundred-bed hospital to the site in 1933. The fifth and final major institution materialized in 1906. Congress, probably influenced by the presence of the

Railroad Yards and Associated Businesses in Atchison, circa 1960. Although not as big an industrial focus as its promoters had hoped, the city's activity along the Missouri Pacific tracks and at the Pillsbury flour and Manglesdorf seed factories was nonetheless impressive. (Courtesy Kansas State Historical Society)

military prison at the post, decided to withdraw land from the southwestern corner of the Fort Leavenworth reservation and construct there a new high-security civil penitentiary.[41]

It is difficult to assess the total impact of the unusual combination of college, hospital, military post, and prisons found in and near Leavenworth. Because the numbers of soldiers and civilian employees at the post have always equaled at least 25 percent of the total city population, the city meets the criterion used to designate an army town in Table 1. When the faculty and students at Saint Mary College are added in, plus the five hundred or so people who typically worked at the military home during the early twentieth century and the seven hundred at the state and federal prisons, this percentage rises to between 35 and 45, depending on the census year. By this measure, Leavenworth qualifies as an institutional town to a greater degree than even Junction City, Lawrence, or Manhattan.[42]

Whereas the construction of a narrow-gauge railroad had provided a clear signal that metropolitan dreams ended in Leavenworth by 1871, contemporary documents suggest that Atchison residents held on to their high initial ambitions until the next decade. This prolongation might have seemed illogical to outsiders as early as 1866. After all, a decision that year to build the Union Pacific Railroad directly from Kansas City to Denver had, as a by-product, turned Atchison's Central Branch project from a main line into a mere spur. Local citizens, however, remained confident for several years thereafter that Congress would subsidize a conversion of their trackage into a through route across the state. Then, when that hope failed to materialize, they logically turned for salvation to the robust Atchison, Topeka, and Santa Fe Railroad that they had helped to found. A bridge over the Missouri River was the only missing link in their eyes. A connection to Chicago via that bridge, in combination with a preexisting Missouri Pacific line from Wyandotte that provided trackage to St. Louis, would position the city well.[43]

The flaw in the Atchison thinking was hubris, including an underestimation of the physical advantages of Kansas City. Local officials thought they could play investors from St. Louis and Chicago against one another to avoid constructing the critical railroad bridge over the Missouri themselves. They also did not foresee the economic logic that led the Santa Fe directorate in 1873 to reposition the company's main line east from Topeka to focus on Kansas City rather than Atchison. The strategy to tie the city's future to railroads persisted even after the Santa Fe effectively abandoned them. Another hope was eastern industrialist Jay Gould, who, by purchasing the Missouri Pacific and the Central Branch lines in the late 1870s, was poised to construct his own through railroad to Colorado. As I have described in chapter 5, Gould did indeed build such a line, but he chose not to base it in Atchison. This decision finally forced local people to plan for an alternate and less expansive future.[44]

Although in retrospect Atchison people can be chided for holding too long to a metropolitan conceptualization of themselves, this behavior was counterbalanced over the next several decades by a series of clearheaded economic strategies. Even though the community was not the focus for any one major rail system, business leaders came to recognize that they nevertheless could profit from the large number of different tracks that passed their way.[45] The count was eight in 1874, including branches of the Burlington and Missouri River Railroad and the Chicago, Rock Island and Pacific. It rose to a very impressive thirteen companies by 1887. To advertise this fact, the local board of trade proclaimed Atchison the "railroad centre of Kansas" and decided that the town's future lay in the utilization of these tracks by a mixture of wholesaling and manufacturing industries.[46]

The most obvious and secure part of the Atchison strategy was to build on existing strengths. After having outfitted mining camps as far away as California and Idaho in earlier decades, the banking systems, warehouses, and grain mills already were in place to supply trade goods to a network of small towns throughout Kansas and southern Nebraska. As the plains region matured, reasoned the Atchison promoters, "the wholesale trade . . . [not only] cannot fail to be remunerative at once, . . . [but will] soon reach an aggregate of business equal to the mammoth establishments of the East." By 1874 this line of activity was well established. The town claimed six jobbers for liquor, five for groceries (including a large operation by the Challis Brothers), four for lumber, three for hardware, two for drugs (including McPike and Fox, which was said to be the largest firm of its kind in the Missouri Valley), two for furniture, and single entries in several other fields.[47]

Lack of census data makes precise assessments of the scale of Atchison wholesalers impossible, but it is clear that they were unchallenged within the state throughout the nineteenth century and well into the twentieth. A local publication claimed that the town's jobbing trade in 1886 totaled $40 million, an amount, it proudly said, that was greater than that generated by all other cities in the state combined. This number also compared favorably with the $60 million produced by the much larger Kansas City, Missouri. One lumber dealer alone (William Carlisle and Company) was said to have sold $13 million in goods, making it the largest in the country for that year. As the decades passed and competitors emerged, some of the Atchison wholesalers either disappeared or moved to larger cities. Enough remained in the middle twentieth century, however, for the town still to qualify as a wholesale specialist (Table 1). Its leading company, Blish Mize, continued to distribute hardware throughout the Midwest much as it did in 1871, for example, when three brothers-in-law had joined forces to purchase the stock of a still older company. Its employees, numbering around 140 throughout most of that long history, were a steadying influence on the local economy.[48]

Flour milling, another venerable, railroad-based business, formed the starting point for Atchison's presence in regional manufacturing. The possibility for grain shipments on Missouri River barges led various businessmen to construct four large elevators by 1882 (including the largest one west of the Missouri at the time), but the lack of through railroads to the west ultimately limited the grain industry's potential there compared with Kansas City, Salina, and similar cities. Blair Milling, begun in 1868, is a survivor. It was overshadowed after 1941, however, by the creation of the Midwest Solvents Company, one of the nation's largest and least-known distilleries. The company is located in Atchison because its founder, Cloud L. Cray, owned McCormick's Distillery in nearby Weston, Missouri, and wanted to expand into the production of industrial alcohol. A largely wholesale focus has meant a low public profile, but the plant does produce considerable whiskey for resale to brand-name distributors. Some 170 people worked for Cray in 1953.[49]

Atchison people carefully watched the campaign that nearby Leavenworth used to attract manufacturers and tried to imitate it to some degree. No suitable coal deposits could be found near their city, however, and so the local recruiting effort never assumed the scale it did down the river. The community in 1880 was proud to count three local carriage makers, one furniture factory,

Locomotive Finished Material Company in Atchison, 1930. LFM has been the city's largest industrial employer for more than a century. Now called Atchison Casting Corporation, it still specializes in heavy steel products. The photograph was taken to show a new addition to its main shop. (Courtesy Kansas State Historical Society)

and a manufacturer of sashes and doors, but all were small and short-lived. Fowler Brothers, a Chicago meatpacking firm, established a branch plant in town in 1878 but relocated it to Kansas City in 1880. Atchison did attract one long-term success, however. Responding to a city offer of $10,000 for anybody who would establish a local iron factory, John Seaton moved his company from Alton, Illinois, in 1872 and soon renamed it Locomotive Finished Materials. Analogous to the Missouri Valley Steel Company in Leavenworth, Seaton's business prospered partly because of a particular pricing structure that made it cheaper to ship ore and coal than it did finished castings. Seaton wisely tailored his products to the region by specializing in engines and boilers for western railroads. He also made an efficient transition from iron to steel products. The plant, employing up to 750 people, still anchors the town's waterfront and its business community. Only the name has changed, first to a division of Rockwell Manufacturing of Pittsburgh, Pennsylvania, in 1956 and then, in the early 1990s, to Atchison Castings.[50]

Atchison, which had been the third-largest city in the state in 1880 with 15,105 residents, never grew much larger. Its population increased slightly through 1910 but then fell sharply as people moved to the nearby and rapidly growing urban centers of Kansas City, St. Joseph, and Topeka. With 12,792 citizens in 1950, Atchison had long since settled into an unusual, seemingly suspended existence midway between a city and a town. St. Benedict's College and Mount St. Scholastica Academy remained to provide an air of sophistication. So did a collection of ornate Victorian houses north of downtown. The workaday life of the community had become increasingly blue-collar, however, with Locomotive Finished Materials, Midwest Solvents, and smaller, related companies providing an increasing proportion of a shrinking base of jobs.

Topeka: The Capital and a Railroad

Among the larger communities in eastern Kansas, Topeka was seen by its early rivals as living a charmed existence. It was not founded as early as Lawrence, could not match the trade connections of Atchison, and stood far behind Leavenworth in size and economic might. Still, partly through happenstance of location and partly through skilled, behind-the-scenes maneuvering by political leaders, townspeople were able to parlay the hosting of a small, free-state convention in 1855 into the acquisition of the state capital in 1861. Then, because the political center of any state inevitably attracts a constellation of businesses and organizations that want to be close to legislators, Topekans found themselves hosting a rich array of insurance companies, publishers, and banks.

Still, as the experiences of Frankfort, Kentucky, and Jefferson City, Missouri, demonstrate, the presence of the capital by itself does not ensure rapid city growth. Perhaps because Topeka leaders knew about these case studies, but more likely simply because of their initial strategic successes, they pushed for other components of economic development. As I discussed in chapter 3, an offer of a thousand acres plus a two-story building convinced the state's Congregational ministers to open what came to be known as Washburn University in 1866. This college was far overshadowed in the newspapers of the time, however, by Cyrus Holliday's project to duplicate the route of the old Santa Fe Trail with a Topeka-based railroad. Holliday, who had invested in this new mode of transportation in Pennsylvania before coming west, had been working on the idea as early as 1860. That year he organized the first statewide railroad conference at his young community and used the occasion to obtain official endorsement for his planned route. Three years later he made the dream a reality by promising an extension northeast to Atchison in exchange for the assistance of Senator Samuel Pomeroy (an Atchison resident) in securing a federal land grant. In 1871, as the railway began to make money with shipments of Texas cattle and Osage County coal, Topekans saw their star begin to rise to an almost unimagined height.[51]

Although city residents agonized as their railroad officials wrestled with the dilemma of how to maximize cattle dollars and yet still build tracks west fast enough to qualify for the land-grant acreage, a local issue gradually became an even greater concern. Once the grant lands were in hand, the pace of development presumably would slow enough for company officials to decide on permanent locations for their general offices and principal repair shops. Topeka, the birthplace, would seem a natural choice, but nothing existed in writing, and Kansans no longer controlled company ownership. Local leaders were aggressive, however. They campaigned for the passage of a $100,000 bond issue to help the company with its construction of these facilities and arranged for the sale of a choice, three-hundred-acre site near the Kansas River and adjacent to the city's eastern edge. The site's availability at this time, which may have been the deciding factor for Topeka, was pure luck. It recently had been vacated by the King Wrought Iron Bridge Manufactory and Iron Works, a large company similar to those that had found success in Atchison and Leavenworth. Zenas King had moved his foundry a few years earlier from Cleveland, Ohio, to Iola, Kansas, but a better financial offer from Topeka had brought the company north in 1872. No more had it started at the new site, however, before mismanagement and the national financial panic of 1873 created bankruptcy. The Santa Fe stepped into the void. As railroad office workers moved into the city in increasing numbers, and especially after the company's lavish new shops opened in 1878 to create a need for eight hundred craftsmen, Topeka experienced a long period of rapid growth. From a population of only 5,790 in 1870, the community grew to 15,452 in 1880 and then doubled to 31,007 by 1890. At the end of this spurt it had become the second-largest urban center in the state (Chart 2).[52]

Topeka's strategy for development during the 1870s focused on providing large subsidies to attract major industry. The King Bridge Works, described by a contemporary observer as "a modern Babylon," would reportedly have given the city "the largest manufactory establishment west of St. Louis." The failure of something so grandiose might have soured boosters on similar ventures had it not been paired with the nearly simultaneous success of their Santa Fe bond issue. In this latter instance, $100,000 had been enough to make railroad officials spurn an offer twice that size from Emporia and bring over a thousand new jobs to the city. Hoping to hit railroad gold a second time, Topekans leveraged themselves financially again the next year. This time the amount was $150,000, and the objective a manufacturer of steel tracks for western railroads. When the Topeka Iron and Steel Company accepted the money and opened a large facility just north of the Kansas River in April 1874, all looked well. Its main customer, the Northern Pacific Railroad, went into receivership almost immediately, however, and the mill operated only sporadically for seven years before burning in 1881.[53]

As local residents watched the tortured adventures of Topeka Iron and Steel during the late 1870s, they became discouraged from further dalliance with big-time, heavy industries that were dependent on the vagaries of a national market. Thereafter, they adopted a more conservative philosophy. First, they would support businesses connected with the city's two major assets: the capital and the Santa Fe railroad. Then, as supplements, they envisioned a series of smaller, locally owned enterprises that could grow with the state. This decision, unusual when compared with aggressive actions in rival cities, turned out to serve Topeka well. By not spending money on brochures touting either their coal reserves (as did Leavenworth) or their railroad nexus (as did Atchison), they unwittingly saved themselves from a gradual loss of many temporarily acquired companies to the even better situated Kansas City. This air of calmness prevailed even in the early 1880s when owners of the Santa Fe railroad recognized a need to run their main line directly into Kansas City from Emporia rather than relying exclusively on the older, indirect path through Topeka and Lawrence. Topeka had established an economic niche that was producing solid, if not spectacular, growth.[54]

Because of the city's location at the junction of the state's two largest railroads, it inevitably attracted a variety of transportation-dependent industries. Wholesalers, for example, were present almost from the beginning, with E. W. Baker and Company (a grocer) the most prominent example during the 1860s. From a single jobber listing in a state gazetteer for 1870, the number grew to eleven a decade later. Although this total paled in comparison to the twenty-eight that operated out of Leavenworth at the time and the twenty-seven out of Atchison, wholesale merchants nevertheless contributed to the local economy and required no subsidies. The development of the W. A. L. Thompson Hardware Company, for example, was exactly what city leaders had hoped would occur. Its founder had come to the state as a young man in 1870 and saved his money for eight years before opening the business. Seventy years later, that company employed twenty-eight salesmen scattered over five states, owned a subsidiary warehouse in Dodge City, and enjoyed a net worth in excess of a million dollars.[55]

Good railroad connections also brought several small foundries and flour mills to Topeka. Three machine shops operated in the 1880s, employing a total of 150 men, while six mills created jobs for 52 additional people. For a short while during the 1890s, Topeka actually was the largest milling center in Kansas. Nine facilities contributed to this cause, led by a branch of the Crosby Mill out of Minneapolis, Minnesota, but the total capacity was only 2,500 barrels of flour per day. When much larger plants began to be constructed in the next decade, owners selected either the better-connected Kansas City or more western cities such as Hutchinson and Salina as their regional sites. Topeka milling gradually declined.[56]

Meatpacking, a fourth railroad-related industry in Topeka, eclipsed the other three in local job provision throughout the first half of the twentieth century. Two plants, employing 75 men apiece, existed as early as 1888: the Kansas Packing Company and the North Topeka Packing House. These soon were supplemented by three others, all with local roots. Burton Hill began to process horsemeat into dog food in 1907, and his Hill Packing Company came to employ 150 men. Larger still was Wolff Packing, which grew out of a small butcher shop established by Charles Wolff in the 1870s. Moving to a large brick facility next to the river, Charles Wolff gradually expanded his slaughtering capacity to five hundred cattle and seven thousand hogs per week and provided jobs for three hundred people. In 1931 the plant was sold to John Morrell and Company of Ottumwa, Iowa, who quadrupled the workforce to thirteen hundred. As such, Morrell was second in employee size locally only to the Santa Fe shops. The third important packer in Topeka was another homegrown business. The Seymour Packing Company, begun in 1893 by Thomas F. Seymour and his brother-in-law George C. Bowman, specialized in poultry and eggs. By 1950 it had become one of the largest companies of its kind in the nation, employing between twelve hundred and nineteen hundred people, pioneering in the drying and freezing of eggs for military use, and establishing five branch plants in Kansas and others elsewhere on the plains.[57]

Even though Topeka's packing plants, wholesale houses, and flour mills provided jobs for many people, their totals were never enough to classify the city as a manufacturing specialist. Overshadowing them and everything else was the older double identity of capital city and railroad town. Outsiders all knew about the capital designation, of course, and although this was important, its impacts were diffuse. The Santa Fe, in contrast, was both big and obvious. In words that were applicable for at least seventy years after he wrote in 1883, local attorney Joseph G. Waters had put the matter succinctly. This railroad, he said, "is the boss of all Topeka things."[58]

The original investment of the Santa Fe in Topeka was modest. In 1869 company officials erected a two-story frame building on Washington Street that served as a combination depot and office. The first shop, erected the next year, was brick but only measured sixty-five by one hundred feet. Engines were serviced in a six-stall roundhouse, but cars had to be repaired in the open air. Cattle money changed all this. A modern office building, erected downtown on Jackson Street in 1883–1884, soon employed 250 people. Even more public attention went to the shops as they rapidly expanded into the largest and most modern facilities of their type in the world. Twelve separate departments existed as early as 1880, including specialties for carpentry, blacksmithing, boiler making, brass working, upholstering, and drafting. Eight hundred men constructed 144 locomotives that year, and the company paid monthly wages in excess of $120,000.[59]

Expansion went on nearly continuously at both the office and the shops through the early 1900s. Two thousand people worked at the shops by 1888,

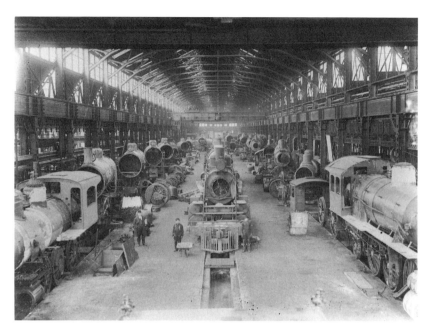

Locomotive Shops of the Atchison, Topeka and Santa Fe Railroad in Topeka, Early 1900s. Although the Santa Fe's shop complex just east of downtown had many buildings and departments during its peak years, this vast room was its core. Engines were built and rebuilt here, including these steam designs intended for passenger service. (Courtesy Kansas State Historical Society)

and in 1901 the company erected a mammoth new machinist's plant and boiler works that measured 152 by 850 feet and featured a unique sawtooth roof design to maximize northern light for the interior. With a monthly payroll that had increased to $250,000 by 1888, it came as no surprise to read and accept matter-of-fact statements that the Santa Fe had done more to build up the community than any other agency, and that "Topeka is a more distinctively railroad center than any other city in the State."[60]

As time moved on, and the Santa Fe's actual trackage focused more and more on Kansas City, Chicago, and Los Angeles, railroad officials contrarily decided to maintain and even increase the company's presence in Topeka. The shops, peaking out at about three thousand employees, were still reported as the world's largest in 1938, while the office staff grew to larger numbers still in a ten-story building downtown. A 1936 report counted 4,430 workers there, "the largest office . . . in the world devoted exclusively to railroads." More astonishing still, perhaps, was the claim that shop employees constituted 40 percent of the city's industrial workforce. When the presence of a special hospital just for railroad workers and a company-sponsored YMCA with fifteen hundred members are added to the mix, the claims for local dominance do not seem at all exaggerated.[61]

The contribution of the Kansas capital to Topeka's economy begins with employees of the state, of course. These numbers reached the low hundreds in the 1880s and then gradually expanded to about a thousand early in the following century. The capital's influence was far greater than this, however. A variety of agencies from the federal government required a presence in the state. They chose Topeka as well. So did state organizations such as the Associated Industries of Kansas, the Kansas Co-operative League, the Kansas Chamber of Commerce, the Kansas Livestock Association, the Kansas Motor Club, and the Kansas Banker's Association. Lobbyists also required hotel space, of course, and partly because of this, Topeka by the 1920s had become known as a convention center for statewide business and fraternal groups. Entrepreneurs opened the large Hotel Jayhawk and Hotel Kansan between 1924 and 1929, more than doubling the previous total of 627 available rooms. By 1935 the number of inns had increased from eighteen to twenty-two, and the city hosted 223 conventions, a remarkable total for a depression year.[62]

Among the broad-based assortment of companies and organizations that sought locations near the capital, special strength developed in two areas. The most predictable of these was the insurance industry, a business always highly regulated by the government. Two such companies, Security Benefit (1892) and Alliance Co-operative (1895), made their homes in Topeka before the turn of the century, but these were not highly touted as a community asset until seven similar establishments joined them between 1909 and 1926. Thereafter, of course, as the group made itself prominent, boosters began to note that Topeka had become one of "the great insurance centers west of the Mississippi." Visibility helped this sense of importance. Stately office buildings erected by three of these companies—Security Benefit, National Reserve, and Preferred Risk Fire Insurance—constituted a third of the photographs used to illustrate the progress of the city in a 1936 article.[63]

In the absence of any historical study, the rise of the insurance industry in Topeka remains poorly understood. No mystery surrounds the city's importance in printing and publishing, however, another industry related to the capital. One obvious anchor was the official state printing office, with its extensive output of annual reports, legislative bills, and the like. Newspapers always are important in any capital, too, but Topekans pushed the general specialty much further. Two longtime contributors, G. W. Crane and Company and Adams Brothers, both began operation before 1890. The former did general work, including books, while the latter specialized in the production of specific business forms. Hall's Lithograph Company also entered the scene early,

The Capper Building in Topeka, circa 1930. From this site at Eighth and Jackson Streets, Capper Publications issued a variety of newspapers and magazines that reached more than four million subscribers during the 1930s and 1940s. Mr. Capper served four years as Kansas governor and thirty as a U.S. senator. (Courtesy Kansas State Historical Society)

but it was a fourth printing entrepreneur, Arthur Capper, who was most responsible for the city's reputation.[64]

Because of its Horatio Alger qualities, Capper's story has been told many times. He came to Topeka in 1884 and obtained a job setting type on the *Daily Capital* newspaper. Nine years later, after moving up to become city editor of that publication and marrying the daughter of former governor Samuel Crawford, he bought a local weekly newspaper of his own. Hard work there enabled him to purchase the *Daily Capital* in 1901, and this in turn led to an expansion into national publications. He converted his old weekly newspaper into a magazine called the *Kansas Farmer* and then purchased companies to publish similar agricultural news in several other states. He also established three national magazines for rural America: *Capper's Weekly, Capper's Farmer,* and, for women, *Household*. By the 1920s, Mr. Capper (actually Senator Capper for four decades beginning in 1918) ran a true empire. Ten different publications had a combined subscription list of four million. Even in 1934, at the middle of the depression, six hundred people worked for him at Eighth and Jackson Streets downtown in what was said to be the fourth- or fifth-largest printing plant in the world. Their output required more than a million dollars to mail each year, 35 percent of Topeka's total postage. Fifteen years later, near the peak of the company's influence, employment reached a thousand people, and publications went into nearly five million homes.[65]

Wichita: The Unexpected Metropolis

When promoters from northeastern Kansas examined their collective commercial and industrial prospects in the early 1880s, they were content. As a group, cities there dominated the entire state and promised to do so for the foreseeable future. An impressive regional railroad network focused on greater Kansas City and attracted entrepreneurs of all sorts. Nearby communities served more specialized roles. Atchison and Leavenworth, also with excellent transportation, concentrated on wholesaling and manufacturing, respectively. Topeka controlled the political sector, Lawrence and Manhattan the state's educational needs. The only threat to this economic hegemony had been major discoveries of coal and lead in southeastern Kansas during the 1870s, but swift new construction and investment by existing railroad companies soon ensured that a sizable proportion of these monies also passed through the hands of northeastern Kansans.

Self-satisfaction in the state's older cities proved temporary. Although no major urban competitors emerged in the mining country and none either in the drier, western half of the state, south-central Kansas was another matter. There, with a success that amazed everybody except local boosters, the 1872 cow town of Wichita turned itself into a real city. From a mere 607 residents in 1870 and only 4,911 in 1880 after the cattle trade had migrated west to Dodge City, the place somehow grew nearly fivefold over the course of the next decade to become the third-largest city in the state (Chart 2). Fifteen years later, after all of Kansas had endured economic depression during the 1890s, Wichita quickly surpassed even fast-growing Topeka in size. In the 1940s, its pace unabated, it transcended Kansas City, Kansas, as well. The former trading post on the Arkansas River, dubbed "the magic city" in the 1880s, was one of the few communities actually to live up to its press clippings.[66]

Calculation of economic specializations explains only a little of the mysterious growth (Table 1). Between 1870 and 1950, Wichita never met the standards to be classified as a manufacturing or railroad town. Neither was it an educational, military, or mining center of note. The city is on the list for wholesaling, however. That label appears from 1930 through the end of the charting period in 1950 but actually can be traced back at least four decades farther. A local publication in 1888, for example, claimed Wichita's jobbing business had recently become the largest in the state. Although people in Atchison surely would have disputed this assertion, it nevertheless provides a good starting point for understanding local strategies for economic development.[67]

The presence of wholesalers during the late nineteenth and early twentieth century implies a good railroad system as the functional basis for Wichita growth. More precisely, the key was the creation of a transportation nexus at a location far enough from Kansas City, Leavenworth, and Atchison (about 190 miles) to be relatively free from competitive pressures. Obtaining and then taking advantage of this circumstance was no easy matter, however, because economic conditions and opportunities changed rapidly throughout the 1870s and 1880s. Wichita had only one outstanding year as a cattle town, for example, before economic depression hit the country and rival shipping points developed. Another early incentive for action came in 1877 when officials of the Santa Fe railroad announced plans to extend their original spur line south from the city. Such a development held potential for opening up more of Indian Territory to local merchants, but it simultaneously would give a competitive edge to entrepreneurs in Arkansas City and other towns close to the border.

In this time of decision, Wichita was fortunate to have two railroad companies positioned on the Kansas-Missouri border with plans to build west. If the owners of both companies could be induced to join the Santa Fe system at Wichita, the town would become the transportation focal point for the region and therefore positioned well to control a large, rapidly evolving frontier to the south and west. This hope became a reality. The St. Louis and San Francisco line, constructed northwest from the Joplin area in 1879 and 1880, was first to arrive. Turmoil in the rival city of Newton over water problems aided in this campaign, as I described in chapter 5. Local lobbyists also were able to convince voters that passage of a bond issue would promote growth. Coal from southeastern Kansas would attract manufacturers, they said, while timber and building stone from the Ozarks would be needed to erect the thousands of homes and businesses soon to appear throughout central and western Kansas. When the second railroad, the Missouri Pacific, arrived in 1883 from Fort Scott in exchange for a $40,000 city subsidy, and Santa Fe workers completed a pair of lines south from town to the border of Indian Territory, Wichita was poised for rapid development (Maps 11 [see p. 24], 14).[68]

Historians of Wichita write about a remarkable boom in real estate values during the late 1880s, one that created fortunes and temporarily ballooned the city to an estimated forty-eight thousand residents. Accompanying the frenzy of tract sales was a concerted effort by the local board of trade to attract manufacturers and other industrialists. Wholesalers, obviously, saw the city's potential early. Two were present in 1880: H. H. Richards (a grocer) and L. Hays (saddlery and hardware). Eleven years later (and just after the onset of hard times), this number had risen to an impressive twenty-eight, including four large lumber dealers. The total, in fact, ranked third highest in the state behind Atchison and Topeka, and second only to Atchison on a per capita basis.[69]

Millers were the first industrialists to see potential in Wichita, and they dominated this sector of the economy through the mid-1880s. City Mills, the largest of five operations present during those years, began business in 1874 and featured a fifty-thousand-bushel elevator and equipment powerful enough to mill 250 barrels of flour daily. This plant also was one of the pioneer processors of hard Turkey Red wheats grown nearby by German Mennonite farmers. Later, as hard-wheat flours grew more popular nationwide after 1900, Wichita acquired several newer and much larger mills because south-central Kansas remained the heart of production for that grain variety. Local capacity for flour making ranked third in the state behind Kansas City and Topeka as early as 1912, and then expanded even more over the next two decades. As output reached 11,300 barrels per day, Wichita's status increased not only to second in the region after Kansas City but also to fifth (and occasionally fourth) in the nation. The Red Star Milling Company led the way in the 1920s with 120 employees. Slightly later the key figure in the industry became Ray H. Garvey. From a Wichita base, this Colby native came to control some one hundred thousand acres of wheat land and a network of large elevators. His local facility held 43 million bushels.[70]

Although numbers for milling capacities and wholesale volumes looked good in booster publications, neither industry offered potential for large employment. Realizing this, the local board of trade in the mid-1880s began an aggressive advertising campaign to solicit manufacturers. The combination of newly arrived railroads and a real estate boom made this task appear easy at first. The board used gifts of overpriced land to "invest" $50,000 in the Burton Car Works, for example. This was a Boston firm that built livestock cars for railroads and that proposed to employ between 1,500 and 2,000 workers. Lesser amounts went to the Peabody Watch Company of La Porte, Indiana; the Gilbert Plow Company; a spice mill; and a wire and nail factory. None of these businesses survived the collapse of the boom in the 1890s. Still, a few initiatives were successful. One was the Union Stock Yards Company, a centralized market developed on the Kansas City model. Meatpacking, a closely related venture, followed shortly thereafter. Although dirty, this industry was thought to be a natural fit for Wichita. The city therefore gave its largest subsidy—$150,000—to Jacob Dold and Sons, an already prosperous firm with plants in Kansas City and in Buffalo, New York. This proved to be a wise choice. Within a year after the 1887 offer, Dold officials had erected a complex of nine buildings and hired 150 men. By 1920, their workforce stood at 425. A second early packer, Whittaker Brothers, was poorly financed but gave way in 1907 to a branch of the large Cudahy company. With 699 workers in 1920, Cudahy claimed the title of Wichita's largest employer for several years, and the packing industry remained a pillar of the local economy through the middle of the century.[71]

Agriculturally related industry dominated Wichita through the 1920s. It was the bellwether group that enabled the city to survive hard economic times during the 1890s and to move into a solid second position behind Kansas City as a regional population center. In many ways the community at this time functioned as "Kansas City West," mirroring the industrial activities of the larger place and acting as its representative and regional holding station on the plains. This role was important, of course, and the envy of other urban centers in the state, for it promised sustained growth into the future. Not even the most enthusiastic local booster could have predicted Wichita's development over the next three decades, however. The idea that this city might increase its population by fifty thousand residents between 1920 and 1930 and then do so again in the 1940s would have to involve legerdemain of some sort, intervention from some new and completely unanticipated source. As things transpired, such magic touched the city twice—first from the ground and then from the sky.

In local folklore, the boom period that began in the 1920s is sometimes incorporated into a more general belief that Wichita residents have a knack for developing timely products with big market potentials. This penchant is sometimes traced back to days of the Chisholm Trail, but it always includes the story of Albert A. Hyde. He created the Yucca Soap Company in 1889 as a way to utilize fat left over from the packinghouses but soon found that many more people were interested in a second product, a salve for colds he called "Mentholatum." Even more important in terms of employment was an idea from William C. Coleman, who had come to town from Oklahoma in 1901 to sell gasoline lamps for use on farms and in small towns that lacked electricity. After finding a way to improve the product's design, he set up his own manufacturing plant in 1907. Success followed almost immediately, and after introducing gasoline stoves and heaters to the line, this factory employed 623 people in 1920.[72]

Although an industrial survey showed that the Wichita economy in 1927 still relied on its tradition of packinghouses (1,531 employees), railroads (1,400 employees), the Coleman company (813 employees), and flour mills (520 employees), this inventory also revealed clear signs of change. Five oil companies now maintained headquarters in town, for example. So did seven airplane factories. Interconnected in ownership, these two new industrial sectors also were growing rapidly. Within two years, in fact, the latter business would operate on such a scale that city promoters would adopt a new slogan: "Wichita—the air capital."[73]

The oil companies that moved into town during the 1920s were associated with the extremely productive wells discovered in the previous decade near El Dorado, forty miles to the east. As I discussed in chapter 6, these fields were the richest in the country during the years of World War I and generated millions of dollars. The largest share of this money came to Wichita. As the closest city of size, it had been the most logical source for investment dollars. Then, as dividends began to flow back to these early financiers at unimagined rates, the effects truly were profound. Three new refineries were the most visible manifestations in the city, and one of these, owned by A. L. Derby, continued to be a major local employer for decades.

Harder to quantify than the refineries, but greater in total impact, was the dispersion of petroleum dollars throughout the broader Wichita economy. Real estate boomed once again, of course, and many people erected mansions. Local money also became prominent in the financing of oil and natural-gas exploration throughout central and western Kansas over the next several decades. Other oil dollars went into many civic and business endeavors. City officials overhauled public sewage, water, and lighting systems and paved the streets. Investors built the Hotel Lassen to provide luxury accommodations for visiting dignitaries, and several local firms grew to become industrial leaders in the production of oil field equipment and in the provision of technical assistance for refineries. The Cardwell Manufacturing Company, begun in 1926, epitomized the former specialty (550 employees in 1951). Koch Industries, started in 1924 by Fred C. Koch and two other petroleum engineers, became (and remains) the leading company in the latter field.[74]

Important as Cardwell, Derby, Koch, and other businesses directly related to oil were to Wichita, their dominance was challenged in the late 1920s by still another industry funded initially by petroleum dollars. This, of course, was the manufacture of airplanes. Oilman Jacob M. (Jake) Moellendick was the connecting link. He observed local barnstorming pilots, decided to learn to fly, and then saw enough commercial possibility for the new machines to justify a major investment. Town boosters inadvertently initiated the process in 1919 when they built a landing field to accommodate pilots who had returned from World War I with a desire to give flying lessons and arrange aerial circuses. When sales from lessons and tickets proved inadequate for these men to make a living, Moellendick decided to utilize their skills in an aircraft manufacturing company. For this he recruited Emil Laird, a successful designer from Chicago, with promises of his own personal investment and the possibility of more from other oil people.[75]

Laird and Moellendick began production in 1920. They built perhaps forty-three airplanes before a quarrel in 1923 led to Laird's resignation, but the company endured. This presence gave Wichita a head start over rival manufacturing centers as improving technical designs led to greater sales over the next several years. In particular, two Laird employees, Walter Beech and Lloyd Stearman, started a competitor company in 1925, and then Stearman split again in 1927 to form still another. By 1929 these three enterprises, plus a fourth major one formed by Clyde Cessna and about nine smaller operations, employed nearly two thousand workers. Together they constructed an estimated thousand airplanes that year, more than were produced in any other city and 26 percent of the national total.[76]

The depression years of the 1930s naturally were difficult for a fledgling industry. Most of the aircraft companies failed, and production fell to some three hundred units annually, but the Cessna and Stearman operations both sur-

Production Line at Beech Aircraft Corporation in Wichita, 1942. Wartime demands led Beech designers to convert their twin-engined Model 18 business plane into a trainer for bomber pilots. Five thousand workers turned out more than 7,400 of these planes, known as AT-10s (plywood) and AT-11s (metal), by 1945. The photograph shows the AT-11 line. (Courtesy Kansas State Historical Society)

vived, and Walter Beech returned after a brief absence to restart a third. These three therefore were active when World War II suddenly transformed their business from a small venture aimed at mail service and business executives into a national defense industry of stupendous proportions. As early as September 1940, even before the United States had officially entered the fighting, Wichita plants had military contracts worth $20 million. Employment soared from two thousand people to more than thirteen thousand in less than a year and then nearly quadrupled again to sixty thousand before 1945. The transformation of the plants and of the city was absolutely astonishing; reporters had trouble finding adjectives vivid enough to describe what they saw.[77]

The massive defense spending that came to Wichita was partly a function of effective lobbying in Washington, D.C., by the newly formed Kansas Industrial Development Commission. This task was comparatively easy, however. Local engineers and workers had more experience than anybody else, and boosters were able to point to additional geographic and demographic ad-

Plant II of Boeing Airplane Company in Wichita, circa 1944. A massive new plant authorized by the federal government in 1940 expanded employment at the Wichita division of Boeing from thirty-five hundred to more than twenty thousand. Their job was to construct B-29s, the military's largest and highest flying bombers. The photograph shows the end of the production line. Each airplane weighed 133,500 pounds, spanned 141 feet across the wings, and generated 8,800 horsepower. (Courtesy Kansas State Historical Society)

vantages. The low humidity of the Great Plains, for example, provided flying weather better than any place other than the desert Southwest. Just as important, the region would be safe. The percentage of native-born residents was extremely high, while the city's position "equi-distant from the two seaboards and from the Canadian and Mexican borders," would make it "invulnerable to attack by foreign aggression."[78]

The Beech and Cessna plants specialized during the war in making twin-engine training planes for pilots who would man the nation's big bombers. This role, although very important, was dwarfed by that assumed at the Stearman factory. Stearman became a division of Seattle's Boeing Airplane Company and, as such, was expanded into a complex that contained nearly sixty-five acres under roof and employed more than twenty thousand people on a twenty-four-hour schedule. The announced mission at this secure, inland location was the manufacture of B-17s, the nation's standard long-range bombers. Actually, the chief product was to be the next generation of these warplanes, massive 133,500-pound machines known as B-29s. Wichita people created more than sixteen hundred such airplanes, including the Enola Gay, which carried the first atomic bomb.[79]

Labor demands during the war years brought new people into Wichita by the thousands. From a base population of 114,996 in 1940, the city probably hit the mark of 200,000 by 1945, creating housing shortages so severe that the federal government had to sponsor three separate tract developments. War's end brought fears of unemployment, of course, and the census of 1950 did indeed reflect a population decline to 168,279 residents. Still, the anticipated problems never really developed. The three airplane plants experimented with converting some of their assembly lines to the manufacture of furniture, home appliances, automobiles, and even houses. More important, Beech and Cessna executives anticipated correctly a substantial postwar demand for small-engine craft and developed specializations in this area. As for Boeing, it shut down, but only for three years. Large-scale employment resumed in late 1948 for the construction of a still larger bomber. Then, as the new B-47s began to emerge, the Department of Defense decided that pilots could best be trained at the site. The government's creation of Wichita Air Force Base next to the Boeing plant in 1949 helped to keep Wichita's unusual economy healthy into another decade.[80]

Hutchinson and Salina: The Secondary Entrepôts

With the emergence of Wichita as the transportation hub for south-central Kansas and later as a center for oil money and airplane manufacturing, the major players in the state's urban system all were in place. Then, from the census of 1890 through that of 1950, it seemed a matter of the rich getting richer. Kansas City, Topeka, and Wichita added population at much faster rates than did other communities that once had claimed equal or near-equal status with them (Chart 2). Although business leaders in some of the smaller cities continued to proclaim imminent greatness in the text of advertising pamphlets, their overall activities suggest that they understood economic planning now had to be on a more modest scale. If their communities no longer could dominate the trade of the entire central plains, for example, they might still be able to supply goods to a region of a dozen or more counties. If they could not attract manufacturers in general, niche production might be a possibility.

Looking again at a chart of population growth, it is interesting to focus on the middle ground that opened up progressively between the three true cities and a grouping of twenty or so smaller places that cluster within the range of 10,000 to 15,000 residents (Chart 2; Maps 18, 19, 20). In the mercantile model of urban growth, this zone is where secondary distribution centers most likely would occur, and the data suggest that both Hutchinson and Salina are good Kansas candidates for this role. Hutchinson, for example, came into its own as a midsize city of 23,298 by 1920, and Salina did much the same a decade later. Expansion rates at both places also remained impressive in subsequent decades, a time when those for most smaller cities in the state remained low or static.

Although the economies of Hutchinson and Salina were by no means identical during the first half of the twentieth century, they did share important elements that conformed to mercantile expectations. First, like Wichita, each was far enough away from Kansas City to escape complete domination from this regional giant. Salina's 172 miles of distance and Hutchinson's 212 granted at least the possibility for them to control trade in portions of central and western Kansas. This potential moved from the theoretical to the actual through decisions by various railroad officials in the 1880s. Through only modest self-promotion by local people, both cities became important transportation junctions. Next to Kansas City, Topeka, and Wichita, in fact, their railroad facilities were the best in the state (Map 15).

Salina's growth, as I have detailed in chapters 4 and 5, began with a decision by Union Pacific men to construct a branch from their main line south to McPherson in 1879 and another west-northwest toward Colby in 1886. These new routes essentially doubled the territory that a local wholesaler potentially could serve (Map 11). To the eight counties traversed by the main Union Pacific line west from the city were now added McPherson to the south, Ottawa to the north, and a row of six counties west from Ottawa. Impressive as this was, it was only half of the good news. A year after the opening of the line toward Colby, financier Jay Gould authorized his Missouri Pacific crews to build west through the row of counties positioned just to the south of Salina and then conveniently added a half-circle connector route north into the city. Via this connector, local merchants began to add residents from another eight counties (Rice west through Greeley) to their lists of potential customers (Map 11).

Because virtually all the agricultural and other traffic that funneled through Saline County from the west was eventually carried on to Topeka and Kansas City by the Union Pacific's main line, no possibility ever existed for Salina to achieve true metropolitan status. Still, for many types of businesses, the opportunity to command the trade of twenty-four or more rural counties was a very attractive proposition. The local population nearly doubled during the decade of railroad expansion, from 3,111 in 1880 to 6,149 in 1890. Drought kept the total at this level until 1900, but then numbers increased by 50 percent each decade through 1930. By 1950 the city had 26,176 residents, and its newspaper, the *Salina Journal,* was the leading daily throughout nearly all of northwestern Kansas.[81]

Salina as it functioned between 1900 and 1950 could serve as a textbook example of mercantile development. Once its railroads were in place and the

Watson Wholesale Grocery Company in Salina, circa 1930. An impressive network of railroads and highways allowed Salina merchants to control trade over most of central and northwestern Kansas. Watson's was the smaller of two grocery jobbers in town, having a third the number of employees as the H. D. Lee Mercantile Company. (Courtesy Kansas State Historical Society)

area to its west fully settled, wholesalers and millers quickly became important parts of the local economy. Three jobbing firms were present as early as 1880 and four in 1891, but major growth in this and other businesses awaited the end of the 1890s drought on the western plains. Although reliable data do not exist before 1930, anecdotal information suggests that Salina jobbers began to predominate over most of northwestern Kansas around the turn of the century. The H. D. Lee Mercantile Company, for example, had become the largest grocery distributor in the region by at least 1920, when it employed 135 people. A decade later the list of wholesale goods going out from the city included automobiles, coal, plumbing supplies, seeds, and nearly every other item in everyday use. Twenty-seven passenger trains and twenty-two freighters passed through town on a daily basis in 1930, and the city easily met the criterion as a specialized wholesale city (Table 1).[82]

Flour milling was a source of pride for Salinans even more than wholesale trade. The number of people employed in each industry was about the same (some three hundred), but the mills were more prominent on the landscape and also brought the city worldwide recognition. The grain business came early to Salina. To promote immigration in the 1870s, officials with the Union Pacific contracted with a large regional landowner, Theodore C. Henry, to advertise his success with wheat cultivation. They even went so far as to promote this section as the state's "golden belt." With good sites for mills available on the Smoky Hill River, Salina had three plants in operation by the early 1880s and expanded this role after the turn of the century as the hard winter wheats of the state became fully accepted on the national market and railroads reduced discriminatory rate practices that favored Minnesota cities. The city never challenged Kansas City as a milling center, but it regularly vied with Wichita for the second position in the state. Then, as more and more flour production moved closer to the grainfields in the 1920s, Salina and Wichita regularly achieved totals that made them the number four and five milling centers for the entire nation. Wheat and Salina went together so much, in fact, that it seemed natural to name the city's minor-league baseball team the Millers. The only change over these decades was a consolidation of operations. From seven separate companies in 1920, none having more than fifty-eight employees, the number dropped to five in 1939: H. D. Lee, Robinson, Shellabarger, Weber, and Western Star.[83]

Salina's advantages in railroad accessibility carried over into nearly every aspect of its economy. Transportation influenced decisions to place three different colleges in the city, for example. The locally sponsored Salina Normal College, built in 1884, lasted only twenty years before burning, but Kansas Wesleyan (a Methodist school) has endured since 1886. Marymount, a similar institution established for Catholic women in 1922, lasted some sixty years as well. In combination with a central location, ease of access also made Salina a popular choice for statewide meetings. The city built a three-thousand-seat auditorium called Memorial Hall with this purpose in mind in 1923. Six major hotels operated by 1930, offering a total of 427 rooms, and civic leaders claimed to have hosted more than twenty thousand people at various conventions during the previous year.[84]

Evidence that city leaders fully recognized the critical role of transportation in local prosperity can be found in their early and enthusiastic support for highways and trucks. The route for what became U.S. Highway 40 parallel to the Union Pacific tracks came through Salina without much controversy. To supplement this, local people hosted a meeting in 1911 to promote construction and marking of an equivalent north-south road. Known first as the Meridian Highway and then as U.S. 81, this route proved critical to the city as trucks began to make inroads into the freight business of the railroads. Together with the ongoing presence of wholesalers, it also meant that Salina became a good spot to establish trucking firms. One of these, the Graves Truck Line established in 1935 by William P. Graves, grew to be a nationally known hauler of general cargo. At its peak, it employed 475 people in thirty-three terminals before closing down in the early 1990s.[85]

Hutchinson's prosperity during the first half of the twentieth century had origins more complex than those at Salina. Salt, of course, provided the initial impetus for that town's growth beyond ordinary county-seat status. As I described in chapter 6, mining on a large scale began in 1888, and the three local producers that survived the early competition—Barton, Carey, and Morton—

employed 772 people in 1920. One of the salt kings, Emerson Carey, also expanded the local industrial sector in the first decade of the century through two other major ventures. His Solvay Process Company utilized local salt to make soda ash and employed up to 500 people before closing in 1921. The other, the Central Fibre Products Company, endured far longer. Its 250 to 300 workers initially used local wheat to make strawboard and later added lines of wallboard and egg-case fillers.[86]

Although Hutchinson's industrialization most certainly depended on salt, good railroad connections were the reason people mined these particular deposits rather than others in the general area. Three tracks formed the core of the system. First, the Santa Fe's main line connected Hutchinson directly with Kansas City to the east. West of the city, this line divided after 1886. To supplement the original trackage northwest along the Arkansas River to Great Bend, railroad officials added a cutoff straight west to Kinsley (Map 11). These two branches efficiently funneled traffic into the city from Dodge City, Garden City, and the other communities of the well-watered Arkansas Valley. It was a trade lucrative enough to attract competition from merchants in nearby Wichita, but connections to Hutchinson were superior. In 1887, construction of a third railroad route enhanced the city's western reach even more. A new Rock Island line added five additional western counties to the trade area, including the towns of Pratt, Greensburg, and Liberal (Map 11).[87]

As was the case in Salina, Hutchinson merchants first took advantage of their extended western hinterland after the drought of the 1890s. The development of flour mills, grain elevators, and warehouses soon followed, again reminiscent of the mix found seventy miles to the north. Because of rivalry with Wichita, Hutchinson's mills never achieved quite the stature of those in Salina, but they nevertheless ranked the city within the top ten flour producers nationally. Local writers in 1908 judged milling to be the equal of salt in the city's industrial structure and praised especially the Hutchinson Mill Company, the William Kelly Milling Company, and Monarch Mills. The daily capacity at these facilities rose from 3,000 barrels per day in that year to 5,700 by 1946, and entrepreneurs simultaneously constructed ever-larger terminal elevators along the railroad tracks. Eight separate operators eventually achieved a combined capacity of 16 million bushels.[88]

Businesses involved in the distribution of goods were smaller individually than either the mines or the mills, but as a group they hired more people after the turn of the century. Thirty-two separate wholesalers existed in 1908. Together they kept four hundred salesmen on the road, employed even more workers locally, and confidently judged their region served—southwestern Kansas and adjacent sections of Colorado, New Mexico, Oklahoma, and Texas—to be "rapidly growing." Census data from 1930 confirm this status (Table 1). The Guymon-Petro Mercantile Company (a grocer founded in 1902) and the Frank Colladay Hardware Company (1913) were local leaders, and Topeka-based Fleming Foods established a local warehouse for IGA grocery stores in 1934. Transportation advantages also drew other entrepreneurs to Hutchinson. The Betts Baking Company began to produce Rainbo-brand bread for the entire state there in 1922, for example, and employed a hundred workers by the 1940s. Similarly, Henry Krause, the inventor of the one-way disc plow used on High Plains farms, moved his manufacturing facilities to town from Meade in 1928. The biggest employer of this type, however, was J. S. Dillon and Sons. John Dillon opened a local grocery store in 1918, but then, over the course of the next three decades, expanded to twenty-four locations throughout central and southwestern Kansas. Three hundred people worked for the company at Hutchinson in 1947, most of them in the general office, warehouse, and bakery.[89]

As the business of distribution increased at Hutchinson, civic leaders again mirrored their Salina neighbors in pushing for good highways to supplement the railroads. In 1910 they organized one of the first road conferences in the state to lay out a "Santa Fe Highway" from Newton west to Dodge City and beyond (now U.S. 50). Another effort gave the city access to the Atlantic and Pacific Highway (now U.S. 54), which ran between Wichita and Liberal. Following these achievements, a new round of businesses established headquarters in town (Map 21; see p. 34). Sixteen trucking firms existed locally in 1951, for example, including the Western Transit Company with thirty-eight vehicles and 110 employees. A group of oil companies also found the city a convenient base for operations in western Kansas, while city hospitals and the *Hutchinson News* expanded their facilities to serve a full quarter of the state. Most emblematic of the trend, however, was a successful effort by boosters to increase the scope and status of their harvest celebration. First it became the Central Kansas Fair, then the official Kansas State Fair.[90]

Smaller Urban Centers in Central and Western Kansas

The tremendous consolidation of economic power within the city limits of Kansas City, Topeka, Wichita, and, to a lesser degree, Hutchinson and Salina constitutes by far the most important change in the urban geography of Kansas between 1900 and 1950. People in these communities, especially those in Kansas City, Hutchinson, and Salina, took advantage of superior railroad connections to grow in classic mercantile fashion. Elsewhere in the state, however, conditions during this time could best be described as status quo, a maintenance of roles laid out in earlier decades by railroad companies, by mine or well operators, and, in a few cases, by other interests. To see how these places functioned and interacted circa 1950, it is best to take a regional approach.

The simplest urban system existed in western Kansas. At one level, nearly this entire region fell within the economic reach of Kansas City.[91] At another scale, as I have just discussed, merchants in Hutchinson and Salina neatly divided the trade area along an east-west axis located just north of Great Bend (Map 15). These three cities, with lesser contributions from Topeka and Wichita, accounted for nearly all of the wholesale trade in the region. The small remainder went through Concordia and Dodge City. Concordia possessed decent railroad connections east to Atchison, southeast to Kansas City, and west a few counties to Stockton (Rooks County) and Lenora (Norton County). With these, the city attracted several small distributors and grew steadily from a population of 3,401 in 1900 to 7,175 in 1950 (Map 14). Neither these wholesalers nor local railroad traffic could compete with the larger and better-positioned Salina in the long run, however, and the community then began a slow decline (cf. Map 15). In fact, Concordia lost not only its jobbers to Salina but also its bishop. Catholic authorities decided in 1944 that a new diocesan headquarters in that latter city could better serve its clientele across northwestern Kansas.[92]

The situation in Dodge City parallels that of Concordia in some ways. Officials with the Santa Fe railroad had relocated their division headquarters for western Kansas there from Coolidge in 1902. This office plus the preexisting railroad shops remained the largest employers in town through 1950, but the company's construction of a spur line southwest to Elkhart in 1912 had also opened up a limited role for small wholesalers (Map 11). The resulting economic boost was never enough to rate specialty status for the city on Table 1, but at least one account claimed that it was sufficient to permit the paving of city streets.[93]

Once the patterns of mercantile trade in the western half of the state are understood, it seems logical to predict that the remainder of the urban network would consist of modest-sized communities where the functions of retail trade and county government are combined. This model holds partially true for the period but is modified to a perhaps surprising degree by a series of economic specializations. Four categories are present. First, railroad division points bolstered populations in eight communities: Belleville, Dodge City, Ellis, Goodland, Hoisington, Liberal, Phillipsburg, and Pratt, while at Colby a junction between the Union Pacific and the Rock Island lines produced another slight bulge above the urban threshold number of twenty-five hundred (Map 20). A second, more limited series of exceptions came via public institutions: Larned with its state hospital for the insane, Norton with the state tuberculosis sanatorium, and Hays with Fort Hays State College. Garden City is a singular third category, a place where 240 factory workers and hundreds more contract farmers with the Garden City Sugar Company formed a population island of nearly eleven thousand. The last urban specialization, mining, is the most extensive. This included Lyons with two salt companies, Hugoton and Liberal with pipeline operations associated with the Hugoton gas deposits, and oil field service businesses and/or refineries in eight principal places: Phillipsburg in the north, Pratt in the south, and Ellinwood, Great Bend, Hays, Lyons, McPherson, and Russell in the extremely productive central part of the state.[94]

Economic specialization makes it almost impossible to fit a model of hierarchical trade centers on the western Kansas world of 1950. Below the population levels of Hutchinson and Salina, in fact, intense rivalry for retail and service business was the rule, a competition partially masked to outside observers by the more visible success of the railroad and other specializations. The situation can best be seen, perhaps, by comparing Pratt with a set of its neighbors to the north. Pratt, a division point for the Rock Island railroad, exemplifies the ideal of a small central-place city. Besides having the railroad and the county seat, it was close enough to the developing south-central Kansas oil fields that it was a logical site for oilmen to place their businesses (Map 12; see p. 25). Once this happened, the town had no immediate rivals and so could expand its retail reach modestly to the north, south, and west beyond the county borders.[95]

If the consolidation of economic functions in Pratt might ease the mind of an urban modeler, the seemingly oversupply of small cities two and three counties to its north would produce theoretical nightmares. There, neither of the railroad division towns—Ellis and Hoisington—was a county seat. The salt-mining operations in Lyons functioned largely as subsidiaries to bigger ones in Hutchinson, while the oil, although abundant, was distributed widely enough to make a single service and refining center impractical. Each city in the area makes perfect sense when evaluated on its own terms, but the urban system as a whole is chaotic.

Consider first McPherson, the easternmost member of the central Kansas agglomeration. Despite being a county seat, this community would seem an unlikely candidate for growth before the age of automobile commuting because of a location midway among the larger transportation centers of Hutchinson, Newton, and Salina. Its steady rise from 2,996 residents in 1900 to 8,689 fifty years later can partly be explained by an extremely fertile agricultural hinterland, one that allowed several local milling companies to persist despite the nearby competition. Entrepreneurship helped as well. One group of people persuaded the state organization of the Farmers' Alliance in 1890 to approve a local insurance firm as its official carrier, thereby creating the long-lived Alliance Insurance Company. Other promoters convinced two religious denominations to establish educational institutions in town: McPherson College (Church of the Brethren, 1888) and Central College (Free Methodist Church, 1914). Oil, the third element fostering growth, became important after 1929. The county led the state in production in 1931, and two years later the Globe Oil Company opened a refinery. This facility changed ownership in 1943 to the National Co-operative Refinery Association and expanded gradually to employ about 350 workers.[96]

West of McPherson the urban organization becomes more confused. Russell, flanked by Hays to its west and Great Bend to its south, formed the core for what became known as the Kansas oil patch. As county-seat towns with decent railroad connections, all three communities became important centers for drilling companies and other oil field service industries once the great extent of mineral reserves became clear in the 1920s. Russell, the city nearest the initial discoveries, expanded first, but oil pools proved larger in Barton County, just to the south. There Great Bend was crowned the unofficial "oil capital of Kansas" and grew rapidly over the next two decades. By 1950, it claimed 12,665 residents and was the largest Kansas community west of Hutchinson. A survey in 1952 found that an almost unbelievable 225 supply and service firms existed locally for the oil fields. Business was brisk enough to merit air service, a television station, expanded public facilities, and a host of new stores and luxury homes. With Hays and Russell, where similar but slightly smaller booms were under way simultaneously, the town shared and rotated an annual celebration called "Oil Appreciation Days."[97]

Looked at from a long-term economic perspective, people in Hays had more reason to smile than their oil-patch colleagues. When the wells eventually would fail, Hays had several advantages that promised a relatively prosperous future. First, Fort Hays State College, which had grown steadily to this point, held the potential to serve fully half the state. Hays also was nicely positioned to command retail trade. Whereas Russell was sandwiched between Hays, Great Bend, and Salina, and Great Bend was uncomfortably close to Hutchinson, Pratt, and Russell, Hays businesspeople could anticipate increasing dominance over some fifteen counties to their north and west. It was a future largely undeveloped as of 1950, but one certainly imagined.

Looking at Map 20 with an eye for other urban rivalries in western Kansas, it is easy to identify four conflicts that might emerge as improvements in automobiles and highways reduced the friction of distance. One of these was due north of Salina, where a railroad division point (Belleville), a prime river site for milling (Beloit), and a significant railroad junction (Concordia) all vied for the same customers. Beloit lost its only competitive advantage as mills converted from water to coal for power, especially since the town's surrounding Smoky Hills were not good grain country (Map 5; see p. 17). Belleville people, undoubtedly aware of the superior rail connections in Concordia, tried to compete by championing early highways. They managed to obtain major east-west and north-south road designations for themselves (the Pike's Peak Highway [U.S. 36] and the Meridian Highway [U.S. 81]), but this was not enough to overcome Concordia's inertia, at least as of 1950.[98]

Three and four counties west of Belleville, the towns of Phillipsburg and Norton formed a second set of small urban rivals in 1950. Both were county-seat towns along one of the Rock Island's main lines, and both had acquired two additional employers of note. Norton grew first because of workers at its state tuberculosis hospital, established in 1914. The community also benefited from being directly north of Dodge City and therefore on the north-south Star Highway promoted by that larger city. As this route was improved and renamed U.S. 283, Raymond E. Blickenstaff reasoned that his hometown might be able to support a trucking firm. Taking advantage of a location midway between Denver and Kansas City, his Ideal Truck Lines eventually grew to a fleet of four hundred vehicles. Phillipsburg, as I related in chapter 6, occupies the northernmost part of the central Kansas oil field and therefore was chosen as the site for a small refinery. The town's second big employer, the Kansas Pipe Line and Gas Company, established its headquarters there because, as a Kansas enterprise with Nebraska customers, it required a base near the state line. This business, later known as Kansas-Nebraska Natural Gas and then as KN Energy, employed 150 or so local people.[99]

Colby versus Goodland, the final rivalry in northwestern Kansas, was a more open and serious affair than the one between Norton and Phillipsburg. This was partly because both towns were dependent on similar railroad economies, and also because the commercial prize available to the winner promised to be trade from at least nine counties in Kansas plus smaller sections of Colorado and Nebraska. In this contest, most early investors favored Goodland over its neighboring county seat to the east. This was because officials of the Rock Island railroad, who essentially had created each place via their track location, had selected Goodland in 1888 as their division headquarters and site for regional shops. The town grew immediately to a thousand people and then increased steadily along with the general development of the High Plains. It had 4,690 residents in 1950.

Colby, a town that has always prided itself on having hardworking entrepreneurs, clearly needed this trait to compete with a larger neighbor. The role of underdog is easy to overstate in this situation, however, because once Rock Island officials decided to make this town a junction between their new line and two preexisting Union Pacific spurs (from Salina and from Oakley), Colby actually was a better location for merchandising than was Goodland (Map 15). While Goodland became known as a blue-collar, shop town, Colby people advertised the accessibility of their community. This was the self-proclaimed "hub of Northwest Kansas," a place whose newspaper "covers its field more thoroughly than any other local paper in eighteen northwest Kansas counties." With companies such as Symns-Shafer (a wholesale grocer), the town's population slowly began to gain on that of Goodland. From a gap of more than fifteen hundred residents in 1920 (1,114 versus 2,664), the margin closed to half that size thirty years later (3,859 versus 4,690).[100]

Southwest Kansas, more sparsely populated than other parts of the state, lacked any obvious rivalries between cities in adjacent counties. This is not to say, however, that simple central-place geometry blanketed the region. Instead, three cities of between seven and twelve thousand residents somehow managed to coexist: Dodge City, Garden City, and Liberal. On a map, these places form an almost perfect equilateral triangle with sides about seventy miles in length (Map 20). This distance means that each city was positioned to control trade from parts of six or more counties and therefore was not a rival of the other two for certain kinds of business. Improving roads and automobiles were expanding the scale of routine travel rapidly at this time, however, and thereby increasing intercity awareness and tensions.

Dodge City, long the self-proclaimed "metropolis of southwestern Kansas," maintained its regional lead in population throughout the 1900 to 1950 period but saw its dominance decline. This change was unexpected because local businessmen made no obvious missteps in the maintenance of traditional economic roles and even had expanded beyond them. The Santa Fe's repair shops continued to employ 125 or so people, for example, and as I noted earlier, several merchants took advantage of the region's only railroad junction to enter the wholesale trade in a modest way. Residents also were proud of their Dodge City Flour Mill (forty-eight employees), which reflected the arrival of large-scale grain farming to the southwestern quadrant of the state, and of St. Mary of the Plains College, the only institution of higher education in that same region. Then, in 1931, boosters added a new enterprise. Noting the trend from cattle to wheat in the area and hoping to attract tourists with an interest in times past, several local men opened the Boot Hill Museum to celebrate the cowboy decades. This venture proved successful, and in 1947 the city's junior chamber of commerce took over its operation.[101]

As Dodge City people pursued traditional economic opportunities, those in Garden City and Liberal opted for specialization. Garden City in 1950 was still the sugar-beet center it had been forty years earlier. In fact, the whole community revolved around a single firm now known as the Garden City Company. Officials at this powerful enterprise financed irrigation systems for individual farmers in eight counties, employed 350 people at the sugar plant itself, and even ran a subsidiary power company that furnished electricity to a large section of western Kansas and eastern Colorado. Residents had begun to talk about diversification, including the use of some irrigated acres to produce grain for feeding cattle, but no such investment had yet materialized. Still, with the sugar business offering reliable yields and steady jobs, Garden City at midcentury was an attractive community, one that 11,037 people called home. Liberal, the final urban center in the region, was without good possibilities for either irrigation or trade. It grew instead as the nearest railroad center to the mammoth Hugoton natural-gas field found just to its north and west. This development, which I described in chapter 6, was based on the acquisition of headquarters for pipeline companies and other minerals industries.[102]

When we shift the regional focus from western to south-central Kansas, the urban patterns reflect much the same processes. The primary exception, of course, is the emergence in this section of a truly dominant central place. Wichita's 168,279 residents in 1950 were the equivalent of five times the population of Hutchinson and more than ten times that of any other area city. Oddly, however, the creation of this giant did not reduce the size of five formerly rival communities in adjacent counties (Map 20). With populations ranging from 7,700 to nearly 13,000 in 1950, in fact, these places generally had managed to double their sizes during the Wichita boom. In a time before widespread automobile commuting, the maintenance of such geometry does not correspond to economic models.

With manufacturing, wholesaling, and several other business functions monopolized by Wichita entrepreneurs, specialization is the only explanation for the viability of Newton to the north, El Dorado to the east, Arkansas City and Winfield to the southeast, and Wellington to the south. The various economic singularities found in these places were so well established, in fact, that they exuded a feeling of permanence. In the case of Newton and Wellington, the role of being division points on the now nationally prominent Santa Fe railroad system had led to continuous growth. Newton, the larger of the two cities, occupied the spot where the railroad's two big routes into Texas joined with the original main line. This nexus produced a large repair facility (three hundred employees versus ninety at Wellington), as well as many auxiliary railroad jobs. A Newton factory made steel rails for shipment to Chicago, San Francisco, and everywhere else on the Santa Fe system. The town also was home to the railroad's food contractor, the Fred Harvey Company. Ninety people worked in the rail mill, while another 150 found jobs at Harvey's dairy farm, ice plant, meat locker, creamery, laundry, and other interconnected enterprises needed to serve twelve million meals per year. Beyond the railroad, the economies of Newton and Wellington also benefited from an average of three flour mills apiece over this time period, products of locations in the heart of the Kansas wheat belt. Finally, Bethel College, an institution sponsored by the General Conference of the Mennonite Church that moved to North Newton in 1892, brought several dozen additional jobs. With 7,747 and 11,590 residents, respectively, Wellington and Newton were prosperous places in 1950.[103]

Although the development of oil fields in western Kansas during the 1920s diverted public attention away from the slightly earlier discoveries just east of Wichita, pumping continued at a good pace in Butler and adjacent counties throughout the period under consideration here. In addition, local companies had laid down pipelines to bring in supplementary supplies from Oklahoma and other nearby sites. As a result, several well-financed refineries survived and even expanded in south-central Kansas. One of these, White Eagle, dominated the small Butler County town of Augusta. After expansions financed by Socony-Vacuum in 1931 and then by Mobil, this plant employed about five hundred workers by midcentury and lifted Augusta's population to 4,483. El Dorado, the original center of the oil boom, maintained this leadership role and supported two additional refineries. One, the El Dorado Company, employed about a hundred workers; the other, a large William G. Skelly operation, hired more than twelve hundred. With Halliburton and similar companies that serviced oil field equipment requiring perhaps a thousand additional local people, El Dorado was able to sustain a population between ten and eleven thousand from 1920 onward.[104]

Arkansas City and Winfield, the two satellite cities southeast of Wichita, had comparatively complex economic structures. Arkansas City, following early but fleeting success as a gateway city to Indian Territory and as a manufacturing center based on power from the Arkansas River, relied after 1920 largely on the railroad and oil businesses. Two refineries began local operations just prior to that year: Midland and Kanotex. Midland (later Shell) employed 127 people in 1920 but soon shut down because of poor financing. Kanotex executives, in contrast, expanded their operation via a licensing agreement with the larger Phillips Petroleum Company of Oklahoma and so provided jobs for some 230 residents. Oil, important as it was in Arkansas City, was exceeded in economic clout by the Santa Fe railroad. After being named headquarters for the company's Oklahoma Division, this city became the site of a major office building and shop complex. Railroad officials also designated the community as a stockyard focus for the line and as the system's primary storage facility for track repair materials. All in all, the Santa Fe directly employed 660 local people in 1950 and at least that many again through indirect influences. Obvious examples included a meatpacking business (Maurer-Neuer), with 450 employees, and a wholesale grocer (Ranney-Davis), whose 65 workers acknowledged the railroad's local importance by selling "Santa Fe" brand foods.[105]

Winfield was the most unusual of the Wichita-area communities. Refinery people had passed this town by, and although it was on one of the Santa Fe's principal lines, it was too close to Arkansas City (twelve miles) to have important railroad functions. Winfield, in fact, had only modest assets in the early 1880s—a beautiful setting in the valley of the Walnut River and the government headquarters for Cowley County. A claim in 1886 that local "energy and liberality" would inevitably lead to prosperity, although echoing thousands of other promotional tracts, actually seems to have been true in this case. Winfield grew to 5,554 by 1910 and then to 10,264 by 1950 largely on the basis of

Officials and Main Building of Southwest Kansas Methodist Episcopal College in Winfield, circa 1897. Institutions traditionally have dominated Winfield's economy. North Hall, photographed here, opened for classes in 1886 and served the campus until 1949, when structural problems led to its razing. The school shortened its name to Southwestern College in 1909. (Courtesy Kansas State Historical Society)

three community efforts. The first, as I described in chapter 6, secured the state school for feeble-minded youth in 1884. This was not a prestigious political plum, of course, but it provided many construction jobs and permanent employment for about seventy-five people. A year later, an offer of $60,000 in cash plus a free site and building stone induced Methodists of the Southwest Kansas Conference to build what became Southwestern College. This second institution, certainly something worth advertising, grew within seven years to become the largest private college in the state. Finally, in 1893, came St. John's College. Sponsored by the Lutheran Church, Missouri Synod, this third institutional employer materialized through the efforts of John P. Baden, a local businessman who donated $25,000 to initiate the fund-raising effort.[106]

Smaller Urban Centers in Eastern Kansas

The penchant of the Kansas experience in 1950 to challenge the ideals of evenly spaced and nicely hierarchical urban systems holds as true for the eastern third of the state as it does in the west. There, greater Kansas City with its half million people was the focus of almost all major mercantile activity. Topeka handled much of the rest. At a smaller scale, however, a massing of economic activities sufficient to control the commerce of as many as six counties occurred only at Emporia (Map 20). The patterns elsewhere look chaotic. Northeastern Kansas, for example, could claim no substantial cities north of the Kansas River and west of the Missouri bluffs. This contrasts with the Kansas Valley proper, where Abilene, Junction City, Lawrence, and Manhattan combined with Kansas City and Topeka to produce sizable urban populations in nearly every county. Finally, and most inexplicable to a casual observer, is southeastern Kansas, where seven cities with populations between 7,000 and 19,500 clustered closely within six adjacent counties.

Because the city scarcity in extreme northeastern Kansas and Emporia's dominance over a large region both have straightforward explanations, they are good places to begin. In the northeastern interior, entrepreneurs simply judged economic potentials as too limited to justify investment. Jefferson County epitomizes this situation, a place where business initiatives beyond local trade always have been stifled by the long-established and dominating presence of Leavenworth immediately to its east, Lawrence to its south, and Topeka to its west. Much the same results were assured in counties farther to the north by the routes of the old St. Joseph and Denver City and the Central Branch railroads, both now operated by the Union Pacific. These transportation corridors enabled merchants in St. Joseph and Atchison, respectively, to control trade (and suppress competition) west at least as far as Washington County.

The two largest communities in the nine-county interior northeast, Hiawatha and Marysville, were bolstered by railroad activities. Hiawatha, with 3,294 people in 1950, served for many years as the division point for freight operations on a Union Pacific run between Atchison and Omaha. Marysville (population 3,866) did the same for both passengers and freight on the line from St. Joseph to Grand Island, Nebraska. Even after this latter headquarters was moved in 1931, the Union Pacific maintained a presence of about 350 workers in town, some at a company stockyard and some in a centralized traffic-control office that had been installed when officials established a cutoff route directly from Topeka to Grand Island (via Marysville) in 1910.[107]

People in Emporia have always attributed their city's steady growth to entrepreneurial efforts. Because the town founders all were politically connected and because the city became home to a United States senator, Preston Plumb, whose service (1877–1891) occurred during a critical time for development, this claim contains more than a grain of truth. Such connections produced two major institutions for the city: Kansas State Teachers College, as I discussed in chapter 2, and the Presbyterian-sponsored College of Emporia in 1882.[108]

The fundamental reason for Emporia's regional commercial dominance and overall growth was not political connections, however. It was transportation. Early railroad planners realized that the Cottonwood River, west from the city, provided an ideal passageway through the Flint Hills (Map 5). With the Santa Fe people taking this route and the Missouri, Kansas and Texas Railroad providing a north-south intersecting line along the town's other river (the Neosho), local growth was assured (Map 10; see p. 23). Two spur lines in the 1870s south from the city into the heart of the surrounding grazing country increased the young city's centrality even more, as did the Santa Fe's decision in 1884 to build a cutoff line from Emporia directly northeast to Kansas City (Map 11). This construction, together with the city's position midway between the emerging metropoleis of Kansas City and Wichita, made Emporia a logical site for crew changes and for large switching yards and stock pens (Maps 14, 15). By the 1920s, the Santa Fe employed some eight hundred local residents, and the city population jumped to more than fourteen thousand. Even without having repair shops or a division headquarters, this was now "one of the most important railroad towns in the state." Finally, because of its transportation focus, meatpackers and other businessmen began to see Emporia as a good location for their operations. All these interconnected activities were large enough, in fact, to regularly overshadow the town's two colleges. Emporia met the statistical requirements for a "college town" in only three of the nine census years between 1870 and 1950 (Table 1).[109]

Neither the model of centrality that works for Emporia's small-scale urban system nor the one of mercantile corridors that helps to explain city absence in extreme northeastern Kansas provides much help in interpreting the nearly continuous chain of cities along the Kansas River upstream from Kansas City. Even in Topeka, that region's largest center, merchants and traders found themselves too close to gigantic Kansas City for efficient operation. Without the obvious plums that came from being state capital and the chance decisions by Arthur Capper and officials of the Santa Fe railroad to maintain important company facilities there, this city would have been many times smaller in 1950 than its 78,791 residents.

Lawrence arguably was the most anomalous community along the Kansas River corridor, a place that somehow supported more than twenty-three thousand people at midcentury despite being only thirty-five miles from Kansas City and even closer to Topeka. Because the two larger cities commanded much better transportation systems, they were also superior locations for manufacturing, wholesaling, and business in general. Lawrence people recognized these handicaps from the start, however, and worked hard to counter them. After failing in plans for a north-south railroad that would rival Topeka's Santa Fe and Kansas City's border-tier lines, city leaders turned to the Leavenworth example of industrialization. Fuel, the key ingredient needed for this activity, was sought in three ways. First, in 1864, local investors recruited Swedish workers to construct a large windmill. Located on a high bluff with sails spanning eighty feet of airspace, this colossus produced power enough for a standard grist operation, a plow factory, and a blacksmith shop. It closed after twenty-one years, however, probably because of an energy output too small and irregular for modern needs. In the meantime, other Lawrence promoters pushed for a railroad spur southwest some forty miles to the same coalfields in Osage County that had been tapped by builders of the Santa Fe. This venture, too, proved short-lived because of low-quality and thin-seamed deposits. The third undertaking, power generation via a dam on the Kansas River, was more successful. First completed in 1874, this structure was advertised to produce 2,500 horsepower. After problems with floods during its early years, several flour mills, a shirt factory, a foundry, and other major manufacturers came to the site beginning about 1880. One of the biggest, Lawrence Paper Company, employed 230 people by 1920. Another, the Consolidated Barb Wire Company, dominated production throughout the entire West before closing in 1898.[110]

Manufacturing, then, rather than the city's venerable but historically small state university, explains most of Lawrence's development to 10,862 residents by 1900. After that date, the emphasis in employment gradually shifted. Although the paper company remained and a new plant, the Kaw Valley Cannery, emerged to process local vegetables (125 workers in 1920), growth at the University of Kansas began to dominate town activities. This school, which did not enroll a thousand students until the early 1890s, reached the mark of four thousand in 1919. As the numbers steadily rose, Lawrence people came to see their major institution less as a curiosity and more as an economic asset. The university hired people in its own right, of course, but also began to attract others through its cultural amenities and technical expertise. This new economy, although modest, was strong enough to raise the city's population over the next four decades to 14,390. Then, in the late 1940s, when masses of veterans from World War II began to descend on America's college campuses, local enrollments increased sharply to more than ten thousand students. Townspeople, suddenly 23,351 in number, now fully embraced the "college-town" label.[111]

West of Topeka, the broken topography of the Flint Hills has always discouraged urban development along a fifty-mile stretch of the Kansas River. Then comes another cluster of cities, first Manhattan at the mouth of the Blue River, next Junction City, where the Republican and Smoky Hill Rivers meet to form the Kansas, and finally Clay Center and Abilene, twenty-five or so miles upstream on these respective tributaries. Each of these communities traditionally has derived some of its economic strength from county-seat functions. Because of riverine sites, they also supported small flour mills. For Clay Center in 1950, this short list of assets plus local retail trade basically exhausted the possibilities for employment, and therefore population remained low. Abilene, because of its position on the main line of the Union Pacific and amid some of the state's best agricultural land, was able to sustain a slightly larger population (5,775 versus 4,528). The success of local businessman Alva L. Duckwall also contributed to this advantage. As his namesake variety store began an expansion into a statewide chain in 1906, he designated centrally located Abilene as the company's distribution center and general office. With fifty-two stores in the system by 1948, this headquarters employed about fifty people.[112]

Junction City and Manhattan, the largest cities in the upriver cluster, have always been textbook examples of the military town and the college town, respectively. Because of their positions on the edge of the sparsely populated Flint Hills and between major transportation centers at Topeka and Salina, people in neither community dreamed of large-scale commercial success. Once Manhattan was named the state's land-grant university in 1863, residents concentrated on its promotion. Junction City people, after briefly entertaining hopes of becoming an important trading center (see chapter 4), similarly focused on adjacent Fort Riley.

Manhattan's history is essentially that of Lawrence minus the flirtation with industry. Even with modest enrollments of several hundred students at Kansas State Agricultural College in the 1890s, the small city qualified as a college town (Table 1). This designation has remained ever since, along with a predictable ratio between student numbers and city population. From a thousand students and 3,438 residents in 1900, these respective numbers rose to 4,902 and 11,659 in 1940. A decade later returning military veterans pushed the student total above 8,000 and the overall population to 19,056.[113]

Developments in Junction City are more complicated than those in Manhattan because better railroad facilities allowed a greater range of economic opportunities. True to its name, this city became a transportation center in the late 1860s and early 1870s when entrepreneurs built what was to become the Missouri, Kansas and Texas Railroad south from the Union Pacific tracks there and Union Pacific leaders themselves constructed a feeder line north along the Republican River (Map 13). These developments attracted several milling operations and were an important factor in the city's acquisition of a divisional headquarters and repair facility for the Union Pacific in 1889. More than two hundred new railroad workers came to town during the 1890s. The biggest and most enduring legacy of the local railroad system, however, was Fort Riley. With trains available to move troops easily wherever they might be needed, army leaders elected not only to keep the post active but also to construct a major training facility there. These decisions, made in the late 1880s, led to ongoing postings of a thousand troops plus huge additional surges in preparation for two world wars. Although personnel totals at Fort Riley did not grow as predictably as the student counts at Kansas State College, they still were enough to boost the population of Junction City regularly upward. The community counts of 8,507 for 1940 and 13,462 for 1950 included several hundred married soldiers who elected to live in town and even more civilian families who made their livings as employees on the post or through military contracts for food and other supplies.[114]

Between the line of cities along the Kansas River and a similarly large cluster in the mineralized southeastern corner of the state, insufficient rural trade existed to support extensive urban development. This was as true in 1950 as it had been in earlier decades, and a tour of the region reveals few surprises. In Miami County, Osawatomie continued with its highly specialized role as home to both a state hospital for the insane (670 employees) and a large repair facility for the Missouri Pacific railroad (700 employees). Its longtime rival, Paola, managed to stay viable through a combination of county-seat functions and a single, large factory. Officials at the Fluor Corporation, a builder of compressor stations for natural-gas pipelines, began to fabricate pipes and other oil refinery products at this site in 1940. The location, they said, was advantageous because it sat midway between the production fields to the southwest and the nation's urban centers. Fluor employed three hundred Kansans in 1950, supporting perhaps a quarter of Paola's 3,972 residents.[115]

Osage City, a third longtime urban presence in east-central Kansas, had not been able to maintain its earlier status. After peaking at 3,469 residents in 1890, this community declined steadily to 1,919 in 1950. The reason was the same as for its founding: coal. Local mines, although well positioned to serve area railroads and cities, could not survive the competition from those farther south near Pittsburg, Kansas. The seams rarely exceeded twenty inches in thickness, and the product produced excessive amounts of ash when burned. Officials with the Santa Fe railroad closed their Osage County mines in the 1890s, and the home-fuel market slowly eroded thereafter.[116]

The two most successful cities in the region just south of the Kansas Valley were Olathe and Ottawa. Both were county seats, but their expansions to 5,593 and 10,081 residents, respectively, were caused primarily by transportation advantages. Each town had received a north-south railroad as a result of promotions initiated in larger cities immediately to their north. They gained a pair of intersecting, east-west links in the 1880s when the Santa Fe established its direct route into Kansas City from Emporia and Jay Gould built his Missouri Pacific main line. Local retailers now had first-rate connections in all directions (Maps 14, 15). Ottawa's expansion exceeded that in Olathe partly because of the early boost from having the shops of the original Leavenworth, Lawrence and Galveston Railroad. These facilities were retained for several decades after the Santa Fe took over this line and employed 262 people in 1920. The city also benefited from employment at a small Baptist school (Ottawa University) and at a hospital run for Santa Fe employees between 1888 and 1930. Finally, slightly greater isolation from Kansas City made Ottawa a better place for rural-oriented companies. The Bennett Creamery, the largest independent facility of its kind in Kansas, was a prime example. In 1940, it bought milk from as far away as Oklahoma, had 140 employees at branches in Chanute, Osage City, and Osawatomie, and sold butter and ice cream throughout the Midwest. Another important but shorter-lived company was Warner Manufacturing, a maker of wire fencing and small gasoline engines, which employed 543 people in 1920.[117]

Olathe could match Ottawa with its own small institution, the Kansas State School for the Deaf, but was too close to Kansas City to attract major manufacturers. The community did claim one nationally famous business, the cowboy-boot factory of Charles H. Hyer and Sons, but this employed only about sixty people. More attuned to the local market was a large grange store owned by the Johnson County Cooperative Association, which sold dry goods, tools, and agricultural supplies to customers from several counties. Olathe's rural world began to change after 1942. That year effective lobbying by Kansas City land developer J. C. Nichols and others on the Kansas Industrial Development Commission produced two major federal war installations that required sites near the Kansas City labor market. The Olathe Naval Air Station trained two thousand people on a continuing basis, while the Sunflower Ordnance Works (nine miles northwest of town) employed twelve thousand people to make explosive powders and rocket propellants. Activity at both sites peaked before the 1950 census, but the impact was permanent. Olathe itself gained 1,500 residents during the 1940s, while Sunflower Village, a new community adjacent to the munitions plant, attracted 3,834 people.[118]

The remaining section of Kansas, the southeast, contained seven different cities with populations greater than seven thousand in 1950. A strong mining heritage suggests a common explanation for this unique pattern, but like most stereotypes, the minerals theory is only partially true. Although coal still was king at Pittsburg at this date, the extraction of lead and zinc fundamental to Baxter Springs and Galena, and natural gas the key to industrialization at several sites, two of the area's larger cities stood almost completely outside any ore-related economy. These places, Fort Scott and Parsons, were the earliest true cities in the region and had grown principally on the strength of railroad connections (Maps 13, 14, 15). Had the various mining activities not occurred, they almost assuredly would have dominated retail trade over several counties apiece and become even larger than their midcentury totals of 10,335 and 14,750, respectively.

As I described in chapter 5, Fort Scott people in the 1870s expected great things for their city. Local coal promised to attract manufacturers, while the intersection of two major railroads, the Missouri River, Fort Scott and Gulf from Kansas City and the Missouri, Kansas and Texas from Parsons to Sedalia, Missouri, boded well for general trade. These advantages were compounded by the town site being approximately a hundred miles from both Kansas City and Sedalia, a distance that guaranteed independence for merchants and good possibilities for railroad division points and shops. None of these things turned out exactly as boosters had hoped. Local arrogance is said to have angered the railroad owners and kept the shops away, while the coal proved inferior to that found a county farther south.

A decade after its failures, a chastened Fort Scott community began to work with the railroad ownerships rather than against them. First, when Missouri Pacific officials constructed a new line east from Wichita in 1883, the city rallied under the leadership of Joseph H. Richards, a railroad attorney. The result was a successful bid not only to be on the route but also to obtain its main shops and division point. This added three hundred jobs starting in 1886. Next, shortly after 1900, when the old border-tier line was sold to the St. Louis and San Francisco Railroad, this company, too, proved receptive to an attractive financial offer. The Frisco's division offices, yards, and shops soon became Fort Scott's largest employer with 440 workers.[119]

Good transportation in Fort Scott naturally attracted other industries. Several foundries from the 1870s remained, but the biggest user of area coal was a brick and tile company called Western Shale Products. Founded in 1910, Western hired sixty people by the next decade and even more in subsequent years. Three food processors each supported similar numbers of workers: a packer, a poultry and egg operation (Edward Aaron, Inc.), and, beginning in 1918, a branch of the Borden Condensed Milk Company. This Borden plant, which purchased 150,000 pounds of milk per day, rivaled the Bennett Creamery in Ottawa as the dairy focus for eastern Kansas. Finally, two important white-collar businesses added diversity to the economy. One was the state's first municipal-owned junior college (1918), which enrolled four hundred students in 1940. The other, the Western Insurance Company (1910), had parlayed local man Oscar Rice's early experiments with automobile insurance into a nationally prominent institution that employed three hundred people in 1950. Together this nicely varied set of businesses produced a steady-state economy for the city.[120]

Fort Scott's principal rival as the railroad center for southeastern Kansas was Parsons. The situations were not completely comparable, however. Whereas Fort Scott was served by three main lines but was headquarters for none, Parsons was the principal city for a single railroad: the Missouri, Kansas and Texas. With the exception of having the state training school for epileptic children (see chapter 6), Parsons in effect was a one-company town. As discussed previously, MKT officials created the city to be their headquarters, shop site, and major junction point. Later adding a hospital for company employees, this railroad employed more than fifteen hundred of the city's residents in 1901, a fifth of the total population.[121]

Parsons rode the success of its railroad company to a population peak of 16,028 in 1920. As in Fort Scott, good transportation also attracted several food-processing companies, including a poultry and egg operation started by Swift and Company in 1903. This firm employed seventy-five people in the 1920s. Efforts to diversify further by the addition of manufacturing industries proved difficult, however, because companies could obtain cheaper fuel either one county to the east (coal) or a similar distance to the north or west (natural gas). The result was a slow population decline over the next two decades to 14,294 in 1940 before war-related activities temporarily halted the slide.[122]

Pittsburg, the largest city in southeastern Kansas, not coincidentally featured the region's most diversified economy. Founded because it sat atop the premier coal deposits in the western Midwest, the community remained an important minerals center in 1950 even though mechanization had dropped the actual number of miners in the county from ten thousand during World War I to some four thousand in the 1930s and to only eight hundred in 1951. These losses in jobs had more than been made up for in other areas, however, many of them also related to coal. As a result, in one writer's words, "Pittsburg's merchants, its professional men, and even its housewives and school children [still] discuss . . . mines as much as western Kansans gossip about prospects and prices for wheat." An annual fall Coal Festival provided further evidence of the local orientation.[123]

Good transportation was one important legacy of the coal. Starting originally with just a spur from Joplin, four fuel-hungry railroad companies built quickly to Pittsburg in the 1880s and 1890s: the Frisco, the Missouri Pacific, the Santa Fe, and the Kansas City Southern. As I discussed in chapter 6, all of these firms invested heavily in the coal business, and the Kansas City Southern people also placed their major shops and a division office in the city. With fifteen hundred local employees at this last-named company in 1915 and still nine hundred in 1951, Pittsburg qualified as a railroad specialty town as well as a mining center (Table 1). A second indirect benefit from the coal business was Pittsburg State College. Maintaining a strong "technology and applied science" program along with more traditional ones in arts and sciences, business, and education, the campus grew steadily after its founding in 1903. At midcentury its 165 faculty members and three thousand students added important variety to the economy.[124]

Pittsburg's residents in 1950 thought more about coal's direct impact on their lives than about its influence through railroads or the college. With quality fuel available at minimal shipping costs, many companies took advantage of the situation. Zinc smelters, an original user, had moved out in the 1890s in search of sites with even cheaper fuel, but foundries and masonry companies found conditions ideal. Pittsburg Pottery, for example, an enterprise dating to 1905, was purchased and expanded in 1941 by three former miners.

Older and larger, the W. S. Dickey Clay Manufacturing Company employed 125 people in 1950 to make sewer pipe for a six-state area. Two foundries were even more prominent in the economy. Both the United Iron Works and the McNally Pittsburg Manufacturing Corporation specialized in coal-mining and ore preparation machinery, and the United plant also designed components for zinc smelters. Intended at first to supply local demand, these companies expanded to hire some two hundred men apiece by 1920. McNally Pittsburg, under the leadership of three generations of McNally men, then evolved onto the national stage. By 1950 it employed eighty-five people in its engineering department alone, ran a branch operation in Wellston, Ohio, and kept a total of 475 people busy erecting large coal treatment facilities around the world.[125]

Spencer Chemical Company, still another industry to grow out of the Crawford County coalfields, affected not only Pittsburg but also Parsons and the communities of Cherokee County in the extreme southeastern corner of the state. The firm's founder, Kenneth A. Spencer, was vice president of the large Pittsburg and Midway Coal Company in 1941 when he conceived a plan to construct a series of military fuel and powder plants in his home region. Working with the Kansas Industrial Development Commission and arguing to officials in Washington, D.C., that this place was relatively safe from enemy attack, that it housed a patriotic and dependable workforce, and that, "more than any other single area in the United States, . . . [it] contained varied and abundant reserves of useful natural resources," he secured major funding. The result was two facilities in southeastern Kansas plus others in nearby Missouri and Oklahoma and the Sunflower Ordnance Works noted previously near Olathe. One of the new plants, the Kansas Ordnance Works, arose twenty-four miles west of Pittsburg and two miles east of Parsons. Its jobs making ammunition were enough to reverse a modest population decline at that latter city. Closer to Kenneth Spencer's heart was the Jayhawk Ordnance Works, located on Spring River, eighteen miles south of Pittsburg and three miles northwest of Galena. The products at Jayhawk—anhydrous ammonia, nitric acid, and ammonium nitrate—were intended for explosives during the war, but Spencer foresaw their easy conversion to agricultural fertilizer use. The organization he created to operate the plant became the Spencer Chemical Company in 1948 and grew rapidly. From a headquarters at Pittsburg, it employed fourteen hundred people in 1953 at Jayhawk and three smaller facilities.[126]

The Jayhawk Ordnance Works, located as far from Pittsburg as it was because of a need for a large water supply, was a boon for several local communities. Baxter Springs and Galena, for example, had long since passed their peaks as lead and zinc mining areas in 1950, with most of that activity now concentrated in northeastern Oklahoma. A lead smelter run by the Eagle-Picher Company at Galena still employed 275 people, but signs of cutbacks or even closure existed. Residents of Baxter Springs, anticipating this eventuality, had been actively recruiting new businesses for several decades. Their biggest coup, a hub for the Yellow Transit trucking company, capitalized on the town's location on U.S. Highway 66, a well-traveled route between Chicago and Los Angeles. This large maintenance shop and relay station employed 150 people. Jobs at the Spencer plant enabled both mining communities to maintain populations just above 4,000. Slightly farther away, the county-seat community of Columbus grew to 3,490.[127]

The final cluster of cities in southeastern Kansas had come into prominence in the 1890s via the state's first boom in oil and natural gas. At a time when these resources were thought to be inexhaustible, first Iola and then Chanute, Coffeyville, Independence, and other places saw their populations increase spectacularly over a fifteen- to twenty-year period. Zinc smelters came to Iola to take advantage of its nearly free supply of gas, glass companies did the same at Coffeyville, and seemingly every community in Allen, Montgomery, and Wilson Counties claimed at least one brickyard and cement company. This story of industrial expansion, which I related in chapter 6, began to ebb as early as 1907 in Iola when pressure in the local gas field fell sharply, indicating an imminent exhaustion of the resource. By 1910 six of the town's smelters were gone, and none of the remaining three lasted beyond 1918. A similar cycle occurred a few years later with the glass companies at Coffeyville and elsewhere.[128]

The decades between 1910 and 1950 were times of economic adjustment in the gas belt. Some cities lost residents, but several others held their own, and one actually expanded. Iola, the birthplace of the boom in 1893, suffered the biggest decline, going from a population of perhaps 14,000 in 1907 to 7,094 in 1950. Iola had decent transportation—a Santa Fe line running north and south (the old Leavenworth, Lawrence and Galveston) intersected by a Missouri Pacific route from Fort Scott into Wichita—but its pool of gas was smaller than others to the south. In 1920, only three major employers remained from the fifteen or twenty of a decade earlier: a brickyard, a cement plant, and a foundry. Another decade reduced this number to two: Lehigh Portland Cement (four hundred employees) and United Tile and Brick (sixty employees).[129]

Iola was able to retain a few of its older companies partly because these particular firms required less fuel than did the smelters and glass factories, and in the case of cement, because it needed an abundant source of water such as the Neosho River (Map 5). Cement and bricks also were products too bulky and too ubiquitous to face long-distance competition. More interesting, perhaps, is the fact that, because Iola was the first sizable city in Kansas to face industrial decline, it was also among the first to recruit assembly plants for clothing and similar products. These operations, seeking locations with reliable and inexpensive workers, were not ideal industries for any town because they were apt to leave in the event that even cheaper labor appeared elsewhere. Still, because jobs were jobs, Iola welcomed as early as 1920 the Brownfield-Sifers

Workers at Lehigh Portland Cement Company in Iola, circa 1925. Started in 1900 during the regional boom in natural gas, this plant was the largest employer in Allen County from 1910 until 1970. Its owners even created their own company town called Bassett just south of Iola. The photograph shows a selection of laborers and office employees but includes none of the firm's African American workers. (Courtesy Kansas State Historical Society)

Candy Company (94 employees), the Iola Button Company (62 employees), and Wheeler-Motter, a maker of overalls (52 employees). An evaporative plant established by the Pet Milk Company promised to be a more permanent part of the economy.[130]

Seven miles south from Iola along the Neosho River sits the town of Humboldt. People there, as well as in Chanute, an additional eight miles downstream, discovered natural-gas fields soon after the original Iola success. Following an experience somewhat parallel with that first boom community, they, too, were each able to retain a large cement plant beyond the halcyon years. At Humboldt, the Monarch company was established in 1907 by Wichita investors and then purchased five years later by a local man. The two hundred jobs at H. F. G. Wulf's plant have formed the economic mainstay for that small community ever since. Chanute's cement company, Ash Grove, had a similar number of workers, but because the gas pool in this region was larger than at Iola and because oil also was abundant, people in northern Neosho County never had to confront the specter of absolute population decline. Chanute's numbers, 10,286 in 1920, remained at this same level over the next three decades.[131]

Because the first large oil strike in southeastern Kansas occurred near Neodesha in Wilson County, that community had received the region's first refinery. Operated by Standard Oil, this facility continued to be a major employer through the 1920s and even to midcentury (844 people in 1920). As mineral explorations had become more sophisticated around 1900, however, most people saw greater potential for oil development twenty miles or so northwest of Neodesha, in the vicinity of Chanute. This finding by itself would have led to substantial growth for that latter community, but Chanute also had the advantage of excellent railroad facilities. Many petroleum entrepreneurs therefore viewed this city as a logical headquarters for regional operations.[132]

The economy of Chanute achieved its apex of strength and balance in the 1930s. A zinc smelter and a glass company had disappeared by that time along with the county's once-cheap natural gas, but the Chanute Brick and Tile Company remained (fifty employees), as did Ash Grove Cement. Moreover, the losses had been more than countered by the oil business. Several exploration and services companies were in town. So were two makers of drilling tools and, most impressively, two small refineries. The oldest of these, the Chanute Refining Company, had opened in 1906 in response to a city offer of free gas, land, and water. Local men built the second in 1931 and then sold it twelve years later to the M. F. A. Oil Company of Columbia, Missouri. These facilities together employed another hundred men and women.[133]

Balancing the minerals industry in Chanute was the railroad. Although the community had been founded because it was the junction of the Leavenworth, Lawrence and Galveston with the MKT, the company that mattered after 1883 was the Santa Fe. After purchasing both the LLG and a smaller, east-west line that linked Chanute with the rest of the Santa Fe system at Winfield, the new conglomerate created a Southern Kansas Division and named Chanute its headquarters. Railroad shops and other facilities soon employed more than five hundred people and formed a bulwark for the local economy for sixty years. This glory ended in 1948. Facing competition from truckers, company officials consolidated operations for southern Kansas at Arkansas City. Chanute people, who already had foreseen this loss, countered by emulating Iola's search for alternative jobs in the apparel business. The Baker Manufacturing Company, a maker of housedresses, had come to town in 1929 and employed 157 people by 1941. Kansas City's Burlington Company built a plant in 1948, hiring another 250 women to cut and assemble overalls.[134]

The Chanute experience between 1910 and 1920 of replacing jobs lost from a natural-gas economy with new positions in the oil and railroad industries was not unique in southeastern Kansas. Forty-five miles to the south, people in Coffeyville discovered that they were the nearest city to what proved to be the second substantial pool of oil in the region. While neighboring towns with only moderately good transportation, such as Caney and Cherryvale, were able to attract one or two factories with this fuel and thereby experience modest growth, Coffeyville became the acknowledged industrial leader in the region. After surging from 5,000 residents in 1900 to more than 12,000 in 1910 via the gas boom, the community kept adding about a thousand new people each decade, for a total of 17,113 residents in 1950.

As I related in chapter 6, Coffeyville's promoters of the gas resource were the most aggressive in the region. Their efforts produced the same range of companies as elsewhere—brick, cement, glass, iron working, and zinc processing—but more of each. They were especially successful at luring glassmakers away from locations in Indiana and Pennsylvania, and at one time had at least ten different companies in residence to make windows, jars, and similar products. Proving that success often feeds on itself, officials with the Missouri Pacific Railroad followed the industrialists into the region by extending tracks from the east. In 1905 they named the city as a division headquarters and thereby brought two hundred additional jobs to town. The final element in the economic transformation, oil, came in 1905 when local recruiters convinced Ohio men to build the National Refinery Company. Its immediate success (two hundred employees early, then three hundred) encouraged owners of the Cudahy Packing Company to diversify into the refining business. Their plant, built in 1908, was sold to Sinclair eight years later. After expansion it became Coffeyville's largest employer, with six hundred jobs.[135]

The Coffeyville economy that reconstituted itself after the gas bonanza was highly oriented to manufacturing. Two refineries remained (the National changed ownership to the Cooperative Refinery Association in the 1940s), as did one major brick maker (Coffeyville Vitrified Brick and Tile). In addition, the city's commercial club recruited two other large, blue-collar employers: the Ozark Smelting and Mining Company and the Acme Foundry and Machine Company. The Ozark people were lead refiners who came from West Plains, Missouri, in 1905. Their emphasis shifted slightly after acquisition by the Sherwin-Williams Paint Company, but the plant continued to employ several hundred residents at top wages. Acme, which was started in 1914 to make iron castings for the brick, cement, and smelter trades, also altered its product mix over the years while maintaining a steady employment for two hundred men. Under the ownership of Ethan L. Graham and then his son Glenn, this company gradually moved into the production of transmission assemblies, valves, pumps, and similar products.[136]

By the 1940s, Coffeyville's reputation as a community with skilled industrial workers began to attract new companies. Ivan Morrow, who had learned his craft at Acme, opened a rival machine shop in 1947. Funk Manufacturing, a maker of small airplanes, came in 1941. Finally, and largest of all with 150 to

300 workers, Dixon Industries moved over from Wichita accompanied by a series of military contracts for airplane components. Coffeyville, a decidedly blue-collar, prounion, and successful community in 1950, was a rare creature on the Kansas scene. Only Kansas City offered a comparable economic structure.[137]

The last remaining city in southeastern Kansas, Independence, is only fifteen miles from Coffeyville. Although never as large as its industrial neighbor, Independence nevertheless was able to maintain a substantial population (10,480 to 12,782) over the period between 1910 and 1950. The two cities shared access to Montgomery County's oil and gas field and had similar initial industrializations. One large cement plant (Lehigh Portland) remained in town at the end of the boom years, along with a slightly newer acquisition that would set a different tone for the future. This pivotal company was Prairie Oil and Gas. Harry Sinclair unexpectedly had relocated its headquarters from Neodesha in 1904. As the enterprise reincarnated itself as the focus of pipeline operations for Sinclair Oil and ARCO, its 250 white-collar jobs gave a decidedly business orientation to the community rather than an industrial one. The company's construction of an elegant, five-story office building in 1916 symbolized this distinction, as did the arrival of many petroleum engineers to staff it.[138]

8

Postindustrial Kansas I: The Interstate Cities since 1950

The railroad-based and regionally oriented economy that had created a relatively stable urban pattern in Kansas for more than half a century began to falter about 1950. Although people did not realize it at the time, a major business restructuring was under way. At its core, this involved agglomerating smaller systems of production into national schemes and then into a global market. The full ramifications of such a massive and complex transformation, which is still ongoing, have yet to be seen, but basic mechanisms and implications are clear. To describe their cumulative effects on traditional patterns of livelihood as far-reaching is understatement.

Seen in retrospect, one of the prerequisites for things to come was a change in transportation that occurred across the United States just after World War II. Once-magical railroad networks ceded their dominance to highways in the movement of people and freight, and in this way options for industrial production became more geographically flexible. Just as fundamental was a switch in employment patterns. The classic industrial structure of the early and middle twentieth century had been based on specialization of labor, economies of scale, and rigid vertical integration within the manufacturing process. Often called Fordism in honor of Henry Ford's pioneering method for manufacturing automobiles, it found expression in Kansas with an array of large aircraft, meatpacking, and milling operations.

Fordist-style manufacturing, and indeed employment in American manufacturing as a whole, peaked around 1970. After that time the nation entered a new and seemingly less predictable state of existence, creating a society that scholars have termed postindustrial. In this realm, automation and computerization have reduced the number of workers needed by manufacturers. Ease of transportation and communication, coupled with an increasingly large and competitive marketplace, also has complicated industrial operation. Centrality still matters, but for some companies, a search for cheaper labor costs and lower taxes has led to relocations away from major cities. Nearly all manufacturers have adopted looser, horizontal networks of production based on obtaining components for their products from independent suppliers rather than controlling the entire process under one roof. This system also contributes to a dispersal of industrial operations.[1]

As manufacturing began to employ fewer Americans, jobs in the service sector of the economy increased to take their place. This, too, is part of the postindustrial process. Some of these gains have been in traditional public arenas such as education, government, and health. Opportunities in the arts and in recreation have expanded as well, but the field's biggest growth has come in specialties needed to support high-technology industry and the new global economy. These fields include financial and legal affairs, research and development, market research, and advertising. Altogether, jobs in the country's service sector have risen steadily, even spectacularly. From 60 percent of the total in 1960, they increased to 71 percent in 1990 and then to 80 percent in 2000.[2]

How exactly has globalization and its related economic changes affected the urban system in Kansas? This question, obviously of concern to residents, also has broader relevance. Most existing case studies about postindustrial society focus either on centers of traumatic readjustment, such as Detroit and the aging mill towns of the South, or on areas that have received hundreds of thousands of new jobs, such as Malaysia and the Mexican-American borderland. Less extreme cases in Kansas City and Wichita, Coffeyville and Liberal, and Derby and Horton can help to provide a more complete portrait of a nation (and world) in a period of major transition. Their stories form the core of this and the next chapter.

A New Urban Pattern

The population of Kansas as a whole grew 41 percent between 1950 and 2000, to a total of 2,688,418 residents. Although this rate was less than the national average, it still represented enough people to fill four and a half new Wichitas or nearly ten new Topekas based on their respective sizes in 1950. Where did all these new Kansans elect to live? At first glance the answer seems obvious (Map 22; see p. 35). Whereas 52 percent of residents had urban addresses at the start of the period (i.e., in cities with populations of twenty-five hundred or more), 70 percent of them did so in 2000. Wichita, the state's largest and most rapidly growing city in 1950, was certainly part of this process. From a base population of 168,279, it more than doubled over the next five decades to a total of 344,284.

Wichita's gains, although impressive, do not accurately reflect the overall nature of the state's recent urbanization. First, rather than claiming simply that big cities have gotten bigger, it is more precise to think in terms of counties. Six such units—Douglas, Johnson, Leavenworth, Sedgwick, Shawnee, and Wyandotte—contained 53 percent of the Kansas population in 2000, or some 1,400,000 residents. Although these same counties also had been important in 1950, they then contained only 34 percent of the state's population, or 632,000 residents. Impressive growth of more than 120 percent in these six units over the last half century represents a gain of some 768,000 people. This number almost exactly equals that for the state as a whole over the same period.

The increasing concentration of the state's population is partially hidden on a traditional charting of city sizes (Chart 3). Although Wichita has indeed grown rapidly since 1950, similar expansion in the Kansas City and Topeka areas goes unrecorded. The missing element, of course, is suburbanization. Commuting ten or more miles to urban jobs by private automobile, a rare phenomenon in 1950, became routine in the decades that followed. This action, in turn, prompted the movement of shopping centers and then industries to suburban locations, thereby diffusing the cities further. The most spectacular Kansas example is Overland Park, an unincorporated place near Kansas City in 1950 that, fifty years later, became the state's second-largest city.

Outside the growth poles of Kansas City, Topeka, and Wichita, the state's smaller urban centers have, on average, stagnated. This generalization hides much variation, however. Whereas a few cities seem nearly unaffected by the changes going on about them, the majority have seen their traditional economies threatened or abandoned. Their strategies for readjustment, whether successful or not, are instructive. Some plans have brought new vigor to their communities. Others were conceived with somewhat lower expectations, and a few were (and are) admittedly short-term efforts designed to stave off seemingly inevitable decline.

Careful comparison of Maps 20 (see p. 33) and 22 reveals an uneven pattern of growth and decline. Eight small cities present on the 1950 version have now fallen below the threshold population of twenty-five hundred: Belleville, Caney, Cherryvale, Council Grove, Ellinwood, Ellis, Horton, and Sunflower Village. Several other, larger places also have lost population, most obviously Coffeyville and other communities in the southeast (Chart 4). All is not gloom, however. Beginning again with small, nonsuburban places, ten new cities can be found on the 2000 map: Burlington, Ellsworth, Hesston, Hillsboro, Lindsborg, Osage City (a return to this level after seventy years), Sabetha, Sterling, Ulysses, and Wamego. Higher on the urban hierarchy, sizable absolute and percentage growth is apparent at Emporia and Manhattan in eastern Kansas; McPherson and Salina in the center; Hays in the northwest; and Dodge City, Garden City, and Liberal in the southwest. Growing cities, like shrinking ones, are spread over nearly every section of the state and feature many different economies. Altogether, they constitute an intriguing pattern, perhaps what one should expect in a world where the rules for survival are constantly being rewritten.

Beyond graphs and maps, an updated functional classification of local economies can help to evaluate the evolving urban system (Table 2). For the sake of comparability, I compiled the new taxonomy using the same measures

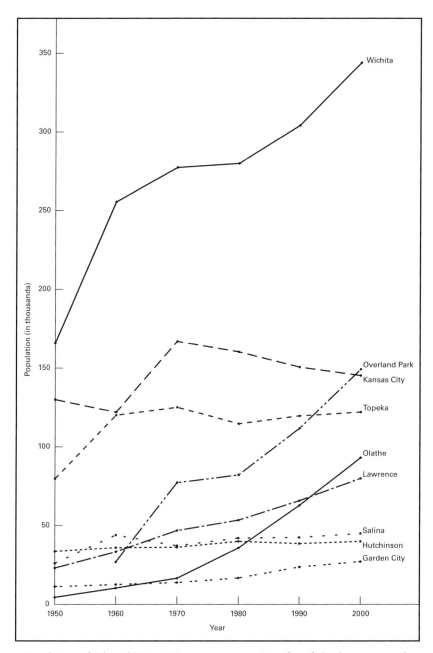

3. Populations of Selected Cities in Kansas, 1950–2000. Data from federal census records.

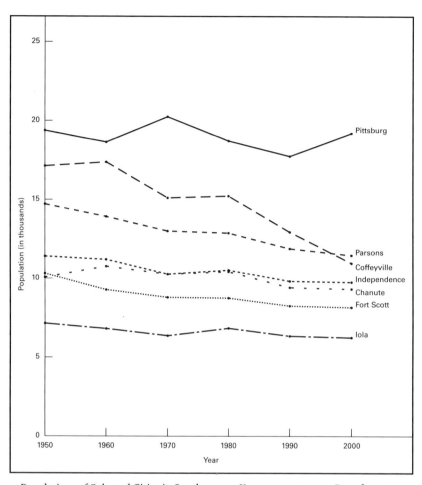

4. Populations of Selected Cities in Southeastern Kansas, 1950–2000. Data from federal census records.

as in earlier years. The only difference has been to double the threshold percentages for employees in particular activities to account for a parallel increase in the workforce that occurred as women began to seek employment outside the home.[3]

Perhaps the most revealing way to examine the city groupings is to compare specializations for three years: 1900, 1950, and 2000 (Tables 1, 2). Whereas few differences exist between the listings for 1900 and 1950, the one for 2000 is quite distinctive. Most obvious, perhaps, is the near disappearance of railroad towns, the largest traditional category. This change began in the 1930s when the gradual introduction of reliable diesel engines lessened the need for repair shops. A severe decline in passenger traffic after World War II hastened

Table 2. A Functional Classification of Cities in Kansas, 2000.

	Qualifying Communities
Army	Junction City
Education	Baldwin City, Colby, Emporia, Hays, Lawrence, Lindsborg, Manhattan, Pittsburg
Health services	Hays, Larned, Marysville, Osawatomie, Parsons, Pratt
Manufacturing	Abilene, Arkansas City, Atchison, Baxter Springs, Chanute, Coffeyville, Columbus, De Soto, Dodge City, Emporia, Fort Scott, Garden City, Gardner, Hesston, Independence, Iola, Liberal, McPherson, Marysville, Neodesha, Parsons, Sabetha, Spring Hill, Wichita, Winfield
Railroad	Herington, Marysville
Wholesaling	Abilene, Lenexa, Mission

Source: Calculated by the author from data in the federal census; Thelma Helyar, ed., *Kansas Statistical Abstract, 2000* (Lawrence: Policy Research Institute, University of Kansas, 2001); and the *Kansas Community Profiles* prepared annually for all cities by the Kansas Department of Commerce and Housing.

this process, and by the 1950s, major companies, including the Rock Island and the Missouri, Kansas and Texas, had declared bankruptcy. Freight traffic continues on Kansas railroads, but only Kansas City and Topeka retain large numbers of railroad workers. Two smaller communities, Herington and Marysville, also have locally important crews. That leaves a dozen or so other cities with abandoned roundhouses and sizable holes in their employment structures.

Another important series of changes has occurred with manufacturing. This is not obvious from a first look at the tables because the total number of Kansas manufacturing towns is substantial in both decades. Only Atchison, Coffeyville, Fort Scott, Iola, Marysville, Neodesha, and Winfield appear on both the 1950 and the 2000 lists, however. Dropping out, most significantly, is Kansas City, along with Augusta, Cherryvale, and El Dorado. Additions are more numerous, perhaps a surprising development given the publicity generated by movements of American manufacturers to foreign countries. Wichita is the biggest of the new cities, followed by Emporia in the Flint Hills and by Dodge City, Garden City, and Liberal in the southwest. A clustering of smaller centers in the southeast also is notable (Map 22).

Although transformations in manufacturing and transportation have affected more of Kansas's specialized communities than the other economic sectors, important changes run nearly across the board. Mining, for example, nearly disappeared as a major employer shortly after 1950. New lead and zinc shafts were opened in south-central Missouri, shutting down the Galena smelter and the rest of the Tri-State District, while Pittsburg coal yielded the regional home-heating market to natural gas and that for power plants to low-sulfur deposits from Wyoming. Wholesale traders, another venerable presence in the region, clearly have taken advantage of ever-better transportation and communication to consolidate their operations. Jobbers in Atchison, Hutchinson, Salina, and Wichita all lost considerable strength over this period to the regional metropolis of Kansas City, or more precisely, to two of its suburbs.

The specializations listed for army, education, and hospitals represent segments of the growing service sector of the economy. Junction City's presence in this regard is nothing new on the regional scene, of course, but expansion is clear in the other two categories. Save Wichita State University, all five other major public universities in the state have grown large enough to strongly influence the economies of their host communities. The same is true for Colby Community College, suggesting a trend that seems likely to spread to Concordia, Pratt, and other communities with similar institutions. Baldwin City and Lindsborg, each home to a small private college, represent another part of the education magnet. Although their colleges are not growing, campus amenities have helped to attract enough new adult residents to push populations above the threshold mark of twenty-five hundred. Finally, and with less notoriety than in college communities, a consolidation of increasingly expensive and specialized health care services has begun to produce "hospital towns." In Kansas, these include the venerable mental-health centers at Osawatomie, Larned, and Parsons plus newer, more general regional foci at Marysville, Pratt, and (most impressively) Hays.

The Kansas Turnpike and a System of Interstate Highways

Before automobiles and trucks could seriously compete with the transportation hegemony of railroads, an integrated system of hard-surfaced roads was a necessity. This achievement by itself would not necessarily change the existing urban hierarchy. For one city to gain a competitive advantage over another, a clear spatial differentiation in highway quality and accessibility also would have to occur. Progress toward the goal of good roads dates from the 1890s, when bicycles first became popular and when the post office department coupled a promise of free rural delivery with a demand for all-weather, accessible routes. These pressures sparked local efforts at grading and ditching, but no substantial advances happened until automobile numbers increased rapidly after 1910. Then, as ownership in Kansas went from 10,490 vehicles in 1910 to 134,000 in 1917, an earlier resistance to funding road improvements vanished. The state legislature assessed an annual registration fee of five dollars beginning in 1913 for this purpose, and federal officials contributed their first funds in 1916.[4]

The highway system that emerged in Kansas between 1910 and 1940 could be described as democratic. Even as state and federal people became more involved in the financing and materials advanced from oiled surfaces to concrete, a grassroots principle that each section of the state be treated alike remained in place. A 1927 law divided Kansas into six equal-area districts, for example, and throughout the 1930s each of these units received the same funding. The result was a map of hard-surfaced roads that resembled a series of spiders. Legs of pavement extended in several directions from each sizable city, but (with the exception of the Kansas River valley) no quality linkages existed between one place and another.[5]

The egalitarian quality of the emerging highway system was not without its challengers. As automobile owners began to tour the country beginning about 1910, various entrepreneurs realized the advantages of marking out routes between cities and of placing their particular towns on as many of these roads as possible. One of the most active Kansans in this regard was A. Q. Miller, editor of the *Belleville Telescope*. In 1911 he helped to organize the Meridian Road to connect a north-south line of county seats from York, Nebraska, to Wellington, Kansas. Two years later he did the same for an east-west route from St. Joseph to Colorado Springs that was called the Rock Island Highway. Miller's pioneering work was critical to having these same pathways designated as federal highways in 1925 (U.S. 81 and U.S. 36, respectively), and he coined the phrase "The Cross Roads of America" in anticipation of growth to come Belleville's way.[6]

The work of Miller and dozens of other Kansas promoters who created the Midland Trail (U.S. 24), the Golden Belt Road (U.S. 40), the King of Trails (U.S. 59), the Great Plains Highway (U.S. 83), and similar routes was not without merit. The presence of the Meridian Road, for example, seems to have inhibited parallel routes one county to the east or west. Still, since anybody could lay out a route, and since federal authorities had guaranteed construction aid for a generous 6,575 miles of roads in the state, it was inevitable that every community of size would eventually be well provided with concrete pavement. Belleville merchants therefore could not gain much ground, if any, on their rivals in nearby cities.[7]

The policy of road equality was first challenged by the needs of World War II. Activity had increased so much at Fort Riley by early 1941 that local traffic was congested. Smaller but still significant problems of access emerged a year later at the Olathe Naval Air Station, airplane factories at Wichita, new air bases at Topeka and Salina, and munitions plants under construction near De Soto, Parsons, and Pittsburg. Selected highway improvements were seen as a national necessity. With the federal government's Defense Highway Act of 1941 providing funds, improvements at these sites were completed quickly, including a seventeen-mile relocation of U.S. 24 around the north edge of Fort Riley.

A second, and more controversial, bypass on U.S. 40 would have connected Topeka directly with Junction City across the lightly populated interior of Wabaunsee County. Angry merchants from St. Marys, Wamego, and Manhattan in the original valley location of that highway delayed this project for five years, but everybody else liked the prospects of shaving twelve miles off the journey from Topeka to western Kansas and of having the federal government finance nearly all the work.[8]

Besides a need to alleviate traffic bottlenecks around defense installations, war conditions also demonstrated the importance of a high-quality road system to link the nation's industrial and military centers. Consequently, Congress authorized the secretary of war to designate a series of existing routes as a national "strategic network" and agreed to finance 75 percent of the costs to upgrade their quality. The plan released for Kansas in late 1941 included parts of eleven different federal and state highways, a total of about fourteen hundred miles (Map 23; see p. 37).[9]

Shortages of materials and higher military priorities delayed implementation of the 1941 program, but as the war ended, the Federal Aid Highway Act of 1944 renewed the call. When asked how Kansas people preferred to spend their share of the $500 million authorized for construction, the state highway commission recommended essentially the same set of routes as it had in 1941. Federal officials, however, had the final say. Judging this list too ambitious, they approved only a third of its mileage at a meeting in 1946. Their choices for inclusion in a new system of "interstate" highways were the combined U.S. 40-83-24 route west from Kansas City to Colorado, U.S. 50 from Kansas City to Newton with an extension south to Oklahoma along U.S. 81 and U.S. 177, and U.S. 66 in the state's extreme southeastern corner.[10]

Although no one has reported exactly how the routes of 1946 were selected, they did meet the basic requirements of linking the state's largest cities and of passing through Fort Riley and the new Smoky Hill Air Force Base (later Schilling) near Salina. Still, no matter how logical in overview, the choices could be debated in particular cases. West of Salina, the choice of continuing on U.S. 40 rather than dropping south to U.S. 50 North (today's U.S. 56) was almost surely determined by the presence of Denver on the former route versus the much smaller Pueblo on the latter. Why this route jumped from U.S. 40 to U.S. 24 at Oakley is more uncertain, although such an alignment had a precedent in the Golden Belt Highway of earlier years, provided service to the relatively large communities of Colby and Goodland, and shortened the trip to Denver by several miles.[11]

Of the possible north-south routes, planners had to give highest priority to connecting the airplane center of Wichita with Kansas City and Oklahoma City. Once this was accomplished via a simple linkage along Highways 50, 81, and 177, the basic needs were covered. Of the remaining candidates for inclusion from the 1941 list, U.S. 81 north of Salina traversed only farming country in Nebraska and the Dakotas. A better case existed for U.S. 73 East (today's U.S. 69), which extended from Kansas City to the vicinity of munitions factories in Cherokee and Labette Counties. This route lacked direct connections with any big city, however. U.S. 75 north from Topeka probably was the hardest route for officials to eliminate because it ran directly to the population and military center of Omaha.

Even though the decisions of 1946 refined the strategic highway plan, the question of whether a particular road received the new interstate designation remained moot for several years. This was because the federal government stuck by its earlier agreement to fund 75 percent of the improvements for the entire "strategic" list of roads. Conditions changed by 1952, however. Demands by Kansas drivers for better highways threatened to outstrip state funds available for construction, especially after expenses from a disastrous recent flood on the Kansas River. In casting about for solutions, highway commissioners hit upon the concept of toll roads. This idea, pioneered in Pennsylvania, was currently being implemented in Indiana, New Jersey, New York, and Ohio. Another project was proposed north from Oklahoma City to the Kansas state line. With financing to come from a bond issue rather than increased taxes, plans for a similar roadway in Kansas faced little opposition.[12]

The legislature created the Kansas Turnpike Authority in 1953. Within a year, this new entity had received reports from two consulting firms and approved a route. Influenced primarily by traffic counts, the proposed connecting link in Oklahoma, and a desire for efficiency, the path extended along a fairly straight line from Kansas City to the Oklahoma border south of Wellington. Kansas City and Wichita were obvious spots on the route, as was Emporia at the halfway point, but some question existed on how best to connect Kansas City with Emporia. The most direct path would have paralleled Highway 50 South through Ottawa, but a slightly longer course passing through Topeka would generate more customers. With traffic counts obviously an issue for bond sales and future toll revenues, this decision was easy. Construction via Topeka began the last day of 1954 and was completed in October 1956 (Map 24; see p. 38).[13]

The Kansas Turnpike—straight, smooth, and uninterrupted by cross traffic—revolutionized transportation in the state. It cut not only twenty-six miles from a trip between Kansas City and Wichita but an even more impressive ninety-five minutes. Although revenues in the 1950s were not as high as expected because the anticipated connecter road in Oklahoma was not completed until 1961, every driver realized that this highway far exceeded the quality of anything that had come before. Seeing that the turnpike would soon affect industrial patterns and regional growth, booster groups immediately made several calls for extensions. One campaign touted the Topeka-Salina

West Lawrence Toll Station on the Kansas Turnpike, circa 1960. Early reports about the turnpike stressed engineered efficiency rather than influences on city growth. Each double lane measured twenty-four feet across, with another fourteen for shoulders. For a mere $3.80 drivers could sample the entire 236-mile length at speeds of more than seventy miles per hour. A football game at the University of Kansas caused the crowd in the photograph. (Courtesy Kansas State Historical Society)

corridor, another the Missouri River valley between Kansas City and St. Joseph, and a third the straight shot from Wichita northwest to Hays via Hutchinson and Great Bend. The Kansas Turnpike Authority made feasibility studies for the latter two routes but concluded that traffic counts would not justify the expense.[14]

Lobbying for new toll roads came to an abrupt end in 1956. Truckers, who opposed the growing fee-based system for travel, joined forces with military planners and others at the national level to advocate limited-access highways financed by some alternate means. Their effort focused initially on the creation of a dedicated trust fund for construction derived from taxes on fuels, tires, and new vehicles. Once this was achieved, the famous Interstate Highway Act of 1956 was an easy sell. This legislation provided 90 percent of the funds for a forty-one-thousand-mile, interconnecting system of ultramodern roads. State highway departments again would suggest routes to be included, but final approval would rest with the United States Department of Commerce.

Existing toll roads also could be brought into the network, but lawmakers left details vague on how to compensate the states involved.[15]

The general form of the Kansas interstate system was never in doubt. In fact, three weeks after President Eisenhower signed the legislation, state and federal authorities had agreed on the same three corridors identified in 1946. People in Baxter Springs and Galena suffered a loss when, to straighten the route between Joplin, Missouri, and Tulsa, Oklahoma, the interstate version of U.S. 66 was relocated just outside the Kansas border. This change was little noticed statewide, however, because bigger negotiations were under way over how to incorporate the Kansas Turnpike into the new system. The result, made with minimal input from the public, was a classic bureaucratic horse trade. To reduce federal costs, state highway commissioners allowed the turnpike section from Kansas City to Topeka to become part of new Interstate 70 and the section from Emporia to the Oklahoma line to become part of new Interstate 35. In exchange, federal officials agreed to build the portion of I-35 between Kansas City and Emporia along the path originally planned in 1946, that is, following U.S. 50 and 50 South through Ottawa. Kansas also would get an additional interstate segment: U.S. 81 from Salina to Wichita. This route, originally I-35 West and now I-135, would connect two important airplane centers and provide interstate access to Newton, a city included on the 1946 plan but excluded from the streamlined path of the turnpike (Map 24).[16]

As freshly poured sections of I-35 and I-70 opened in the late 1950s to join the existing turnpike mileage, urban leaders rapidly came to realize that proximity to these limited-access highways would be fundamental to the next generation of regional growth. This conclusion, obvious in retrospect, nevertheless caught entrepreneurs off guard at the time. As a group, they had grown up in a railroad world and not been active in planning the new network. Instead, dispassionate engineers had dictated the path of the Kansas Turnpike, while guaranteed funding from the highway trust fund had helped to insulate the interstate network from local and regional lobbies.[17]

Armed with a growing awareness of the transportation revolution, members of the Kansas business community saw a clear pattern of winners and losers when they looked around their state. No complaints came from Kansas City, Topeka, and Wichita, of course, but all the area north of I-70 essentially was disenfranchised. So was the state's entire southwestern quadrant. Still, because these latter two areas were largely rural, construction decisions there could not be questioned. Hutchinson people had a more legitimate complaint. Their city, with 33,575 citizens in 1950, was the fourth largest in the state. They had no major military facility, however, and so highway engineers had not been able to justify a thirty-mile, westward shift in their route from Salina to Wichita. Residents of McPherson and Newton, on the other hand, were fortunate beneficiaries of this decision.

Southeastern Kansas contained the largest number of people unserved by the new highways. Leaders from this region were bitterly disappointed that U.S. 69 had not been selected as an interstate corridor, and they were therefore among the first Kansans to move the debates over new highway construction into the arena of regional politics. Their chance came in 1956. When federal officials solicited proposals for how to allocate an additional thousand miles of interstate highway, a lobbying group called the U.S. 69 Association pushed for inclusion of their corridor. The Kansas Highway Commission supported this idea and sent it on, but hopes for federal acceptance dimmed when it was learned that Missourians had submitted a similar proposal for U.S. 71, a parallel route only one county to the east. Predictably, perhaps, officials in Washington, D.C. took no action on either idea.[18]

A possibility for reviving the southeastern plan came in 1959 as an indirect result of turnpike revenues not being as high as expected. Reasoning that the still-unbuilt portion of I-35 between Ottawa and Emporia might eventually reduce turnpike use even further, state officials offered their federal counterparts a chance to abandon this part of the construction and to bring the portion of the turnpike between Topeka and Emporia into the interstate system instead. Jack L. Goodrich, a member of the Kansas Highway Commission from Parsons, saw opportunity in this proposal. He argued that, even though the basic exchange had merit, it was foolish for Kansans to give up a federally funded road and receive nothing in return. Why not pressure federal planners for a substitute interstate link to connect Ottawa with Tulsa (Map 24)? Such a new route, he said, "would serve more Kansans than if U.S. 69 were put on the interstate system."[19]

Goodrich's suggestion was opportune, logical, and very popular in his hometown, but it necessarily angered people in Fort Scott and Pittsburg. Debate raged for a year and followed predictable lines of thought. To the view from U.S. 69 that previous designations made by federal officials "should not be disturbed," the Ottawa–South Interstate Highway Association countered that a direct road from Kansas City to Tulsa would "traverse the heart of southeast Kansas," benefit the state as a whole, and stand a better chance for funding than the alternative. After all points were aired, the Kansas Highway Commission in July 1959 formally endorsed the Goodrich proposal, substituting it for the U.S. 69 plan in a new recommendation to Washington.[20]

The "Ottawa-South" interstate never was built. It lingered on the federal list of projects for four years but gradually fell victim to changing values and rising costs. The highway trust fund continued to generate dollars, of course, but public sentiment began to favor investment in loop roads around major cities such as I-435 and I-635 in Kansas City and I-235 in Wichita. When offered a chance in 1963 to rescind their earlier recommendation in exchange for a promise to complete the original short stretch of I-35 remaining between Ottawa and Emporia, the state's commission agreed.[21]

The Ottawa decision was the last major action taken on the Kansas interstate system. Although lobbying efforts persisted and a feasibility study was conducted in 1966 for turnpikes from Wichita to Hays, from Wichita to Joplin, and along U.S. 69, nothing materialized except a plan to gradually raise these pathways (and a few others) to interstate quality through the state's regular funding process. Progress continues in this way as I write, but an incremental approach clearly has been too little, too late from the perspective of competitive city growth. Broadly speaking, anticipation of substantial population increases for any sizable city along the state's interstate system would have been easy to justify for the decades between 1950 and 2000. Similarly, stagnation or decline would be a reasonable expectation for any community outside this high-speed network of asphalt and concrete.[22]

Kansas City and Johnson County: A Dispersed Regional Center

Building on the axiom that capitalist economies concentrate power and money, it follows that this concentration should increase as modern technology lessens the restraints of time and distance. Metropolitan Kansas City has been a beneficiary of this process, becoming one of twenty-five or so American cities to achieve the status of regional metropolis. With a current total population of 1,776,062 (about 40 percent of whom live in Kansas), this community has sphere of influence that encompasses the entire state except for a handful of counties in the far northwest that look instead to Denver. This dominance, which is as true for the service sector as it is for wholesale or other trade, can be seen through measures as diverse as shopping behavior and allegiances to professional sports teams.[23]

From a Kansas perspective, the most interesting aspect of greater Kansas City over the past five decades has not been overall growth so much as a reconfiguration of the economic and social roles played by the residents of Wyandotte and Johnson Counties. Johnson County, as anybody familiar with the Midwest knows, went from being a typical farming region to the second-largest population center in the state during these years. Wyandotte County's concurrent experience, continuous economic adjustment in the face of new market conditions, is less flamboyant but equally compelling. I begin with the older center.

People in Kansas City, Kansas, felt good about themselves in 1950. Their traditional milling, packing, and soap-making industries continued to employ thousands of workers. An even larger number of jobs existed at their new, two-thousand-acre Fairfax Industrial District. As one of the first and finest places designed exclusively for manufacturing and distribution, Fairfax was garnering national praise seemingly every day for its efficient and attractive layout, on-site parking, and easy access to rail, water, and air transportation. Businesses actually competed with one another for the chance to locate there. These included the "largest bakery in the world" (Sunshine Biscuit Company), the biggest single employer in the state (a forty-one-hundred-person assembly plant for Buick and Oldsmobile cars), and eight of the nation's leading manufacturers of farm implements. With 125 companies in all, twenty-five thousand workers, and an annual payroll of $100 million, Fairfax was a true community jewel.[24]

Despite all its advantages and inertia, the economy of Kansas City, Kansas, was not totally healthy at this time. Even as community leaders symbolically congratulated themselves in 1951 by opening a major downtown convention center (the fourteen-story Town House hotel), adjacent retail businesses on Minnesota Avenue were beginning to experience revenue declines. Many more people had automobiles than before the war and were using them to patronize bigger stores that existed just across the state line. Then, that same June brought one of the worst floods in state history. It damaged every enterprise in the bottomlands, including Fairfax, the various railroad yards, the mills, and the packers. Although Fairfax investors quickly authorized $12 million to improve levees and drainage systems and most businesses reopened, the flood caused many executives to rethink long-term business strategies.[25]

The packers were the most concerned. Their multiple-story plants, modern in 1890, were not well suited to newer conveyor belt styles for processing animals. The old sites also had inadequate parking for a new generation of automobile-driving workers. These issues, minor at first, grew in importance throughout the 1950s and 1960s. Then came an even bigger complication. In place of the traditional arrangement whereby thousands of midwestern farmers each fattened a few head of cattle, entrepreneurs in more remote (often irrigated) locations began to construct large feedlots that could handle hundreds of animals at one place. As this trend developed, it became obvious that new, one-story packing plants could be built profitably near these feedlots. This possibility turned into a reality in the late 1960s. Within a span of twenty years, Kansas City went from a leadership role in the business to complete noninvolvement, from eleven packing plants to none.[26]

The other traditional mainstays of the Kansas City economy, although faring much better than the packers, also have done little to promote local population growth in recent decades. The two soap makers, Colgate-Palmolive and Procter and Gamble, remain in their familiar Armourdale locations. The city also has retained its status as a national center for milling and railroads, and as a regional one for shipping and wholesaling. These industries all have in-

Buick-Oldsmobile-Pontiac Plant in Kansas City, 1960. General Motors came to the Fairfax Industrial District in 1946 and quickly became the city's largest industrial employer. It maintains this status today. (Courtesy Kansas State Historical Society)

creased the volumes of goods they produce or handle. Because of consolidation and mechanization, however, they were able to make these gains while simultaneously shedding employees. The two soap companies, for example, which had hired a total of 1,210 people in 1919, operated with half that number in 1995. The Union Pacific Railroad absorbed the Missouri Pacific system over this same period but still cut back four hundred workers locally to a new level of five hundred. The only notable exception to this employment trend occurred at the Santa Fe yards in Argentine. Several major expansions of facilities there, including a gravity assembly system for freight trains, an extension of its classification yard westward past Turner to a total of nine miles, and a new office building and diesel-service center, have greatly increased the city's role within this railroad's overall management scheme. Such actions also have enabled local yard employment to hold steady near its 1920 level of thirteen hundred workers.[27]

Although losing the packing industry and seeing contractions by other long-time employers hurt Kansas City pride, increased mechanization and efficiency were occurring in every industrial center across the country. The real problem locally was not losses in the old industries but threats of relocation from the newer manufacturers in Fairfax plus an inability to compete effectively within the metropolitan area for most parts of the growing service

The Great Flood of July 1951 in Kansas City. Taken from bluffs in the Strawberry Hill neighborhood of Kansas City, Kansas, this view looks southwest across the Central Industrial District. A portion of the intercity viaduct (now Interstate 70) is visible on the extreme left. The Armour packing plant occupies the middle foreground, with the stockyard just beyond the photograph to the right. Experts estimated total losses in Kansas at more than $2 billion. (Courtesy Kansas State Historical Society)

economy. These were complex issues. Because lofty office buildings filled with advertising firms and insurance executives had never been central to the identity of this proudly blue-collar city, no one was surprised that such companies continued to stay away. A pair of decisions made by General Motors officials in 1955 were much more worrisome; they suggested that Kansas City had weaknesses even in its traditional appeal to manufacturers.

Because General Motors maintained two large assembly plants in greater Kansas City (the one at Fairfax and another on the Missouri side in Leeds), it came as no big surprise when the company announced that both a large training center for mechanics and a manufacturing plant for batteries would be constructed in the area. Neither of these facilities went to a traditional industrial site, however. The training center, one of six in the nation and intended to instruct five thousand people per year, was placed along a highway, not a railroad line. Its location was the small Johnson County suburb of Shawnee. Even more disturbing from the perspective of Kansas City, Kansas, was the battery plant. This employer of 350 was the type of industry that a few years before would have fought to get into Fairfax. Instead its owners chose a site on the edge of Olathe, another Johnson County position even farther from the central city. H. D. Dawson, the general manager for the Delco Remy Division of General Motors, noted that this site was served by the Santa Fe railroad and then added simply that the company "prefers to operate in smaller communities." Clearly, however, the situation was much more involved than this.[28]

The biggest issue for the battery plant was transportation, in terms of both automobiles for workers and trucks for supplies and product shipment. This was ironic from the perspective of Fairfax promoters because they had always prided themselves on this very point. The problem was that, while Fairfax was designed with excellent railroad, airplane, and barge service, its road access had always been marginal. The construction of the Seventh Street Trafficway in 1933, which bridged both the Kansas and Missouri Rivers and was designed as the new route of U.S. Highway 169, had been undertaken with the new Fairfax in mind. It served the area well in the years before World War II, but it traversed the heart of the city and was designed as a conventional versus a limited-access route. Soon it was clogged with traffic.

To alleviate the growing complaints of Fairfax workers and suppliers, city officials began to lobby as early as the 1950s for an extension of the Kansas Turnpike and/or interstate system into their district. They found sympathy for their cause but a competition for dollars with an older city plan that would build a route parallel to the elevated Seventh Street Trafficway along Eighteenth Street into Argentine. When the Eighteenth Street project received priority (it opened in 1956), Fairfax began to suffer. Workers living in new suburban developments resented congested drives, and companies heard similar complaints from truckers who carried increasing percentages of their supplies and products. A solution finally arrived in 1975 with the opening of Interstate 635, which connected Interstates 29, 35, and 70 in a loop roughly parallel with Fortieth Street. The project included a two-mile spur road of the same quality (Kansas Highway 5) that ran directly into Fairfax.

Highways 5 and 635 were godsends for Kansas City in that they were essential to the continued profitability of Fairfax and to the retention there of the 3,350 employees who today work at the big General Motors assembly plant as well as the 550 at Sunshine Biscuit, the 525 at Owens-Corning Fiberglas, and the slightly lesser numbers at Midwest Conveyor, Sealy Mattress, and a hundred other factories. They were not enough to completely reinvigorate it, however. The same argument held for Armourdale with its two soap companies. The biggest industrial bonus associated with I-635 occurred between its junctions with I-35 and I-70 in the Kansas bottomlands. This area, called Turner, was adjacent to the western section of the Santa Fe yards. It became the city's newest industrial center in the 1970s when two of the area's major grocery wholesalers relocated there (Fleming Foods and Associated Grocers—a total of 2,025 employees) along with much of its trucking industry (Overnight Transportation, Roadway Express, the U.S. Post Office's regional bulk-mail center). Besides having an enviable combination of highway and railroad access that was critical for these types of business, this site was also much closer than Fairfax to suburban developments in both Wyandotte and Johnson Counties.[29]

When one compares the total employment picture for the various manufacturing, transportation, and wholesale industries that formed the traditional core of the Kansas City, Kansas, economy from the 1950s to the present, the result is a slight loss. This finding is encouraging in some senses, for jobs in these sectors have plunged much more steeply nationwide. Still, the city has experienced a slow erosion of its tax base since 1980 and is now unable to meet the criteria for designation as a manufacturing specialty place (Table 2).[30]

Despite its struggles to cope with ongoing industrial trends, Kansas City still managed a modest population increase between 1950 and 2000, from 129,553 to 146,866. Some of this was the result of expanding political boundaries, but the city's participation in selected aspects of the growing service economy helped as well. The shining star in this regard has been hospitals. Through a series of consolidations, the Bethany Medical Center downtown and the Providence–St. Margaret Health Center on the west side have emerged as two of the most comprehensive municipal facilities in the state. Each has more than a thousand employees. Even more impressive, and serving a region far beyond the metropolitan area from its Rosedale base at Thirty-ninth Street and Rainbow Boulevard, is the University of Kansas Medical Center. Placed there because of a land donation in 1894, this unit has grown from a single building in 1905 to a sprawling research, specialty care, and training center. Its total of 5,800 employees increases every year and now dwarfs even that at the General Motors plant.[31]

Beyond hospitals and local government, the two other major service employers in Kansas City in recent years have been its community college (750 people) and a regional office of the Environmental Protection Agency placed downtown through the political influence of Robert Dole, the longtime United States senator (700 people). Although these two institutions have been important to a city that has struggled financially, neither offers the possibility for substantial additional growth. Local officials thought they saw a panacea when Kansas voters approved pari-mutuel wagering in 1986. They acted quickly and obtained a license to build a large horse and greyhound track called the Woodlands. Their site was adjacent to a just-opened Interstate 435 (the outer loop road for the metropolitan area) and so seemed ideal. The track opened to great expectations, and more than 550 people worked there in 1995. Thereafter, competition from riverboat gambling in nearby Missouri eroded its profitability.[32]

Even as prospects for the racetrack faded, its presence inspired local planners and officials to look further into the recreation industry as an economic engine. They merged their city and county governments in 1997 to allow for better coordination and explored possibilities for a Disney-style family theme park. Eventually, however, they signed an agreement with a Florida firm to build another state-of-the-art racetrack, but this time for automobiles. Called the Kansas Speedway, it focused on a seventy-five-thousand-seat stadium set on fifteen hundred acres near the intersection of Interstates 70 and 435. The city-county retained 400 of these acres for a major retail "tourist district." As I write in 2003, the shopping area is still in development, but the Speedway itself, which opened in 2001, has been judged a total success. With it Wyandotte County may at last be able to enlarge upon its longtime reputation as an important but aging center for manufacturing and transportation.[33]

If a person tried to envision a portion of the metropolitan area that could stand as the antithesis of Kansas City, Kansas, that conception would match closely the reality of northeastern Johnson County. Whereas the former area is ethnically rich, venerable by Kansas standards, and (with its gritty industrial base) solidly working-class, the latter place is almost entirely white, recently developed, and high income. Where businesses exist, they tend toward highway-oriented shopping malls, corporate office parks, light industries, and distribution centers. The state of Kansas, it seems, has somehow ended up with two extreme segments of a modern metropolitan area, leaving the slightly larger, middle section of life to develop on the Missouri side of the border.

Johnson County so typifies suburban development in the United States that its general characteristics need little recounting here. Its principal distinction has always been the high social status of its new communities relative to others in the metropolitan area. Many commentators also note how unusual it is for such wealth to have emerged across a state line from the region's principal retail and commercial center. These two facts are easy to explain, for they both were creations of Jesse C. Nichols, a man whose reputation is legendary in Kansas City.

Nichols pioneered the development of modern shopping centers in the United States between 1914 and the mid-1920s with his Country Club and Brookside Plazas. These sites, about four and six miles south of downtown Kansas City, respectively, are on the Missouri side of the state line but less than a mile from Kansas. He was equally innovative in planning upscale residential communities in this general area, doing so on a scale large enough to afford the donation of gateway statuary, land for churches and schools, and similar amenities. His Kansas developments began in 1908 with the purchase of a 230-acre tract near what is today Sixty-third Street and State Line Road. To induce Missourians to cross the political border, he decided to create there the region's most exclusive residential community. First he ringed the area with three golf courses. Then he laid out large lots, limited construction to stone and brick, and added a series of deed restrictions to protect property values. This was Mission Hills.[34]

The strategy of promoting conspicuous displays of wealth worked exactly as planned. As several hundred affluent people began to build in Mission Hills in the years after 1910, tens of thousands of middle-class Kansas Citians became envious and vowed to live as close to this Elysium as possible. J. C. Nichols, of course, was more than willing to accommodate. In 1941 he began to develop five square miles to the south and west of Mission Hills as Prairie Village. Later that decade came the smaller Fairway, Westwood, and Westwood Hills adjacent to his jewel on the north. Roeland Park followed in 1951, the year after the developer died and just before his company decided to concentrate in the future on subdivisions rather than entire communities.[35]

J. C. Nichols never envisioned his new cities as anything other than residential settings. He would interrupt the houses occasionally for a park, a school, or a small shopping center, of course, but industries, railroads, and highways were not part of the agenda. This philosophy was that of most suburban developers, of course. Still, the same amenities of trees, quiet, and fresh air that made for comfortable housing also appealed to certain kinds of businesses, particularly those in the service sector and others that would not offend residents with their smoke or noise. The concept of placing corporate and industrial enterprises in suburban positions took several decades to incubate across the country, but it gathered momentum in the 1970s. When it did, a previously homogeneous landscape began to develop variety. Many local governments did not embrace the new possibility, wanting their towns to remain pure residential havens. Others would have welcomed the additional tax revenue provided by businesses but lacked the prerequisite of abundant land lying adjacent to the interstate highway system. A few places, however, had both the will and the means to merge the two roles. In so doing they created a new phenomenon that has come to be known as edge city.[36]

About half of the Johnson County suburbs have remained almost totally residential: all six of the Nichols creations plus Leawood, which the Kroh Brothers company began to develop directly south of Prairie Village in 1948. Significantly, no railroad lines or interstate highways cross these places with the exception of a single mile of I-435 in Leawood. The possibilities for industrialization were much greater in Mission, Merriam, Lenexa, Olathe, and Overland Park. The first four of these lie in a northeast-to-southwest arc along the route of the old border-tier railroad, now part of the Burlington Northern Santa Fe system. When this same pathway was selected for Interstate 35 in 1970, a major opportunity presented itself to interested officials. These possibilities grew still larger late in that decade when plans were announced for the area's final interstate link, I-435, which would intersect I-35 at nearly a right angle and cross through four miles of Lenexa and six of Overland Park. Lenexa, at the junction of the two roads, became especially favored once the newer highway opened in 1985.[37]

Jack Marshall, the mayor of Lenexa from 1965 to 1970 when his town had fewer than five thousand residents, was one of the first people in the region to see the future. By convincing his city council to aggressively seek light industry and distribution firms, he initiated a chain reaction. First came a master zoning plan that led to the creation of thirteen industrial parks by 1981 and then additional incentives through industrial revenue bonds. With this program already in place as the interstates were completed, Lenexa had a head start on the competition that it has never relinquished. Company officials who want a location in the Kansas City area continue to choose Lenexa in large numbers, collectively raising property evaluations there from $2 million in 1960 to an amazing $414 million in 2002. The city's thirty-one industrial parks now include, on the distribution side, a mammoth J. C. Penney catalog center with 1,400 employees and major regional operations for United Parcel Service (825 jobs), the Sports Apparel Corporation (575), Puritan Bennett medical gases (500), and Ford automobile and truck parts (230). Among the larger manufacturers attracted to the city are Deluxe Check Printers with 1,000 employees, Johnson Controls (plastics) with 485, Gill Studios (industrial printing) with 450, and the Coca-Cola Bottling Company of Mid-America with 325. Their collective contribution to city coffers has permitted Lenexa's homeowners to enjoy extremely low property taxes, a factor that pushed the population to 40,238 by 2000.[38]

The successful blending of residential life with interstate industry in Lenexa quickly found imitators in Olathe and Gardner, the next two communities south along I-35. These were a bit farther from Kansas City proper, of course, but could offer abundant land at lower prices. In addition, Olathe promoters touted access to the Santa Fe's main line while Gardner people promoted the amenities of a refurbished airport designed expressly for industrial and executive use (the former Olathe Naval Air Station, now renamed New Century Airport).

Olathe's transition from sleepy county seat to industrial suburb actually began with the acquisition of two companies long before the new highway: the Delco Remy battery plant discussed previously (1956) and the King Radio Corporation in 1959. The King facility, unlike the Delco one, came to town without acclaim. It originally was a small endeavor set up by a local man who had become dissatisfied with the quality of communications equipment then available for light aircraft. This market niche proved extremely profitable, and Edward King Jr. decided to expand the company where he had started it. By 1982 he had 1,239 employees in Olathe and an equal number at nearby branch factories in Lawrence, Ottawa, and Paola. With 2,000 people currently on the job at the main plant, his company (now a division of Honeywell) remains the biggest private employer in the city.[39]

Although the King operation is unusual for a suburban community in being homegrown, it is otherwise typical of the industry drawn to such interstate corridors. More recent acquisitions for Olathe include distribution centers for Sysco Food Services (500 employees), Dillard's department stores (200), and Aldi Foods (100). TransAm Trucking (500 employees) is also present, along with Garmin International (a maker of global positioning systems with 1,000 employees) and Osborn Laboratories (test services and testing kits with 400). The local population, 5,593 in 1950, boomed to 37,258 by 1980 and to 92,962 in 2000. Surprising even its residents, this total currently ranks Olathe as the fifth-largest city in Kansas.[40]

Gardner, six miles southwest of Olathe, doubled its size to 9,396 people in the last decade and seems poised to do so several times again before 2010. This old community, known primarily during its first hundred years as the junction between the Oregon and Santa Fe Trails, received a temporary boost in 1942 when the United States Navy established an air station nearby to train pilots. This was a short-lived affair, and the 850-acre site was isolated enough that in 1966, when the military decided to relinquish ownership, Johnson County officials were hesitant to accept it as a gift. They did so, however, renaming it the Johnson County Industrial Airport. Then, primarily through the promotional work of Joseph Dennis, the new facility's first director, business leaders began to see the advantages there of a combination of inexpensive land, interstate access, and air freight. The biggest tenants, the North Supply wholesale equipment division of the Sprint telecommunications company (1,023 employees) and the bar-code maker Graphic Technology (550 employees) are enough by themselves to qualify Gardner as a manufacturing specialty city (Table 2).[41]

Johnson County's largest edge city, Overland Park, differs from its neighbors in both history and present orientation. Its origins go back to entrepreneur William Strang Jr., who, after a major flood in 1903, saw the value of upland locations for future urban expansion. He therefore purchased six hundred acres near today's intersection of Seventy-ninth Street and Metcalf Avenue. Because his site was eight miles from downtown Kansas City, Missouri (too far to commute by horse and buggy), he built an interurban railroad. Strang's Overland Park development attracted several thousand residents but lost much of its allure in the 1940s and 1950s to the newer and more stylish residential communities of J. C. Nichols. To compete, local planners decided on a policy of aggressive land acquisition with an eye to attracting corporate offices and light industry. Thinking like their neighbors in Lenexa, they hoped to keep residential property taxes low and thereby to have an advantage in their competition with Leawood and Prairie Village.[42]

Annexation proved easy in the early 1960s because no one else was doing it. From thirteen square miles in 1961, Overland Park quickly increased to forty-four, extending north to 47th Street and, more impressively, south to 143rd.

This expansion was successful in that it eliminated future westward growth by Leawood, Mission, and Prairie Village, and that it included many potential business sites. The process also was frustrating, however, in that the I-35 corridor established in 1970 lay almost entirely within other municipal jurisdictions. Fortune smiled fifteen years later when the route of the new I-435 passed directly through Overland Park's heartland.

Because of the timing of the two interstate routes, Lenexa entrepreneurs had a head start in pursuing distributors and industrialists. Rather than compete directly for this market, Overland Park people decided to court a different clientele. Seeking corporate offices, they labeled themselves "executive country" and worked primarily with Thomas Congleton, who recently had developed a successful industrial park along I-35 at the Lenexa–Overland Park border. Congleton's new vision was called Corporate Woods, a lavishly designed environment in which office buildings would be set amid the serenity of trees, grass, and water. It was, quite simply, the suburban residential dream applied to business America.[43]

The original Corporate Woods development, set on 294 acres adjacent to I-435, was in place as the interstate opened and has been successful beyond everybody's expectations. Seventeen buildings were open by 1981, housing, among others, the premier Kansas City engineering firm of Black and Veatch (2,600 employees in 2002), the Farmers Insurance Group (920 employees), and the Employer's Reinsurance Corporation (600). Favorable press reports soon spawned additional developments, including a complex called Executive Centre, major hotels, and, most impressive, the world headquarters of Sprint Corporation, the telecommunications giant. Sprint, a company with local roots, began construction in 1997 for a "campus" similar in look and scope to Congleton's model. Set on 247 acres near 119th Street and Nall Avenue, the facility is designed to contain twenty main buildings, eighteen parking garages, six thousand trees, and a lake. As of 2002, eight thousand Sprint employees worked there, and there were plans to add another six thousand. These positions, plus continued annexations south around both Leawood and Olathe, have helped Overland Park to achieve Kansas's fastest rate of urban growth over the past two decades. With 149,080 residents, the young city was second in size only to Wichita in the 2000 census.[44]

Wichita and Its Satellite Communities

The nature of modern-day Wichita is perhaps best revealed by its inclusion on a list of manufacturing-specialist cities (Table 2). This listing is unusual in two ways. First, economic specialization of any sort is uncommon in larger cities because of reduced chances there for dominance by a single company or institution. Yet, somehow, it has occurred in a city that is large already (344,284) and that continues to grow rapidly. The stability of Wichita's specialty is also noteworthy. During the same set of decades that has seen mass relocation of most industrial production in the United States, Wichita has retained and actually solidified its emphasis on aircraft and other basic manufacturing. The city's slogan of "air capital" fits as well now as it did eight decades ago. The only difference is that this industry has now expanded to include neighboring communities as well.

The manufacturing orientation of much of Wichita's industry today stands in contrast to the city's other traditional roles as a processor of agricultural products and marketing center for a large portion of central and western Kansas. These three emphases have always been complementary in several ways, creating diversification and therefore stability for the economy. Over the past several decades, however, the city's export-oriented manufacturing businesses have expanded faster than its regional and processing ones. This development, in conjunction with a sense of metropolitan self-sufficiency that necessarily accompanies population growth, has reduced long-standing ties between Wichita and its hinterland. A symbolic demonstration of this change occurred in 1996 with an announcement that the city's newspaper would henceforth restrict daily circulation to a thirteen-county area.[45]

The recent history of milling and meatpacking in Wichita parallels that described for Kansas City and epitomizes the city's loss of connection to rural Kansas. Although milling remains important, with the conglomerates Cargill and Cereal Food Processors having operations, both are highly mechanized. Cudahy, the last of the old-line packers, closed in 1975, followed five years later by the local stockyard. Farmland Foods remains with 415 employees, but agribusiness in Wichita has only a small fraction of the workforce it once did.[46]

In terms of regional trade beyond agriculture, Wichita's experience over the past several decades has been mixed. The greater size of Kansas City, together with that community's superior rail, highway, and air transportation, has allowed its entrepreneurs to gain increased control over the wholesale and distribution business for the state. The picture is much brighter for Wichita in retail trade, or at least it was until 1996, when stores lost the advertising power of the city's *Eagle* newspaper over most of the state. Throughout the 1960s, 1970s, and 1980s, however, whether defined by television market saturation or actual trade statistics, the city's merchants dominated the market for luxury and specialized purchases from Oklahoma all the way to the Nebraska and Colorado borders. Only on the east was this reach somewhat limited by the combined competition of Kansas City, Joplin, Topeka, and Tulsa (Map 25; see p. 39).[47]

If Wichita people recently have focused less on their connections to Kansas and more on those to the broader world, who could blame them? This ten-

Aerial View of Downtown Wichita, 1961. Wichita gained nearly ninety thousand people during the 1950s and assumed a genuine metropolitan character. This view looks to the northwest. The Romanesque towers of the old city hall, a landmark since 1892, still stand at 204 South Main in the lower left of the photograph. The Lassen Hotel, noted in chapter 7, is one block east and two blocks north. (Courtesy Kansas State Historical Society)

dency is a mark not only of postmodern society but also of the particular economic history of their city. The transforming event locally was World War II and its immediate aftermath. As I described in the previous chapter, military needs pushed employment at the three local airplane factories from two thousand to sixty thousand and remade a rather quiet mercantile community of

115,000 into a more freestanding and nationally oriented metropolis of nearly twice that size. Everybody expected a partial collapse of this boom-time economy after 1945, followed by greatly increased diversification. Instead, following only a brief adjustment, business growth resumed along almost the same trajectory as it had during the war years. The city's population hit 250,000 in 1960 and 300,000 in 1990. And, despite continuing fears of overspecialization, its singular economic orientation still shows no signs of long-term stagnation.

If World War II initiated a larger and more specialized economic role for Wichita, what happened during the late 1940s at the gigantic Boeing aircraft plant is key to understanding the modern city. When a few years of world peace gave way to the Korean War and then to a persistent tension with the Soviet Union, American military strategists saw a need for continuous upgrades to its fleet of bombers and other airplanes. The question of how to allot the new production contracts associated with these decisions, an issue much debated a decade earlier, was now easily answered. Wichita's superb facilities were already in place and its workers trained and reliable. The first orders for B-47s, a revolutionary design with swept-back wings and jet engines, arrived at the city in September 1948, and from then on the pace only quickened.[48]

Historian Craig Miner has observed that, once the initial B-47 contract was issued, an ongoing, central role for the city in the nation's military-industrial complex became "almost automatic." One of the major auxiliary benefits of the Boeing rebirth occurred at the adjacent Wichita Air Force Base. Because this would be the most logical place to train pilots for the new jet aircraft, the government poured $40 million into its enlargement (it was rechristened McConnell in 1954). The city received not only construction dollars from all this but also some fifteen thousand additional residents and a new municipal airport. This last-named perk, on Wichita's southwest side, was a consequence of the old airport grounds being needed for the McConnell expansion.[49]

As employment at Boeing reached 26,500 in 1956, when production of B-47s was giving way to that of even larger B-52s, it was easy to overlook the smaller manufacturing operations at Beech and Cessna. These also were great successes, however. With more than five thousand employees apiece, the two companies combined to produce nearly twenty-five hundred light commercial planes in 1955, 72 percent of the industry total. They also were participating in the development of practical jet designs and selling many of their models to the military for training purposes. Boeing's size also robbed attention from still another important component of Wichita's aircraft operation: a group of subcontractors who had materialized to make everything from instruments to valve assemblies. These machine shops of various sizes numbered in the hundreds by the mid-1950s and handled contracts worth scores of millions of dollars annually.[50]

In 1956, nearly one in three of Wichita's industrial employees could be found in the three airplane plants, a ratio that was matched in major cities of the time only by the steel plants in and near Pittsburgh, Pennsylvania. Taking all workers into consideration, local dominance by the aircraft industry peaked around 1967, when it employed about 23 percent of the city's labor force. This was shortly after the "big three" companies had been joined by a fourth: Learjet, a maker of luxury business planes. This company's owner, eccentric inventor William Lear, had come to Wichita in 1962 without being recruited and set up operations on a sixty-four-acre site near the municipal airport.[51]

By the early 1980s, Lear was employing more than 2,000 people, a nice complement to the 4,000 or so at Cessna, the 7,000 at Beech (purchased by Raytheon Corporation in 1979), and the 14,000 at Boeing. The largest subcontractor in the city, IFR Systems, makers of electronic test instruments, employed an additional 540 workers. These impressive numbers represented about 15 percent of the local labor force at the time, less than in 1967 but still remarkable for a large American city. Today production expansion at all these facilities continues to outpace labor savings achieved via the use of robots and other technology, creating gains in employment. Boeing still is king, with a labor force of 18,500. Cessna follows with 10,100, then Raytheon with 9,755, and Bombardier-Learjet (a 1990s merger) with 4,000.[52]

Although paling before the influence of the aircraft plants, other aspects of Wichita's long-standing entrepreneurial spirit also contribute importantly to the local economy. Oil, which created the initial investment capital for the area's airplane business, continues to provide new dollars from fields in adjacent Butler County. Most local petroleum-related business, however, is now tied less directly to actual oil fields than in decades past. The Derby Refining Company, an obvious symbol of the industry, shut down in the 1980s. Two other prominent city businesses—Koch Industries and Vulcan Chemicals—continue the tradition in a more oblique fashion.

Koch Industries is an outgrowth of a local engineering firm from the 1920s that built refineries. Fred C. Koch and his sons were extremely successful over the years, expanding their business until in 1998 it generated an incredible $30 billion in annual revenue. Now with subsidiaries far beyond oil and into chemical industries, financial services, and beef production, Koch's gross income is larger than that of the Disney Corporation, Coca-Cola, or Boeing. It is a private, efficient, and low-key operation, however, and employs only a relatively small number of people (2,070) at its eight-story headquarters on East Thirty-seventh Street North.[53]

Vulcan Chemical, highly visible on Wichita's southwest side and often in the news because of pollution issues, is superficially the antithesis of Koch's quiet corporate presence. The two companies actually have much in common. Vulcan makes chlorine-based chemicals such as caustic soda, carbon tetrachloride, and hydrochloric acid. The company (then known as Frontier) came to Wichita in 1950 partly to obtain a central location from which to supply oil drillers throughout Kansas and Oklahoma with the acid they required for their work. Wichita also offered convenient access to the rock salt (NaCl) needed as a basic raw material. Company workers drilled a series of brine wells for this near the town of Clearwater, tapping the same deposits that have been exploited for 120 years at nearby Hutchinson. Seven hundred people work at the plant today, a number that has been constant for decades.[54]

Beyond the airplane factories and the Koch operations, Wichita's next largest industrial employers are a pair of descendants from William C. Coleman's original gasoline lantern factory of 1901. Leaders at this company have been able to reposition their products several times over the last century to grow successfully and gain international status. As the rural market for their lanterns decreased in the 1930s, for example, they developed a line of home furnaces. During World War II they created a pocket stove used widely by American troops; then, around 1960, they entered and soon dominated the growing market for outdoor camping and recreation products. Although officials moved their corporate offices to Denver in 1995, the main factory remains on Wichita's north side along with 1,530 employees. In addition, the city has retained Evcon Industries, a former (before 1990) division of Coleman. Evcon's thousand employees specialize in heating and air-conditioning units for manufactured homes and recreational vehicles.[55]

The list of entrepreneurs whose investments collectively have helped to form Wichita's special economic character is long. To the names of the various airplane pioneers, Fred Koch, and William Coleman, one certainly should add Ray Garvey and his son Willard. The elder Garvey, as I noted in chapter 7, gained wealth in the grain business. His son became Wichita's most important developer of real estate and the creator, in 1986, of the twenty-one-story Epic Center, the city's tallest building. Other names only slightly less hallowed in the community include Dan and Frank Carney, who founded Pizza Hut; J. Robert Day, the developer of Shepler's stores for western apparel; Dan and Robin Foley, who created the Taco Tico franchise; and Tom Devlin, who did the same with Rent-A-Center. Believing in this spirit of commerce and wanting to nurture it further, officials at Wichita State University unabashedly have poured large amounts of money into their business programs. This includes the creation in 1977 of the nation's first "Center for Entrepreneurship." As population growth continues strong in Wichita even as it declines elsewhere on the Great Plains, residents still have good reason to envision their community as an exceptional, "magic city."[56]

Although the phrase "military-industrial complex" in today's society necessarily raises the specter of world destruction, Wichita people would argue that it is much better to be inside this system than on its periphery. They say this not so much out of feelings of excessive patriotism as from an appreciation of the financial benefits this business has brought into their community.

Boeing, Cessna, and Raytheon rank behind only Overland Park's Sprint Corporation as the leading employers in the state. The forty thousand or so jobs they provide are all unionized and therefore offer excellent salaries and benefit programs. Even though their business is cyclical and therefore prone to temporary layoffs for younger employees, this problem is more than counterbalanced by assurances that wages will remain high and that the companies will be permanent fixtures in the city. These last two items are key because they are more and more unusual in a globalizing economy that now forces most manufacturers to seek out cheaper locations on an ongoing basis. Spending for engineering expertise and military equipment, of course, is among the few categories exempt from such mundane considerations.[57]

Not surprisingly, the zone of Boeing-induced prosperity extends beyond the political borders of Wichita. A combination of commuters by the thousands and subcontracting firms by the dozens has now made the airplane business central to the suburbs of Andover, Derby, and Haysville and to the formerly self-sufficient communities of Augusta and Wellington. The same process is under way as well at Arkansas City, El Dorado, Newton, and Winfield, places slightly farther away whose residents once thought of Wichita as their economic rival.

Because Derby and Haysville are almost entirely residential communities, the process of economic change is easiest to see in Augusta and Wellington. Like most of Wichita's satellite towns, these both have histories as specialized places. Augusta was heavily dependent on its oil refinery (operated most recently by Mobil), and Wellington on the shops and a division office of the Santa Fe railroad. As diminishing mineral reserves and aging machinery at Augusta and new railroad realities at Wellington produced the threat and then the reality of plant and shop closures, townspeople sought alternative jobs. They pursued initially, and with some success, branch facilities of major manufacturing companies that wanted better distribution of their products in the Great Plains or Midwest. Typical of these efforts was the recruitment of the National Furniture Company to Wellington in 1950. National was the nation's third-largest maker of living room chairs and sofas. From its headquarters in Evansville, Indiana, officials shipped springs and frames to Kansas. Fabrics came from southern mills. The plan proved profitable for a while, with jobs for a hundred Wellington residents, but the company eventually saw more advantages in consolidation than dispersal and retreated to Indiana.[58]

The loss of National Furniture and similar companies after relatively short tenures made people in Wellington and Augusta more appreciative of the alternative and generally more lucrative opportunities provided by the aircraft business. The process began with mechanically minded, local men taking jobs in Wichita. After gaining experience and contacts in the plants there, some of them decided to set up independent machine shops back home. In Wellington,

G and C Precision Corporation began operations in 1950, followed the next year by Clark Manufacturing and Oxwell Incorporated, all doing sheet-metal work. Lamar Electro-Air Corporation (repair and overhaul of engines) and Precision Machining added to this core during the 1960s. Today, all five companies remain in business on Wellington's north side and employ a total of about 750 people. Together they dominate the local economy and help to maintain the city's population at about nine thousand residents. In Augusta, the pattern is similar. The community's biggest company, Sigma Tek, employs 110 or so people to manufacture aircraft instruments. The next largest, DJ Engineering, is a machine shop. These companies, together with easy commuting to the Wichita plants along a four-laned U.S. Highway 54, have enabled the town's population to grow from 4,483 to 8,423 over the last fifty years. Few people miss their refinery anymore, a place that has been dark and shuttered for more than twenty years.[59]

Nine miles west of Augusta on Highway 54, and therefore nearly adjacent to both the Kansas Turnpike and the aircraft plants, is Andover. The population there began to grow in the mid-1980s, just after turnpike officials installed a new exit at this point. It reached 4,204 in 1990 and 6,698 in 2000. Some people have compared Andover's future to that of Overland Park, seeing there a potential to combine commercial development with high-quality residential life. This vision has materialized to a degree through the creation of Terradyne by developer Roe Messner, a sprawling combination of golf course, resort, and expensive houses. The proximity to the aircraft factories has made Andover equally attractive to specialized manufacturers, however. Raytheon itself maintains a facility there that employs 400 people. Sherwin-Williams is present, too, with 215 workers who make paint and other surface coatings for the air industry.[60]

Because the huge Boeing plant sits on Wichita's southeastern flank, it comes as no surprise that the satellite cities on this side of the metropolitan area have the greatest concentrations of subcontracting companies. Within this spatial arc, however, the specific locations for shops have been determined largely by the state of preexisting economies. If residents of Augusta and Wellington, with their declining industrial bases, were the most aggressive in pursuing new airplane-related opportunities, people in El Dorado represent the other extreme. As the focus for the huge Butler County oil field, El Dorado historically has commanded the largest share of regional refineries, as well as ancillary employment in pipelines, oil field-service industries, and bulk fuel trucking. All these activities remained strong into the second half of the twentieth century when the aircraft subcontractors were establishing themselves elsewhere. They continue today in a somewhat diminished fashion. After a series of mergers and sales that changed the name of the large Skelly refinery to Getty, Texaco, Equilon, and now Frontier, it still employs 460 people.

Although the workforce at El Dorado's two refineries is now less than half what it was in 1950, the city has been able to maintain its population at about twelve thousand. Commuting to Wichita jobs explains much of this stability, of course. Important also was a community decision in 1964 to erect a modern campus for a small junior college that had operated through the local high school since 1923. Butler County Community College, the product of this process, now employs 306 people and is well positioned to serve the entire Wichita region. A similar metropolitan-wide role is played by El Dorado State Park, a facility built around a reservoir impounded on the Walnut River in 1973. This lake attracted nearly a million visitors in 2000, second among state parks only to Hillsdale near Kansas City. Finally, two more recent, though polar opposite, sources of jobs include a state correctional facility near the lake that has grown from a small honor camp in the 1980s into a full-fledged prison (460 employees), and the Pioneer Balloon Company, a success in the quirky business of manufacturing balloons and marbles (200 employees). The phrase "economic diversity" is definitely applicable to El Dorado today.[61]

The two remaining cities south of Wichita, Winfield and Arkansas City, fall into intermediate positions on the continuum of aircraft subcontracting. As longtime rivals with one another for trade within Cowley County and beyond, leaders in each town had put together different but equally successful economic strategies between the 1880s and the 1950s. These raised and then maintained the two populations at just over ten thousand residents. Arkansas City was a blue-collar community, heavily dependent on an oil refinery (Kanotex, later Total), the Santa Fe railroad (headquarters for the Oklahoma Division), and meatpacking (Maurer-Neuer, later John Morrell). Life at Winfield, in contrast, revolved around its state mental hospital, two church-sponsored colleges, and the Gott Corporation, a homegrown manufacturer of water coolers. The Binney and Smith Company, Winfield's equivalent to the National Furniture Company at Wellington, enriched this mix in 1952. A maker of crayons and paints based in Easton, Pennsylvania, Binney and Smith had established a branch in Los Angeles and wanted another in the Midwest. When they selected Winfield because of its sophisticated yet small-town environment and did so without any local recruitment, residents beamed. This decision not only confirmed their own cultural values and identified their town with a famous product but also brought in 350 jobs.[62]

The economic system in Cowley County received its first major setback in 1960 when a consolidation within the Santa Fe system moved Arkansas City's division headquarters to Amarillo, Texas. Another round of problems came in the 1980s with the demise of both the small St. John's College in Winfield (80 employees) and the large Morrell packing plant in Arkansas City (1,030 employees). Despite having good name recognition for the "Rodeo" brand that had been invented there, Morrell could not compete with the higher volume and increased automation at newer plants. Finally, in 1997, came a third episode of closings, large enough this time to generate sympathetic news coverage across the state. First came an announcement that, to save tax dollars, the Winfield State Hospital would shut down. This meant 800 jobs. Then officers at Total Petroleum declared their facility at Arkansas City too antiquated to continue (272 jobs), and Binney and Smith's owners decided that a consolidation of their operations back in Pennsylvania now made economic sense (345 jobs).[63]

The events of 1997 were a classic illustration of the locally callous nature of postindustrial economics. Cowley County lobbyists were able to salvage a little from the disaster by a conversion of one part of the hospital into a veterans' home and another into a prison. They also were grateful that Southwestern College and the Gott Corporation continued on, the latter even after its purchase in the late 1980s by Rubbermaid. Salvation for both Arkansas City and Winfield has proved to be Strother Field, a World War II air base midway between the two cities that was deactivated in 1945 and converted into a quality industrial park. Although not all of its twenty or so tenants were and are connected to Wichita's air industry, several of the larger ones are. Cessna people placed a facility there in 1966 that employed more than 500 people before business recession in the late 1970s forced its closing. General Electric remains, however, with a thousand-worker operation that overhauls airplane engines. Other important and longtime companies at Strother include Gordon-Piatt (a maker of gas and oil burners with 205 employees) and KSQ Blow-molding (a plastics firm with 289 workers).[64]

At Newton, the last of Wichita's major satellite communities, recent economic history has a starting point similar to that at Wellington. Both places were heavily dependent on the Santa Fe railroad and therefore faced crises when that company reorganized after World War II. As the railroad's shops and its Fred Harvey dining-supply business closed, city leaders looked for replacement jobs. Since downtown Wichita lay only twenty-five miles south on Highway 81, one might have expected an immediate influx of airplane subcontractors. This did not occur, in part because Boeing and the other plants sit on the far (i.e., southern) side of Wichita and its traffic. People in Augusta and Wellington had a locational advantage.

Airplanes turned out to be unneeded at Newton in the 1950s because area promoters were quick to find alternative industry. The key idea originated with George Sharp, a former mayor and owner of a distributorship for mobile homes. He knew that a Michigan-based manufacturer of these products was planning to open a branch plant in the southern Midwest. With an offer to owner Gurden Wolfe of a city-constructed building and assurance that former railroad shop workers would have the mechanical skills needed, "Great Lakes Homes" began to be made locally in 1955. This move provided seventy jobs immediately and initiated a chain reaction. A second homes manufacturer, the American Coach Company, also decided to expand to Newton in 1956. So did the Fred V. Gentsch and the Stevens Spring Companies, both makers of interior furnishings for such homes. Part of the reason for all this was that Norman Wolfe, the brother of Gurden, owned American Coach, but economies of scale were involved as well. With four major factories present by 1983, Newton people advertised themselves as the "mobile homes capital of the Midwest." Their city's population in 1980, 16,332, was 4,000 more than the mark from thirty years earlier.[65]

Impressive as the manufactured-home business was, a totally unexpected company was able to match its employment at Hesston, a small town seven miles northwest of Newton on Highway 81. Called simply the Hesston Manufacturing Company, this was a 1948 creation of Lyle E. Yost, a local farmer and banker who had an idea for a grain auger that could unload combines even as they traveled across fields. When he coupled this product with a second, the "swather," that could mow, rake, and condition hay in one operation, Yost's company swelled to five hundred employees by the early 1960s and then to a thousand. The town of Hesston rapidly became known as the community with more jobs than people.[66]

The economic situation in the Newton area today is not as simple as it was in the 1980s. The Hesston Corporation has changed ownership and is now called Hay and Forage Industries, but it continues to employ the same number of workers. The excellent wages paid at Hesston drew employees away from the mobile-home industry, however, and several firms moved elsewhere. The remaining local companies connected with this business, in fact, are specialty operators such as the Hehr Glass Company (85 employees) and the cabinetmaker Norcraft Incorporated (561 employees) that have expanded beyond their original market. Gradually, but inevitably, Newton people find themselves drawn into the orbit of Wichita aircraft. More residents commute to jobs there each year, and three small, air-related businesses now call Newton home: Avcon Industries, Executive Aircraft, and Midwest Aircraft Services.[67]

Topeka, Leavenworth, and Lawrence: Specialized Roles under the Dominion of Kansas City

A hundred years ago the citizens of Leavenworth learned the impossibility of vying with greater Kansas City for control of regional manufacturing and trade. Superior transportation and economies of scale in that larger community forced the once-premier city of Kansas into an increased reliance on its military base and other specialized institutions for which Kansas City did not compete. Much this same story is true for the university town of Lawrence, situated only slightly farther from the metropolis. Contemporary strategies for growth at these two places, however, can best be appreciated by pairing them with the post-1950 experience of Topeka, Kansas's traditional third city and a place more recently affected by the neighborhood giant.

Topeka's business leaders in 1950 were a contented group. Starting from an exceptionally solid employment core of the state capital and the Atchison, Topeka and Santa Fe Railroad, they had amassed a diversified economy. The capital had attracted a variety of state and federal agencies, as well as headquarters for statewide utilities such as Southwestern Bell and the Kansas Power and Light Company. Insurance and printing businesses came for the same reason. The Santa Fe company, with its corporate headquarters and a major repair shop in town, was both a major employer in its own right and the centerpiece for an efficient transportation network that attracted millers, wholesalers, and meatpackers. This combination, together with Forbes Air Force Base (a creation of World War II that had become an important focus for the Strategic Air Command), had produced impressive population growth rates. Between 1900 and 1960, the increases averaged 14,500 per decade. By the end of this period, Topeka's total (119,484 residents) nearly equaled Kansas City for second place in the state (Charts 2, 3).

According to an adage, success always carries with it the seeds of complacency. Whether or not this was the case for Topeka in the 1960s and 1970s is hard to say, but for a variety of reasons, losses in employment during those years started to nearly equal gains. The city's population actually fell between 1970 and 1980. Then, despite several expansions of its political borders over the next decades, Topeka's total for 2000 was still only 122,377 residents, hardly more than its count from forty years earlier. Looking back, it is interesting to note that this economic stagnation was little discussed locally until the 1980s. Part of the reason was because people in the 1960s were still digesting the flush times of the previous decade and had confidence that the city's central position in the ongoing construction of the interstate highway system would soon yield renewed prosperity. Another contributor to the nonchalance was a general business recession that spread across the country in the 1970s. Topekans bided their time, secure in their beliefs that economic recovery would come and that the state capital and the Santa Fe railroad would never leave them.

An attitude of laissez-faire proved justified to some degree. The country's recession was indeed temporary, and both the capital and the railroad remain as major employers. The problem was with some of the industries traditionally tied to government and transportation. Starting with government-related business, the number of state employees in Topeka rose from about 1,000 in 1900 to a peak twelve times that size in the mid-1990s before falling back to 8,798 in 2002. The pattern for federal agencies in the city is similar, with a maximum of 1,860 in 1986 and 1,447 in 2002. Statewide organizations continue to locate in Topeka, although most maintain only modest offices. Still, Southwestern Bell, which moved to town in 1950 from Kansas City, employs 943, and Westar Energy (the renamed Kansas Power and Light Company) 709.[68]

Unlike the experience with utilities and other state organizations, however, Topeka has lost many of its once vaunted insurance and printing industries. Insurers, like other businesspeople, have merged in order to compete on larger scales. As this has happened, home offices have tended to relocate from state capitals to metropolitan centers. National Resource Life, for example, one of the eight companies that made insurance "one of the community's biggest businesses" as recently as 1958, moved to Kansas City in the early 1980s as part of a consolidation. At present, Topeka's only remaining large employers in this field are Blue Cross/Blue Shield and the Security Benefit Group, with 1,834 and 600 employees, respectively. The city's decline in the printing business has a simpler explanation. Capper Publications, its kingpin company, was heavily dependent on magazines directed toward rural America. Although the organization adjusted somewhat to changing times, including the creation of Topeka's first television station (WIBW, Channel 13) in 1953, it dissolved soon afterward. Again, only a remnant of this past continues, most prominently with the 1,200 people who work at Jostens School Products Group. This Minnesota-based company, in Topeka since 1969, is the nation's largest printer of school yearbooks.[69]

The history of transportation-dependent business since 1950 runs parallel with that of insurance. Although Topeka has good rail and highway connections, Kansas City enjoys better ones. This reality, plus the latter community's greater size and its possession of the region's only international airport, made it a logical choice when manufacturers, millers, and wholesalers sought sites for consolidation. These businesses could operate in Topeka, but most could do so more efficiently in Kansas City. Meatpackers were an exception, of course, nearly all of whom forsook their former centers in favor of sites closer to feedlots. Topeka has retained one of its traditional packers, however. Hill's Pet Products (750 employees) has been making dog food in the city since 1907.[70]

Because consolidation and relocation were and are slow, ongoing processes, such losses of individual companies were easy for citizens to rationalize. Topekans first began to grasp what really was happening to them in 1976 when two separate events focused attention on the overall local economy. The most obvious was a government decision to close Forbes Air Force Base, which cost the city seven thousand jobs. The other, subtler incident was the opening in Salina of a large convention, sports, and concert facility called the Bicentennial Center. This construction was a direct threat to another of Topeka's taken-for-granted roles, that of meeting place for all manner of statewide organizations. When discussion over these issues was augmented in 1980 by the census bureau's announcement of a population loss of almost ten thousand people, realistic assessments and new strategies for growth rapidly materialized.[71]

The first order of business was retention of the Santa Fe offices and shops. Despite the heavy investments this company had made in Topeka, this was no easy task. Ever since 1884, when the Ottawa cutoff into Kansas City had been established, the city literally had been off the main track of the railroad and therefore dependent more on goodwill than business logic for its continuing pool of jobs. A merger of the company with the Southern Pacific Railroad in 1983 now tested this goodwill further. To keep the main office in town, desperate city leaders not only provided a free, cleared city block for a new building on Quincy Street between Ninth and Tenth but also agreed to construct an adjacent parking garage and to purchase the Santa Fe's older offices. The result was a $40 million complex completed in 1985 that housed two thousand workers. In addition, all parties agreed that the Santa Fe shops, a six-hundred-employee unit that specialized in the construction and repair of freight cars, would continue to serve the new combined company.[72]

With the railroad happy, Topeka planners next set their sights on improvements to infrastructure. The city was too close to Kansas City ever to justify a true international airport. But to avoid losing more companies such as the Brock Hotel Corporation (owners of many Holiday Inns and the ShowBiz pizza franchise), which left for Dallas in 1980 specifically to obtain better air connections, leaders pushed successfully to build a modern terminal at the old Forbes site south of town. They also began to develop a three-hundred-acre industrial park there along the Santa Fe tracks where the General Services Administration once had operated. To counter the challenge posed by Salina's Bicentennial Center, the community decided to convert the Shawnee County Fairgrounds along Topeka Boulevard between Seventeenth and Twenty-first Streets into a $20 million convention facility. Called the Expocentre, it opened in 1986 to a wide range of activities including minor-league football and hockey.[73]

The final element in Topeka's strategy for economic rejuvenation was salesmanship. Becoming more aggressive in the use of industrial revenue bonds and tax exemptions, the pitch of Topeka business leaders was not to be better than Kansas City but to offer, in the words of Mayor Doug Wright, "a refuge" from the big city. They provided, as demonstration, a list of major manufacturers that had been successful in Topeka for decades. These included a Goodyear tire plant that began operations in 1945 (1,700 employees in 2002); DuPont's cellophane factory in suburban Tecumseh since 1958 (now owned by the Belgian UCB Group, 200 employees in 2002); and the homegrown Volume Shoe Company, a pioneer in self-service stores since 1959 (now renamed Payless ShoeSource, 1,550 employees in 2002). The sales talk was that Topeka combined good business facilities with land that was more abundant and less costly than in Lenexa. Amenities were present, too: a relaxed atmosphere spiced with access to university life in nearby Lawrence and Manhattan.[74]

The strategy that Topeka people have employed since the 1980s can be called a partial success. Traffic at the refurbished Forbes Field has proved inadequate for the needs of first Frontier and then United Airlines, and so shuttle vans to Kansas City International have become a fact of life. Another of Topeka's significant institutions, the Menninger Foundation, has fallen victim to changing funding patterns for hospitalization. This pioneer in the treatment of mental illness has been forced to cut its employment from twelve hundred to three hundred and now faces imminent consolidation and relocation. On the positive side of the ledger, the Santa Fe, following still another merger to create a mammoth Burlington Northern Santa Fe Railway in 1995, announced that the conglomerate would retain Topeka as a major business center. More than thirteen hundred Topekans currently work in its offices (the specialization is information technology) and another 670 in the shops. The most recent positive development comes from the Target Corporation, a respected, Minnesota-based retailer. In 2002 its officers selected Topeka over Olathe and Wichita as the site for a 650-employee distribution center that will serve ninety of its expansive stores.[75]

If Topeka, a sizable place in its own right, can be affected negatively by competition from Kansas City's airport and businesses, the potential for similar change in the smaller and more proximate cities of Lawrence and Leavenworth is great indeed. A prediction of population stagnation or decline at these sites is not guaranteed, however, because of at least two countering prospects. First, since they are located closer than Topeka to the big city, leaders may well have better anticipated the competition and found ways of dealing with it earlier. Also, with both sites little more than twenty-five miles from Kansas City, the communities could now be functioning as suburbs and thereby experiencing Olathe-like growth.

Population figures show that Lawrence and Leavenworth have followed quite different paths since 1950. In that initial year they were of similar size, Leavenworth with 20,579 residents and Lawrence with 23,351. Leavenworth then grew modestly, but steadily, until 1990 before falling back over the next census period from 38,495 people to 35,420. Lawrence's experience has been far more dramatic. It doubled its 1950 count within twenty years and then nearly did so again by 2000. Among the state's larger cities, in fact, Lawrence's percentage increases over this period have been exceeded only by Olathe, Overland Park, and Wichita (Chart 3).

Because the population trends in Leavenworth approximate those for Topeka, it is reasonable to expect similar causes. To a degree, this is true. Both communities rely heavily on institutional economies whose employment, although steady, has expanded only modestly in recent years. Leavenworth is even more firmly identified with its institutions than is Topeka, however, not so much on purpose but because of difficulties in attracting and retaining other major employers. Fort Leavenworth leads the way today as it has since the birth of the city, currently employing 3,144 military personnel and 2,279 civilians. This is followed by 900 workers at the Eisenhower Veterans Affairs Medical Center and its outpatient pharmacy, 692 at the state prison, 560 at the federal penitentiary, and a combination of 706 students and 169 employees at St. Mary College. Taken altogether, the list constitutes an impressive 24 percent of the city population.[76]

As I discussed in chapter 7, Leavenworth boosters realized by the 1940s that relatively poor railroad transportation doomed their previous efforts with manufacturers. Their furniture makers and coal miners all had left by that time. Forty years later, so had the Missouri Valley Steel Company, which as recently as the 1950s employed three hundred workers under federal contracts to construct tugboats and other small ships. Only one remnant of this past survives, in fact. A hundred people at the Great Western Manufacturing Company still make milling equipment as they have since 1858. Beyond this, the field of major private businesses is now limited to two branches of Kansas City–based Hallmark Cards (850 employees) and Heatron, a maker of industrial heating elements (218 employees).[77]

Hope for industrial renewal flickered briefly for the city in the 1970s and early 1980s with the openings of the new Kansas City International Airport just ten miles east of Leavenworth and then Interstate 435 an even shorter distance in the same direction. Such dreams proved illusory, however. Although a quality bridge exists over the Missouri River, the state of Missouri has never provided good access routes to it. The road to I-435, for example, is a crooked fourteen miles instead of a beeline eight, and the airport lies another ten miles beyond. The business that Leavenworth people anticipated as theirs went instead to neighboring Platte County, Missouri.[78]

Not as important as transportation matters, perhaps, but still a significant cause of Leavenworth's stagnation are two interrelated issues of image. The biggest is nationwide and strongly negative name recognition for its federal penitentiary. Before the 1990s, townspeople played down this fact of local life and stressed instead the glories of Fort Leavenworth and its internationally famous Command and General Staff College. Since population data suggested that this tactic was not working, the local visitor's bureau recently decided to reverse the stance and attempt to poke fun at their circumstance. They now post ads in which a man in a convict's suit asks travelers, "How about doin' a little time in Leavenworth?"

Prison imagery, whether light-handed or not, undoubtedly has hurt Leavenworth in its ongoing competition to provide suburban housing for newcomers to the greater Kansas City area. Nearly as important in this regard is the intervening presence of Wyandotte County between Leavenworth and either Kansas City, Missouri, or Overland Park. Wyandotte, with its blue-collar, ethnic population, of course, also presents a negative image to many new suburbanites. As a result, expensive tract housing is now being erected thirty miles east, south, and west of Kansas City, Missouri, while physically more attractive and accessible settings lie untouched between Kansas City, Kansas, and Leavenworth.[79]

Future hopes for Leavenworth probably lie with the people who know the city best: officers in the United States Army. Their view of this area is reflected in the nickname "Little Pentagon." Fort Leavenworth is where up-and-coming military people go for advanced training in command and strategy, an elite posting at which officers find quality golf courses, housing, medical care, and even a fox-hunting club. For this reason, and especially since the 1960s with the opening of the Munson Army Hospital and the Smith Dental Clinic, Leavenworth has become home to many military retirees. These people and their families, often still relatively young, provide a substantial infusion of money into the community. They buy and restore Victorian houses left over from the city's glory years and volunteer for a wide variety of civic projects.[80]

From one perspective, the problems that faced planners in Lawrence in 1950 loomed even larger than those at Leavenworth. Their city not only was deep in the shadow of Kansas City's economic might but also was even closer to competition from Topeka. Few business endeavors, it would seem, had much of a chance. Two flaws exist in this reasoning. First and foremost, of course, is that it ignores the presence of the University of Kansas. This institution, unlike the public agencies in Leavenworth and Topeka, was just beginning a major growth phase at this time. Bolstered by returning war veterans, enrollments had doubled their 1940 total to more than ten thousand students. Substantive increases would continue over the next decades as more people saw a need for higher education. Currently, some twenty-six thousand students and fifty-three hundred instructors and other support individuals reside in or near Lawrence. A second local college, Haskell Indian Nations University, augments these numbers. By adding bachelor-level degrees in the 1990s to its earlier programs, Haskell now serves nine hundred students and employs more than two hundred staff members.[81]

A second big asset for Lawrence has been transportation. Because of the community's position on the direct route between Kansas City and Topeka, it had long enjoyed good railroad service. This same circumstance brought the Kansas Turnpike to town in the mid-1950s and then, thirty years later, a second interstate-quality highway (K-10) that connects the city's Twenty-third Street directly east to Interstate 435 in Johnson County.

In 1950, with its students and support staff constituting almost half the community's population, Lawrence fit the label of "college town" almost perfectly.

Things became more complicated in the following decades, initially because the combination of good transportation and university appeal attracted a small group of manufacturers and distributors. To join the Lawrence Paper Company (the only holdover from the city's nineteenth-century industrialization) came first a phosphates plant built by the FMC Corporation (1951), then a nitrogen fertilizer factory by the Co-operative Farm Chemicals Association (1954) and a branch of Hallmark Cards (1957). These, led by Hallmark's 800-person staff, added some 1,540 industrial jobs to the city. With the exception of the Co-op plant (a recent closure), these companies continue to employ similar numbers.[82]

The 1960s and 1970s saw gradual expansion of Lawrence's industrial base, particularly the additions of Packer Plastics (a maker of injection-molded products such as drinking cups), a pet-food factory owned by Quaker Oats, a branch plant of Olathe's King Radio, and, biggest of all, a distribution center for the Kmart Corporation designed originally to serve a third of the country. Again, these companies all have stayed to become long-term members of the community. In 2002, Kmart employed 950 people, Packerware/Berry Plastics 400, Honeywell (i.e., King) 330, and Del Monte (Quaker's new owner) 130.[83]

Important as industrial jobs and increasing university enrollments were to Lawrence's rapid growth, they have been matched at least since the 1980s by two other phenomena. First, as the initial big generation of university graduates began to retire at this time, many decided to relocate to their old college towns. Concurrent with this trend, improvements to Highway 10 made the idea of commuting to jobs in Johnson County a thirty-minute proposition. Like the retirees, some of the people who were transferred into the Kansas City area began to choose Lawrence as a home. The city was nearly as close to their jobs as more traditional suburbs and offered university amenities as a bonus. A variation on this theme involves couples in which one person finds a job in Topeka and the other in Kansas City. Lawrence then becomes the midway choice for residence.[84]

Recent developments in the continuing economic expansion of Lawrence include aspects of edge-city development and niche retailing. Industrial promoters have now dubbed Highway 10 Kansas's "smart corridor," and growth along it is fast-paced. On its eastern end, the city of De Soto is poised to explode with suburban population during the next decade. Several industrial parks line its route as well, including East Hills in Lawrence. There, National Computer Systems and Amarr Garage Doors have added 800 and 573 new jobs, respectively. More noticeable to most community residents is a revival of downtown Lawrence as a retail center. This development, surprising given the accessibility of shopping malls in Topeka and Overland Park, is largely boutique in nature. In general, Lawrence people travel to the malls to purchase ordinary goods, while their counterparts in Kansas City and Topeka venture to

Lawrence for bookstores, chic restaurants, international fashions, and clothing accessories. Again, the town's college atmosphere underlies this trend.[85]

Interstate Cities in Western and East-Central Kansas

Beyond developments in and near Kansas City, Topeka, and Wichita, one of the most notable changes in the Kansas urban system since 1950 has been the role of interstate highways in creating a new order of central places. Limits to travel and trade caused by what geographers refer to as the friction of distance had been relatively constant in the state throughout the railroad decades until World War II. A combination of better automobiles and the superhighways changed all this, however, and initiated a chain reaction of city growth and decline.

The first stage of the new order saw populations increase in almost every county seat during the 1950s. Merchants attributed this development to personal charms, but it was much more a matter of rural villages shutting down economically as their former customers sought out better selection in somewhat larger towns. In the 1960s and 1970s (I-70 was completed across Kansas in June 1970), the scale changed again. Now it was county-seat towns themselves that declined. Part of the reason was the general migration to larger cities, but just as important was a new willingness of people to drive thirty or fifty miles for an even greater range of goods and the ability to do so in the time they once had spent going ten.[86]

The new pattern of trade and services for Kansas and most other states was easy to predict. Interstate cities would become focal points, drawing in customers from a corridor at least three counties wide. In the case of central and western Kansas, this reach from I-70 extended across the entire northern half of the state. Not every community along the highway could grow in this fashion, of course. Mostly it was a matter of big places getting bigger and of the spacings of active cities increasing across the High Plains. Looking at the still-emerging system, the first node west from Topeka is the dyad of Manhattan and Junction City (Map 22). Then comes a similar-sized jump of fifty miles to the traditional distribution point of Salina, followed ninety miles farther on by the oil and college center of Hays. The last and smallest focus on the route, Colby, sits just more than a hundred miles beyond Hays. This pattern leaves in its interstices the once-important cities of Abilene, Russell, and Goodland. Each of these places has struggled economically as their neighbors have boomed. They are victims of the new geography almost as much as towns far more isolated.

Despite the importance of the transportation model just discussed, the success of two of the region's interstate communities, Manhattan and Junction

City, has come largely from other sources. These places, the state's purest examples of a college town and a military town, respectively, prosper in accord with their institutions. Enrollments at Kansas State University have followed the same pattern as those at the University of Kansas. From 5,000 students in the early 1950s, the number rose to 20,600 in 2000, with a faculty and support staff of 4,790. This total exceeds 50 percent of the city's population of 44,831. In a similar way, ten thousand soldiers and 2,300 civilian employees at Fort Riley together account for 65 percent of the population in Junction City (18,886).[87]

The creation of economic diversity is more difficult at Manhattan than at Junction City because Manhattan actually lies eight miles north of the interstate. Owing to this handicap, Manhattan business strategies have been modest but deliberate. First, and logically, they have built on the university's reputation in agriculture to recruit three related organizations. The U.S. Department of Agriculture's Grain Research Station is there with 125 employees, and so is the American Institute of Baking with 105 people. Larger still is the home office of Kansas Farm Bureau (506 employees), a lobbying organization dating back to 1918 that now works primarily in insurance and marketing. Beyond these ventures, Manhattan has three other sizable industries: the locally owned Steel and Pipe Supply Company that started in 1948 (144 employees), Parker Hannifin (a maker of hydraulic hoses with 200 employees), and the McCall Pattern Company (277 employees). McCall's, which caters to do-it-yourself seamstresses, sought out Manhattan in 1969 because of an existing labor pool from spouses of college students and soldiers.[88]

Junction City, with four exits on I-70, has considerable potential for industrialization. So long as adjacent Fort Riley remained strong, however, the town's business leaders were not active recruiters. This situation began to change in the late 1980s when military leaders said the post, despite having 101,000 acres, was too small for battalions to maneuver in modern tanks. The possibility for local disaster increased further when Congress announced a series of base closings to reflect the reduced size of today's military forces. To date Fort Riley has escaped closure, but troop strength is down 40 percent from its earlier level of 18,000. Industrial recruitment is therefore vigorous in Junction City. So far the biggest acquisitions have been Provell (a telemarketing firm that utilizes military spouses, 330 employees), a sausage plant operated by Armour Swift Eckrich (530 employees), and a Footlocker Distribution Center for shoes and other athletic apparel (755 employees).[89]

An important but sometimes overlooked part of both the Junction City and Manhattan economies consists of military retirees. This population, although twenty-three hundred strong, is less visible than a similar group in Leavenworth because it is less affluent. Fort Riley is a post dominated by enlisted personnel, not officers, but the same amenities of good health care, commissary food, and recreational opportunities attract many former soldiers. Enlisted people cite Junction City's welcoming atmosphere, low cost of living, and tolerance of diversity (including of interracial marriages). Officers are more likely to choose neighboring, and somewhat more expensive, Manhattan for their retirement. They feel the access there to both the fort and the university is ideal.[90]

Salina, with 45,679 residents the largest I-70 city in Kansas west of Topeka, is much more commercial and industrial than the highly specialized Manhattan and Junction City. Its growth, as I have developed previously, was tied closely to being a railroad center and therefore a good point for the assembly and distribution of goods. Salina never was independent of Kansas City in this regard but played an important secondary role within the broader system. Faster speeds of travel and economies of scale over the past fifty years have now eroded much of this traditional industry. Milling, for example, is nearly gone, as are the wholesaling and trucking businesses.

The ebbing of older industries almost surely would have raised outcries from city leaders were it not for the presence of a newer employer that more than made up for these losses. This was the Smoky Hill Air Force Base, a World War II installation that, after closing briefly in 1949, was reactivated as one of eight permanent bomber stations for the Strategic Air Command. As its name was changed to Schilling in 1957, it added intercontinental ballistic missiles to its repertoire, employed nine thousand people, and accounted for a third of the county's annual payroll.[91]

Military officials abruptly closed Salina's number one employer in 1965 and, in so doing, prompted anxious discussion about the city's future. Population dropped 13 percent at the 1970 census, to 37,714, but a gift of the abandoned base for industrial or other use and the near completion to the city's doorstep of Interstates 70 and 35 West (today's I-135) rendered the future far from bleak. The immediate goal of job replacement led to incentives that attracted branches of Beech Aircraft (900 workers) and Westinghouse Electric (310 workers) in the late 1960s, and then General Battery in 1975 (308 workers). A slightly older ball-bearing factory owned by Federal Mogul and a local restaurant called Tony's expanded as well. The factory grew to 300 employees, while Tony's met with even greater success by marketing frozen pizza nationally. Its size increased exponentially after 1969 to 1,200 employees by 1982.[92]

With employment stabilized by a group of industrialists who liked Salina's combination of good transportation, inexpensive housing, and intermediate size, city officials then concentrated on retail trade and public services. The community had derived great pride and profit from its unofficial designation as capital of northwestern Kansas, but it realized that improved traffic flow on the new highways could both help and hinder its status. On the plus side, I-70 would logically enhance the city's reach to the west, enabling more shoppers to come from farther away. This was countered, however, by conditions to the east and south. The better road would now tempt people in Junction City, for example, to spend money in Topeka (sixty-five miles) instead of Salina (forty-eight miles). Even more frightening was the possibility that Salina residents themselves might start to drive ninety miles south to Wichita's trendy stores and gigantic malls rather than keeping all their dollars at home.

The double-edged nature of Salina's new position has played out slightly more positive than negative so far, with a modest population increase of four hundred people between 1980 and 1990 and nearly three thousand more during the most recent decade. No large industries have joined the cadre attracted over a generation ago, but most members of this group have expanded. Tony's Pizza is still the employment leader with 2,000 workers, followed by Exide Corporation (formerly General Battery) with 710, Philips Lighting (formerly Westinghouse) with 600, and Raytheon (formerly Beech) with 493. In addition, a local maker of farm and landscape equipment, Great Plains Manufacturing, has expanded greatly since the 1980s and now employs 650.[93]

Beyond traditional industry, the efforts of Salina leaders to promote their central location have been mixed. Frustration marks their recent experience with retail trade and distribution. No major warehouse has come in to replace a Western Auto distributorship that closed in the 1980s, and no truck line has replaced Graves, the locally owned company that was the largest in the state. The city's plan to fortify retail trade was Central Mall, a $30 million enclosed shopping center erected twenty blocks south of downtown in the 1980s. This and a later, ambitiously named Mid-State Mall succeeded mainly in killing the original business center downtown. The shopping allures of Wichita, Topeka, and (more recently) Hays become stronger every year.[94]

Success in the fields of health and job services and the arts has come much more easily. Job training springs from an 1892 business curriculum established as part of the city's Kansas Wesleyan College. This unit, now independently operated as Brown Mackie College, was joined in the 1960s by Kansas Technical Institute created at the old Schilling base. Known currently as Kansas State University, Salina, this program enrolled 926 students in 2000. Hospitals, once mostly small operations found in nearly every county, have been consolidating at extremely rapid rates in recent decades as sophisticated (and expensive) new equipment becomes available. Salina's two traditional facilities, Asbury and St. John's, merged in 1995 to form the Salina Regional Medical Center. Later that same year, Concordia's St. Joseph Hospital joined the group. The new alliance employed 1,204 people in 2000, a total that ranked Salina statewide behind only the health agglomerations in Kansas City, Wichita, and Topeka.[95]

Finally, in the arts, the key event for Salina was a decision to replace the city's old Memorial Hall with a new Bicentennial Center in 1976. This act, symbolic as well as functional, has enabled the community to regain prestige through the sponsorship of a minor-league basketball franchise and the national junior-college basketball tournament for women. The building also attracts large audiences with music concerts and, every June, serves as the main venue for the city's popular Smoky Hill River Arts Festival. This latter celebration, feeding originally off the strong arts tradition in the neighboring town of Lindsborg, is now strong in its own right and attracts up to one hundred thousand people each summer.

Although it pales in comparison with the loss of Schilling Air Force Base, another reason why economic development became important in Salina during the 1970s was a growing competition that city leaders felt from Hays, the next urban center west along the I-70 corridor. Hays, by virtue of its state college and rich reserves of oil, had long been the largest community in northwestern Kansas. With a population of 8,625 in 1950, it was a trading hub for sixteen or so counties but still functioned under Salina's umbrella for wholesaling and specialized retail purchases. This relationship gradually changed as population in Hays increased at a rate nearly double that of Salina over the next fifty years. With 15,396 residents in 1970 compared with Salina's 37,714, a dependent relationship became, in part, a rivalry.

The rapid growth of Hays has several causes. First, as a college town, it saw its student count rise from just over a thousand in 1950 to 5,130 two decades later before leveling off in more recent years. Another source of people, money, and amenities was the county's oil fields, which have rivaled those in neighboring Barton and Russell Counties as the most productive in the state since the 1940s. Hays in this way acquired the public services, cultural opportunities, and discretionary capital it needed to attract retirees and businesspeople alike. These things, in turn, led to a booming retail trade and the establishment of Travenol Laboratories, a maker of plastic medical apparatus that moved to town in 1967 and soon offered jobs to a thousand area people. All in all, according to a former president of the chamber of commerce, it was "growth without much public effort."[96]

Hays planners first became proactive in 1972 when they gambled on the construction of the first large shopping mall in northwestern Kansas. Its size, 230,000 square feet, was far too large for the city itself, but the center succeeded by spreading its customer base into at least twenty counties. This move was encouraged, in part, by the recent completion of Interstate 70, as well as a realization that oil production, which had peaked about 1960, could not sustain the economy for many more years.[97]

The wisdom of trade expansion proved itself in the mid-1980s when, in rapid succession, Travenol Laboratories closed and world oil prices plummeted from thirty dollars per barrel to ten. These events prompted a new office of economic development and, with it, three major initiatives that together have positioned the city well. First, on the industrial front, they attracted a Japanese maker of large stationary batteries to move into the Travenol building. They also constructed a factory shell on speculation and leased it to a direct-mail company from Mount Pleasant, Iowa, that had outgrown its labor supply. These plants, now known as EnerSys and Hays International Mailing Services, employed 390 and 121 people, respectively, in 2002 and have eased the trauma of Travenol's demise.[98]

The second and third initiatives in Hays came in the service sector. Just as local people had bolstered the city's retail potential a decade ahead of their Salina neighbors, they did the same with medicine. From a situation with two small hospitals and twenty doctors in the 1960s, an aggressive policy of recruitment and construction pushed physician numbers to thirty in 1982 and to sixty-six in 2000. The two older hospitals merged in 1991 to create a modern Hays Medical Center, a facility designed to serve a population of 130,000 and that includes the only heart-surgery unit in the western half of the state. With 1,200 employees in 2002 plus another 776 at three related institutions (Developmental Services of Northwest Kansas, Central Plains Laboratory, and the High Plains Mental Health Center), Hays now qualifies as a specialized health services community (Table 2). Less important than the hospital, but nevertheless significant for the local economy, was the opening in 1999 of a spectacular new public facility for one of the nation's best collections of Cretaceous-age fossils. The Sternberg Museum of Natural History at Fort Hays State University, based largely on locally collected specimens, now sits next to the interstate highway and has become the region's main tourist destination. With 20,013 residents in 2000, Hays has coped well within the postmodern world.[99]

Because of the tremendous expansion of health, education, and retail functions at Hays, economic opportunities for other cities in northwestern Kansas are severely limited. Stores with stocks of everyday and perishable goods and businesses that cater to agricultural and other local resources would seem to offer the best possibilities. The chances for success, of course, logically should increase in direct proportion to distance from the Hays competition.

Two cities long have vied to be the northwesternmost urban focus in Kansas: Colby and Goodland. Goodland, home of a division point and repair shop for the Rock Island railroad until 1953, was traditionally the larger of these places. The loss of several hundred jobs when this transportation company consolidated and then declared bankruptcy hurt the economy badly. Local people, looking for an alternate employer, thought they had found one later that decade when geological studies revealed good supplies of groundwater in surrounding Sherman County. Investors soon converted this water into sugar-beet production and induced the Great Western Sugar Company to construct a refinery in 1968. The plant employed seventy-three people by 1982, but it closed shortly thereafter because its corporate owners feared future water shortages and changes in national sugar policy.[100]

Natural History Museum at Fort Hays State University in Hays, circa 1958. Specimens collected from the rich Cretaceous fossil beds of western Kansas form the core of this nationally known collection. Opened to the public in 1914, it attracted so many visitors that the university recently acquired a huge new facility for it next to Interstate 70. The Sternberg Museum is now a major contributor to the Hays economy. Note the exhibit of a six-foot fish swallowed by a larger creature that hangs on the back wall in the photograph. (Courtesy Kansas State Historical Society)

As population at Goodland fell from 5,708 to 4,983 during the 1980s, its longtime rival eclipsed the city in size for the first time. Colby, its citizens with the mind-sets of underdogs, had countered Goodland's two large employers with an economic policy based on trade, diversity, and political activism. As the junction of the Rock Island and Union Pacific railroads, Colby always had enjoyed good transportation. Perhaps because of this tradition, local people lobbied early and successfully to relocate the route planned for I-70 several miles closer to their city.[101]

Besides their highway coup, Colby residents in the 1960s also engineered the construction of a community college. This initiative, conceived originally by Representative Don Smith to be a branch of Fort Hays State College, was ready for action when state politicians passed enabling legislation for a comprehensive junior-college system in 1965. The institution was popular from the start. It now enrolls more than a thousand students and, with a support staff of about 150, constitutes the city's largest industry. Colby, in fact, now deserves the label of college town and has built an institution that evokes civic pride on a par with that found in Manhattan and Lawrence.[102]

Despite its successes, Colby has not yet eliminated Goodland as a rival. Wal-Mart, the retail giant, has placed stores in both locations. Goodland also offers a vocational-technological school to counter Colby's college, and each community has a small hospital and, oddly, a processing plant for sunflower seeds. Colby people remain the more aggressive in their approach to the changing world, however, which probably bodes well for the future. A fairly recent effort to establish a factory-outlet center at the interstate exit already has failed like many other such facilities around the country. Another venture at the same general location is a success, however—the county's Prairie Museum of Art and History. At the 2000 census, Colby's population of 5,450 exceeded the count in Goodland by 502.

A population increase in Goodland of only 6 percent between 1950 and 2000 marks it as one of the interstitial cities along the I-70 corridor, places too close to more favored sites to capture much more than local trade. Of the many examples of such bypassed communities, Abilene and Russell deserve special comment. Moderate successes there in the past had given boosters hope for brighter futures.

Russell's history is simpler than that of Abilene. Never having the advantages of Salina's railroad connections or Hays's college, it was a small county seat until major oil pools came into production in the 1920s. The 1950 census captured Russell at its apex, with 6,483 residents and a well-developed system of city services. Things held steady for another decade, but then the wells began to ebb and, along with them, the oil field service companies that had employed many townspeople. Today three small enterprises (EOTT Energy, Shield's Oil Producers, and Kaw Pipeline) continue this business with a total of ninety-two workers. A few other residents cater to tourists at the Oil Patch Museum established near the interstate exit.[103]

Seeing the success that Newton enjoyed with the manufacture of mobile homes, Russell people approached a Nebraska maker of travel trailers, King of the Road. Their pitch was that former oil field workers could provide ideal labor for a branch of this plant. When company officials agreed, they established a 120-person operation that prospered in Russell from the 1980s until 2000. One factory, however, was not enough to stem continuing loss of retail trade to Hays, and conditions worsened when that company's officers decided to consolidate operations back in Nebraska. Russell's population fell to 4,696 in 2000 and then looked to decline again the next year with the closing of the city's second-largest manufacturer, a gluten plant owned by Farmland Industries of Kansas City. As I write, however, the economic situation has improved somewhat. Seeing in the gluten facility an opportunity for synergy, Wichita-area investors purchased it and then built an ethanol plant on an adjacent property. Ethanol demand is high at the moment because of a California requirement for 100 percent ethanol-blended gasoline by the year 2004. By using waste from the gluten plant as raw material for alcohol production and then selling the alcohol refuse as cattle feed, both facilities promise to prosper. The complex, called U.S. Energy Partners, currently employs 120 local people, enough to stabilize Russell's economy for the immediate future.[104]

With 5,775 residents in 1950, Abilene not only was slightly smaller than Russell but also faced physical conditions that seemed certain to stymie additional growth. Junction City lay only twenty-five miles in one direction, Salina twenty-seven in the other. Abilene has been lucky with enterprises generated by two of its native sons, however. Directly and indirectly, these have been responsible for modest city growth to 6,543 people by the year 2000. The easier of the personal impacts to measure is that of Alva L. Duckwall, who founded a variety store in 1901 and then expanded it into an eleven-state chain. His company, now called Duckwall/Alco Stores, currently operates 260 units from a general office and warehouse in Abilene. With 401 jobs in the city, it traditionally has been the community's largest employer.[105]

The local influence of Abilene's second famous son, Dwight Eisenhower, is subtler than that of Duckwall. The town obviously basked in pride when "Ike" led the Allied forces to victory in World War II and then served two terms as president of the United States. Visitors became common in the 1960s after the creation of the official Eisenhower Center to house the presidential library, boyhood home, chapel, tomb, and museum. The most interesting aspect of this development from Abilene's perspective was an economic strategy that emerged from it. Planners reasoned that, since people were coming to see one museum, why not build others to keep them in town longer? Two obvious themes lent themselves to implementation: a reconstruction of the cattle-driving landscape they called Old Abilene Town and a hall of fame for greyhound racing that built on Abilene's longtime national leadership in the breeding and training of these animals. A third successful entry has been the Museum of Independent Telephony, underwritten largely by Sprint Corporation in honor of that company's beginnings in Abilene as the United Telephone Company.[106]

An economic development program based on a network of small museums makes Abilene unique in Kansas and perhaps in the Midwest, but the concept is not all that different from the more common practice of ethnic heritage tourism that has flourished since the 1980s in many communities around the country, including nearby Lindsborg. In all cases, success hinges on having major population centers fairly close at hand and good access via an interstate highway. Abilene people, however, have been able to take their strategy one level higher. In 1995, the Russell Stover Company, a candy manufacturer from

Kansas City, selected this community for its fifth production plant. Officials cited not only good transportation as a reason but also the advertising potential of a highway location and the number of museum visitors who might want to add a factory tour and candy purchase to their day's experience. Abilene gained 501 jobs from the Stover addition. Smaller in scale but similar in vision was an even more recent decision by owners of Kansas's most famous restaurant, the Brookville Hotel. Forced out of their small namesake town by an inadequate sewer system, they rebuilt in Abilene directly at the interstate exit.[107]

Kansas demonstrations of the power of interstate highways to concentrate retail and service trade, and therefore population growth, are not limited to the I-70 corridor and the Wichita area. The segments of the Kansas Turnpike between Topeka and Wichita, Interstate 35 between Kansas City and Emporia, and Interstate 135 between Salina and Wichita, each 90 to 140 miles in length, all feature a predictable urban success story at their approximate midpoints (Map 24). For the turnpike this has been Emporia, whose population increased 71 percent between 1950 and 2000. Ottawa (an 18 percent increase) and McPherson (a 58 percent increase) occupy the equivalent positions on I-35 and I-135, respectively.

Although Emporia, McPherson, and Ottawa all possess college campuses, they otherwise are quite distinct communities whose economies have taken separate paths. McPherson, the smallest of the three in 1950 with 8,689 residents, provides a good starting point because of several issues it shares with Hays, Russell, and Salina. Like Hays and Russell, McPherson experienced oil prosperity starting in the late 1920s. Unlike its neighbors, however, McPherson people secured a refinery, one that remains active and that has helped to establish a local philosophy in favor of industrialization. McPherson's similarity with Salina comes from the double-edged nature of its interstate development. In addition to making the town more accessible for area shoppers, the highway also fostered easier conditions for McPherson residents themselves to travel the thirty-seven miles north to Salina or the fifty-eight south to Wichita.

Seeing that possibilities to increase retail sales locally were modest at best, McPherson planners decided instead to concentrate on industry (Table 2). Their strategy, in addition to touting good transportation, revenue bonds, and a reliable workforce fresh from the farm, was to offer the cheapest electricity in Kansas. The local board of public utilities instituted a symbiotic relationship with the Kansas Power and Light Company. By this, the city would provide power to the larger utility during times of peak demand in exchange for the right to purchase base-load power back at low rates. To promote their plan, city officials flooded their streets with bright illumination and proclaimed themselves the "light capital" of the state. The tactic proved successful because, beginning in 1960, quality companies have come and stayed.[108]

The list of manufacturers attracted to McPherson is impressive and includes branches of Kit Manufacturing (a maker of mobile homes, 1960), Johns-Manville (fiberglass insulation, 1974), and Sterling Drugs (pharmaceutical products, 1977). With a total of just more than a thousand employees in 2002, they, along with the venerable NCRA refinery (511 employees), form the core of the city's economy. Supplementing these companies is another longtime and homegrown institution, Farmer's Alliance Mutual Insurance (261 employees), and several makers of plastic extrusion products. This latter group, led today by a branch of the CertainTeed Corporation with 304 workers, began almost by chance when Keith Swinehart came to town in 1959 and entered the market for PVC pipe just as this product came into widespread use for plumbing. Vanguard Plastics (polybutylene pipes and 147 employees) and American Maplan (extrusion tools and 135 employees) complete the local inventory. Together they have boosted the city's population to 13,770 and made McPherson a destination for many commuters.[109]

People in Ottawa have never been as proactive toward growth as those in McPherson. The city's population of 10,081 in 1950 was mainly a result of infrastructure provided by outsiders. First had come a north-south railroad promoted in Lawrence and then the main line of the Santa Fe, built by that company as part of a cutoff to connect Emporia directly with Kansas City. This transportation network, which brought with it a small Santa Fe repair shop, made Ottawa a convenient location for several rural-oriented industries such as creameries and windmill plants. The city's population rose to 9,018 by 1920 and held there almost without variation for the next seventy years. As the older companies faded from the scene, they were replaced, almost one for one, by apparel makers, including the H. D. Lee Company in 1941 (work pants, 224 employees) and Mode O'Day in 1945 (dresses, 97 employees). These stayed until the 1980s before seeking cheaper labor elsewhere. The only local manufacturer to make good after midcentury was Ottawa Steel Products, a producer of hydraulic front-end loaders for tractors. It evolved into the Ottawa Truck Corporation and a specialization in vehicles for freight terminals. This company employed 288 people in 2002.[110]

Ottawa's steady-state existence lasted into 1994, when Olathe-based AlliedSignal (the former King Radio) closed its local branch. This cost the city 450 jobs and prompted planners to become proactive. They realized that, with Kansas City suburbs now less than twenty miles away along I-35, companies that wanted access to the big urban market but not the high-priced land of Lenexa could well consider Ottawa. The promotion of these assets began to yield dividends in 1995. First a small maker of dinnerware, Heartland China, selected the city over Lawrence and Overland Park. Later that year came a true bonanza: a Wal-Mart distribution center just northwest of the city that hired 1,250 workers. After a hiatus of several years in which labor markets adjusted to the new

Line Workers at the Iowa Beef Processors Plant in Emporia, circa 1975. After an animal is killed, its carcass is hung on a chain that moves past a series of workers who skin, gut, and trim. The jobs are repetitive and difficult, with safety closely tied to the speed of the chain. Packing plants hire thousands of workers but since the early 1980s have had annual turnover rates in excess of 70 percent. (Courtesy Kansas State Historical Society)

job openings, this trend has continued. A distribution facility for American Eagle Outfitters (250 employees) is the most recent addition. These new positions, together with expansions at Ottawa Truck and at several small makers of heavy steel products, have changed the community's business climate. Over the course of twenty years, Ottawa has undergone a transition from employment self-sufficiency through increased reliance on commuting to jobs elsewhere and now to being a commuter destination of its own. The local population soon should expand rapidly beyond the 11,921 reported at the last census.[111]

Emporia, much farther from larger, competing cities than either McPherson or Ottawa, better fits the model of growth via an interstate node. Transportation, of course, always has been central to this city's development, and as local railroads began to reduce employment in the 1950s, it is not surprising that people realized this significance and started early to explore alternatives. Part of their plan was to promote the community's most important institution, Kansas State Teachers College (now Emporia State University), as a more personal educational experience than that available at the three larger state universities. Commercially, the strategy was three-pronged: to encourage retail expansion with a new Flint Hills Mall, to attract agriculturally related industries, and to encourage local entrepreneurs.[112]

Emporia's blend of initiatives has produced steady growth over the past half century and a current population of 26,760. Looked at more closely, however, retail and university expansions have been less than planners hoped. With the stores, the cause is much the same as at Salina: an increase in shoppers from adjacent counties countered by losses from the city itself to turnpike-accessible Topeka and Wichita. The university's enrollment, possibly influenced by a declining appeal of low-paying teaching jobs for its graduates, also has stagnated over this time. Even so, Emporia State's 5,200 students and support staff of 1,669 remain central to local life.

Industry is chiefly responsible for the city's recent growth. One of the first companies, and by far the biggest, to show local interest was Iowa Beef Processors (IBP). In the 1960s, as now, this organization was a leader in reconceptualizing the meatpacking industry. It sought out smaller cities close to where cattle were raised and built one-story, modern plants. From this perspective, Emporia was an ideal choice. The surrounding Flint Hills are premier grazing country, of course, and a series of small feedlots existed along the pasture margins. With two interstate highways in town as additional incentive, IBP erected a large facility on the city's southwestern side in 1969. More than 1,640 people worked there in 1982 and 2,590 in 2002.[113]

Joining IBP at Emporia's transportation nexus in agricultural country has been a second meat company (Fanenstil Packing, 50 employees); a branch of a large national bakery (Dolly Madison, 767 employees); and two pet-food plants (Midwest Menu and Safeway, 225 total employees). More intriguing, perhaps, is the presence of an unusual number of locally developed factories. Rivaling Wichita people in their spirit of entrepreneurship, Emporians long prided themselves on the Didde Corporation. This maker of web offset presses and collators was born in Carl Didde's print shop in 1943 and employed 725 people during the 1980s. Although the company fell victim to mismanagement and sold out in 2001, two similar endeavors created by former employees remain: Kansa Corporation (newspaper insertion equipment) and Glendo Corporation (engraving equipment), with a total of 120 employees. In other fields, local ideas to make headlight-aiming devices and a new type of pressure vessel for oil refineries have led to 340 and 76 jobs, respectively, with Hopkins Manufacturing and Saunders Custom Fabrication.[114]

9
Postindustrial Kansas II: Life beyond the Exit Ramps

As Emporia, Hays, Salina, and similar cities along the interstate corridors began to expand their trade areas during the 1950s and 1960s, it was inevitable that other, more isolated urban centers would suffer. Such relatively bypassed places are concentrated in three sections of the state: the far north, fifteen or so counties in the southeast, and the entire southwestern quadrant. Because planners in each of these regions obviously deal with many common issues, an economic analyst might be tempted to create a simple model of population decline for them based on distance from interstate nodes. This would be a mistake, however, because differing resource bases and modes of entrepreneurship have yielded widely varying fortunes in the various cities. The outcomes provide useful cautions for single-minded theorists and can best be understood region by region.

A New Role for the Southwest

Southwestern Kansas differed from the north and the southeast during the 1950s, 1960s, and 1970s in having widespread prosperity based on oil and natural gas. Although these minerals are not totally ubiquitous, they underlie every community of consequence except Dodge City, Larned, and Scott City and are central to regional thinking (cf. Maps 12, 22; see pp. 25 and 35). Residents in the easterly, oil-oriented portion of this region do not attribute recent hard times to problems of highway access, for example. Their more immediate concerns focus on aging wells and fluctuating petroleum prices. These issues led to economic crisis in the mid-1980s when the market dropped suddenly from more than thirty dollars per barrel down to ten after leaders in Saudi Arabia decided to increase their production dramatically in 1986.[1]

Although prices recovered somewhat from their nadir, twenty thousand Kansans lost jobs during the downturn, and the industry never regained its earlier stature. The biggest impact, potentially, should have been on the three cities most central to production and drilling: Great Bend, Hays, and Russell. Despite continuing linkages among these places, however, the negative impact fell disproportionately on Great Bend. No commentator ever stated the reason directly, but clearly it was Interstate 70 that enabled Hays and Russell to escape major association with oil's problems. Hays, of course, emerged as a poster city for commercial prosperity, while writers accorded Russell blander, but still not negative, words of steady-state existence. Great Bend, in contrast, faced the reality of decline. From a boomtown mentality in the 1950s, self-confidence had fallen so far by 1987 that someone posted a sign: "Last One Out, Turn Off the Lights."[2]

With no superhighway that could be promoted to potential new companies, population in Great Bend fell in both the 1980s and the 1990s from a peak of 16,608 to 15,345. This decline of 8 percent almost certainly would have been greater, however, had city leaders not decided to invest some of their oil money in economic diversification during the heady 1960s and 1970s. They upgraded their hospital to form the Central Kansas Medical Center in 1964 and bid successfully the following year for one of the new junior colleges authorized by state legislators. As in Salina, they also developed a modern industrial park at a B-29 bomber base that had operated briefly during World War II. These efforts, advertised just before the interstate highway system was fully in place, attracted several firms that aimed for a regional market and have proved themselves durable: Doonan Trailer Corporation (grain and oil conveyances, 120 employees in 2002); Great Bend Industries (hydraulic cylinders, 240 employees); and Great Bend Manufacturing (front-end loaders, 148 employees). Somewhat surprisingly, officials at the Fuller Brush Company of Hartford, Connecticut, also decided to relocate their manufacturing and warehousing facilities there. Praising the centrality of Kansas, company officials invested $9 million in Great Bend in 1972 and hired 660 workers.[3]

The problems posed by a forty-mile distance to an interstate highway were countered for industrialists in the 1980s by an abundance of labor made available by the oil crisis. More recently, the economic balance has shifted slowly toward the negative side. The Fuller people have decided to remain (a current workforce of 345), as have the three regional manufacturers. Employee numbers at the community college (600) and the medical center (650) equal the industrial totals. Additional growth at any of these units is unlikely, however, because of competition only sixty miles away at the larger Hays and Hutchinson. College officials already have made overtures to merge with Fort Hays State University, and the hospital may be forced to do the same. In manufacturing, Marlette Homes and several other firms have abandoned the city for better-connected sites, and no replacements are at hand. Finally, and most obviously, Great Bend's retail trade area has been reduced to only two counties: its own and adjacent Pawnee.[4]

If Great Bend soon may be judged a superfluous city as its oil fields deteriorate and retail sales stagnate, the futures of its tributary communities of Hoisington and Larned seem darker still. Hoisington especially faces challenges.

Founded as a railroad town, it achieved prosperity almost entirely from its designation as a division point for the Missouri Pacific. From a peak population of 4,248 in 1960, Hoisington faded along with the railroads and could count only 2,975 residents by 2000. A brick and tile plant started in 1954 by banker Ray Smith to take advantage of quality local clays has helped to stem the exodus somewhat, as has a factory for telephone cable established in 1975 by the Detroit-based Essex Group. With 260 employees apiece, they support nearly the entire population. Cheyenne Bottoms, one of the Midwest's premier sites for migratory waterfowl, lies just east of town. Increased tourism there in recent years has added several additional jobs, but the potential for expansion seems small.[5]

Larned people, although also familiar with population decline (from 5,001 in 1960 to 4,236 in 2000), are better positioned than those in Hoisington to maintain at least a limited viability. The town serves as the seat of Pawnee County, has access to good groundwater along the Arkansas River, and, most important, maintains its longtime institution: Larned State Hospital. Mechanization has reduced employment at Doerr Metal Products (the city's only manufacturer) to nineteen, but these jobs have been more than replaced by the development of two large feedlots near the river. Ward's, founded in 1962 (39 employees), and Pawnee Beefbuilders, dating from 1972 (27 employees), sell cattle to packers upstream in Dodge City and Garden City. The state hospital remains the community's economic mainstay. More than 840 people worked at its campuslike setting four miles west of town in 2002 and another 345 in special buildings there that have been converted into prisons for juvenile and insane offenders. So long as the hospital remains, Larned's future is assured.[6]

The combination of relatively poor transportation and a declining minerals economy that defines Great Bend today is reproduced on a larger scale at Hutchinson. With 40,787 people in 2000 and status as the tenth-largest city in the state, Hutchinson at first would seem a poor candidate for decline. The city has not grown in the last two decades, however, and has seen its size ranking fall steadily from the number four spot it occupied in 1950. The problem is related somewhat to the oil economy and to mechanization at the city's salt companies, but mostly to a location that simultaneously is too far from an interstate highway (thirty-two miles) to attract new industry and too close to expanding Wichita (forty-nine miles) to avoid damaging competition for retail and wholesale businesses.

An economic survey in 1980 characterized Hutchinson as a city where initiatives are studied carefully before implementation. Ongoing obstacles to success have now made this trait a requirement. Of the elements that promoted local growth early in the twentieth century—milling and wholesale trade via one of the state's best railroad networks, large-scale salt production, and an unlikely pair of public institutions (a prison and the state fair)—the mills and warehouses gradually declined along with the railroads. Town leaders assessed changing conditions accurately, however, and tried to use the community's size and accumulated capital to diversify their pool of jobs.[7]

Because of competition to the east from Newton and Wichita, Hutchinson retailers always looked westward for their trade. This effort, which was aided by the ability of the local newspaper to reach distant communities via the rail system, produced customer loyalty to the city as far away as the Colorado border. Such a pattern continued strong into the 1980s but has since eroded badly. The cause lies partially with Hutchinson residents themselves. Their own increasing preference for Wichita retailers contributed to the closing of the long-lived Peagues department store and several specialty shops.[8]

Because the services portion of any economy is more flexible than retail sales, Hutchinson entrepreneurs decided to make special efforts there. Plans to expand the city hospital were hampered by the presence of specialized facilities at Wichita, but ideas for more singular and culturally oriented offerings have been able to reverse this traffic flow and draw in customers from throughout south-central Kansas. That this approach could work had long been demonstrated by the success of the Kansas State Fair each September. To this, promoters added the men's basketball tournament of the National Junior College Athletic Association every March and the equally popular Mennonite Relief Sale and Auction in April. The city now has achieved even greater fame and visitor numbers with the Kansas Cosmosphere and Space Center (a multimillion-dollar science museum opened in 1980 with funding from the Carey family) and the promotion of their unique Prairie Dunes Golf Course in the nearby sand hills as a site for the U.S. Open Championship for women.

Attracting and maintaining industrial jobs has been the most challenging task for Hutchinson promoters. With inexhaustible salt deposits and well-equipped plants, their basic industry remains strong. One firm, IMC Salt (formerly Carey) closed its local plant in 1999 in favor of a newer one nearby at Lyons, but the big evaporative facilities of Morton and Cargill (formerly Barton) employ more than a hundred people apiece. Other venerable companies that continue include Krause Plow, which has specialized in wheat-farming equipment since 1924 (300 employees); Consolidated Manufacturing, a rebuilder of automobile engines since 1943 (283 employees); and the headquarters and distribution center for Dillon's, a major chain of grocery stores that started in Hutchinson in 1918 (800 employees).[9]

To attract a wider range of manufacturers, Hutchinson leaders followed the example of Great Bend in converting an old army airfield into an industrial park. This conversion occurred only after many other cities had done the same, however, and so success has been only moderate. Plants for Detroiter Mobile Homes, Rainbo Bread, and Far-Mar-Co flour have closed, but replacements exist in Doskocil (a sausage maker, 400 employees); Lowen Corporation

Corn and Watermelon Exhibit at the Kansas State Fair in Hutchinson, circa 1965. With an annual attendance that surpasses four hundred thousand, this fair is advertised as "the largest event" in Kansas. Hutchinson leaders extol it as part of a general push for tourism. When not displaying produce in September, the 280-acre fairground hosts more than 250 flea markets, trade shows, auto races, and similar activities. (Courtesy Kansas State Historical Society)

(decals, 229 employees); and Collins Industries (small buses, ambulances, 452 employees). Officials with these operations praise the local workforce but disparage city inaccessibility. The industrial solution that has worked for other Wichita-area communities—a tie to the aircraft juggernaut—is hampered somewhat by location (the entire Wichita metropolitan area sits between Hutchinson and the aircraft plants) but has nevertheless produced the city's biggest employer. Eaton Corporation, which began life in 1942 as Cessna's first branch outside Wichita, gradually has been transformed into a separate division that makes hydraulic power systems for airplanes and other uses.[10]

Eaton's connections to Wichita suggest that its 981 jobs will be permanent, a great satisfaction for Hutchinson's planners in their ongoing efforts to recruit. The future for the city likely will be population stability, however, with industrial uncertainty balanced by an increase in commuting to Wichita. That short drive became routine in the 1990s when Highway 96 was improved to a four-lane trafficway. A transition toward a suburban existence will not be easy for this city with its independent past, of course. One indication is that congressional redistricting completed in 2002 continues Hutchinson's long-standing separation from Wichita and its affiliation instead with the state's western counties.[11]

Just as Great Bend's tributary cities of Hoisington and Larned have found existence away from the interstate system to be challenging, so too have Lyons and Ellsworth, towns respectively one and two counties north of Hutchinson (Map 22). Both of these places are county seats and have histories in salt mining that raised community expectations for additional growth. One mine continues at Ellsworth (53 employees) and two at Lyons (158 employees total), as does a modest amount of area oil production. Retail sales have not been good for decades, however, leaving residents with options for accepting decline or recruiting the type of industry that is not welcome in more prosperous places.[12]

Ellsworth people opted for risk over decline when they voted in the late 1980s to invite construction of a medium-security state prison. Most citizens are happy with the decision. No escapes have occurred, and the addition of 235 new jobs to the community has converted slow population decline into modest gain (a total of 2,965 in 2000). At Lyons, twice as large as Ellsworth in 1950 with 4,545 residents, two separate possibilities for growth have materialized. A tentative selection of local salt beds by the Atomic Energy Commission as the nation's storage vault for nuclear waste was greeted with enthusiasm in 1970. Officials promised forty new jobs to replace those lost by a recent company closure, and people knew the area to be seismically stable. When the project was revoked after subsequent testing revealed too many wells in the area that might allow radiation to leak, people actually got mad at the surveyors. Lyons's second venture was a large processing plant for hogs proposed just west of town by the Seaboard Corporation of Kansas City. This idea was greeted with more skepticism than the nuclear vault, but still as a positive event. Concentrated animal production needed to supply the plant certainly would generate odor and pollution, but the builders promised more than two thousand jobs. Seaboard planned to open the facility in 2001 but canceled construction after reassessment of national market conditions. Lyons residents, their air and soil still pristine, have accordingly endured a population slide of 19 percent between 1960 and 2000, down to a total of 3,732.[13]

Pratt, the remaining sizable community in the eastern portion of southwestern Kansas, has shared the ups and downs of the oil industry with Great Bend and Lyons and the tribulation of losing the division point on a major railroad with Hoisington. These events are basic to understanding a local loss of 1,408 people since 1960. Despite this trend, however, Pratt seems poised for growth over the next several years. Unlike most of their competitors in the region, the city's merchants and service providers are able to command a fairly large trade area, especially to the south and west (Map 22). This has made the small community of 6,570 a natural choice in recent years for McDonald's, Pizza Hut, Wal-Mart, and similar franchises. It also has given leaders incentive to enlarge two basic service institutions: a hospital into the Pratt Regional Medical Center and a nearly moribund vocational school into Pratt Community College. With 410 and 140 employees, respectively (8 percent of the city's population), these units serve an eight-county area and have become central to Pratt's economic existence.[14]

Two secondary location issues contribute to Pratt's economy as well. Although the city is near the southwestern edge of the Kansas oil region (Map 12), it is nonetheless midway between concentrations of wells in the Great Bend–Hays area and in north-central Oklahoma. Several drilling-supply companies including Halliburton and R. C. Williams consequently find this a convenient site from which to service the entire realm. Feedlot operators also like the city. A market is nearby in Dodge City, and their other major requirement of water is met in the northern half of Pratt County by lowlands near the Arkansas River. Pratt Feeders, four miles north of the city on U.S. 281, has a capacity of twenty thousand head and employs thirty-five people. The Pratt Livestock Exchange, the busiest auction barn in south-central Kansas, hires another seventy-five.[15]

The economic struggles of Great Bend, Hutchinson, and Pratt in recent decades are textbook examples of what has happened nationwide to once-vibrant places that lost much of their trade areas and industrial bases to cities located on the interstate system. If an outside observer were armed with this local knowledge and then asked to predict a demographic scenario for extreme southwestern Kansas, the task would seem ridiculously easy. In this most arid and isolated section of the state (the nearest superhighways are ninety miles to the north and south), population decline would almost have to be the rule. Jobs connected with the large natural-gas field centered at Hugoton might limit problems to a degree, but since output there peaked in the 1970s (paralleling the experience of oil in central Kansas), our observer also could argue that company contractions would contribute to an overall exodus of jobs and people.

The facts for southwestern Kansas, of course, stand in marked contrast to any model of declension. Between 1950 and 2000 each of the three largest cities in the region—Dodge City, Garden City, and Liberal—more than doubled its population. Although their rates of expansion have not rivaled Wichita or Overland Park, they far exceed interstate-rich Salina and match the university towns of Hays and Manhattan. Dodge City, the longtime regional trade center and division point for the Santa Fe, grew from 11,252 residents to 25,176. Even more interesting, that community's two somewhat smaller and more specialized rivals increased at even faster rates. The former sugar-beet center of Garden City actually surpassed its neighbor in population as early as 1970 and counted 28,451 residents in 2000. Liberal, the smallest of the three in 1950, grew fastest of all: 176 percent. With 19,666 people in 2000, it has become a near-equal business competitor in the region.

Although unprecedented, the widespread prosperity in southwestern Kansas is no mystery. It clearly derives from the meatpacking industry. The intriguing aspect of this economic transformation is its history, beginning as it does with the failure of another agribusiness endeavor. A series of national political maneuverings in the mid-1950s abruptly made the production and refining of sugar beets unprofitable in Garden City. As area farmers struggled to find alternative uses for the thousands of irrigated acres developed as beet fields, Earl Brookover hit upon the idea of converting his fields to corn and then feeding this grain to local cattle. Brookover, a native of nearby Scott County, opened the region's first commercial feedlot near Garden City in 1952. His business started small but grew during the droughty years later that decade when a scarcity of grass prompted ranchers to try grain fattening.[16]

Another jump in scale, with many new feedlots joining the original, came in the 1960s. This involved the widespread adoption of center-pivot technology as a way to utilize groundwater for spray irrigation and therefore to avoid the necessity of ditches. To employ center-pivot systems effectively, a person needed both abundant groundwater and inexpensive fuel for the pump. These two conditions are met ideally in the cluster of counties south from Garden City and into Oklahoma and Texas. There the panhandle and Hugoton natural-gas fields underlie the unconsolidated sands of the Ogallala formation, one of the largest aquifers in the nation.[17]

As local people developed the center-pivot and feedlot industries (between 1964 and 1969 irrigated land in southwestern Kansas increased by over 175,000 acres), it was inevitable that major meatpacking companies would see potential for themselves in this new regional mix. Existing plants at Kansas City and similar sites were outdated and a source of smells increasingly abhorred by sophisticated neighbors. By moving, the companies could obtain new and often subsidized facilities in communities that welcomed their business. Such sites also were cost efficient because of local feedlots. Within a decade, nearly the entire beef industry had relocated to rural plains communities. Kansas, which had accounted for only 5 percent of the national slaughter market for cattle in 1965, rose to 18 percent by 1987.[18]

The first slaughterhouse to relocate in southwestern Kansas was Producers' Packing Company, which came to Garden City from Iowa in 1965. Welcomed from the beginning, the company cosponsored an annual celebration called Beef Empire Days with the city's chamber of commerce and so established a resonance with the region's past. This event soon became a three-day symbol of area identity and, as such, helped convince larger companies that southwestern Kansas was a good location for their investments. Entrepreneurs in

Liberal, concerned over declining local employment on the Rock Island railroad, were the first to broker such an arrangement. The National Beef Packing Company began operations there in 1969. This move transformed the community. From an initial employment of 150, National expanded to 900 workers by 1979 and then to 1,900 by 1982.[19]

The presence of cheap and abundant cattle on the High Plains proved to be incentive enough to overcome the inconvenience of a mediocre highway network. Labor supply was potentially a more serious problem. If companies were to build modern plants that required up to three thousand employees, could these numbers of workers ever be found? The experience of National Beef during the 1970s answered this question definitely in the affirmative. Immigrants recruited from Vietnam, Mexico, and elsewhere easily bridged the gap between need and local supply and proved to be excellent employees. With this last doubt removed, officials at Iowa Beef Processors (the largest of the new generation of packers) built a huge plant ten miles west of Garden City in 1980 and hired on two thousand hands. Later that same year, Excel Corporation of Wichita erected a slightly smaller facility near Dodge City.[20]

Liberal, the city that pioneered big packing in the region, did so because its business community had experienced the joys of a boom in the 1950s and wanted desperately to maintain that momentum. The initial surge had been based on natural gas. Drillers in those years were finding new wells with potential yields of up to a hundred million cubic feet per day at depths greater than known before. Because of this, the area's major pipeline company, Panhandle Eastern, expanded its staff size to 430 by 1961 and created a subsidiary operation called Anadarko Petroleum to concentrate on exploration and development.[21]

Gas-based monies allowed construction of a new courthouse, two grade schools, and the other public facilities needed to accommodate a population surge that nearly doubled the city's size by 1960 (from 7,134 to 13,813). Leaders also actively broadened the economic base through financial incentives and the creation of an industrial park at an air base abandoned by the army in 1946. A maker of steel truck beds called Tradewind Industries was the first success. It moved to town in 1947 from nearby Perryton, Texas, with thirty-two employees. Four years later, the scope of growth increased dramatically with the acquisition of Beech Aircraft's first plant outside Wichita. Having no more space at their headquarters, Beech executives moved their spar shop for military planes across the state and into existing hangars at the old base. Soon they hired 350 workers.[22]

Economic conditions and population numbers held steady in Liberal during the 1960s. Planners lobbied successfully for expanded hospital facilities (the Southwest Medical Center) and an entry in the state's new network of junior colleges (Seward County Community College). They worried, however, that a decline in drilling activity implied falling production of oil and gas in the near future. This explains the decision to pursue a major packing company at that time. Some fifteen years later, by 1983, such initiative was justified. Officials at Anadarko Petroleum had moved their headquarters to Houston in 1972, and those at Panhandle Eastern were now considering similar action. In addition, the Beech executives decided to reconsolidate their operations at Wichita. These changes hurt Liberal's diversity badly, but National Beef continued to prosper. In 2002 the main plant employed 3,000 people, with another 110 working at its trucking subsidiary (National Carriers) and 75 at its Supreme Feeders cattle company. Without anyone intending to do so, Liberal became a company town. Its beef monopoly at present is challenged only slightly by employment at the medical center (450), the college (150), and a swine-breeding operation owned by the corporate giant DeKalb (310). The ethnicity of the community also has changed tremendously via this process. Hispanic residents now constitute 42 percent of the total and are beginning to move from packing employment into the wider life of the city.[23]

Garden City's move into the world of big-time meatpacking was similar to that in Liberal. Having lost their signature sugar company in 1955, residents needed a way to stave off population decline. They therefore readily embraced a change to irrigated corn, large feedlots, and then giant packers. With the Hugoton gas field lying mostly to the south, the economic future of Garden City in 1960 actually was more tenuous than that of Liberal. The town did hold several advantages from a packer's perspective, however. These included vast irrigation acreage (225,000 in 1972), equally large feedlots (a capacity of one million head within sixty miles of the city in 1980), and a demonstrated favorable attitude toward all aspects of the cattle business.[24]

Garden City's agribusiness economy grew steadily between 1950 and 1980, led by the success of its original packer. Though the company changed names from Producers' to Farmland to Valagri to Monfort, its employment reached thirteen hundred by 1982. Such incremental gains then gave way to an immense increase. In 1980, officials at Iowa Beef Processors (now IBP) built a hundred-million-dollar plant near the city that required two thousand workers for initial operation. With the plant's completion and eventual expansion to twenty-nine hundred employees, Garden City rapidly acquired the status of a poster community for the modern packing industry and its multinational workforce.[25]

As in any other city heavily reliant on a narrow economic base, Garden City leaders have tried to lessen this dependency. One successful project was a new shopping mall in 1980 that enlarged the town's retail trade at the expense of rival communities. Efforts in the 1990s concentrated on alternate uses for their irrigated cropland, particularly on the recruitment of dairies to supply milk for cities in Arizona and other desert locations. These new employers proved critically important after December 25, 2000. A fire that day destroyed the city's original packing plant (now owned by ConAgra) and threw twenty-three hundred people out of work. As I write in 2003, no reconstruction has begun. Company officials say they await news from insurers, but the city meanwhile badly misses the plant's $43 million annual payroll. Besides IBP, the only other large employer in town is St. Catherine's Hospital with 625 workers.[26]

Because the Garden City economy has always been agriculturally based, its residents made the transition to meatpacking life and its multinational culture relatively smoothly. People in Dodge City, in contrast, were somewhat more resistant. This attitude probably was to be expected from a citizenry used to regional leadership and a steady, diversified economy. As recently as the 1960s that city had prided itself on a broad spectrum of regional roles: a railroad center, a leader in wholesale and retail trade, a major tourist destination for the old cattle trade as re-created on the television series *Gunsmoke,* and the site of the area's only four-year college: St. Mary of the Plains. The power of this quartet then waned. New stores in Hays and Wichita began to erode the trade area, and Dodge lost most of its Santa Fe jobs to Newton on the east and La Junta on the west. Tourism, once on the level of 250,000 visitors per year, fell to 110,000 by 1984 as memories of the television series began to fade and the nonprofit group that owned the principal "Boot Hill" site decided not to advertise extensively. Finally, St. Mary's, which long had suffered financial problems, closed in the early 1990s.[27]

Although Dodge City is east of the big Ogallala Aquifer, the valley of the Arkansas River there still holds considerable potential for irrigation wells. Some eighty thousand acres in Ford County had been so developed by 1972, which attracted several large feedlots and one small slaughterhouse: HyPlains Dressed Beef. As bigger packers became interested in the region, Dodge City entrepreneurs did not promote as actively as their neighbors. Excel Corporation out of Wichita did build locally in 1980, but without great fanfare. Possibly as a result of this nonchalance, relations between immigrant workers and older residents in Dodge City were less cordial than in Garden City or Liberal.[28]

By 2002, the communities of Dodge City and Garden City had evolved similar economic structures. The HyPlains plant was sold to National Beef and expanded to employ 1,010 workers, while Excel (now owned by the Cargill conglomerate) hired 2,560. After these two giants, local manufacturing employment fell off rapidly to pallet maker Arrowhead West (120) and the Crust Buster/ Speed King factory for farm equipment (77). As for the service sector, the acquisition of a community college assuaged the loss of St. Mary, and the local Western Plains Regional Hospital has earned a reputation for heart care to complement a cancer specialization at its Garden City equivalent. The longtime rivalry between the two cities actually is less now than in the recent past. Residents still measure themselves against one another, but they face similar

Re-creation of Front Street in Dodge City, 1958. Tourism became an important part of the city's economy in the 1950s and 1960s. Capitalizing on the popularity of the television series *Gunsmoke,* which was supposed to take place locally, townspeople constructed an 1880s facsimile of downtown a block west of the real thing. The Long Branch Saloon, where a costumed Miss Kitty talked with Marshal Dillon, drew a quarter-million visitors annually until television tastes began to change in the 1970s. (Courtesy Kansas State Historical Society)

economic and social challenges for the future. For retail trade the bigger competitors for both are Hutchinson and Wichita.[29]

In the shadows of the major cities, three other communities in southwestern Kansas exceeded the urban threshold of 2,500 people in 2000 (Map 22). All have done so on the basis of local natural resources. Scott City, thirty-six miles north of Garden City, owes its growth to 3,855 residents primarily to a northern extension of the same groundwater formation that supplies the general region with such abundance. Scott County consequently is a state leader in irrigation acreage (mostly for corn) and feedlot capacity. The feedlots themselves employ more than 300 people.[30]

Hugoton and Ulysses have locations central to both natural gas and groundwater. Although better transportation connections and greater initial size at Liberal and Garden City allowed those places to profit most from these resources, many cattle feeders prefer to operate in smaller places, as do drilling companies and field-service people. Hugoton, with, 3,708 residents in 2000, still proclaims itself "gas capital of the world" and retains a regional office of Northern Natural Gas (a company that started there). Ulysses, a county to the north, has grown somewhat larger, to 6,960 people. Its success comes through jobs in pipeline, feedlot, and irrigation enterprises plus positions at two specialized, minerals-based plants: BP America for liquid petroleum (138 workers) and Columbia Chemicals for carbon black (50).[31]

The Bypassed Northern Tier

Traffic on Highway 36 westward across Kansas from St. Joseph, Missouri, used to be nearly as heavy as that on its parallel routes of U.S. 40 out of Kansas City and U.S. 30 along the Platte River of Nebraska. Gradually, however, as the latter two highways were converted into Interstates 70 and 80, respectively, activity along and near the middle route began to stagnate. The distance of sixty or so miles north or south to the new corridors of development no longer was enough to insulate local merchants from competition.

Without mineral resources or other special circumstances to provide alternative jobs, cities in extreme northern Kansas generally experienced population declines after 1950. Predictably, the losses on a percentage basis were greatest in specialized railroad communities, especially those near other, more diversified centers. The larger cities in the region, the ones that traditionally served roles in wholesaling, also declined substantially. The only areas to avoid economic problems, in fact, were those within easy commuting distance of the interstate system.

Atchison, always the urban focus of the northern tier, lost 2,560 people between 1950 and 2000. Left with only 10,232 residents by the most recent census, this onetime rival with Kansas City has now devolved to a point where many people consider it a small town. Some analysts might even conclude that the community is lucky not to have experienced even greater losses. Its highway connections southwest to Lawrence and Topeka are poor and those southeast to Kansas City only marginally better.

A study based on highway access, although important, misses Atchison's major assets. For one thing, the community's two biggest employers, Atchison Castings and Midwest Grain Products, have been in place since 1872 and 1938, respectively. Both have invested heavily in their sites and employ highly skilled workers. Since neither operation requires a rapid highway distribution network, they happily keep their 1,058 combined jobs in Atchison. Midwest Grain, in fact, recently has built a $6 million addition for the production of wheat proteins. Moreover, the city's blue-collar reputation has attracted a state-sponsored vocational-technical school and several additional companies of this type, including a corn products plant run by ConAgra (122 employees), the National Pipe Company (85 employees), and Fargo Assembly, a maker of electrical wiring systems (80 employees).[32]

Atchison's second important asset is a large stock of spacious Victorian houses. This heritage, when combined with the presence of Benedictine College and the proximity of Kansas City (fifty-five miles), is giving the town increasing cachet as a retirement community. A property-tax rebate program for substantial home improvements encouraged this trend in 1995. An even larger accelerant has been the phenomenal success of Nell Hill's and G. Diebolt's, two home decor businesses established in downtown storefronts by local resident Mary Carol Garrity. With shoppers coming in large numbers from Omaha, Des Moines, and beyond, Atchison's impressive architectural stock is showcased to an appreciative audience. The city's long decline is probably at an end.[33]

In theory, the northern-tier communities best able to survive in recent times would be those spaced far enough apart to assure at least modest regional trade. Starting with the obvious node of Atchison and St. Joseph and assuming an interval of a hundred or so miles, this reasoning would predict viability for Marysville (96 miles from Atchison); Concordia or Beloit (70 and 101 miles, respectively, from Marysville); and Phillipsburg or Norton (85 and 111 miles, respectively, from Beloit). Health for any other community presumably would require special circumstances.

The distance model works reasonably well across the region (Map 22). Marysville, with 3,271 people in 2000, had declined 15 percent from its total of fifty years before but was still much larger than county seats adjacent on the east and west. Similar situations exist for Beloit (4,019 residents), Concordia (5,714), Norton (3,012), and Phillipsburg (2,668). Exceptions to the model's predictions are perhaps more interesting to contemplate. Why have economic pressures not produced single dominant cities instead of the continuing rivalries at Beloit and Concordia and at Norton and Phillipsburg? How have the interstitial communities of Clay Center, Hiawatha, Holton, Sabetha, and Wamego all managed to retain populations over 2,500?

The successes of Clay Center (4,564 residents), Holton (3,353), and Wamego (4,246) are easiest to explain. The first two of these combine roles as county seats with locations only thirty-odd miles from larger interstate population centers: Junction City–Manhattan in the case of Clay Center, and Topeka for Holton. In addition, both communities have attracted and retained a small pair of agricultural manufacturers. Holton is home to Oldham's Farm Sausage (a 150-employee plant now owned by A. S. E. Deli Foods) and a similar-sized maker of rotary mowers (the Alamo Group of Kansas). In Clay Center, two producers of grain-handling equipment (Hutchinson-Mayrath and GT) together employ 192 residents. Wamego lacks the designation of county seat but counters by being only fourteen miles from Manhattan. The community also had the good fortune for a local blacksmith shop to evolve into a major manufacturer of earthmoving equipment. Neil Balderson began his expansion in 1929 with a contract to build snowplows for the Kansas Highway Department. He later marketed his products through the large Caterpillar Company of Peoria, Illinois, and the two entities then merged. Some 340 people now work at the plant.[34]

Along the hundred-mile route that separates Atchison and Marysville, three small cities have managed to achieve populations of twenty-five hundred in at least one census year. Of these, the experience at Horton best fits the model of overall decline for such interstitial sites. This city, of course, was a creation of the Rock Island company, which planned it as a major railroad division point and repair center. Although the city was never on a major highway or the seat of county government, the company's employment pushed Horton's population to 4,031 by 1930. Since then, as Rock Island executives allowed first their shops and then their tracks to deteriorate, townspeople have tried to create alternative employment. Even though innovative in at least two instances, these plans have largely failed. The city's current population of 1,967 represents a decline of 51 percent from its peak.

The first attempt to rejuvenate Horton was the sale of several old shop buildings to the Hammersmith and Son Manufacturing Company in 1965. This has proved to be the city's only lasting success, as the steel fabricator continues to employ about forty-five workers. Two more elaborate development schemes followed. In the 1980s, Tony Pizzuti of the Bank of Horton saw growth potential in the processing of applications for federal student loans. By offering quick, reliable service, the bank soon handled 750 applications per day and employed a hundred extra people to do so. Although this loan business declined locally around 1990 as other banks began to enter the competition, a new opportunity for the town appeared in the form of a private prison. Kansas and other states at that time were under a federal order to reduce inmate crowding. The situation prompted Horton entrepreneurs to reason that a private facility might simultaneously meet this need and save taxpayers money. Promoters had gone so far as to purchase land for a thousand-cell building and to arrange financing with a Texas company when state officials decided instead to construct a similar-sized public facility at El Dorado. A location near Wichita, they said, made more sense geographically than having a second prison in northeastern Kansas. With this defeat, the people of Horton finally seemed to accept a much-reduced economic role for their community.[35]

The recent history of Sabetha demonstrates that the economic success desired at Horton can be more than chimeric. Although both of these communities lie on the geographic periphery of their respective counties and currently lack major railroad connections, Sabetha has managed to add 416 citizens over the past fifty years and to qualify officially as an urban place. This completely unpredictable growth to 2,589 residents is the result of several local people starting viable companies in their hometown. The principal successes are Wengers Manufacturing (an innovator in extrusion work since 1935) with 186 employees, and Keim Transportation (a trucking firm since 1957) with 284.[36]

The experience of the final interstitial city in the northeast, Hiawatha, falls in between that of its neighbors. With the relative advantages of being a county seat and having decent railroad (Union Pacific) and highway (U.S. 36) connections, the town has held its population at just above three thousand for more than eighty years. Hiawatha retailers have profited to a degree by the decline of nearby Horton, and several franchise businesses, including Wal-Mart, have decided to build. Countering this retail success has been increasing difficulty in attracting and holding manufacturers to a relatively isolated location. The community relied for decades on the Flair-Fold Corporation, a maker of shutters, frames, and doors. After this company quit business in 1999, the loss of its 340 jobs quickly made Brown County's unemployment rate the highest in the state. A replacement facility called Kansas Mill Work did not stay long, and at present, Hiawatha's industrial base consists of only two tool-and-die makers that employ a total of eighty-six workers. Commuting seventeen miles to jobs in Sabetha has become common.[37]

Marysville arguably occupies the best business location in Kansas along Route 36. It has good water via the Blue River, service from one of the Union Pacific's main lines, two federal highways, and sufficient isolation from St. Joseph and other larger cities to ensure viability for small retailers and service companies. Although the population at the current time (3,271) is smaller than it has been since 1920, the outlook looks reasonably bright. Besides its locational advantages, Marysville has been fortunate to have two large and long-time employers. The Union Pacific has made this a changeover point for train crews and home to one of its centralized traffic-control offices. In addition to directly providing four hundred jobs in this way, the railroad also is responsible for a new motel-diner complex built in 1999 with crewmen in mind. The motel (the Oak Tree Inn) features special shades that completely darken the rooms to accommodate the needs of shift workers, while the restaurant is open continuously.[38]

Landoll Corporation, the second major company in Marysville, resembles Wengers in Sabetha in being homegrown and innovative. Dan Landoll, the owner-founder, expanded from a small welding shop in 1962 into an employer of five hundred. He started by manufacturing tillage equipment, then added specialized trailers, and finally deicing equipment for airplanes. By the late 1980s he was selling to all major airlines as well as the U.S. military. The company's business formula, which stresses high-value, low-volume products and highly skilled labor, has been imitated by many other small-town entrepreneurs.[39]

In contrast to the seemingly effortless dominance of Marysville over its competing towns, the situation farther west in north-central Kansas always

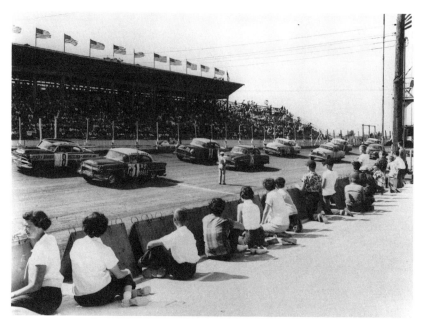

Automobile Racing at the North Central Kansas Free Fair in Belleville, 1958. Belleville's residents host the state's largest free festival every August as a means of bolstering their community's pride and economy. Racing on the city's half-mile track remains the featured attraction today, but the stock cars of 1958 have given way to smaller, customized designs that compete in the Midget Nationals. (Courtesy Kansas State Historical Society)

has been one of intense rivalry. People in Belleville, Beloit, and Concordia all achieved their initial successes by combining the business of county government with additional employers of note. At Belleville it was a division point for the Rock Island railroad. Beloit claimed an excellent mill site on the Solomon River, while Concordia became the nexus for area railroads. None of these assets could be transferred to the needs of the late twentieth century, of course, and common sense would predict that only one of the three places would make the economic transition successfully.

Belleville, the smallest of the three cities in 1950, was also the one positioned most poorly for growth. With Concordia only seventeen miles to the south, it never commanded a large trade area. Then, when its mainstay railroad declared bankruptcy in 1975, the community's only remaining asset was the intersection of U.S. Highways 36 and 81. Local people did what they could. Promoting this junction as the "crossroads of America," they began to host a Kansas travel information center and a Crossroads Car Festival while continuing sponsorship of their long-standing North Central Kansas Free Fair. They also built on the success of a local manufacturer of hospital supplies (Scott Industries) by recruiting a second complementary company (M-C Industries, now Precision Dynamics Midwest). These efforts, all well conceived, unfortunately have done little more than slow the town's decline. The local population, now 2,239, has fallen in each of the last three censuses.[40]

As Belleville stagnated, most observers thought that Concordia, rather than Beloit, would assume dominance in the region. It was the larger community (7,221 versus 4,121 in 1970, for example) and was well positioned to pull trade away from Belleville. In addition, townspeople had been active in adding public services. To complement an excellent hospital built and run by the local Sisters of St. Joseph, Cloud County voters in 1965 agreed to finance one of the state's new community colleges. Finally, Concordia offered a decent array of small industries led by the sheet-metal fabricator C-E Process Equipment with 150 employees and Boogaart's Food Distribution with 100.[41]

Today, Concordia's leadership in the region is far less certain than it was a generation earlier. Even though its principal commercial firms are still present (C-E Process is now Alstom Power, and Boogaart's is now F & A Food Sales) and the community college has expanded to 143 employees, the city's population has fallen dramatically. The latest census shows only 5,714 people in town, a decline of 21 percent since 1970.[42]

Concordia's difficulty lies not with its college, its industries, or even its traditional competition with Belleville and Beloit. Instead, the problem stems from increasingly easy access to Salina, forty-eight miles to the south on Highway 81. Just as officials with the Catholic Church saw fit to relocate their diocesan center from Concordia to that larger city in 1944, so did the Sisters of St. Joseph with their hospital in 1985. Local retailers have felt the pull to the south greatly as well, especially in recent years as the state gradually upgrades this section of Route 81 to interstate quality. As I write, downtown Concordia is not a healthy business location.[43]

As first Belleville and then Concordia suffered major population losses, Beloit managed to gain relative status simply by holding its own. The city's size in 2000, which totaled 4,019, was only 2 percent less than it had been fifty years earlier. Some of this vitality, ironically, comes from being more isolated than Belleville or Concordia. Not only is Beloit farther from Salina, but it is also well positioned on U.S. Highway 24 to attract retail trade from Osborne County and other points to the west.

Beloit combines a decent retail market with a mixture of manufacturing and service industries. Its longtime state institution, a reformatory for girls, is now coeducational and an employer of 133. To this the city has added the North Central Kansas Technical College (a staff of sixty-six) and a solid, homegrown industry called Sunflower Manufacturing. Sunflower makes tillage machinery and grain drills and has grown steadily since its founding in 1965. Two hundred people now work there, and the company has joined with the college to train new workers in agricultural equipment technology. People in Beloit and Concordia now see each other as allies at least as much as antagonists in the struggle to remain vital. The first major sign of cooperation came in the 1960s when the communities finally allowed the road between them (Kansas Highway 9) to be paved.[44]

The dynamics between Norton and Phillipsburg, the second city rivalry along Kansas's northern tier of counties, are reasonably straightforward. These towns have enjoyed recent population stability at levels near three thousand because of different, but balancing outside circumstances. With Norton this long has been the presence of a state institution, whereas Phillipsburg has capitalized on a position at the northern tip of the central Kansas oil field.

Until 1990, Norton's institution was a state mental hospital, a descendant of the tuberculosis sanatorium that had been in town since 1914. Changing public attitudes and improved medication then promised closure for this and several similar installations. A concurrent demand for additional prison space saved the institution, however, and so today the Norton Correctional Facility (minimum security) employs almost exactly the same number of people (266) as did the hospital. Phillipsburg grew beyond normal county-seat size because oil field developers in the 1930s and 1940s wanted access to good railroad transportation. This desire, coupled with Phillipsburg's Rock Island facilities and vigorous local promotion by newspaperman McDill (Huck) Boyd, produced three sizable companies for the town: the offices of the Kansas-Nebraska Natural Gas Company (later KN Energy), a small oil refinery, and Tamko Asphalt Products (roofing shingles). Although the refinery has now been downgraded to a fuel storage terminal, this triad of companies continues operations today. Tamko is the largest at 240 employees.[45]

With their anchor industries in place to provide stability, leaders in both Norton and Phillipsburg then attempted to corner the area's regional trade. Norton advertises itself as the "service center for northwest Kansas" and is proud to host both the Northwest Kansas Library and the divisional office of the Kansas Department of Transportation. Phillipsburg people counter by promoting themselves as the "market center of north-central Kansas," complete with an Alco department store. The two communities also balance almost evenly with other industries. To Norton's New Age Industrial Corporation (aluminum extrusion products, 120 employees), Phillipsburg can offer the headquarters of the Kyle Railroad, a company with 108 workers that Huck Boyd and others convinced to take over this section of the abandoned Rock Island tracks in 1984.[46]

Coping Strategies in the Southeast

Anyone who attempts to generalize about the fate of American communities beyond the interstate highways will find frustration in Kansas. Experiences vary widely from the far-northern section of the state to the southwest, for example, and from the southwest to the southeast. In the small cities of the north, population figures have held relatively constant. Most of these places never had big industrial or transportation employments to lose, and hometown entrepreneurs helped to counter declining retail sales by the creation of small manufacturing plants with products aimed at regional markets. Southwest Kansans, in contrast, confounded every outsider's expectations for these years by doubling their population through a successful promotion of their region's water and gas supplies to the nation's leading meat producers. Recent history in the state's southeastern counties is different still. Although the overall trends there are more in line with textbook expectations of decline, local planners crafted several strategies that have yielded considerable success.

Before focusing on events at the core of this region southward from Fort Scott and Iola, several recent urban developments just to their north deserve comment. First, in Miami County near Kansas City, both Osawatomie and Paola have grown in recent decades, and once-tiny Louisburg reached the urban threshold in 2000. These increases, as well as those at Spring Hill just to their north, are almost all products of expanding job opportunities in Lenexa and Overland Park and the improvement of U.S. Highway 69 in this area to interstate standards.

Louisburg is almost purely a residential community, whereas Paola and Osawatomie display aspects of both suburban and older, independent economies. Osawatomie people continue to rely heavily on the city's venerable state mental hospital (572 employees), which has now been modified to have one building serve as a correctional facility (31 employees). Commuting also is surprisingly common despite a location more than forty miles from Johnson County jobs. The numbers involved approximate the positions lost when the local repair shops for the Missouri Pacific Railroad closed. As a result, the population has remained at just more than four thousand. Paola, unlike Osawatomie, has managed to retain all its earlier major employers: the county government, Lakemary Center (a school for mentally retarded children begun in 1969 through a gift from the town's long-standing Ursuline convent), and Taylor Forge. Taylor, a new name for the Fluor Corporation, has been in the city since 1930, making pressure vessels and fittings for the petrochemical industry. Its 150 employees and several hundred new commuters have pushed the city population to 5,011.[47]

Like Louisburg, two communities south of Topeka recently have achieved urban status. In the case of Osage City, this process actually was redemptive, for nearly 3,500 people once had lived there during its coal-mining peak of the 1880s. The town's resurgence back from a low of 1,919 residents began in 1962 with the recruitment of a manufacturer of paper napkins, the Osage Products Company. This business prospered enough to be purchased by Kansas City–based Hallmark Cards in 1973, but consolidation closed the plant twenty years later. Many of its two hundred workers then joined a growing group of commuters to Topeka, thirty-two miles away. Others transferred to the remaining industrial employer in town, Kan-Build. This venture, a pioneer in the creation of customized modular homes, came to Osage City because of the work ethic its outside owners perceived in small-town Kansas. After control passed to local people in 1989, the company prospered. It now employs 230 carpenters and other skilled workers.[48]

Burlington, a county-seat town thirty miles south of Osage City, is too far from Topeka for easy commuting. Its recent appearance on the urban map (2,790 in 2000) is a direct result of a nuclear power station built nearby by two state utilities. The Wolf Creek reactor, which currently employs nine hundred people, began operation in 1985. Its location takes advantage of abundant cooling water from the adjacent John Redmond Reservoir on the Neosho River. The site also offers a good balance between seclusion (for safety) and proximity to urban markets.[49]

Finally, two other small but specialized urban places exist three counties farther west. Herington and Hillsboro both lie in the triangle framed by Interstate 70 and 135 and the Kansas Turnpike. Neither is a county seat, but the former had 2,563 residents in 2000 and the latter 2,854. Herington, of course, was historically a major junction point on the Rock Island railroad. With the decline of the nation's rail system in general and the bankruptcy of the Rock Island in particular, many observers expected this city's fate to differ little from that of Horton, another Kansas hub for that company. Herington's population has indeed declined every decade since 1940 but now has stabilized. Executives at Union Pacific, the new owner of the city's tracks and yards, have designated this as the switching station for intermodal traffic moving between Chicago and St. Louis on the east and various California destinations on the west. With 150 Union Pacific employees in town, Herington still lives up to the "railroader" name displayed on its school uniforms.[50]

Hillsboro, a county south of Herington, is close enough to the interstate cities of McPherson and Newton (twenty-five miles each) to allow easy commuting. In addition, the town long has been the cultural center for the Mennonite Brethren church. This tight-knit Anabaptist group with 3,740 members in the state has sponsored Tabor College in Hillsboro since 1908. A total of 140 people work at the school, another 85 at two small farm-trailer factories, and fifty at the Barkman Honey Company (a packager of the Busy Bee brand).[51]

Burlington, Herington, Hillsboro, and Osage City, much like the communities of far-northern Kansas, represent small peaks of population in sparsely settled areas. This pattern changes abruptly in the state's nine southeasternmost counties, where a combination of mining activities and older industrial development supported 220,361 people in 1950 (an average of 24,484 per county). Although these numbers had been even higher in earlier decades, they nevertheless represented a valuable labor pool that might be touted to a new generation of manufacturers. A continuing modest production of oil and gas in the area gave planners hope as well. Fuel, they thought, could be as important a lure for industry as interstate transportation. Finally, optimists reasoned that the earlier, 1900 to 1920 period of boom and bust had honed local skills in the art of economic reorientation.

The context for understanding recent challenges faced by southeastern Kansans begins with the period between 1920 and 1950. The halcyon days of world status in zinc smelting and glassmaking were gone by this time, but the region's railroads, mines, brick and tile makers, and cement plants continued to provide employment to thousands of people. Then, and seemingly all at once, these enterprises declined precipitously. Chaos was greatest in the specialized zinc towns of Baxter Springs and Galena, the coal center of Pittsburg, and the railroad division and repair hubs of Fort Scott and Parsons, but all area communities were affected. Each decade brought further declines in the output of oil and gas wells along the belt from Iola to Coffeyville, for example, as well as drops in efficiency at aging area refineries and smoky factories that had been established at the turn of the century.

A charting of population trends for the larger regional cities provides a good platform for analysis (Chart 4). The most obvious point is somewhat paradoxical. Although each community has endured a net loss over the past fifty years, these changes also have produced the first truly dominant city in the region. At Pittsburg, people somehow have managed to maintain their economy while their traditional rivals in Coffeyville and Parsons suffered some of the biggest population declines in the state (36 and 22 percent, respectively). The graph also suggests that the economic crisis was more intense for particular places at different times. The 1950s, for example, were especially difficult in Fort Scott, the 1980s in Coffeyville.

Looked at from afar, it is tempting to judge the economic strategies of southeastern Kansans as failures. Too much focus on downward-trending charts ignores the arguably even more important process of job recruitment, however. This ongoing, second wave of industrialization is subtle but impressive. More broadly based than its predecessor, it has enabled the seven communities listed on Chart 4 to retain between 64 and 99 percent of their populations. It also has unfolded in stages, with many of the pioneering efforts made in Iola, the first area community to suffer losses after the original period of expansion. Case studies based on this historical sequence of adjustment provide a convenient framework for discussion.

The history of Iola exemplifies that of southeastern Kansas. As the site of the state's first major discoveries of natural gas in 1893, the community attracted industries by the dozens and at one time claimed a population of fourteen thousand. Then, in 1907, pressure in the field fell sharply. Although surrounding Allen County still produces oil and gas, the sense of limitless supplies vanished with the pressure drop, and factories and people began to move elsewhere. Iola's decline, a modest 11 percent between 1950 and 2000 (Chart 4), becomes a much bleaker 55 percent if the period is extended back another forty-three years.[52]

As I discussed in a previous chapter, Iola managed to retain into the 1950s only two of its gas field industries: Lehigh Portland Cement and United Tile and Brick. These both struggled, however, and disappeared before 1980. The reaction of city officials to their first round of job losses had been a somewhat panicked solicitation of any and all new employers in an attempt to slow population exodus. Out of this came several plants that sought cheap labor to make work clothes and a condensery for Pet Milk. With time, it became obvious that these factories were less than ideal for the city's long-term health. Clothing companies would move again if they could find cheaper help, and the condensed milk business was in decline nationally.

Iola promoters in the early 1970s decided to adopt a new strategy, one that soon would become a model for the entire region. Their strongest asset, they reasoned, was a sizable, semiskilled, and highly reliable labor force set in a community atmosphere of low-key, family life. A current brochure from the economic development office at a neighboring community summarizes the case well: "Fort Scott's labor force is marked by strong work ethics, low absenteeism and high productivity born of strong rural roots. Industries benefit from doing business in a right-to-work state and profit from fair wage rates while employees enjoy a comfortable living environment where the average cost of an existing single-family home is 48 per cent lower than the national average."[53]

By being aggressive while most other cities were not, Iola was able to attract new factories from two major companies: Midland Brake in 1973 and Gates Rubber in 1975. Owners of both corporations favor small towns for their manufacturing plants, reasoning that lower operating costs and "a choice labor supply" overcome slightly higher shipping costs. With 500 and 425 employees, respectively, Gates and Midland (now Haldex) are by far the biggest companies in Iola and draw in commuters from Chanute and elsewhere. Their success also influenced a 1997 decision by another such manufacturer, Kansas City–based Russell Stover Candies, to establish its second Kansas factory in town. Stover now employs 250 local people.[54]

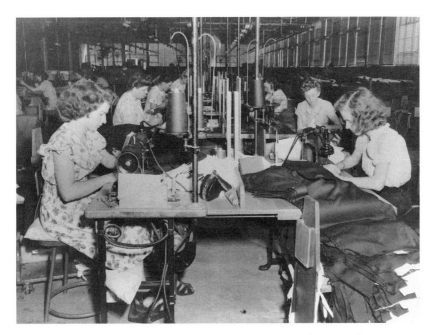

Reliance Manufacturing Company in Cherryvale, circa 1941. Clothing makers came to many communities in southeastern Kansas at the end of the gas-based economic boom. They provided abundant jobs (here, sewing pockets on trousers) but paid low wages and often moved on to sites that promised even cheaper help. This particular plant had three incarnations—Miller Manufacturing, Reliance, and Cherryvale Manufacturing—before closing completely about 1970. (Courtesy Kansas State Historical Society)

Following Iola, Neodesha became the second community in the region to rethink its economy. This Wilson County town has never been larger than 3,943 people, but a refinery established there in 1897 by a young Standard Oil Company gave it a definite industrial tone. When city officials learned in 1946 that Standard planned to close this facility, they had little choice but to think ambitiously if they were to survive. The result was a private development corporation capitalized at $100,000 and funded by $25 shares sold in the community. Its leaders bought and improved real estate for subsidized sale or lease to prospective industrialists.[55]

The companies targeted in Neodesha were mostly Kansas based and much smaller than those later pursued at Iola. This record is instructive. Some of the initial ventures, such as a sawmill and a dairy, did not last long. Another but grander failure was an attempt to rescue the financially troubled Independent Plow Company. This plant briefly employed 250 people before going bankrupt a second time. Persistence ultimately paid off, however. Airosol, a fledgling maker of insecticides in aerosol containers, came in 1949 and today employs 100 people. The 1960s and early 1970s produced three more ongoing successes: M-E-C (steel storage tanks and dryers for agriculture), Prestige (oak cabinetry) and Cobalt Boats. Wages paid to the thousand workers at these factories are high enough that no one misses the smelly old refinery.[56]

Fort Scott provides a third early example of economic reorientation in the region. Here the spur for change was closure of repair yards by both the Frisco and Missouri Pacific railroads plus, somewhat earlier, the city's single entries into the brick and cement businesses. Redevelopment plans have never been as coordinated as at Iola or Neodesha, however, perhaps because the Fort Scott losses were spread over a period of two decades. The result has been a potpourri of new companies, some well suited to the area for the long haul and others likely to relocate.

The most tenuous of Fort Scott's larger enterprises include Key Industries (a maker of work clothes) and the insurance companies Great-West and Med Plans 2000. Officials at Key feel pressure to find less expensive pieceworkers overseas, while the insurance companies find it hard to attract young executives to a nonmetropolitan area. Insurance has a strong history in Fort Scott, of course, with the rise of locally owned Western Casualty to national status. An even larger company called Lincoln Life acquired Western in 1985, however, and closed the Fort Scott office five years later. The availability of local skills in this field and open office space helped to attract the two present firms, but problems remain.[57]

Local companies better suited to southeastern Kansas include two medium-sized specialists in aluminum extrusion (a total of 385 workers) and two others in the printing of business forms. The latter industry is led by Ward/Kraft, which has grown from 25 to 450 employees since its beginnings in 1972 and added a branch in Fredericktown, Ohio. Fort Scott also has been home to a 170-person plant of the Dayco Corporation since the mid-1980s. This maker of molded belts and hoses for automobiles expanded to Fort Scott from nearby Springfield, Missouri, partly to escape that state's more restrictive laws for unionization. Another, and nearly concurrent, connection to Missouri is the warehouse headquarters for Valu Merchandisers, a subsidiary of Kansas City–based Associated Wholesale Grocers that was created to handle health and beauty products. Because the parent company's network of retail stores fans out from distribution centers in Kansas City and Springfield, Fort Scott was a central location for this 219,000-square-foot facility and its 218 jobs.[58]

If Fort Scott has been less efficient in industrial recruitment than earlier-starting Iola, it atones with business related to tourism. In 1978, federal authorities completed restoration of the community's namesake fort and opened it as a national historic site. Local people were quick to establish an adjacent visitor's center, design a walking tour of historic homes and buildings, and subsidize a trolley to run a circuit from the fort to another military legacy, Fort Scott National Cemetery. Visitor numbers began modestly, but now total more than 75,000 annually at the fort. To date the biggest benefit to the town has been several score new residents who, after having seen the city's large stock of Victorian homes on a tour, decided to relocate.[59]

Officials at cities south of Fort Scott and Iola enjoyed the luxury of planning their futures in a relatively unharried environment. Many traditional employers in this section of the state endured into the 1980s, creating a situation in which decline was widespread but not yet critical. Three interesting cooperative efforts emerged from these circumstances. The first, in 1957, was called Mid-America, Incorporated. Although chronically underfunded, this was the nation's first multiple-county organization for economic development, and it played an important role in lobbying the state government about regional issues for its forty-three-year existence.[60]

Mid-America supported the southeast's second and third major cooperative endeavors. In the 1960s it helped to pass legislation for a statewide system of junior colleges. Seeing both general and vocational education as important for economic development, communities in the southeast endorsed this concept overwhelmingly. Of the nineteen campuses authorized for state aid, six were in this region. With the exception of Pittsburg, which already had a state college, this meant a new campus for every other significant city in the region. Bids even were accepted from both Coffeyville and Independence, making Montgomery County unique in the state in having two such institutions. Today all six programs continue to prosper, with full-time enrollments ranging from 668 to 1,151.[61]

Better highway transportation was the region's third and most important cooperative endeavor. Pushes for new construction grew intense in the 1960s as people realized the obvious benefits of superhighways, but intercity rivalries prevented a united front. A plan to upgrade U.S. 69 through Fort Scott and Pittsburg, for example, faced competition from a parallel promotion for U.S. 169 through Chanute and Coffeyville. The location for a new east-west connector between Wichita and Interstate 44 near Galena was even more controversial. Should it take a southerly route through Independence and Coffeyville, a northerly one through Chanute and Pittsburg, or a compromise path via Neodesha, Parsons, and Columbus?

The state legislature authorized a toll road east from Wichita in 1974 but never provided funding. Hopes rose again in 1987 and were finally realized two years later with agreement for a two-lane freeway along the middle route via Parsons. Proponents predicted 19,300 new jobs in the area, 34,600 new residents, and the permanent banishment of a new and unwelcome image for the region as "the Appalachia of Kansas." Slow-paced construction dampened such forecasts, but as I write in 2003, the new "super two" highway (U.S. 400) is complete except for a bypass around Parsons and a diagonal across Cherokee County. Although it is still too soon to measure actual business gains against earlier expectations, people remain justifiably optimistic.[62]

The less-than-urgent view toward industrial change prevalent until recently in Chanute, Coffeyville, Fredonia, Independence, and Parsons is best understood by reference to other cities in the region. A contrast can be seen, for example, between the hustlings in Neodesha during the 1960s and 1970s and feelings of arrogance in Independence because it housed the national office of ARCO Pipeline. At Fredonia, the smallest city of the group, such thinking holds true even to the present. The 2,600 people there seem blessed. Jobs associated with government (the seat of Wilson County) and cement (the long-standing Lafarge Corporation) have now been augmented by easy access to Neodesha's factories eleven miles away on new U.S. 400.[63]

The economic anchor at Chanute since 1908 has been Ash Grove Cement, which still employed 144 people in 2002. Its presence cushioned the losses of the Santa Fe's division offices in 1948 (500 jobs) and then, four decades later, of the Mid-America Refining Company (83 jobs) and several other oilfield-related operations. The equivalent steadying force for Parsons, twenty-three miles to the south, has been its state hospital (600 employees in 2002) and, more ephemerally, the federal government's Kansas Ordnance Works (1,453 in 1987, but 717 in 1993 and only 205 in 2002). Employment at this ammunition plant, a contractor-operated legacy from World War II, has always been irregular. Its useful life now appears over, however, since government officials have announced plans to sell the 13,727-acre site. Parsons's other big loss since 1950, of course, was the shops and business offices of its founding company, the Missouri, Kansas and Texas Railroad. Following consolidation in 1957 and then sale to the Union Pacific, local facilities were systematically closed. A thousand area residents lost their jobs.[64]

People in Coffeyville and Independence were able to retain more of their traditional industries into the 1990s than anyone else in the region. Several factors underlie this stability, the most important being the relatively large size of local oil and natural-gas deposits. This output contributed to the longevity of all three of Independence's traditional industrial employers: Lehigh (now Heartland) Cement (125 workers in 2002); Union Gas Company (150 in the 1980s before its closing); and ARCO Pipeline (200 in 1995). At Coffeyville the wells remain critical for the refinery run by Farmland Industries (333 employees in 2002) and important for big power users such as the Acme Foundry (284 in 2002) and Sherwin-Williams Paints (250 in 2002).[65]

Among the strategies developed to replace lost industrial jobs, the most informal have been at Chanute and Parsons. Such stances likely are the result of job losses in these cities since 1950 being small and drawn out over a period of years. Initial recruiting at both places produced an eclectic mix of new man-

Farmland Industries Refinery in Coffeyville, circa 1986. The survival of this plant on the city's east side is evidence that oil reserves in surrounding Montgomery County are still among the largest in southeastern Kansas. National Refining, the original company at this site, began operations in 1905. Ownership passed to the Co-operative Refinery Association in the 1940s, and that company renamed itself Farmland Industries in 1966. (Courtesy Kansas State Historical Society)

ufacturers similar to that at Fort Scott. In Chanute one found specialists in ductwork (H. K. Porter Company, 130 employees in 1982); industrial boilers (Chanute Manufacturing, 170 employees); and radar units (Kustom Signals, 371 employees). At Parsons it was gym clothes for schools (Broderick Company, 161 employees in 1982); storage tanks (Peabody Tec, 196 employees); and concrete products (Superior, 119 employees). Since about 1970, however, the paths of the two cities have diverged.[66]

The older Chanute companies have gradually yielded center stage to an integrated and locally conceived set of factories based on the manufacture of recreational trailers. The initial and most central of these is Nu-Wa Industries, a pioneer in this business nationally beginning in 1962. Its success inspired an imitator called Custom Campers (which recently has become a subsidiary of Nu-Wa), and the two now employ more than 400 area people. Nu-Wa's growth also has prompted a pair of existing businesses to refocus their product lines. The 112 employees at Young's Welding have moved from oil field work to the creation of trailer frames, while the 324 people at Hi-Lo Tables (now Hi-Lo

Industries) now emphasize trailer cabinetry. Through such expansion and conversion, Chanute has become a focused industrial community. Although specialization is always risky, Nu-Wa's concentration on quality rather than volume is a formula followed by many successful Kansas manufacturers.[67]

Parsons people are envious of Chanute's manufacturing focus but have been unable to foster a similar concentration for themselves. Their three older companies remain (albeit with new names), but factories added during the 1980s each specialized in a different realm: machinery fabrication, printing, and louvered shutters. Employment at the fourteen local industrial companies in 2002 totaled 1,903, an impressive number in some ways but worrisome to planners, since none of the companies have strong enough ties to local raw materials or markets to guarantee continuity. This concern has been balanced by hopes that the city's prime position on the region's new east-west express highway would attract additional companies and help to retain others. So far the results have been good. No major employers have left in the last decade, and truck traffic through town has tripled in volume. The city also is headquarters for a major new transportation firm, Wichita SEK Transit. Its 180 jobs are the best sign yet that, unlike the situation in the 1970s, the world is no longer "passing by" this once vibrant place.[68]

As noted earlier, the petroleum-blessed economies of Coffeyville and Independence traversed the 1950s, 1960s, and 1970s relatively unscathed. Coffeyville did lose a refinery and a set of railroad shops, and Independence a brickyard, but an impressive series of major industrial employers remained. The 1980s brought serious problems, however. International conditions lowered oil prices drastically and forced a series of consolidations and closures. Coffeyville initially suffered more than its rival community, particularly in its oil field service businesses, but a 1995 decision by officials at ARCO Pipeline to move their headquarters from Independence to Houston, Texas, more than evened the score. These losses prompted the most intensive process of reevaluation and recruitment ever seen in southeastern Kansas. Taking a countywide approach, people boosted funding for their vocational-technical school in Coffeyville and created the Montgomery County Action Council in 1985. Even more impressive, they approved a one-cent sales tax intended solely for industrial aid.[69]

Offering low rates for electricity from a municipally owned plant and generous tax incentives, Coffeyville has attracted four new companies. One of these, Aptus, was received with initial skepticism in 1985 because its business involves the processing of toxic materials. The others—Western Publishing (puzzles and games), American Insulated Wire, and Amazon.com (a book warehouse)—earn general praise. Amazon has employed about five hundred people since its opening in 1999 (about as many as the three other firms combined) and seems particularly well suited to the area. Echoing the sentiments made earlier at Gates Rubber in Iola, company officials have stated that they admire the work ethic in small Kansas cities and needed a central location to complement their existing shipping centers in Delaware and Nevada.[70]

Independence, the last city in the southeast to suffer major employment losses, saw its bubble burst in the 1990s. Besides ARCO, the Emerson Electric Company of St. Louis closed its local motor factory after twenty-six years. Elimination of 500 jobs at these two companies in just four years left only the DANA Corporation (a maker of replacement automobile parts with 1,010 workers) and Heartland Cement (125 workers) as large industrial employers. Adding to the tension was knowledge that DANA is an international conglomerate with little or no local loyalty. Another relocation or consolidation therefore could occur at a moment's notice.[71]

Independence was a community united by its economic desperation when officials at Wichita's Cessna Corporation announced plans to build a second major plant for their single-engine airplanes. Company leaders wanted to locate in a medium-sized Kansas community where they would have less competition for workers than in their headquarters city. Over several other competing sites—Emporia, Hays, Manhattan, and Pittsburg—Independence won with an incentive package worth $36 million. Besides the money, Cessna people said they liked the worker training available at the county's technical and community colleges, the abundance of quality housing in the city, and a certain civic pride evident in negotiations. The new plant opened in 1997 and, two years later, employed 875 people. With them, as expected, came big jumps in retail activity, including a new men's store, two department stores, and a Wal-Mart. Not as many of the Cessna workers live in town as projected (speculators drove up housing prices), but people still judge the project an overwhelming success. The Wichita tie makes Cessna a good bet for employment stability, a commodity that is increasingly scarce in the contemporary world.[72]

The final set of cities in southeastern Kansas includes the state's traditional coal center of Pittsburg, the lead and zinc towns of Baxter Springs and Galena, and several satellite communities. Since mining in this region essentially ceased in the late 1950s, one might expect recent censuses to reveal mass emigration. Instead, decline has been much less than the regional average. Galena, the hardest-hit city, lost 18 percent of its population over these years (to 3,287 in 2000), but both Baxter Springs and Pittsburg have held nearly steady with 4,602 and 19,243 residents, respectively. In fact, the greater Pittsburg area actually has gained at least 2,000 people when one considers increases in the neighboring communities of Frontenac and Girard.

The continued economic viability of Baxter Springs and Galena is difficult for many people to imagine. Not only are the mines closed, but federal officials also have been warning residents since at least 1983 about an ongoing danger of lead poisoning. Gigantic piles of mining debris cover forty square miles in this area, and from them lead leaches into water supplies. Tests are conclusive enough for the Environmental Protection Agency to have designated this a Superfund site, but many local people remain unconcerned. They prefer to concentrate on the advantages of staying put.[73]

Besides inexpensive housing, both Galena and Baxter Springs offer commutes of less than fourteen miles to jobs in Joplin, Missouri. They also retain a larger industrial base than most outsiders expect, much of it taking advantage of the best highway access in this quadrant of the state. These two communities were the only Kansas stops on the famous Route 66. Now a modern equivalent of that highway called Interstate 44 passes just to their south and contains an exit no more than seven miles from either town. Yellow Freight Systems, one of the nation's leading truckers, established a terminal in Baxter Springs in 1932 and then expanded it to employ more than five hundred people. Although this center closed recently, a smaller carrier called Bingham Transportation (200 employees) has taken over some of its business. Both towns also have clothing plants—Vogel in Galena (75 employees) and King Louie in Baxter Springs (219 employees). This is one of three shirt factories for King Louie, a Kansas City–based company that keeps its entire operation close to home for efficiency. Finally, each of these towns retains some mining ties. A German conglomerate called Liebherr makes specialized mining trucks at Baxter Springs (163 employees). Eagle-Picher, the last operator of the Galena smelter, runs the old Spencer fertilizer plant north of town (72 employees).[74]

The continuing viability of Columbus and Girard, the seats of Cherokee and Crawford Counties, respectively, could be regarded as another anomaly in the southeastern region. These towns always have been in the shadow of mining communities, yet they maintain populations near three thousand. Much of the explanation lies in the creativity and generosity of a man named Omer F. Witt. Witt, a printer, was working in Parsons during the 1940s when he decided to manufacture printing equipment on his own. He established plants in Girard and Fort Scott and, as he grew successful, offered machines to area people at half the regular price. Through his efforts, southeastern Kansas became a national center for commercial printing. Fort Scott has two such operations today, Parsons and Pittsburg one apiece, but the biggest local impacts have been in Columbus and Girard. Calibrated Forms is the largest employer in Columbus (325 workers), while five firms (277 employees) in the slightly smaller Girard make this community the self-proclaimed "printing capital of the nation."[75]

Any survey of southeastern Kansas appropriately can conclude with the emerging regional hub at Pittsburg. This city's path to recent success—holding steady while its rivals decline—often has been attributed to economic diversity. Certainly planners there have sought a broad range of businesses, and a chamber of commerce director in 1982 went so far as to call the mix at that

Machine Shop at the McNally Pittsburg Manufacturing Company in Pittsburg, circa 1965. Expanding from a small boiler plant in the 1890s, McNally Pittsburg gradually became one of the world's leading makers of coal-preparation equipment and the city's largest employer. This fifty-two-thousand-square-foot shop was a source of pride when it opened in 1955. It featured a sixty-foot ceiling in the main bay, two overhead cranes capable of lifting twenty tons, and an interior railroad siding. More than a hundred men worked here on a variety of boring and jig mills, radial drills, planers, and lathes. (Courtesy Kansas State Historical Society)

time nearly "recession-proof." When the local employment structure is examined in detail, however, change is the dominant theme. The famed stability applies only to population totals.[76]

In 1950, Pittsburg's mines and the shops of the Kansas City Southern Railroad each still employed more than eight hundred people. In addition, the Spencer fertilizer plant south of town accounted for another thousand jobs, and two foundries and two clay-products companies for eight hundred more. Although Pittsburg State College was present as well and added cultural amenities and technical training, the community epitomized blue-collar life. Thirty-five years later, this list of employers had been pruned substantially. Clemons Coal was the sole mining company to survive, and it employed only eighty people. The railroad had consolidated its shops to Louisiana, the Spencer plant had changed hands twice and was down to a hundred workers, and both the United Iron Works and Pittsburg Pottery had quit business.

Pittsburg State and Mount Carmel Hospital continued to grow, but public attention focused on even more rapid expansion at the remaining foundry, now called McNally Pittsburg, Incorporated. Mac-Pitt decided to specialize in the manufacture of coal-washing equipment and in so doing became "one of the state's most far-reaching and desirable businesses." It employed 750 people at its headquarters and 2,500 around the world.[77]

In a move that proved to be disastrous from a local perspective, a Swedish company called Svedela purchased Mac-Pitt at its pinnacle in 1988. Svedela, in turn, sold it to Metso Minerals of Finland, whose officers decided that maintenance of an operation in faraway Kansas was not in their best financial interest. They therefore closed the corporate office in Pittsburg in 1994 and the plant in 2002. As the course of these events became clear, residents were forced to reassess the city's circumstances and prospects. The findings, to the surprise of many, proved surprisingly upbeat. While closing the door on heavy industry, the community simultaneously and seemingly without effort was opening another toward an equally rewarding concentration of service industries. Most obviously, the 1,175 employees and 6,800 students at its university now formed a large enough mass to create a genuine college-town atmosphere and economy. In addition, the 640 people at the Mount Carmel Regional Medical Center were on the verge of similar accomplishments for health services.[78]

Pittsburg State University is without doubt the key to the city's current economy. Its direct employment and social contributions are substantial, of course, and campus leaders also opened the $28 million Business and Technology Institute in 1997. This center provides new impetus to the college's long-standing support for industrial development and has helped to attract and retain many small manufacturers. The biggest of these, Superior Industries, relocated its automobile wheel factory from Arkansas in 1989 and generated 790 jobs. A few years earlier Ohio's Sugar Creek Packing Company had hired 240 people in Frontenac at one of its four bacon-processing plants.[79]

The most significant development for Pittsburg's future is an evolving symbiosis with Joplin, thirty miles to the southeast. Joplin, with more than forty-five thousand people, has long been the shopping center for much of southeastern Kansas. Although this success obviously has hurt Pittsburg's chances for growth in retail sales, Joplin's lack of a four-year public college until recently has allowed Pittsburg State to extend its educational hegemony into an urban area that contains more than a hundred thousand people. The key to this expansion was legislative approval for in-state tuition rates to apply to a select list of counties in Missouri and Oklahoma. In this and other ways, Pittsburg people increasingly and profitably are reconceptualizing their location. From seeing themselves on the periphery of Kansas, they now envision their city as part of the core for a growing tristate urbanized area.[80]

10
Conclusion

The details of Pittsburg's transformation from mining camp to college town, the recent economic difficulties at Great Bend, and the mercurial rise of Overland Park are all intriguing stories in their own way. To focus exclusively on these particulars, however, is to lose perspective on the urban system as a whole. In concluding this look at the state's cities, I want to return to the larger concern and pose several rhetorical questions for consideration. What forces, for example, have been most important in determining urban success and failure in Kansas, and how have these agents changed over time? Has the ongoing process of development been impossibly complex, or can we replicate its essentials through static or dynamic models? Finally, if such modeling can be achieved, how does the Kansas experience compare to existing theory?

Throughout the territorial and early statehood period of the 1850s and 1860s, I judge the match between theory and reality to be very close. Indeed, it seems as if Professors Vance, Muller, Meyer, and Wyckoff might have had Kansas in mind as they developed the mercantile model of long-distance trade described in chapter 1. The frontier cities they hypothesize compare well to the actual experience at Atchison, Leavenworth, and other communities in the elbow region of the Missouri River. Entrepreneurs there strove mightily to connect the mining camps, military outposts, and remote valley settlements of the west with the established cities of Chicago and St. Louis to the east.

The transition from waterways and overland trails to railroads did not alter the basic mercantile objectives of early Kansans. Control of the process did change, however, as the expense of steel tracks required major financial inputs from Boston, New York, and other eastern locations. Although railroads eventually connected nearly every community in Kansas, the main lines quickly came to focus on sites favored by eastern capitalists. Again this meshes with theoretical expectations. By the mid-1870s Kansas was firmly connected to the national urban system, with Kansas City as its undisputed entrepôt.

If the rise of Kansas City seems inevitable in hindsight and on a theoretician's diagram, closer inspection of the historical record provides much grist for alternate scenarios. Leavenworth people, for example, were the first Kansans to grasp the tremendous potential for growth inherent in a transcontinental railroad. Had they achieved their dream in the 1850s before eastern interests were fully organized against them (as they nearly did), Kansas City would now be that city's suburb instead of the other way around. Atchison, home to one of the state's first United States senators, also maintained a realistic hope for metropolitan status into the early 1860s by securing a branch of that first rail line to the Pacific (Map 7; see p. 19). Another power play by outside capitalists crushed this ambition in 1865. By changing the route of the Union Pacific's main line to run west from Junction City to Denver instead of northwest to Nebraska, Atchison's track was left as little more than a cul-de-sac.

The mercantile model also applies to several Kansas settings beyond the Missouri elbow. As I described in chapter 4, many entrepreneurs in the 1860s believed that railroad technology could allow a city west or south of the older riverine cluster to usurp the entrepôt position for the developing frontier. Before the assertion of Kansas City's hegemony, rational plans existed for both Lawrence and Topeka to become the principal western junction for railroads out of Chicago and St. Louis. Farther south, people in Fort Scott believed in a similar possibility for themselves. So, too, did the railroad creators of Parsons.

Finally, mercantilism provides an initial guide to how the state's urban network adjusted after the ascendancy of Kansas City. In this variation on the model, Kansas City assumes the former role of St. Louis; other, mostly newer cities vie to become its subsidiaries. The railroad network of 1869 reveals a preliminary version of this system (Map 8; see p. 20). As people constructed new tracks branching away from the main line of the Union Pacific, they simultaneously created a series of junction and terminal points that (at least temporarily) served the needs of wholesale merchants well. These places included Emporia, Fort Scott, Junction City, Lawrence, Ottawa, and Topeka. Waterville enjoyed a similar advantage as the terminus for the Union Pacific branch out of Atchison.

A second but still temporary expansion of mercantilism existed between 1870 and the early 1890s in extreme southern Kansas. Great potential existed there for gateway cities so long as residents of adjacent Indian Territory restricted the promotion of railroads and cities across their rich and extensive lands. Entrepreneurs developed Baxter Springs, Chetopa, and Coffeyville with such aspirations firmly in mind. Farther west, in the ranching country near Arkansas City, Caldwell, Kiowa, Englewood, and Liberal, the motive was similar except that warehouses progressively gave way to stockyards. The trading monopolies held by these border cities ended soon after the first Oklahoma land rush in 1889. Although traffic to newly established Guthrie and Oklahoma City continued to funnel through Kansas City, no need remained for intermediate stops at Chetopa or Kiowa. People on the border either had to leave or seek other livelihoods.

The form assumed by the region's mercantile system between about 1890 and 1950 can be assembled from charts of the state's mature railroad network (Maps 14, 15; see pp. 27 and 28) plus a listing of cities with wholesaling concentrations

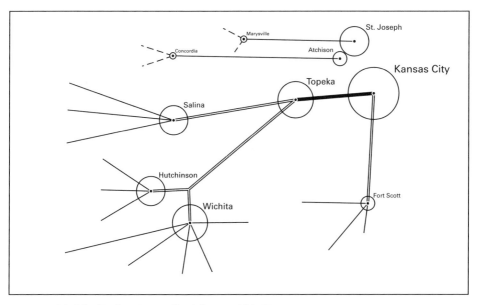

5. A Model of Major Mercantile Linkages within Kansas, 1890–1950.

(Table 1). Most goods flowed westward from Kansas City to Topeka during this period (Chart 5). There, in a split following the tracks of the Union Pacific and Santa Fe railroads, one branch extended on west to Salina and another southwest to Hutchinson and Wichita. From these three cities, jobbers exploited additional trellises of rail lines to reach at least two-thirds of the state. Most of the remaining communities were served either directly by Kansas City merchants or through older trading networks operating in reduced capacities out of St. Joseph via Marysville, Atchison via Concordia, and Sedalia (Missouri) via Fort Scott.

Because the mercantile system in Kansas after 1865 relied almost exclusively on railroads for transportation, and because railroad owners realized much of their revenue from the shipment of trade goods, it seems logical that the two groups would promote and develop a common set of cities. Such reasoning works for Kansas City but, contrarily, not for the rest of the state. Indeed, close inspection reveals that the needs of individual railroad corporations actually were quite different from those of wholesalers.

The ideal position for a distributor at this time was amid a web of rail lines. Any given railroad entrepreneur, in contrast, thought more in linear terms: the extension of a single main track as far as possible and then the addition of several spurs into untapped territory. Although it is true that a railroad company occasionally would split its main line and thereby create a junction of its own, such places were rare in nineteenth-century Kansas. Newton is the principal example from the Santa Fe system (with Emporia added later), while Herington and Horton on the Rock Island line and Parsons on the Missouri, Kansas and Texas essentially complete the list.

It is significant that Newton, Parsons, and similar "company towns" never became wholesale centers (Tables 1, 2; Chart 5). Warehouse merchants require good transportation, of course, but also desire multiple carriers within this system as a guard against monopolistic prices. In this, a jobber's objective during the railroad years was similar to that of city leaders. A key component of Wichita's early success, for example, was the recruitment of the St. Louis and San Francisco Railroad to open trade areas beyond those provided by the Santa Fe. For Salina, the equivalent coup was the Missouri Pacific to use as a supplement to the Union Pacific; for Hutchinson it was the Rock Island to augment the Santa Fe. Topeka, of course, hosted the Union Pacific, the Santa Fe, and the Rock Island, while Kansas City eventually cornered all these and several more (Map 11; see p. 24).

Individual railroad companies, with power to designate communities as division headquarters, repair centers, and junction points for branch lines, obviously were capable of creating urban networks on their own. This was especially true for the owners of the Union Pacific and the Santa Fe, who, by constructing large portions of their lines ahead of the settlement frontier, had the opportunity to create initial cities across half the state. These corporations largely squandered their chance for urban promotion, however. As I detail in chapter 4, the financial allures of their land grant and the cattle trade so dominated the thinking of Union Pacific people that they positioned their division points west from Kansas City with nothing more glamorous in mind than water, perfect spacing, and cheap property. The selection of Wamego, Brookville, and Ellis instead of nearby and larger Junction City, Salina, and Hays, respectively, introduced complications to the emerging urban network across a third of the state and thereby slowed its maturation. The situation was much the same along the Santa Fe. No sooner was Newton named a division point and repair center than the company undercut the community's emerging centrality by building a spur track south and developing a cow town at Wichita. Farther west, railroad officials gave Dodge City the next set of shops and a big stockyard but allotted the division point to Coolidge. Urban promotion, although potentially lucrative, was only one of several ways a railroad could make money. Owners of these two land-grant companies clearly concentrated on alternative schemes.

The only full-scale effort by a railroad company to create a regionally dominant city in Kansas occurred in the southeast. Railroads built ahead of settlement in this portion of the state just as they did in the west, and the Missouri, Kansas and Texas (MKT) company had the power to control development once it had obtained the exclusive right to build on south through Indian Territory.

When this company's officials decided to construct an additional route as a more direct link to eastern markets, they saw potential for a new city at the junction of their original and newly established tracks. Parsons, the "infant wonder" created in 1870, served briefly as the mercantile hub for a significant portion of Kansas, Missouri, and Indian Territory.[1] Had the mineral discoveries that created urban rivals at Pittsburg and Galena been delayed only a few years, Parsons almost surely would have grown large enough to maintain its regional control. As it happened, of course, southeastern Kansas soon filled with a series of specialized communities, none of which was able to prevail over the others.

Foci for the state's mercantile system during the railroad decades theoretically should be at intersections between the main lines of the pioneering Union Pacific and Santa Fe companies and those of their slightly later competitors (Map 11). Such sites would provide efficient circulation at competitive prices. Along the Union Pacific artery from east to west this logic predicts development at Lawrence, Topeka, Junction City, and Salina. The Santa Fe equivalents are Lawrence, Topeka, Emporia, and Hutchinson plus (along a second important Santa Fe corridor south from Newton) Wichita and Wellington. When Lawrence and Wellington are eliminated as jobbing sites because of their proximity to the larger Kansas City and Wichita, respectively, and Junction City similarly devalued because the MKT track there was little used, the remaining cities all could serve the needs of wholesalers well. In fact, they formed the skeleton structure for the longtime mercantile reality of the state (Chart 5).

To predict the Kansas trading network at a smaller scale, one logically should only have to plot intersections between competing railroad lines outside the Union Pacific–Santa Fe axis. North of the Union Pacific, these places include Marysville (UP and St. Joseph and Denver City); Belleville (Rock Island and UP); Concordia (Central Branch and UP); and Colby (Rock Island and UP). Herington (Rock Island and Missouri Pacific) and McPherson (Rock Island and UP) have similar assets in central Kansas, as do Fort Scott (MKT and Frisco) and Chanute (MKT and Santa Fe) in the southeast. Comparing this list with actual development, the model works for Concordia, Fort Scott, Marysville, and (at a smaller scale) Colby (Tables 1, 2; Chart 5). Herington and McPherson are both too close to Salina to have attracted jobbers, and similarly, Belleville to Concordia. Chanute's competitive position always was weakened by lack of traffic on its branch of the MKT.[2]

Mercantile models provide fundamental explanations for the Kansas urban scene from the early years up through the last big railroad decade of the 1940s. Today, however, when businesses rely on fast, long-distance deliveries and minimize warehouse stocks to save money, much of the mercantile world has been transformed to scales larger than the state level. Cities in Kansas still participate in these operations, of course, but most area activity has become restricted to interstate highway sites in the greater Kansas City area. Abilene, with the retention of its homegrown Duckwall/Alco warehouse and Coffeyville with its piece of a new-style national network for the bookseller Amazon.com are notable exceptions.

Theorists of urban systems have been right to concentrate on trade. It explains basic patterns. Still, when a mercantile model of Kansas is compared with maps of city population for various years, discrepancies are numerous (cf. Chart 5; Maps 16–20, 21 [see pp. 29–33]). These residual places, almost by definition specialists in economic activities other than trade, obviously should be accounted for in any reasonably complete urban model. The question becomes whether it is possible to do so.

Of the various functional specializations, railroad towns have been the most common in the state. Following the locational logic for division points and shops discussed in previous chapters, one can generalize that railroad communities, beyond their blue-collar employments, also competed for retail trade with smaller mercantile cities. The clearest example, perhaps, is in the northwest, where the Rock Island town of Goodland has long vied with Colby for the trade dollars of area residents. Phillipsburg (Rock Island) versus Norton, Belleville (Rock Island) versus Concordia, Ellis (UP) versus Hays, Hoisington (MP) versus Great Bend, Osawatomie (MP) versus Paola, Herington (Rock Island) versus Abilene, Horton (Rock Island) versus Hiawatha, Arkansas City (Santa Fe) versus Winfield, and Osage City (Santa Fe) versus Burlingame all exhibit a similar pattern. At another scale, so does Newton (Santa Fe) versus Hutchinson and Wichita.

An adversarial relationship between railroad and more commercially oriented centers is not inevitable, of course. Company officials awarded shop sites and division headquarters for many reasons, sometimes giving them to already prospering places. Kansans sometimes forget, for example, that the most important railroad centers in the state have always been Kansas City and Topeka. At a smaller scale, major railroad activity has reinforced general economic dominance over sizable areas for Dodge City, Emporia, Fort Scott, Liberal, Marysville, Pratt, and several other communities (Table 1).

Complex decision-making processes and large numbers of individual cases make it difficult to incorporate railroad towns into a general urban model. This issue is most problematic for the years before 1950, when the collective impact of these communities was to complicate the mercantile network rather than to reinforce it. More recently, as railroad consolidations have drained population from several communities, the older mercantile ideal comes closer to reality again. Horton has nearly disappeared as a business center in northeastern Kansas, for example, and a similar process can be seen at work in Belleville, Ellis, and Hoisington.

Compared with the abundance of railroad communities, the number of cities with major public institutions is small enough to accommodate within an urban model. This is not to say that locating a new state's capital, university, and prison is a straightforward process. Certainly it was not in Kansas. American history makes it clear, however, that citizens usually placed these institutions in separate cities. Moreover, the communities selected typically were somewhat isolated from the mercantile-industrial mainstream as it existed at the time of statehood. The basic idea, as I detail in chapter 3, was to spread the public largesse. This was supplemented by a belief that students, criminals, and even a state's public servants all would function best in bucolic environments.[3]

Transforming general location principles for public institutions into an urban model is not difficult. The historical record suggests that residents likely would place their capital along the principal transportation artery of the time and between the largest population cluster and the physical center of the state. For Kansans this would mean a site on the Kansas River somewhere between Lawrence and Junction City. University location would follow a similar rationale. The prison, in contrast, usually went to a city with industrial potential. Convicts at the time were seen as a valuable labor source for manufacturers.

Viewed dynamically, a model for institutional cities would project differential growth for the three major types. The capital, a power center by definition, should expand most rapidly through the attraction of federal offices, banks, and service operations of many kinds. University towns, in contrast, catered to only a limited clientele throughout the nineteenth century and the first half of the twentieth century. They would therefore remain relatively small until recent decades. Development at a prison town is hardest to predict. An early start for manufacturing might initiate a positive cycle of transportation improvements and more industry. On the other hand, even better transportation would exist elsewhere at the state's mercantile center, and the negative connotations of prison life could hurt a city's public image. At Leavenworth, of course, the latter scenario triumphed over the former.

Cities created by minerals development constitute the most intractable problem for modelers of the Kansas urban system. The deposits were large, and their discoveries occurred after county seats and most railroad routes already had been laid out. Throughout half the state, new economic opportunities with coal, lead and zinc, natural gas, oil, salt, and/or irrigation agriculture disrupted fully functioning mercantile and governmental systems. Had this mining potential been smaller or known from the outset of settlement (as was the case in Colorado and Montana), the current urban network would be simpler. As events happened, however, the interruption to the settlement fabric was near a maximum.

As a given mineral resource became known, its nature and location sometimes were such that processing and service activities could be established at existing railroad and county-seat communities. This was a common scenario in central and western Kansas. Consider first the Hugoton gas field. Easy shipments via pipelines allowed its developers to place their offices and service centers at Liberal, a transportation center at the edge of the field. Seventy miles to the north, topographic conditions made irrigation feasible along the Arkansas River in Finney County. Since a major railroad already paralleled that waterway, it again was natural to concentrate business operations in the valley's largest community: Garden City. Although developments at these two communities did not disrupt urban geographies within their home counties, they very much affected trading patterns at a larger scale. Dodge City people in the 1880s had confidently believed that railroad jobs and marketing experience would ensure their town's dominance over the entirety of southwestern Kansas for decades to come. Ten years later Garden City emerged as a significant competitor; Liberal did the same in the late 1920s. In this way, simplicity in the region's urban network yielded to a complexity that persists to the present.

Oil and salt, the major minerals of central Kansas, allow flexibility for processing and worker residence similar to that at the Hugoton gas field. Salt is bulky, of course, but the Kansas deposits are so vast that mines could be positioned directly beside railroad tracks. Oil production led to sizable population increases for the county-seat towns of Great Bend, Hays, Lyons, McPherson, Phillipsburg, and Pratt. Salt did the same for Lyons and especially Hutchinson. Again, although these developments left the urban hierarchy within each county intact, they created regional trade rivalries where none had existed before. The locations of Russell and Great Bend were too close to the college community of Hays for anybody to have predicted substantial growth before the good fortune of oil in the 1920s. The same was true for McPherson in relation to the mercantile center of Salina.

Minerals-induced activity at Hutchinson is more complex. Before the owners of several salt companies decided to concentrate their mines there, Hutchinson seemed destined to a population of no more than two thousand. Nearby Newton was the area's railroad hub, and equally close Wichita had captured most of the mercantile market. Then, as a direct result of the salt bonanza, officials of the Rock Island railroad elected to build through the community. With these tracks to supplement the earlier Santa Fe ones, Hutchinson suddenly became an excellent site for wholesale trade and therefore a serious urban rival for its two neighbors to the east.

Southeastern Kansas, the remaining center for minerals production, displays the most complicated urban pattern in the state. City competition in this area began even before the mines. When competing railroad companies decided to bypass existing county seats and establish their own cities, Labette County obtained Parsons and Chetopa to compete with Oswego, Montgomery

County secured Coffeyville as a rival for Independence, and Neosho County gained Chanute to vie with Erie. Then, later in the 1870s, came the coal mines at Pittsburg and zinc shafts at Galena. Because these bulky minerals were best processed on site, another set of specialized communities emerged in Crawford and Cherokee Counties. Even this did not end the region's urban complexity, however. In the 1890s, the state's first big oil and natural-gas discoveries elevated Neodesha into a rival for Fredonia in Wilson County and Humboldt into one for Iola in Allen County. Wealth from these wells also bolstered the populations of Chanute, Coffeyville, and Independence.

Mineral development alone would have produced a tangled web of cities and functional distinctions in the southeast. The situation there was magnified many times, however, by the potential of two of these products—coal and natural gas—to attract manufacturers in need of inexpensive fuel. The result was an industrial tradition that became deeply entrenched. Outside of Kansas City and Leavenworth, in fact, nearly all of the specialized manufacturing cities in the state have been and remain in the southeastern quadrant (Tables 1, 2). The only variation in the pattern was an expansion several counties to the west between 1910 and 1920 after new oil fields led to refineries at Arkansas City, Augusta, El Dorado, and Wichita.

It is appropriate to conclude this book with thoughts about how the Kansas urban system might operate in the future. Two recent and ongoing processes, the transitions from railroad to superhighway transportation and from regional to globalizing economies, have in common a change in scale. A similar and parallel adjustment with important consequences is taking place with retail trade and service industries. A typical county-seat town today, although still able to attract customers to its grocery stores, churches, and gasoline stations, is increasingly unable to do so for even slightly more expensive or specialized items such as household appliances or accounting services. To sample such goods and services in quantity and to obtain better prices on them, most people now are willing to drive an hour or more.

Long-distance shopping behavior does not immediately affect the populations of either the smaller community or the larger one, but the impacts multiply over the course of ten or twenty years. The reality of fewer customers for the smaller town leads to closures of automobile dealerships and dental offices, for example. In turn, the absence of such stores necessitates trips to the larger city even for people who do not particularly want to go. The children of such travelers are affected, too. Growing up with regular exposure to the movie theaters, trendy clothing stores, and other allures of the larger city, they naturally are tempted to move there permanently as they mature. A cycle of growth for a few places and decline for many more is thereby set in motion.

Cities best positioned for growth from this perspective are those that combine quality offerings in higher education and health services with a good selection of retail stores. Hays is the state's exemplar in this regard. Its university and health center dominate their respective fields across twenty or more counties, essentially the entire northwestern quarter of Kansas (Map 26; see p. 40). Colby, with an active community college, could be considered its satellite. The size differential between these two cities and others in the northwest seems sure to increase in the foreseeable future.

The power of a combination of education, health, and retail services to promote growth in recent years is easy to see and has become a goal of several other cities in the state. Emporia and Pittsburg, with their respective state universities as anchor points, are making progress in this direction. Both cities face obstacles not found at Hays, however. Emporia businesspeople control only a three-county trade area and have little chance for further expansion because of bigger competitors at Topeka only fifty-eight miles to the northeast and at Wichita eighty-seven miles to the southwest. The situation for Pittsburg appears even worse at first inspection because five neighboring communities have populations of more than eight thousand, and even larger Joplin sits just across the Missouri line. None of Pittsburg's erstwhile rivals in Kansas show potential for growth, however. As for Joplin, the college town's future relationship with that larger city may be analogous to that of Lawrence with Kansas City. If so, both the specialized community and the more general commercial center may prosper.

Further variations on the Hays model can be seen at the state's premier college towns of Lawrence and Manhattan, of course. More intriguing, they also occur at several smaller cities where quality services in health or education are offered at locations accessible to the rest of the modern retail and services package. People in Marysville and Pratt, for example, have worked hard to build their local hospitals into superior facilities able to serve multiple-county areas (Table 2). These health centers, in turn, have helped to keep their retail sectors vital. The two cities increasingly function in satellite roles similar to that of Colby. For basic services, Marysville supplies the northern sector of the Topeka trade area and Pratt a portion of the Wichita realm (Map 26).

Specializations in higher education have paid increasingly handsome dividends for communities across the nation ever since soldiers returning from World War II decided that college degrees would help their futures. More recently, additional growth for these towns has come from retirees and other seekers of cultural amenities. Lawrence and Manhattan, where population increases show no sign of abatement, function in similar ways. They offer good health services and shopping selections but do so in symbiotic relationships with larger communities. Manhattan people rely on nearby Topeka for major department stores, for example, while Lawrence people often shop in metropolitan Kansas City. At a smaller scale, Lindsborg with its Bethany College has a similar relationship with Salina; Baldwin City (Baker University) with Kansas City and Lawrence; and North Newton (Bethel College) with Newton.

Only in southwestern Kansas does a pattern of single-city dominance analogous to that of Hays fail to appear. Dodge City, Garden City, and Liberal all are contenders for the role. Their residents have invested heavily in community colleges, hospitals, and shopping facilities, and each city now controls economic activity over several counties apiece. Which of them might prevail as trade areas continue to expand, however, is not yet certain. The three communities each grew substantially between 1990 and 2000, but at nearly identical rates. Still, Garden City has several advantages over its rivals. Its position, nearly equidistant from Amarillo, Colorado Springs, and Wichita, is ideal to avoid competition with larger metropolitan areas. It also has the area's best water supply, sharing the Ogallala Aquifer with Liberal and the Arkansas Valley with Dodge City. These basic assets plus an aggressive public attitude to embrace multiculturalism and to diversify the economy with dairying and other ventures should lead to its gradual ascendancy in the region.

The easiest predictions for the Kansas urban system involve its core cities of Kansas City and Wichita. Greater Kansas City, atop the regional hierarchy, is firmly enmeshed into larger national and international business networks and should continue to serve as the mercantile linchpin for an area much larger than Kansas. To the chagrin of many people in Atchison, Lawrence, Leavenworth, Ottawa, and even Topeka, it is obvious that these once independent cities operate increasingly as specialized neighborhoods within the growing metropolitan complex.

Wichita plays a somewhat complicated role in the state and national economy. Although the community functions within the orbit of Kansas City at higher levels of mercantile activity, it also is the primary supplier of goods and services to south-central Kansas and the source of specialty items for people from Salina, Hays, Garden City, and beyond. Indeed, nearly all Kansans living southwest of Emporia regard Wichita as "their" city. In another type of competition, metropolitan Wichita itself continues to expand and now extends south to include Arkansas City and Winfield, east to El Dorado, north to McPherson, and northwest to Hutchinson. The future of Boeing and other Wichita-based airplane companies is the most difficult aspect of the city to project. Their business is cyclical, of course, but has proved durable because military buyers trust these experienced contractors. Local people expect continued success for their signature industry, and I have no reason to doubt their optimism.

If the future urban system of Kansas indeed resembles that shown on Map 26, with vital trading centers at Hays and Garden City in the west, Salina and Wichita in midstate, and Emporia, Joplin-Pittsburg, St. Joseph, Topeka, and of course Kansas City in the east, what fate awaits the state's other sizable communities? For many the answer will be prosperity via secondary or satellite roles within this mercantile and services system. Colby, Dodge City, and Liberal are well positioned for this in the west. Concordia, Hutchinson, Pratt, and Winfield–Arkansas City should assume similar roles in the central region, as should Atchison, Coffeyville-Independence, Fort Scott, Lawrence, Manhattan, and Marysville in the east. Lawrence and Manhattan will thrive doubly, of course, because they also possess statewide reputations as education centers.

Growth centers are not particularly hard to identify in Kansas or anywhere else. The challenge lies in predictions for cities in less favorable locations. Declines are inevitable, but this judgment should be tempered by the historical findings reported throughout this book. Remarkably few instances exist where once-thriving communities have fallen below the urban threshold population of twenty-five hundred. The railroad towns of Belleville, Ellis, and Horton have suffered such a fate, of course, as have Caney, Cherryvale, Humboldt, and several other communities developed during the state's early industrial spree based on natural gas. Elsewhere (and even in the places just mentioned to a degree) resilience has been the rule. People make emotional as well as financial investments in their communities. Although economic realities have made it impossible to maintain the hundreds of hamlets that once blanketed the state, people who live in slightly larger towns often have found inventive ways to preserve their vitality.

Historically, the cities facing the biggest challenges have been those specialized in mining or railroads. Where such communities occupy positions close to larger cities, as do Ellis and Frontenac, their citizens could become commuters. At Caney, Horton, and other isolated sites, however, residents could not cope with the losses and so began to move elsewhere. The next specialized cities to face threats likely will be Coffeyville, El Dorado, and McPherson. The oil refineries there are old, and so are Kansas wells. Great Bend and Russell are other cities in trouble. With area oil production in decline, how can these communities survive? Diversified manufacturing has worked reasonably well for Coffeyville and constitutes the best hope for Great Bend and Russell. The alternative is more commuting, a solution already common in El Dorado and McPherson.

Among the state's smaller cities, the most interesting are those that manage to prosper without the obvious advantages of community colleges, interstate highways, major institutions, or easy commuter destinations. Beloit fits into this category. So do Neodesha and Sabetha. In all these places, a visitor gets the sense that a few key people simply willed their communities into continued life through the creation of home-owned industries suited to the regional economy. Although I do not look for substantial growth at these three cities or their slightly larger counterparts such as Goodland, Iola, and Parsons, neither do I expect them to decline further. Community pride is a powerful force.

Appendix

Populations of Cities in Kansas, 1860–2000

Populations of Cities in Kansas, 1860–2000.

	1860	1870	1880	1890	1900	1910	1920
Abilene	—	—	2,360	3,547	3,507	4,118	4,895
Andover	—	—	—	—	—	—	—
Anthony	—	—	345	1,806	1,179	2,669	2,740
Arkansas City	—	—	1,012	8,347	6,140	7,508	11,253
Atchison	2,616	7,054	15,105	13,963	15,722	16,429	12,630
Augusta	—	—	922	1,343	1,197	1,235	4,219
Baldwin City	—	—	325	935	1,017	1,386	1,137
Baxter Springs	—	1,284	1,177	1,248	1,641	1,598	3,608
Bel Aire	—	—	—	—	—	—	—
Belleville	—	—	238	1,868	1,833	2,224	2,254
Beloit	—	—	1,835	2,455	2,359	3,082	3,315
Bonner Springs	—	—	—	—	609	1,462	1,599
Burlington	118	960	2,011	2,239	2,418	2,180	2,236
Caney	—	—	—	542	887	3,597	3,427
Chanute	—	—	887	2,826	4,208	9,272	10,286
Cherryvale	—	—	690	2,014	3,472	4,304	4,698
Chetopa	—	960	1,305	2,265	2,019	1,548	1,519
Clay Center	—	—	1,753	2,802	3,069	3,438	3,715
Coffeyville	—	—	753	2,282	4,953	12,687	13,452
Colby	—	—	—	516	641	1,130	1,114
Columbus	—	402	1,164	2,160	2,310	3,064	3,155
Concordia	—	—	1,853	3,184	3,401	4,415	4,705
Council Grove	—	712	1,042	2,211	2,265	2,545	2,857
Derby	—	—	—	—	—	235	247
De Soto	—	—	—	—	—	337	345
Dodge City	—	—	996	1,763	1,942	3,214	5,061
Edwardsville	—	—	—	—	—	—	203
El Dorado	—	—	1,411	3,339	3,466	3,129	10,995
Ellinwood	—	—	352	684	760	976	1,103
Ellis	—	—	689	1,107	932	1,404	1,876
Ellsworth	—	—	929	1,620	1,549	2,041	2,065
Emporia	—	2,168	4,631	7,551	8,223	9,058	11,273
Eudora	—	—	572	618	640	640	627
Eureka	—	—	1,127	2,259	2,091	2,333	2,606
Fairway	—	—	—	—	—	—	—
Fort Scott	262	4,174	5,372	11,946	10,322	10,463	10,693
Fredonia	—	—	923	1,515	1,650	3,040	3,954
Frontenac	—	—	—	600	1,805	3,396	3,225
Galena	—	—	1,463	2,496	10,155	6,096	4,712
Garden City	—	—	—	1,490	1,590	3,171	3,848
Gardner	—	—	203	515	475	514	514
Garnett	237	1,219	1,389	2,191	2,078	2,334	2,329
Girard	—	—	1,289	2,541	2,473	2,446	3,161
Goodland	—	—	—	1,027	1,059	1,993	2,664
Great Bend	—	—	1,071	2,450	2,470	4,622	4,460
Hays	—	370	850	1,242	1,136	1,961	3,165
Haysville	—	—	—	—	—	—	—
Herington	—	—	—	1,353	1,607	3,273	4,065
Hesston	—	—	—	—	—	—	—
Hiawatha	—	—	1,375	2,486	2,829	2,974	3,222

1930	1940	1950	1960	1970	1980	1990	2000
5,658	5,671	5,775	6,746	6,661	6,572	6,242	6,543
—	—	—	186	1,880	2,801	4,204	6,698
2,947	2,873	2,792	2,744	2,653	2,661	2,516	2,440
13,946	12,752	12,903	14,262	13,216	13,201	12,762	11,963
13,024	12,648	12,792	12,529	12,565	11,407	10,656	10,232
4,033	3,821	4,483	6,434	5,977	6,968	7,848	8,423
1,127	1,096	1,741	1,877	2,520	2,829	2,961	3,400
4,541	4,921	4,647	4,498	4,489	4,773	4,351	4,602
—	—	—	—	—	—	3,695	5,836
2,383	2,580	2,858	2,940	3,063	2,805	2,517	2,239
3,502	3,765	4,085	3,837	4,121	4,367	4,066	4,019
1,837	1,837	2,277	3,171	3,884	6,266	6,413	6,768
2,273	2,379	2,304	2,113	2,099	2,901	2,735	2,790
2,794	2,629	2,876	2,682	2,192	2,284	2,062	2,092
10,277	10,142	10,109	10,849	10,341	10,506	9,488	9,411
4,251	3,185	2,952	2,783	2,609	2,769	2,464	2,386
1,344	1,606	1,671	1,538	1,596	1,751	1,357	1,281
4,386	4,518	4,528	4,613	4,963	4,948	4,613	4,564
16,198	17,355	17,113	17,382	15,116	15,185	12,917	11,021
2,153	2,458	3,859	4,210	4,658	5,544	5,510	5,450
3,235	3,402	3,490	3,395	3,356	3,426	3,268	3,396
5,792	6,255	7,175	7,022	7,221	6,847	6,152	5,714
2,898	2,875	2,722	2,664	2,403	2,381	2,228	2,321
294	256	432	6,458	7,947	9,786	14,691	17,807
384	383	518	1,271	1,839	2,061	2,291	4,561
10,059	8,487	11,262	13,520	14,127	18,001	21,129	25,176
228	243	274	513	619	3,364	3,979	4,146
10,311	10,045	11,037	12,523	12,308	10,510	11,495	12,057
1,115	2,059	2,569	2,729	2,416	2,508	2,329	2,164
1,957	2,042	2,649	2,218	2,137	2,062	1,814	1,873
2,072	2,227	2,193	2,361	2,080	2,465	2,757	2,965
14,067	13,188	15,669	18,190	23,327	25,287	25,512	26,760
599	603	929	1,526	2,071	2,934	3,006	4,307
3,698	3,803	3,958	4,055	3,576	3,425	2,974	2,914
—	—	1,816	5,398	5,227	4,619	4,173	3,952
10,763	10,557	10,335	9,410	8,967	8,893	8,362	8,297
3,446	3,524	3,257	3,233	3,080	3,047	2,599	2,600
2,085	1,766	1,569	1,713	2,223	2,586	2,628	2,996
4,736	4,375	4,029	3,827	3,712	3,587	3,308	3,287
6,121	6,285	10,905	11,811	14,790	18,256	24,097	28,451
493	510	676	1,619	1,839	2,392	4,277	9,396
2,768	2,607	2,693	3,034	3,169	3,310	3,210	3,368
2,442	2,554	2,426	2,350	2,591	2,888	2,794	2,773
3,626	3,306	4,690	4,459	5,510	5,708	4,983	4,948
5,548	9,044	12,665	16,670	16,133	16,608	15,427	15,345
4,618	6,385	8,625	11,947	15,396	16,301	18,632	20,013
—	—	—	5,836	6,531	8,006	8,364	8,502
4,518	3,804	3,775	3,702	3,165	2,930	2,685	2,563
526	403	686	1,103	1,926	3,013	3,012	3,509
3,302	3,238	3,294	3,391	3,365	3,702	3,603	3,417

Populations of Cities in Kansas, 1860–2000 (continued)

	1860	1870	1880	1890	1900	1910	1920
Hillsboro	—	—	133	555	754	1,134	1,451
Hoisington	—	—	—	446	789	1,975	2,395
Holton	—	—	—	2,727	3,082	2,842	2,703
Horton	—	—	—	3,316	3,398	3,600	4,009
Hugoton	—	—	—	136	54	105	644
Humboldt	—	1,202	1,542	1,361	1,402	2,548	2,525
Hutchinson	—	—	1,540	8,682	9,379	16,364	23,298
Independence	—	435	2,915	3,127	4,851	10,480	11,920
Iola	—	—	1,096	1,706	5,791	9,032	8,513
Junction City	—	—	2,684	4,502	4,695	5,598	7,533
Kansas City	—	2,940[a]	10,311[b]	45,324[c]	60,566[d]	88,291[e]	108,851[f]
Kingman	—	—	—	2,390	1,785	2,570	2,407
Lansing	—	—	—	—	—	—	—
Larned	—	—	1,066	1,861	1,583	2,911	3,139
Lawrence	1,645	8,320	8,510	9,997	10,862	12,374	12,456
Leavenworth	7,429	17,873	16,546	19,768	20,735	19,363	15,912
Leawood	—	—	—	—	—	—	—
Lenexa	—	—	—	—	—	383	472
Liberal	—	—	—	—	426	1,716	3,613
Lindsborg	—	—	466	968	1,279	1,939	1,897
Louisburg	—	—	499	760	665	603	556
Lyons	—	—	509	1,754	1,736	2,071	2,516
Manhattan	—	1,173	2,105	3,004	3,438	5,722	7,989
Marysville	171	300	1,249	1,913	2,006	2,260	3,048
McPherson	—	—	1,590	3,172	2,996	3,546	4,595
Medicine Lodge	—	—	373	1,095	917	1,229	1,305
Merriam	—	—	—	—	—	—	—
Mission	—	—	—	—	—	—	—
Mission Hills	—	—	—	—	—	—	—
Mulvane	—	—	215	724	667	1,084	1,239
Mulberry	—	—	—	—	—	997	2,697
Neodesha	—	—	924	1,528	1,772	2,872	3,943
Newton	—	—	2,601	5,605	6,208	7,862	9,781
Norton	—	—	634	1,074	1,202	1,787	2,186
Olathe	—	1,817	2,285	3,294	3,451	3,272	3,268
Osage City	—	—	2,098	3,469	2,792	2,432	2,376
Osawatomie	—	—	681	2,662	4,191	4,046	3,293
Oswego	—	1,196	2,351	2,574	2,208	2,317	2,386
Ottawa	—	2,941	4,032	6,248	6,934	7,650	9,018
Overland Park	—	—	—	—	—	—	—
Paola	—	1,811	2,312	2,943	3,144	3,207	3,238
Park City	—	—	—	—	—	—	—
Parsons	—	—	4,199	6,736	7,682	12,463	16,028
Phillipsburg	—	—	309	992	1,008	1,302	1,310
Pittsburg	—	—	624	6,697	10,112	14,755	18,052
Plainville	—	—	39	347	378	1,090	1,004
Prairie Village	—	—	—	—	—	—	—
Pratt	—	—	—	1,418	1,213	3,302	5,183
Roeland Park	—	—	—	—	—	—	—
Rose Hill	—	—	—	—	—	—	—

1930	1940	1950	1960	1970	1980	1990	2000
1,458	1,580	2,150	2,441	2,730	2,717	2,704	2,854
3,001	3,719	4,012	4,248	3,710	3,678	3,182	2,975
2,705	2,885	2,705	3,028	3,063	3,132	3,196	3,353
4,031	2,872	2,354	2,361	2,177	2,130	1,885	1,967
1,368	1,349	2,781	2,912	2,739	3,165	3,179	3,708
2,558	2,290	2,308	2,285	2,249	2,230	2,178	1,999
27,085	30,013	33,575	37,574	36,885	40,284	39,308	40,787
12,782	11,565	11,335	11,222	10,347	10,598	10,030	9,846
7,160	7,244	7,094	6,885	6,493	6,938	6,351	6,302
7,407	8,507	13,462	18,700	19,018	19,305	20,642	18,886
121,857	121,458	129,553	121,901	168,213	161,087	151,521	146,866
2,752	3,213	3,200	3,582	3,622	3,563	3,196	3,387
—	—	—	1,264	3,797	5,307	7,120	9,199
3,532	3,533	4,447	5,001	4,567	4,811	4,490	4,236
13,726	14,390	23,351	32,858	45,698	52,738	65,608	80,098
17,466	19,220	20,579	22,052	25,147	33,656	38,495	35,420
—	—	1,167	7,466	10,645	13,360	19,693	27,656
452	502	803	2,487	5,549	18,639	34,110	40,238
5,294	4,410	7,134	13,813	13,862	14,911	16,573	19,666
2,016	1,913	2,383	2,609	2,764	3,155	3,077	3,321
616	677	590	862	1,033	1,744	1,964	2,576
2,939	4,497	4,545	4,592	4,355	4,152	3,688	3,732
10,136	11,659	19,056	22,993	27,575	32,644	43,081	44,831
4,013	4,055	3,866	4,143	3,588	3,670	3,360	3,271
6,147	7,194	8,689	9,996	10,851	11,753	12,422	13,770
1,655	1,870	2,288	3,072	2,545	2,384	2,453	2,193
—	—	—	5,084	10,955	10,794	11,819	11,008
—	—	—	4,626	8,125	8,643	9,504	9,727
—	—	1,275	3,621	4,198	3,904	3,446	3,593
1,042	940	1,387	2,981	3,185	4,254	4,683	5,155
1,596	1,175	779	642	622	647	555	577
3,381	3,376	3,723	3,594	3,295	3,414	2,837	2,848
11,034	11,048	11,590	14,877	15,439	16,332	16,700	17,190
2,767	2,762	3,060	3,345	3,627	3,400	3,017	3,012
3,656	3,979	5,593	10,987	17,917	37,258	63,402	92,962
2,402	2,079	1,919	2,213	2,600	2,667	2,683	3,034
4,440	4,145	4,347	4,622	4,294	4,459	4,590	4,645
1,845	1,953	1,997	2,027	2,200	2,218	1,870	2,046
9,563	10,193	10,081	10,673	11,036	11,016	10,667	11,921
—	—	—	—	77,934	81,784	111,790	149,080
3,762	3,511	3,972	4,784	4,622	4,557	4,698	5,011
—	—	—	2,687	2,529	3,778	5,081	5,814
14,903	14,294	14,750	13,929	13,015	12,898	11,919	11,514
1,543	2,109	2,589	3,233	3,241	3,229	2,828	2,668
18,145	17,571	19,341	18,678	20,171	18,770	17,789	19,243
1,058	1,232	2,082	3,104	2,627	2,458	2,173	2,029
—	—	—	25,356	28,378	24,657	23,186	22,072
6,322	6,591	7,523	8,156	6,736	6,885	6,687	6,570
—	—	—	8,949	9,760	7,962	7,706	6,817
—	—	—	273	387	1,557	2,399	3,432

Populations of Cities in Kansas, 1860–2000 (continued)

	1860	1870	1880	1890	1900	1910	1920
Russell	—	—	861	961	1,143	1,692	1,700
Sabetha	—	—	849	1,368	1,646	1,768	2,003
Salina	—	918	3,111	6,149	9,688	15,085	20,155
Scott City	—	—	—	229	212	918	1,112
Shawnee	—	—	—	—	—	—	—
South Hutchinson	—	—	—	321	225	387	639
Spring Hill	—	—	502	573	580	605	555
Sterling	—	—	1,014	1,641	2,002	2,133	2,060
Tonganoxie	—	—	426	673	848	1,018	971
Topeka	759	5,790	15,452	31,007	33,608	43,684	50,022
Ulysses	—	—	—	198	—	—	103
Valley Center	—	—	71	167	343	381	486
Wakeeney	—	—	418	439	394	883	1,003
Wamego	—	—	—	1,473	1,618	1,714	1,585
Weir	—	—	376	2,138	2,977	2,289	1,945
Wellington	—	—	2,694	4,391	4,245	7,034	7,048
Wichita	—	—	4,911	23,853	24,671	52,450	72,217
Winfield	—	—	2,844	5,184	5,554	6,700	7,933

Source: Federal census records.

[a] This number is for Wyandotte, the major parent community of modern Kansas City, Kansas.

[b] The total of the populations of Wyandotte (6,149), Rosedale (962), and the original Kansas City, Kansas (3,200).

[c] Includes the populations of Argentine (4,732) and Rosedale (2,276).

[d] Includes the populations of Argentine (5,878) and Rosedale (3,270).

[e] Includes the population of Rosedale (5,960).

[f] Includes the population of Rosedale (7,674).

1930	1940	1950	1960	1970	1980	1990	2000
2,352	4,819	6,483	6,113	5,371	5,427	4,783	4,696
2,332	2,241	2,173	2,318	2,376	2,286	2,341	2,589
20,155	21,073	26,176	43,202	37,714	41,843	42,299	45,679
1,544	1,848	3,204	3,555	4,001	4,154	3,785	3,855
553	597	845	9,072	20,946	29,653	37,962	47,996
669	915	1,045	1,672	1,879	2,226	2,444	2,539
566	489	619	909	1,186	2,005	2,191	2,727
1,868	2,215	2,243	2,303	2,312	2,312	2,191	2,642
1,109	1,114	1,138	1,354	1,717	1,864	2,347	2,728
64,120	67,833	78,791	119,484	125,011	115,266	119,883	122,377
1,140	475	2,243	3,157	3,779	4,653	5,474	5,960
896	700	854	2,570	2,551	3,300	3,624	4,883
1,408	1,852	2,446	2,808	2,334	2,388	2,161	1,924
1,647	1,767	1,869	2,363	2,507	3,159	3,706	4,246
1,115	1,038	819	669	740	705	730	780
7,405	7,246	7,747	8,809	8,072	8,212	8,517	8,647
111,110	114,966	168,279	254,698	276,554	279,272	304,017	344,284
9,398	9,506	10,264	11,117	11,405	10,736	11,931	12,206

Notes

1. Systems of Cities on the Plains

1. Stanley Vestal, *Queen of Cow Towns: Dodge City* (New York: Harper, 1952).

2. Edward K. Muller, "From Waterfront to Metropolitan Region: The Geographical Development of American Cities," in *American Urbanism: A Historiographical Review,* ed. Howard Gillette Jr. and Zane L. Miller (New York: Greenwood Press, 1987), pp. 105–6.

3. All textbooks in urban geography contain sizable sections on central-place theory. Classic studies include John E. Brush, "The Hierarchy of Central Places in Southwestern Wisconsin," *Geographical Review* 43 (1953): 380–402; Walter Christaller, *Central Places in Southern Germany,* trans. Carlisle W. Baskin (Englewood Cliffs, N.J.: Prentice-Hall, 1966); Brian J. L. Berry, *A Geography of Market Centers and Retail Distribution* (Englewood Cliffs, N.J.: Prentice-Hall, 1967).

4. Richard C. Wade, *The Urban Frontier: The Rise of Western Cities, 1790–1830* (Cambridge, Mass.: Harvard University Press, 1959).

5. Charles F. J. Whebell, "Corridors: A Theory of Urban Systems," *Annals of the Association of American Geographers* 59 (1969): 1–26.

6. James E. Vance Jr., *The Merchant's World: The Geography of Wholesaling* (Englewood Cliffs, N.J.: Prentice-Hall, 1970).

7. Donald W. Meinig, "American Wests: Preface to a Geographical Interpretation," *Annals of the Association of American Geographers* 62 (1972): 159–84.

8. Edward K. Muller, "Selective Urban Growth in the Middle Ohio Valley, 1800–1860," *Geographical Review* 66 (1976): 178–99; Muller, "Regional Urbanization and the Selective Growth of Towns in North American Regions," *Journal of Historical Geography* 3 (1977): 21–39. See also David R. Meyer, "The Rise of the Industrial Metropolis: The Myth and the Reality," *Social Forces* 68 (1990): 731–52; and Robert D. Mitchell and Warren R. Hofstra, "How Do Settlement Systems Evolve? The Virginia Backcountry during the Eighteenth Century," *Journal of Historical Geography* 21 (1995): 123–47.

9. Andrew F. Burghardt, "A Hypothesis about Gateway Cities," *Annals of the Association of American Geographers* 61 (1971): 269–85; John R. Borchert, "America's Changing Metropolitan Regions," *Annals of the Association of American Geographers* 62 (1972): 352–73; Michael P. Conzen, "Capital Flows and the Developing Urban Hierarchy: State Bank Capital in Wisconsin, 1854–1895," *Economic Geography* 51 (1975): 321–38; Conzen, "A Transport Interpretation of the Growth of Urban Regions: An American Example," *Journal of Historical Geography* 1 (1975): 361–82; Conzen, "The Maturing Urban System in the United States, 1840–1910," *Annals of the Association of American Geographers* 67 (1977): 88–108; Allan R. Pred, *Urban Growth and the Circulation of Information: The United States System of Cities, 1790–1810* (Cambridge, Mass.: Harvard University Press, 1973); Pred, *Urban Growth and City-Systems in the United States, 1840–1860* (Cambridge, Mass.: Harvard University Press, 1980); David R. Meyer, "A Dynamic Model of the Integration of Frontier Urban Places into the United States System of Cities," *Economic Geography* 56 (1980): 120–40; William K. Wyckoff, "Revising the Meyer Model: Denver

and the National Urban System, 1859–1879," *Urban Geography* 9 (1988): 1–18; Anne Mosher Sheridan, "The Development of Nucleated Settlement Systems: The Case of the Upper Mississippi Valley, 1800–1860" (master's thesis, Pennsylvania State University, 1983); Timothy R. Mahoney, *River Towns in the Great West: The Structure of Provincial Urbanization in the American Midwest, 1820–1870* (Cambridge.: Cambridge University Press, 1990); Lawrence A. Brown, Rodrigo Sierra, Scott Digiacentro, and W. Randy Smith, "Urban-System Evolution in Frontier Settings," *Geographical Review* 84 (1994): 249–65.

2. The River Towns to 1860

1. Nineteenth-century writers employed the term *le grand detour* at two different scales. I use it in their larger sense. Others applied it only to the specific river bend where Atchison was established.

2. Elvid Hunt and Walter E. Lorence, *History of Fort Leavenworth, 1827–1937*, rev. ed. (Fort Leavenworth, Kans.: Command and General Staff School Press, 1937), pp. 12–15. The name Little Platte River is confusing. In 1827 it referred to the Missouri stream now known simply as the Platte.

3. Ibid., pp. 17, 19.

4. Walter A. Schroeder, "The Presettlement Prairie in the Kansas City Region (Jackson County, Missouri)," *Missouri Prairie Journal* 7, no. 2 (December 1985): 3–12.

5. Perry McCandless, *A History of Missouri,* vol. 2, *1820–1860* (Columbia: University of Missouri Press, 1972), pp. 116–17; Hunt and Lorence, *History of Fort Leavenworth,* p. 56.

6. Daniel Fitzgerald, "Town Booming: An Economic History of Steamboat Towns along the Kansas-Missouri Border, 1840–1860" (master's thesis, University of Kansas, 1983), pp. 12, 45–48; Hunt and Lorence, *History of Fort Leavenworth,* p. 36.

7. Hunt and Lorence, *History of Fort Leavenworth,* p. 65; Fitzgerald, "Town Booming," pp. 47–49.

8. John M. McNamara, *Three Years on the Kansas Border* (New York: Miller, Orton, and Mulligan, 1856), p. 20.

9. Eugene T. Wells, "St. Louis and Cities West, 1820–1880: A Study in Historical Geography" (Ph.D. diss., University of Kansas, 1951), p. 320.

10. Ibid., pp. 320–22; William A. Bowen, *The Willamette Valley: Migration and Settlement on the Oregon Frontier* (Seattle: University of Washington Press, 1978), pp. 13–14.

11. *St. Joseph Gazette,* June 6, 1845; Wells, "St. Louis and Cities West," pp. 322–23; Fitzgerald, "Town Booming," pp. 33–39. The two rivers Oregon travelers avoided by a departure from St. Joseph were the Wakarusa and the Kansas.

12. *St. Joseph Gazette,* March 9, May 18, and June 15, 1849.

13. Fitzgerald, "Town Booming," pp. 40–42, 50.

14. Frank S. Popplewell, "St. Joseph, Missouri, as a Center of the Cattle Trade," *Missouri Historical Review* 32 (1937–1938): 444–45.

15. Hunt and Lorence, *History of Fort Leavenworth,* pp. 61–62, 75.

16. Ibid.

17. Wells, "St. Louis and Cities West," pp. 289–91.

18. Ibid., pp. 291–93; David Dary, *The Santa Fe Trail: Its History, Legends, and Lore* (New York: Knopf, 2000), pp. 116–21; Charles J. Latrobe, *The Rambler in North America, MDCCCXXXII–MDCCCXXXIII* (New York: Harper and Brothers, 1835), vol. 1, pp. 104–5.

19. Wells, "St. Louis and Cities West," p. 302.

20. Walker D. Wyman, "Kansas City, Mo.: A Famous Freighter Capital," *Kansas Historical Quarterly* 6 (1937): 3–4; Wells, "St. Louis and Cities West," pp. 313–15.

21. Wells, "St. Louis and Cities West," pp. 359–60; Charles N. Glaab, "Business Patterns in the Growth of a Midwestern City: The Kansas City Business Community before the Civil War," *Business History Review* 33 (1959): 156–64.

22. Fitzgerald, "Town Booming," pp. 58–59; Wells, "St. Louis and Cities West," p. 308.

23. *St. Joseph Gazette,* June 26, 1846.

24. *St. Joseph Gazette,* January 15, 1847.

25. *St. Joseph Gazette,* November 6, 1846.

26. Fitzgerald, "Town Booming," p. 37; Popplewell, "St. Joseph," p. 446.

27. Fitzgerald, "Town Booming," p. 53.

28. Ibid., pp. 69–75; Wells, "St. Louis and Cities West," pp. 338–47.

29. Robert E. Riegel, "The Missouri Pacific Railroad to 1879," *Missouri Historical Review* 18 (1923–1924): 3–12; John L. Kerr, *The Story of a Western Power—The Missouri Pacific: An Outline History* (New York: Railway Research Society, 1928), pp. 3–7.

30. H. Craig Miner and William E. Unrau, *The End of Indian Kansas: A Study of Cultural Revolution, 1854–1871* (Lawrence: Regents Press of Kansas, 1978), pp. 1–8.

31. Ibid.; James C. Malin, *The Nebraska Question, 1852–1854* (Lawrence, Kans.: James C. Malin, 1953), pp. 128–36.

32. Miner and Unrau, *End of Indian Kansas,* pp. 9–12.

33. Ibid.; Paul W. Gates, *Fifty Million Acres: Conflicts over Kansas Land Policy, 1854–1890* (Ithaca, N.Y.: Cornell University Press, 1954), p. 17.

34. James R. Shortridge, *Peopling the Plains: Who Settled Where in Frontier Kansas* (Lawrence: University Press of Kansas, 1995), pp. 15–20.

35. *St. Joseph Gazette,* June 7, 1854; Fitzgerald, "Town Booming," pp. 95–96.

36. *St. Joseph Gazette,* January 18, 1850; Walter Williams, *History of Northwest Missouri* (Chicago: Lewis Publishing Company, 1915), pp. 353–55; Fitzgerald, "Town Booming," pp. 86–88.

37. Fitzgerald, "Town Booming," pp. 88–96.

38. William G. Cutler, ed., *History of the State of Kansas* (Chicago: A. T. Andreas, 1883), pp. 472, 494–95.

39. Ibid.; Daniel Fitzgerald, *Ghost Towns of Kansas: A Traveler's Guide* (Lawrence: University Press of Kansas, 1988), pp. 17–20.

40. Wyman, "Kansas City," p. 4; Fitzgerald, "Town Booming," pp. 106–8. Although the name City of Kansas remained officially in place until 1889, people began to use the less formal phrase Kansas City almost immediately. I do the same in this book.

41. *Kansas Public Ledger,* July 4, 1851. The *Ledger* was the first newspaper published in Westport.

42. Fitzgerald, "Town Booming," pp. 98–100.

43. Louise Barry, *The Beginning of the West: Annals of the Kansas Gateway to the American West, 1540–1854* (Topeka: Kansas State Historical Society, 1972), pp. 1226, 1228.

44. Fitzgerald, "Town Booming," p. 101.

45. *Kansas Weekly Herald,* September 15, 1854.

46. *St. Joseph Gazette,* August 9, 1854; Wells, "St. Louis and Cities West," p. 536.

47. Raymond W. Settle and Mary L. Settle, *War Drums and Wagon Wheels: The Story of Russell, Majors, and Waddell* (Lincoln: University of Nebraska Press, 1966), pp. 38–39.

48. Ibid., pp. 40–45. David G. Taylor, "Boom Town Leavenworth: The Failure of a Dream," *Kansas Historical Quarterly* 38 (1972): 393; Horace Greeley, *An Overland Journey from New York to San Francisco in the Summer of 1859* (San Francisco: H. H. Bancroft and Co., 1860), pp. 38–39.

49. Fitzgerald, "Town Booming," pp. 116–17; Taylor, "Boom Town Leavenworth," p. 394.

50. Cutler, *History of the State,* p. 419.

51. William A. Dobak, *Fort Riley and Its Neighbors: Military Money and Economic Growth, 1853–1895* (Norman: University of Oklahoma Press, 1998), pp. 21–43.

52. *Kansas Weekly Herald,* February 23, 1855.

53. Ibid., November 15, 1855. For background, see Gates, *Fifty Million Acres,* pp. 48–71.

54. Taylor, "Boom Town Leavenworth," pp. 395–97; Taylor, "The Economic Development of Leavenworth, Kansas, 1854–1870" (master's thesis, University of Kansas, 1966), pp. 25–41; Taylor, "The Business and Political Career of Thomas Ewing, Jr.: A Study of Frustration" (Ph.D. diss., University of Kansas, 1970), pp. 16–43.

55. William T. Sherman, *Home Letters of General Sherman,* ed. M. A. De Wolfe Howe (New York: Scribner's, 1909), pp. 156–57.

56. Settle and Settle, *War Drums and Wagon Wheels,* pp. 51–90, 126; Taylor, "Economic Development of Leavenworth," p. 74.

57. Taylor, "Economic Development of Leavenworth," pp. 51–74.

58. Taylor, "Boom Town Leavenworth," p. 400; Wyckoff, "Revising the Meyer Model," pp. 6–7; Edward Langsdorf and Robert W. Richmond, eds., "Letters of Daniel R. Anthony, 1857–1862," *Kansas Historical Quarterly* 24 (1958): 214.

59. Cutler, *History of the State,* p. 459; Gates, *Fifty Million Acres,* p. 21.

60. Daniel Fitzgerald, *Faded Dreams: More Ghost Towns of Kansas* (Lawrence: University Press of Kansas, 1994), pp. 6–14; Cutler, *History of the State,* p. 459.

61. Peter Beckman, "The Overland Trade and Atchison's Beginnings," in *Territorial Kansas: Studies Commemorating the Centennial* (Lawrence: University of Kansas Publications, 1954), p. 148.

62. James Sutherland, comp., *Atchison City Directory and Business Mirror for 1860–61* (Indianapolis: James Sutherland, 1860), p. 7.

63. Cutler, *History of the State,* pp. 370–71; Wells, "St. Louis and Cities West," pp. 555–56; Beckman, "Overland Trade," pp. 150–51.

64. Cutler, *History of the State,* p. 371; Fitzgerald, "Town Booming," pp. 119–20; Walker D. Wyman, "Atchison: A Great Frontier Depot," *Kansas Historical Quarterly* 11 (1942): 301.

65. John N. Holloway, *History of Kansas: From the First Exploration of the Mississippi Valley, to Its Admission into the Union* (Lafayette, Ind.: James, Emmons and Company, 1868), p. 109.

66. Beckman, "Overland Trade," pp. 151–52.

67. *Atchison Freedom's Champion,* February 20, 1858, and February 11, 1860; Beckman, "Overland Trade," p. 153; Wyman, "Atchison," p. 303.

68. Cutler, *History of the State,* p. 379; Beckman, "Overland Trade," p. 157. Although East Atchison remains a popular name for the station area, officially it was (and is) called Winthrop.

69. Peter Beckman, "Atchison's First Railroad," *Kansas Historical Quarterly* 21 (1954): 153–65; Taylor, "Economic Development of Leavenworth," pp. 61–74; *Atchison Union,* February 25, 1860; Beckman, "Overland Trade," pp. 154–58; Wyman, "Atchison," p. 303.

70. Beckman, "Overland Trade," p. 154; Wyman, "Atchison," pp. 301, 304–5.

71. *Manufactures of the United States in 1860 Compiled from the Original Returns of the Eighth Census* (Washington, D.C.: Government Printing Office, 1865), pp. 164–65, 296, 302–3.

72. Sutherland, *Atchison City Directory,* pp. 95–96; Sutherland, *Leavenworth City Di-*

rectory and Business Mirror for 1860–61 (Leavenworth, Kans.: James Sutherland, 1860), pp. 203–5.

3. Lawrence, Topeka, and the Placement of Public Institutions, 1854–1866

1. Samuel A. Johnson, *The Battle Cry of Freedom: The New England Emigrant Aid Company in the Kansas Crusade* (Lawrence: University of Kansas Press, 1954), pp. 3–34.

2. Ibid., pp. 7–19; Alice Nichols, *Bleeding Kansas* (New York: Oxford University Press, 1954), pp. 12–14. Edward Everett Hale's book was *Kanzas and Nebraska: The History, Geographical and Physical Characteristics, and Political Position of Those Territories; An Account of the Emigrant Aid Companies, and Directions to Emigrants* (Boston: Philips, Sampson and Co., 1854).

3. Johnson, *Battle Cry,* pp. 30–33.

4. Ibid., p. 53.

5. William G. Cutler, ed., *History of the State of Kansas* (Chicago: A. T. Andreas, 1883), pp. 312–13.

6. *Kansas Free State,* February 7, 1855. This newspaper, which had begun publication in Lawrence in January, was edited by Josiah Miller, a native of South Carolina, and R. G. Elliott, who came to Kansas from Cincinnati (see Cutler, *History of the State,* p. 316).

7. Cutler, *History of the State,* pp. 314, 317–19; Johnson, *Battle Cry,* p. 155.

8. This map is reproduced in David A. Dary, *Lawrence, Douglas County, Kansas: An Informal History* (Lawrence, Kans.: Allen Books, 1982), p. 28.

9. Johnson, *Battle Cry,* p. 250; Clifford S. Griffin, *The University of Kansas: A History* (Lawrence: University Press of Kansas, 1974), pp. 14–15, 21–23.

10. Even this total of three thousand emigrants overstates the case, because only one-half to two-thirds of these remained as permanent settlers in Kansas (see Johnson, *Battle Cry,* p. 296); Albert R. Greene, "The Kansas River—Its Navigation," *Transactions of the Kansas State Historical Society* 9 (1905–1906): 338–43.

11. Cutler, *History of the State,* p. 330; James Sutherland, *Lawrence City Directory and Business Mirror for 1860–61* (Indianapolis: James Sutherland, 1860), p. 49.

12. Johnson, *Battle Cry,* p. 81; Cutler, *History of the State,* p. 539.

13. Roy D. Bird and Douglass W. Wallace, *Witness of the Times: A History of Shawnee County* (Topeka: Shawnee County Historical Society, 1976), p. 12; Frye W. Giles, *Thirty Years in Topeka: A Historical Sketch* (Topeka: G. W. Crane and Company, 1886), p. 51.

14. Cutler, *History of the State,* pp. 539–40.

15. James R. Shortridge, *Peopling the Plains: Who Settled Where in Frontier Kansas* (Lawrence: University Press of Kansas, 1995), pp. 24–27.

16. Cutler, *History of the State,* pp. 539–41, 557.

17. Bird and Wallace, *Witness of the Times,* pp. 24, 110, 189–93; Giles, *Thirty Years in Topeka,* pp. 81–84, 108; Peter MacVicar, *A Historical Sketch of Washburn College* (Burlington, Kans.: Republican-Patriot Printers, 1886), pp. 2–6. The Episcopalian seminary, which opened in 1860, became known as the College of the Sisters of Bethany in 1872. It closed in 1928. The Congregational school held its first classes in 1866 under the name Lincoln College. Two years later it was rechristened Washburn College to honor Ichabod Washburn, an industrialist from Worcester, Massachusetts, who had donated $25,000 to the endowment fund.

18. Cutler, *History of the State,* pp. 315–16.

19. Ibid., p. 101.

20. Homer E. Socolofsky, *Kansas Governors* (Lawrence: University Press of Kansas,

1990), pp. 33–39; Franklin G. Adams, "The Capitals of Kansas," *Transactions of the Kansas State Historical Society* 8 (1903–1904): 331–33.

21. Adams, "Capitals of Kansas," p. 332.

22. Ibid., p. 333; Cutler, *History of the State*, pp. 86, 101. Reeder owned eighty acres, or a quarter, of the Pawnee town site (see George W. Martin, "The Territorial and Military Combine at Fort Riley," *Transactions of the Kansas State Historical Society* 7 [1901–1902]: 368).

23. William A. Dobak, *Fort Riley and Its Neighbors: Military Money and Economic Growth, 1853–1895* (Norman: University of Oklahoma Press, 1998), p. 141; Cutler, *History of the State*, p. 101; Adams, "Capitals of Kansas," p. 332.

24. Martin, "Territorial and Military Combine," p. 370; Cutler, *History of the State*, p. 105; Robert W. Richmond, "The First Capitol of Kansas," *Kansas Historical Quarterly* 21 (1955): 321–25.

25. Cutler, *History of the State*, p. 351; Adams, "Capitals of Kansas," pp. 338–39. The Isacks name is sometimes spelled Isaacs in the literature.

26. Cutler, *History of the State*, p. 533; Bird and Wallace, *Witness of the Times*, p. 18; Adams, "Capitals of Kansas," p. 339; C. K. Holliday, "The Cities of Kansas," *Kansas Historical Collections, 1883–1885* 3 (1886): 397.

27. Sara Walter and Iona Spencer, *Territorial Days: A Story of Historic Lecompton, Kansas* (Lecompton, Kans.: Lecompton Historical Society, 1980), pp. 3, 16.

28. Adams, "Capitals of Kansas," pp. 342–43.

29. Ibid., pp. 346–48.

30. Cutler, *History of the State*, pp. 106–8; Dary, *Lawrence, Douglas County*, p. 54.

31. Adams, "Capitals of Kansas," pp. 342–44.

32. G. Raymond Gaeddert, *The Birth of Kansas* (Lawrence: University of Kansas Publications, 1940), pp. 32–35, 58–65; Adams, "Capitals of Kansas," pp. 344–45; Cutler, *History of the State*, pp. 173–76.

33. Cutler, *History of the State*, pp. 173–76.

34. Gaeddert, *Birth of Kansas*, p. 116; Giles, *Thirty Years in Topeka*, p. 110.

35. Gaeddert, *Birth of Kansas*, pp. 118–20.

36. David G. Taylor, "Boom Town Leavenworth: The Failure of a Dream," *Kansas Historical Quarterly* 38 (1972): 397–99.

37. David G. Taylor, "The Economic Development of Leavenworth, Kansas: 1854–1870" (master's thesis, University of Kansas, 1966), p. 161.

38. Frank M. Gable, "The Kansas Penitentiary," *Kansas Historical Collections, 1915–1918* 14 (1918): 379–80; Theodore L. Heim, "The Administration and Organization of the Kansas State Penitentiary" (master's thesis, University of Kansas, 1960), pp. viii–ix.

39. Heim, "Administration and Organization," pp. viii–ix; Cutler, *History of the State*, p. 282.

40. Griffin, *University of Kansas*, pp. 13–15.

41. Gaeddert, *Birth of Kansas*, p. 120; Winifred N. Slagg, *Riley County Kansas: A Study of Early Settlements, Rich Valleys, Azure Skies and Sunflowers* (Manhattan, Kans.: Winifred N. Slagg, 1968), pp. 77–79.

42. Gaeddert, *Birth of Kansas*, p. 120.

43. Ibid., pp. 120–21.

44. The land amounts for the Morrill grants were based on thirty thousand acres for each member of a state's congressional delegation. Kansas, with a single member in the House of Representatives, received the minimum of ninety thousand acres.

45. Gaeddert, *Birth of Kansas*, pp. 120–21; Griffin, *University of Kansas*, pp. 24–26; Frank W. Blackmar, *The Life of Charles Robinson: The First Governor of Kansas* (Topeka: Crane and Company, 1902), p. 134. Mr. Russell's railroad interest was in having St. Joseph rather than Atchison named as one of the eastern terminals for the first transcontinental railway. A bill about this was then under consideration by Congress. If the line were to go west from St. Joseph, it would pass through Doniphan County (see George W. Glick, "The Railroad Convention of 1860," *Transactions of the Kansas State Historical Society, 1905–1906* 9 [1906] 479).

46. Cutler, *History of the State*, pp. 845–49, 853, 859.

47. Ibid.; Gaeddert, *Birth of Kansas*, p. 122.

48. Cutler, *History of the State*, p. 297.

49. Harold M. Liberman, "Early History of the Kansas School for the Deaf, 1861–1873" (master's thesis, University of Kansas, 1966), pp. 61, 75.

50. Lowell Gish, *The First 100 Years: A History of Osawatomie State Hospital* (Topeka, Kans.: Boys' Industrial School, 1966), p. 21.

51. Cutler, *History of the State*, p. 1229, quoting an older resident of Wyandotte named D. B. Hadley; Frank H. Betton, "The Genesis of a State's Metropolis," *Kansas Historical Collections, 1901–1902* 7 (1902): 114–15. The process by which "Wyandot," the name of the Indian people, was transformed into "Wyandotte," the name of the city and county, is a mystery. One authority attributes it to a clerical error (see Sondra Van Meter McCoy and Jan Hults, *1001 Kansas Place Names* [Lawrence: University Press of Kansas, 1989], p. 214).

52. Betton, "Genesis," pp. 114–15; Horace Greeley, *An Overland Journey from New York to San Francisco in the Summer of 1859* (San Francisco: H. H. Bancroft and Co., 1860), p. 20.

53. Cutler, *History of the State*, p. 1231; Betton, "Genesis," pp. 115–17.

54. Cutler, *History of the State*, p. 1238; Betton, "Genesis," p. 120.

55. Gish, *First 100 Years*, p. 21.

56. Cutler, *History of the State*, p. 211; Greeley, *Overland Journey*, p. 26; Lowell Gish, *Reform at Osawatomie State Hospital: Treatment of the Mentally Ill, 1866–1970* (Lawrence: University Press of Kansas, 1972), pp. 15–17.

57. Cutler, *History of the State*, pp. 887, 1232; Gish, *Reform at Osawatomie*, p. 43.

58. Liberman, "Early History," pp. 37–84.

59. Ibid., pp. 96–111; *Olathe Mirror*, February 15, 1866; Cutler, *History of the State*, p. 631.

60. Cutler, *History of the State*, pp. 272, 282.

4. Railroad Promotion and a Reconceptualization of Urban Kansas, 1863–1880

1. Henry Nash Smith, *Virgin Land: The American West as Symbol and Myth* (Cambridge, Mass.: Harvard University Press, 1950), pp. 13–48.

2. Nelson Trottman, *History of the Union Pacific: A Financial and Economic Survey* (New York: Ronald Press, 1923), p. 5; William Cronon, *Nature's Metropolis: Chicago and the Great West* (New York: Norton, 1991), pp. 295–309.

3. George W. Glick, "The Railroad Convention of 1860," *Transactions of the Kansas State Historical Society, 1905–1906* 9 (1906): 467–80.

4. Ibid.

5. *Leavenworth Herald*, February 26, 1859; Frank S. Popplewell, "St. Joseph, Missouri, as a Center of the Cattle Trade," *Missouri Historical Review* 32 (1937–1938): 446–50.

6. Daniel Fitzgerald, "Town Booming: An Economic History of Steamboat Towns

along the Kansas-Missouri Border, 1840–1860" (master's thesis, University of Kansas, 1983), pp. 199–200; Charles S. Gleed, "The First Kansas Railway," *Transactions of the Kansas State Historical Society, 1897–1900* 6 (1900): 357–59; Peter Beckman, "Atchison's First Railroad," *Kansas Historical Quarterly* 21 (1954): 155–57.

7. David G. Taylor, "Thomas Ewing, Jr., and the Origins of the Kansas Pacific Railway Company," *Kansas Historical Quarterly* 42 (1976): 155–79.

8. Ibid., pp. 155–67; Paul W. Gates, *Fifty Million Acres: Conflicts over Kansas Land Policy, 1854–1890* (Ithaca, N.Y.: Cornell University Press, 1954), pp. 109–33.

9. Wallace D. Farnham, "The Pacific Railroad Act of 1862," *Nebraska History* 43 (1962): 141–67. The term "Leavenworth ring" is from Albert Castel, *A Frontier State at War: Kansas, 1861–1865* (Ithaca, N.Y.: Cornell University Press for the American Historical Association, 1958), p. 84.

10. Farnham, "Pacific Railroad Act," pp. 141–67.

11. Imre E. Quastler, *The Railroads of Lawrence, Kansas, 1854–1900* (Lawrence, Kans.: Coronado Press, 1979), p. 61.

12. Taylor, "Thomas Ewing, Jr.," pp. 173–76; William R. Petrowski, *The Kansas Pacific: A Study in Railroad Promotion* (New York: Arno Press, 1981), pp. 73–85; Alan W. Farley, "Samuel Hallett and the Union Pacific Railway Company in Kansas," *Kansas Historical Quarterly* 25 (1959): 1–6.

13. Farley, "Samuel Hallett," pp. 1–6; John D. Cruise, "Early Days on the Union Pacific," *Kansas Historical Collections, 1909–1910* 11 (1910): 536. Business leaders in Kansas City, Missouri, also probably influenced Hallett's decision to move his company to Wyandotte. On September 14, 1863, members of Kansas City's city council offered him control of a company they had chartered to build what proved to be a critical connector railroad from their community to the Hannibal and St. Joseph line at Cameron. Because Hallett died the next year, this control was never realized (see Charles N. Glaab, *Kansas City and the Railroads: Community Policy in the Growth of a Regional Metropolis* [Madison: State Historical Society of Wisconsin, 1962], pp. 116–17).

14. Farley, "Samuel Hallett," pp. 1–16.

15. Petrowski, *Kansas Pacific*, p. 80; Albert R. Greene, "The Kansas River—Its Navigation," *Transactions of the Kansas State Historical Society* 9 (1905–1906): 353–58.

16. Farley, "Samuel Hallett," pp. 5–6. Farley quoted the description of Wyandotte from the October 1862 issue of the *Congregational Record*, a Lawrence publication.

17. Cruise, "Early Days," p. 534; John Speer, *Life of Gen. James H. Lane* (Garden City, Kans.: John Speer, 1896), pp. 272–75.

18. North Lawrence Civic Association, "Early History of North Lawrence" (mimeographed, 1930), pp. 8–16; William G. Cutler, ed., *History of the State of Kansas* (Chicago: A. T. Andreas, 1883), p. 544.

19. James R. Shortridge, *Our Town on the Plains: J. J. Pennell's Photographs of Junction City, Kansas, 1893–1922* (Lawrence: University Press of Kansas, 2000), pp. 28–29; John B. Jeffries, "An Early History of Junction City, Kansas: The First Generation" (master's thesis, Kansas State University, 1963), pp. 55–63, 130–33.

20. Cronon, *Nature's Metropolis*, pp. 295–309.

21. Petrowski, *Kansas Pacific*, pp. 66, 96.

22. Ibid., pp. 96, 136–52. The junction of the Kansas and Nebraska railroads was eventually made at Cheyenne, Wyoming, and not Utah.

23. Ibid., pp. 147–57; Shortridge, *Our Town on the Plains*, pp. 29–30.

24. Petrowski, *Kansas Pacific*, pp. 96–97; Arthur M. Johnson and Barry E. Supple,

Boston Capitalists and Western Railroads (Cambridge, Mass.: Harvard University Press, 1967), pp. 223–34.

25. Gates, *Fifty Million Acres,* pp. 136–40; George L. Anderson, "Atchison and the Central Branch Country, 1865–1874," *Kansas Historical Quarterly* 28 (1962): 3–4.

26. Petrowski, *Kansas Pacific,* pp. 149–50.

27. Anderson, "Atchison and the Central Branch Country," p. 4; Petrowski, *Kansas Pacific,* p. 157.

28. Anderson, "Atchison and the Central Branch Country," pp. 6–7, 9. The phrase "Great Railroad Centre of Kansas" is closely associated with John A. Martin, editor of the *Atchison Champion* and the chief booster of the city at this time. The Lawrence writer's name is unknown. His article in the *Lawrence Journal* was reprinted in the *Atchison Champion* on May 17, 1867.

29. Quastler, *Railroads of Lawrence,* pp. 72–73.

30. Ibid., pp. 90–101.

31. Imre E. Quastler, "Charting a Course: Lawrence, Kansas, and Its Railroad Strategy, 1854–1872," *Kansas History* 18 (1995): 27–30.

32. Glenn D. Bradley, *The Story of the Santa Fe* (Boston: Richard G. Badger, 1920), pp. 54–61.

33. Ibid., pp. 68–80.

34. V. V. Masterson, *The Katy Railroad and the Last Frontier* (Norman: University of Oklahoma Press, 1952), pp. 11–14.

35. Eugene T. Wells, "St. Louis and Cities West, 1820–1888: A Study in Historical Geography" (Ph.D. diss., University of Kansas, 1951), pp. 578–87; *Kansas Press,* December 5, 1859.

36. Masterson, *Katy Railroad,* pp. 17–22.

37. Glaab, *Kansas City and the Railroads,* pp. 125–37.

38. H. Craig Miner, "Border Frontier: The Missouri River, Fort Scott & Gulf Railroad in the Cherokee Neutral Lands, 1868–1870," *Kansas Historical Quarterly* 35 (1969): 105–6; Miner, "The Kansas and Neosho Valley: Kansas City's Drive for the Gulf," *Journal of the West* 17 (October 1978): 75–85.

39. Glaab, *Kansas City and the Railroads,* pp. 138–68.

40. Ibid., pp. 125–68; George L. Anderson, "Atchison, 1865–1886, Divided and Uncertain," *Kansas Historical Quarterly* 35 (1969): 30–45; Quastler, "Charting a Course," pp. 18–33; David G. Taylor, "Boom Town Leavenworth: The Failure of a Dream," *Kansas Historical Quarterly* 38 (1972): 389–415.

41. Glaab, *Kansas City and the Railroads,* pp. 43–46.

42. Ibid., pp. 138–68; Miner, "Kansas and Neosho Valley," pp. 82–83.

43. Miner, "Kansas and Neosho Valley," pp. 82–83; Anderson, "Atchison, 1865–1886," p. 32; Quastler, *Railroads of Lawrence,* pp. 174–75. Also in Kansas City by 1870 were the Missouri Valley Railway, whose line extended through St. Joseph to Council Bluffs, Iowa, and the North Missouri Railway that connected to St. Louis on the north side of the Missouri River.

44. John C. Hudson, *Plains Country Towns* (Minneapolis: University of Minnesota Press, 1985).

45. Masterson, *Katy Railroad,* p. 28.

46. Ibid., pp. 51–54, 116, 177; Harold J. Henderson, "The Building of the First Kansas Railroad South of the Kaw River," *Kansas Historical Quarterly* 15 (1947): 236; Miner, "Border Frontier," pp. 105–29; Quastler, *Railroads of Lawrence,* p. 170.

47. Cutler, *History of the State,* pp. 607, 609; B. Smith Haworth, *Ottawa University: Its History and Its Spirit* (Ottawa: Ottawa University, 1957), pp. 6–11.

48. Robert E. Hosack, "Chanute—The Birth of a Town," *Kansas Historical Quarterly* 41 (1975): 468–87.

49. Ibid.

50. Cutler, *History of the State,* pp. 827–28, 837; Harvey F. Nelson, "An Economic History of Chanute, Kansas" (master's thesis, University of Kansas, 1939), pp. 11–29.

51. Jimmy D. Skaggs, *The Cattle-Trailing Industry* (Lawrence: University Press of Kansas, 1973), pp. 103–4.

52. Cutler, *History of the State,* p. 1161; Miner, "Border Frontier," pp. 127–28; Masterson, *Katy Railroad,* pp. 58–68.

53. Cutler, *History of the State,* p. 1161; Claude H. Nichols, comp., *The Baxter Springs Story* (Baxter Springs, Kans.: Baxter Springs Centennial, 1958).

54. Cutler, *History of the State,* p. 1565.

55. Ibid., pp. 1565, 1581.

56. Ibid., p. 1574; L. Wallace Duncan, comp., *History of Montgomery County, Kansas* (Iola, Kans.: L. Wallace Duncan, 1903), p. 128.

57. Masterson, *Katy Railroad,* pp. 27–28. The importance that officials attached to the decision to build into Missouri is reflected in their choice for a new company name—the Missouri, Kansas and Texas Railroad—at this same time.

58. Cutler, *History of the State,* p. 1473.

59. Ibid., pp. 1473–74.

60. Ibid., p. 1456; Masterson, *Katy Railroad,* pp. 79, 84.

61. Masterson, *Katy Railroad,* pp. 96, 153; Cutler, *History of the State,* p. 1457; *Parsons: Its Past, Present and Future (Special Edition)* (Parsons, Kans.: Parsons Palladium, October 16, 1901), p. 1.

62. Masterson, *Katy Railroad,* pp. 96–97; Cutler, *History of the State,* p. 1465; Mrs. Tommie J. Crispino, ed., *The Centennial Story of Parsons, Kansas* (Parsons, Kans.: n.p., 1971), p. 57.

63. Mary L. Barlow, comp., *The Why of Fort Scott* (n.p., 1921), pp. 9–57.

64. Leo E. Oliva, *Fort Scott: Courage and Conflict on the Border* (Topeka: Kansas State Historical Society, 1984), p. 67; Cutler, *History of the State,* pp. 1073, 1075; William G. Calhoun, comp., *Fort Scott: A Pictorial History* (Fort Scott, Kans.: Historic Preservation Association of Bourbon County, 1978), pp. 14, 26.

65. Cutler, *History of the State,* p. 1072; *Fort Scott Monitor,* October 6, 1869. The decision of Levi Parsons to cross the border-tier tracks at Fort Scott, rather than elsewhere, was determined largely by his earlier choice to develop the town of Parsons as his headquarters instead of Chetopa. Had he chosen Chetopa, the road to Sedalia likely would have taken a different angle and intersected the border-tier line farther south, at Girard, the seat of Crawford County. Girard has better coal than Fort Scott, and its citizens actively courted Mr. Parsons for the prize. Girard was a company town of James Joy and other border-tier officials, however, and this factor may have dissuaded Parsons. Had the selection been in Girard's favor, the future urban development of southeastern Kansas almost certainly would have been different. Fort Scott would have been smaller, and Pittsburg probably never founded. For more on Girard, see Edward F. Keuchel Jr., "The Railroad System of Fort Scott: A Study of the Impact of Railroad Development upon a Southeastern Kansas Community, 1867–1883" (master's thesis, University of Kansas, 1961), pp. 80–95.

66. Charles W. Goodlander, *Early Days of Fort Scott: Memoirs and Recollections* (Fort Scott, Kans.: Monitor Printing Company, 1900), p. 129; Keuchel, "Railroad System," pp. 22, 29, 72, 84–85.

67. Keuchel, "Railroad System," pp. 85–87, 107. The poem appeared in the *Fort Scott Monitor* on January 1, 1871. It was unsigned but has been attributed to Ware by historian James C. Malin (see Keuchel, "Railroad System," p. 87).

68. Ibid., pp. 101–9.

69. Gates, *Fifty Million Acres*, p. 251; Petrowski, *Kansas Pacific*, pp. 144, 177; George L. Anderson, *Kansas West: An Epic of Western Railroad Building* (San Marino, Calif.: Golden West Books, 1963), pp. 69–79.

70. Anderson, *Kansas West*, p. 27; Leo E. Oliva, *Fort Wallace: Sentinel on the Smoky Hill Trail* (Topeka: Kansas State Historical Society, 1998), pp. 6–15.

71. Oliva, *Fort Wallace*, pp. 12–13, 44.

72. Thelma J. Curl, "Promotional Efforts of the Kansas Pacific and the Santa Fe to Settle Kansas" (master's thesis, University of Kansas, 1961), pp. 40–48.

73. *Leavenworth Daily Conservative,* January 24, 1867 (quoted in Joseph W. Snell and Robert W. Richmond, "When the Union and Kansas Pacific Built through Kansas," *Kansas Historical Quarterly* 32 [1966]: 335).

74. Cutler, *History of the State,* pp. 697–98.

75. Ibid., pp. 700–701; Harry Hughes and Helen C. Dingler, *From River Ferries to Interchanges: A Brief History of Saline County, Kansas from the 1850's to the 1980's* (Salina, Kans.: Harry Hughes and Helen C. Dingler, 1988), pp. 15, 49–55; *Junction City Weekly Union,* November 30, 1867 (quoted in Snell and Richmond, "When the Union," p. 337).

76. Cutler, *History of the State,* pp. 1274–77; Robert R. Dykstra, *The Cattle Towns* (New York: Atheneum, 1979), p. 36.

77. Curl, "Promotional Efforts," pp. 73–79, 154–55; H. Craig Miner, *West of Wichita: Settling the High Plains of Kansas, 1865–1890* (Lawrence: University Press of Kansas, 1986), pp. 38–51.

78. The Richardson comments, from a letter written July 21, 1872, are cited in Miner, *West of Wichita,* pp. 39–40.

79. Cutler, *History of the State,* pp. 1284–86.

80. Leo E. Oliva, *Fort Hays: Keeping Peace on the Plains* (Topeka: Kansas State Historical Society, 1980), pp. 2–19.

81. Ibid.; Cutler, *History of the State,* p. 1291.

82. Oliva, *Fort Hays,* pp. 35–40; Cutler, *History of the State,* pp. 1291–92; *Lawrence Kansas Weekly Tribune,* June 8, 1868 (quoted in Snell and Richmond, "When the Union," p. 346).

83. Cutler, *History of the State,* p. 1290; James R. Shortridge, *Peopling the Plains: Who Settled Where in Frontier Kansas* (Lawrence: University Press of Kansas, 1995), pp. 101–4.

84. Oliva, *Fort Wallace,* p. 41; Cutler, *History of the State,* p. 1297; Snell and Richmond, "When the Union," p. 347.

85. Oliva, *Fort Wallace,* pp. 31–41. The quotation is attributed by Oliva to a Missouri newspaper correspondent who visited the fort in May 1870.

86. Ibid., pp. 31, 122.

87. Cutler, *History of the State,* pp. 706–7, 1007, 1294; Hughes and Dingler, *From River Ferries,* p. 56; Jennie Martin, *A Brief History of the Early Days of Ellis, Kansas* (n.p., 1903), pp. 7–8; Maureen Winter, ed., *Indians to Industry: A History of Hays and Ellis County* (Ellis, Kans.: Ellis County Star, 1967), pp. 90–92.

88. Dykstra, *Cattle Towns*, pp. 15–21.

89. Ibid.; Joseph G. McCoy, *Historic Sketches of the Cattle Trade of the West and Southwest* (Glendale, Calif.: Arthur H. Clark, 1940), pp. 113–16.

90. McCoy, *Historic Sketches,* pp. 116–20; Cutler, *History of the State,* p. 688; George L. Cushman, "Abilene, First of the Kansas Cow Towns," *Kansas Historical Quarterly* 9 (1940): 240–58.

91. Cushman, "Abilene," pp. 240–58; Dykstra, *Cattle Towns,* p. 358.

92. Dykstra, *Cattle Towns,* pp. 37–40; Cutler, *History of the State,* p. 1280.

93. Joseph W. Snell and Don W. Wilson, "The Birth of the Atchison, Topeka and Santa Fe Railroad," *Kansas Historical Quarterly* 34 (1968): 117.

94. Cutler, *History of the State,* p. 1555; Snell and Wilson, "Birth of the Atchison," pp. 130, 134.

95. D. Lane Hartsock, "The Impact of the Railroads on Coal Mining in Osage County, 1869–1910," *Kansas Historical Quarterly* 37 (1971): 431.

96. Ibid., p. 433; Oscar A. Copple and Joyce M. Hitchings, *History of Osage City, and Osage County* (n.p., 1970), pp. 21–22; Cutler, *History of the State,* pp. 1542–43, 1555.

97. Cutler, *History of the State,* p. 1542.

98. Ibid, p. 846; George A. Hamm, "The Atchison Associates of the Santa Fe Railroad," *Kansas Historical Quarterly* 42 (1976): 353; Snell and Wilson, "Birth of the Atchison," pp. 118–19, 141, 330.

99. William T. Moran, *Santa Fe and the Chisholm Trail at Newton* (Newton, Kans.: Moran, 1970), pp. 9–10.

100. Ibid., pp. 26, 43, 50; Snell and Wilson, "Birth of the Atchison," p. 336.

101. The favorable financial news that accompanied Newton's creation perhaps explains the town's name. It honors the Massachusetts home of several major railroad investors. See Sondra Van Meter and Jan Hults, *1001 Kansas Place Names* (Lawrence: University Press of Kansas, 1989), p. 143.

102. Dykstra, *Cattle Towns,* pp. 48–50; L. L. Waters, *Steel Trails to Santa Fe* (Lawrence: University of Kansas Press, 1950), p. 47.

103. Dykstra, *Cattle Towns,* p. 50; Moran, *Santa Fe,* p. 42.

104. Cutler, *History of the State,* pp. 772–73; Dykstra, *Cattle Towns,* p. 52; Moran, *Santa Fe,* p. 67.

105. Moran, *Santa Fe,* pp. 80–82, 94; *Fiftieth Anniversary Number* (Newton, Kans.: Newton Kansan, August 22, 1922), pp. 42–44.

106. The story of Newton's loss of the railroad shops and division point in 1879 has an ironic ending, because through another economic adjustment by Santa Fe officials, the town was able to reclaim both of these prizes in 1897. As I will discuss in detail later in the text, a relocation of a portion of the railroad's main line in 1886 left the big-bend section of the tracks in a bypassed position. This isolated section included Nickerson, the replacement site for the shops. Railroad people then privately approached Newton leaders with a proposal to return the facilities to the city if a better water supply could be procured. This opportunity prompted local fund-raising and then the drilling of a well field in the Little Arkansas Valley near Halstead. The Santa Fe furnished the right-of-way for a pipeline back to Newton, and in exchange, the city council provided building lots for transferred employees and free water to the company for seven years. See *Fiftieth Anniversary Number,* pp. 42–43.

107. Dykstra, *Cattle Towns,* pp. 41–42, 55; H. Craig Miner, *Wichita: The Early Years, 1865–80* (Lincoln: University of Nebraska Press, 1982), pp. 1–17.

108. Miner, Wichita: *The Early Years,* pp. 19–49.
109. Ibid., pp. 19–49, 141; Dykstra, *Cattle Towns,* pp. 53–55.
110. Snell and Wilson, "Birth of the Atchison," pp. 347–48, 354.
111. Cutler, *History of the State,* pp. 1371–72.
112. Ibid.; Snell and Wilson, "Birth of the Atchison," p. 337.
113. Dykstra, *Cattle Towns,* p. 56; Myron C. Burr and Elizabeth Burr, eds., *The Kinsley-Edwards County Centennial, 1873–1973* (Kinsley, Kans.: Nolan Publishers, 1973), p. 4; Snell and Wilson, "Birth of the Atchison," p. 348; Cutler, *History of the State,* pp. 765, 1368.
114. C. Robert Haywood, *The Merchant Prince of Dodge City: The Life and Times of Robert M. Wright* (Norman: University of Oklahoma Press, 1998), p. 38; Frederic R. Young, *Dodge City: Up through a Century in Story and Pictures* (Dodge City, Kans.: Boot Hill Publications, 1972), pp. 19–20.
115. Haywood, *Merchant Prince,* pp. 37–38.
116. Robert M. Wright, *Dodge City, the Cowboy Capital, and the Great Southwest* (Wichita: Wichita Eagle Press, 1913), pp. 76–77; Dykstra, *Cattle Towns,* pp. 60–61, 79; Cutler, *History of the State,* p. 1560; C. Robert Haywood, *Trails South: The Wagon-Road Economy in the Dodge City–Panhandle Region* (Norman: University of Oklahoma Press, 1986).
117. Waters, *Steel Trails to Santa Fe,* p. 156; Agnesa Reeve, *Constant Frontier: The Continuing History of Finney County, Kansas* (Garden City, Kans.: Finney County Historical Society, 1996), pp. 11, 14. The community of Sherlock was later renamed Holcomb.

5. Later Railroads and Railroad Towns, 1877–1910

1. Glenn D. Bradley, *The Story of the Santa Fe* (Boston: Richard G. Badger, 1920), p. 140; Joseph W. Snell and Don W. Wilson, "The Birth of the Atchison, Topeka and Santa Fe Railroad," *Kansas Historical Quarterly* 34 (1968): 355.
2. Bradley, *Story of the Santa Fe,* pp. 142–43; Imre E. Quastler, *The Railroads of Lawrence, Kansas, 1854–1900* (Lawrence, Kans.: Coronado Press, 1979), pp. 226–27, 258–65.
3. Bradley, *Story of the Santa Fe,* pp. 228, 235.
4. H. Craig Miner, *The St. Louis–San Francisco Transcontinental Railroad: The Thirty-fifth Parallel Project, 1853–1890* (Lawrence: University Press of Kansas, 1972), pp. 5–13, 40–66, 94–105.
5. Ibid., pp. 107–13.
6. Ibid.
7. Ibid., pp. 113–21.
8. William G. Cutler, ed., *History of the State of Kansas* (Chicago: A. T. Andreas, 1883), p. 904; William E. Bain, *Frisco Folks: Stories and Pictures of the Great Steam Days of the Frisco Road (St. Louis–San Francisco Railway Company)* (Denver: Sage Books, 1961), pp. 26, 36, 69; T. F. Rager and John S. Gilmore, *History of Neosho and Wilson Counties, Kansas* (Fort Scott, Kans.: L. Wallace Duncan, 1902), p. 907.
9. Bradley, *Story of the Santa Fe,* pp. 227–28, 236.
10. Ibid., pp. 228–32; Quastler, *Railroads of Lawrence,* p. 269; Robert R. Dykstra, *The Cattle Towns* (New York: Atheneum, 1979), p. 70.
11. Bradley, *Story of the Santa Fe,* pp. 145–46, 248; Keith L. Bryant Jr., *History of the Atchison, Topeka and Santa Fe Railway* (New York: Macmillan, 1974), pp. 192–93. The point of departure for the branch toward New Mexico apparently was never debated in the Santa Fe boardrooms, but that of the line to northern Texas was. With the acquisition of the old LLG system in 1880, that company's border city of Coffeyville became a possible choice instead of Arkansas City. Arkansas City lay almost directly north of the

stipulated destination in Denison, however, whereas a route from Coffeyville would have had to be angled and longer. A track from Arkansas City also would be farther away from the existing Katy railroad line and therefore better able to control a sizable trade area of its own in Indian Territory once that area would be opened for general settlement. See L. L. Waters, *Steel Trails to Santa Fe* (Lawrence: University of Kansas Press, 1950), p. 77.

12. Cutler, *History of the State*, pp. 1496–97.

13. Ibid., pp. 1597–98; James F. Clough, *The Story of Arkansas City* (Arkansas City, Kans.: n.p., 1961); Mrs. Bennett Rinehart, *Blaze Marks on the Border: The Story of Arkansas City, Kansas, Founded 1870–1871* (North Newton, Kans.: Mennonite Press, 1970), p. 168.

14. Eldie F. Caldwell, ed., *Illustrated Southern Kansas*, rev. ed. (Lawrence, Kans.: E. F. Caldwell, 1886), pp. 35–46; Winfield Commercial Club, *Souvenir of Winfield, Cowley County, Kansas, 1904* (Winfield, Kans.: n.p., 1904), p. 3E. The western segment of the Kansas City, Lawrence and Southern Railroad was also known as the Southern Kansas and Western Railroad.

15. Nelson Trottman, *History of the Union Pacific: A Financial and Economic Survey* (New York: Ronald Press, 1923), p. 120; William R. Petrowski, *The Kansas Pacific: A Study in Railroad Promotion* (New York: Arno Press, 1981), pp. 248–49.

16. Petrowski, *Kansas Pacific*, pp. 177–78; George L. Anderson, *Kansas West: An Epic of Western Railroad Building* (San Marino, Calif.: Golden West Books, 1963), pp. 70–79.

17. Cutler, *History of the State*, pp. 1003, 1007.

18. James R. Shortridge, *Our Town on the Plains: J. J. Pennell's Photographs of Junction City, Kansas, 1893–1922* (Lawrence: University Press of Kansas, 2000), pp. 31–33, 98. The reorganization of division points by the Union Pacific in 1889 that affected Brookville, Junction City, and Wamego left the facilities at Ellis and Kansas City intact.

19. Cutler, *History of the State*, pp. 700–702.

20. Ibid.; *Salina Facts* (Salina, Kans.: Salina Chamber of Commerce, 1930); Harry Hughes and Helen C. Dingler, *From River Ferries to Interchanges: A Brief History of Saline County, Kansas from the 1850's to the 1980's* (Salina, Kans.: Harry Hughes and Helen C. Dingler, 1988), p. 16.

21. Arthur M. Johnson and Barry E. Supple, *Boston Capitalists and Western Railroads* (Cambridge, Mass.: Harvard University Press, 1967), pp. 248–49.

22. Cutler, *History of the State*, pp. 1019–20, 1318.

23. Ibid., pp. 1016, 1018; *Hand-book of Concordia and Cloud County, Kansas* (Chicago: C. S. Burch Publishing Co., 1888), p. 3; Mrs. E. F. Hollibaugh, *Biographical History of Cloud County, Kansas* (n.p., 1903), pp. 150–56, 164, 173, 201–16; Janet P. Emery, *It Takes People to Make a Town: The Story of Concordia, Kansas: 1871–1971* (Salina, Kans.: Arrow Printing Company, 1970), pp. 2–3, 94–95; James R. Shortridge, *Peopling the Plains: Who Settled Where in Frontier Kansas* (Lawrence: University Press of Kansas, 1995), pp. 113–14.

24. Cutler, *History of the State*, pp. 1017, 1023–24; George L. Anderson, "Atchison, 1865–1886, Divided and Uncertain," *Kansas Historical Quarterly* 35 (1969): 39. Because the Central Branch and the Kansas Pacific came under joint ownership in 1879 with a purchase of Jay Gould, the Solomon Valley Branch was never extended beyond Beloit. This circumstance avoided the construction of parallel railroads along the Solomon similar to what had happened in the Republican Valley and therefore the emergence of a series of competing towns. See Johnson and Supple, *Boston Capitalists*, pp. 241–45.

25. James L. Ehernberger and Francis G. Gschwind, *Smoke above the Plains: Union Pacific, Kansas Division* (Calloway, Nebr.: E. and G. Publications, 1965), pp. 13–14.

26. Cutler, *History of the State*, pp. 917–19; Charles Brown, *Marysville as It Is* (Marysville, Kans.: Press of P. Springer, 1887), pp. 3–4; *Marshall County, A County of Wealth, Beauty, and Historic Interest (Special Issue)* (Topeka: Kansas Commercial News, January-February, 1902), pp. 1–3; *Seventieth Anniversary Edition* (Marysville, Kans.: Marshall County News, February 27, 1931).

27. John L. Kerr, *The Story of a Western Power—The Missouri Pacific: An Outline History* (New York: Railway Research Society, 1928), pp. 1–27; Robert E. Riegel, "The Missouri Pacific Railroad, 1879–1900," *Missouri Historical Review* 18 (1923–1924): 173–84.

28. Riegel, "Missouri Pacific," pp. 173–84; Atchison Merchants' and Manufacturers' Bureau, *The Advantages of Atchison, Kansas, as a Commercial and Manufacturing Centre* (Atchison, Kans.: Merchants' and Manufacturers' Bureau, 1887), pp. 16–17.

29. Anderson, "Atchison, 1865–1886," pp. 39–40.

30. *Miami County, Kansas 1987* (Paola, Kans.: Miami County Historical Society, 1987), p. 30; Anna January, *Historic Souvenir of Osawatomie and Environs* (Kansas City: R. P. Company, 1910).

31. A. Bower Sageser, "Building the Main Line of the Missouri Pacific through Kansas," *Kansas Historical Quarterly* 21 (1955): 326–30; *Salina Facts.*

32. Cutler, *History of the State*, p. 768; *Biographical History of Barton County, Kansas* (Great Bend, Kans.: Great Bend Tribune, 1912), pp. 194–98; *Fiftieth Anniversary Edition* (Hoisington, Kans.: Hoisington Dispatch, March 9, 1939), p. 1.

33. William E. Hayes, *Iron Road to Empire: The History of 100 Years of Progress and Achievements of the Rock Island Lines* (New York: Simmons-Broadman, 1953), pp. 72–114.

34. Ibid., p. 116; Anderson, "Atchison, 1865–1886," pp. 39–44.

35. Cutler, *History of the State*, pp. 711–12; A. N. Ruley, *A. N. Ruley's History of Brown County* (Hiawatha, Kans.: Hiawatha World, 1930), pp. 195–96.

36. Oliver P. Byers, "Early History of the El Paso Line of the Chicago, Rock Island and Pacific Railway," *Kansas Historical Collections, 1919–1922* 15 (1923): 573; History and Literature Club, *History of Horton and Surrounding Neighborhoods* (Horton, Kans.: Horton Pride, 1974), pp. 17–18; Ollie Krebs, ed., *Horton, Kansas, 1886–1986* (Horton, Kans.: Horton Headlight, 1986), pp. 3, 33.

37. Hayes, *Iron Road to Empire*, p. 117.

38. I. O. Savage, *A History of Republic County, Kansas* (Beloit, Kans.: Jones and Chubbic, 1901), pp. 81–85; *New Year's Edition* (Belleville, Kans.: Belleville Telescope, December 29, 1905), pp. 16–17; Ernest W. Burgess and J. J. Sippy, *Belleville Social Survey* (Belleville, Kans.: Telescope Print, 1914), p. 16.

39. *Progress and Historical Edition* (Phillipsburg, Kans.: Phillips County Review, May, 1952), sec. 1, p. 8.

40. Book Committee, *They Came to Stay: Sherman County and Family History* (Goodland, Kans.: Sherman County Historical Society, 1980), vol. 1, pp. 146–49, 269.

41. Byers, "Early History," pp. 574, 577–78; Hayes, *Iron Road to Empire*, p. 117.

42. For slight variations on the Herington-Low encounter, see Byers, "Early History," p. 574; *Portrait and Biographical Record of Dickinson, Saline, McPherson, and Marion Counties* (Chicago: Chapman Bros., 1893), pp. 163–64; *Fiftieth Anniversary Edition* (Herington, Kans.: Herington Times-Sun, August 24, 1937), sec. B, pp. 1–2; and Karen Edson, ed., *Herington: A Century of Pride* (North Newton, Kans.: Mennonite Press, 1987).

43. H. Craig Miner, *Wichita: The Magic City* (Wichita: Wichita-Sedgwick County Historical Museum, 1988), p. 58.

44. Byers, "Early History," p. 574; Bradley, *Story of the Santa Fe*, p. 254.

45. *Hutchinson, the Salt City in the Heart of the Great Kansas Wheat Belt* (Hutchinson: News Company, 1908); Willard Welsh, *Hutchinson: A Prairie City in Kansas* (Wichita: McCormick-Armstrong Company, 1946), pp. 18, 65–66.

46. Byers, "Early History," p. 574; *Pride, Progress Edition* (Pratt, Kans.: Pratt Daily Tribune, 1954), pp. 46, 82.

47. Byers, "Early History," p. 574.

48. Ibid.; *Kansas Centennial Futurama Edition* (Liberal, Kans.: Southwest Daily Times, March 18, 1961), sec. 7, p. 1.

49. Hayes, *Iron Road to Empire,* pp. 151–64, 213.

50. Waters, *Steel Trails to Santa Fe,* p. 252.

6. Mining, Irrigation, and Newer Institutional Towns, 1876–1950

1. Walter H. Schoewe, "The Geography of Kansas, Part IV, Economic Geography: Mineral Resources," *Transactions of the Kansas Academy of Science* 61 (1958): 359–60.

2. Paul W. Gates, *Fifty Million Acres: Conflicts over Kansas Land Policy: 1854–1890* (Ithaca, N.Y.: Cornell University Press, 1954), pp. 153–93.

3. Elmer Coe, *Fort Scott as I Knew It* (Fort Scott, Kans.: Monitor Binding and Printing Company, 1940), p. 86; Schoewe, "Geography of Kansas, Part IV," pp. 370–71, 375–78.

4. William E. Powell, "Coal and Pioneer Settlement in Southeasternmost Kansas," *Ecumene* 9 (1970): 10; William G. Cutler, ed., *History of the State of Kansas* (Chicago: A. T. Andreas, 1883), p. 1121.

5. Arrell M. Gibson, *Wilderness Bonanza: The Tri-State District of Missouri, Kansas, and Oklahoma* (Norman: University of Oklahoma Press, 1972), pp. 3–40; H. Craig Miner, *The St. Louis–San Francisco Transcontinental Railroad: The Thirty-fifth Parallel Project, 1853–1890* (Lawrence: University Press of Kansas, 1972), pp. 47–66; Federal Writers' Project, Works Progress Administration, *Missouri: A Guide to the "Show Me" State* (New York: Duell, Sloan and Pearce, 1941), pp. 235–37.

6. Cutler, *History of the State,* p. 1128.

7. Gibson, *Wilderness Bonanza,* pp. 21, 34, 123–24.

8. Fred N. Howell, "Some Phases of the Industrial History of Pittsburg, Kansas," *Kansas Historical Quarterly* 1 (1932): 281–82; Gibson, *Wilderness Bonanza,* p. 37.

9. A. J. Georgia et al., *A Twentieth Century History and Biographical Record of Crawford County, Kansas* (Chicago: Lewis Publishing Company, 1905), p. 117; Howell, "Some Phases," p. 283; Frederick N. Howell, "Pittsburg, Kansas, and Its Industries" (master's thesis, University of Kansas, 1930), p. 55.

10. Ibid., p. 94; Georgia, *Twentieth Century History,* pp. 110–14.

11. William E. Powell, "The Cherokee-Crawford Coal Field of Southeastern Kansas: A Study in Sequent Occupance," *Midwest Quarterly* 22 (1981): 123–24; Powell, "European Settlement in the Cherokee-Crawford Coal Field of Southeastern Kansas," *Kansas Historical Quarterly* 41 (1975): 150–55; Howell, "Some Phases," pp. 289–94; Federal Writers' Project, Works Progress Administration, *The Pittsburg Guide* (Pittsburg, Kans.: Pittsburg Chamber of Commerce, 1941), p. 18.

12. Federal Writers' Project, *Pittsburg Guide,* p. 11; *Pittsburg Diamond Jubilee 1876–1951 (Official Program)* (Pittsburg, Kans.: n.p., 1951), pp. 32–33.

13. Howell, "Pittsburg, Kansas," p. 130; Federal Writers' Project, *Pittsburg Guide,* p. 21; William T. Bawden, *A History of Kansas State Teachers College of Pittsburg, 1903–1941* (Pittsburg, Kans.: Kansas State Teachers College, 1952).

14. Howell, "Some Phases," p. 284.

15. Cutler, *History of the State,* pp. 1151–52, 1165–67; Pens-Lens View Company, *Pens-Lens Views of the Galena-Empire City Mining Camp* (Carthage, Mo.: Pens-Lens View Company, 1899), p. 28; Nathaniel T. Allison, *History of Cherokee County, Kansas, and Representative Citizens* (Chicago: Biographical Publishing Company, 1904), pp. 159–62; Irene G. Stone, "The Lead and Zinc Field of Kansas," *Kansas Historical Collections, 1901–1902* 7 (1902): 243–60.

16. Gibson, *Wilderness Bonanza,* pp. 28, 37; Claude H. Nichols, comp., *The Baxter Springs Story* (Baxter Springs, Kans.: Baxter Springs Centennial, 1958).

17. Cutler, *History of the State,* p. 1152; Allison, *History of Cherokee County,* pp. 116, 124, 173–74; Georgia, *Twentieth Century History,* p. 46.

18. Robert Taft, "Kansas and the Nation's Salt," *Transactions of the Kansas Academy of Science* 49 (1946): 258–62; Frank Vincent, "History of Salt Discovery and Production in Kansas, 1887–1915," *Kansas Historical Collections, 1915–1918* 14 (1918): 358–59; *Hutchinson Daily News,* September 28, 1887.

19. Taft, "Kansas and the Nation's Salt," pp. 237–41, 265; Schoewe, "Geography of Kansas, Part IV," pp. 438–40.

20. Vincent, "History of Salt," pp. 358–78; Taft, "Kansas and the Nation's Salt," p. 266.

21. Vincent, "History of Salt," pp. 362–65; *Hutchinson, the Salt City in the Heart of the Great Kansas Wheat Belt* (Hutchinson, Kans.: News Company, 1908); Taft, "Kansas and the Nation's Salt," pp. 269–70.

22. Taft, "Kansas and the Nation's Salt," pp. 269–70; *Hutchinson, the Salt City;* Schoewe, "Geography of Kansas, Part IV," p. 440.

23. Cutler, *History of the State,* pp. 676–77; L. Wallace Duncan and Charles F. Scott, eds., *History of Allen and Woodson Counties, Kansas* (Iola, Kans.: Iola Register, 1901), p. 56.

24. Earl K. Nixon, "The Petroleum Industry of Kansas," *Transactions of the Kansas Academy of Science* 51 (1948): 381–86; H. Craig Miner, *Discovery! Cycles of Change in the Kansas Oil and Gas Industry, 1860–1987* (Wichita: Kansas Independent Oil and Gas Association, 1987), pp. 13–23.

25. Miner, *Discovery!* pp. 29–40, 57–77.

26. Ibid., pp. 79–95.

27. Ibid., pp. 109–13; Angelo Scott, "How Natural Gas Came to Kansas," *Kansas Historical Quarterly* 21 (1954): 233–46; *Third Industrial Edition* (Independence, Kans.: South Kansas Tribune, December 30, 1903).

28. Duncan and Scott, *History of Allen,* pp. 57–58; John G. Clark, *Towns and Minerals in Southeastern Kansas: A Study in Regional Industrialization, 1890–1930,* Special Distribution Publication No. 52 (Lawrence: State Geological Survey of Kansas, 1970), pp. 23–24.

29. Clark, *Towns and Minerals,* pp. 26, 65; Duncan and Scott, *History of Allen,* p. 58.

30. Clark, *Towns and Minerals,* p. 51.

31. *Centennial Edition* (Iola, Kans.: Iola Register, May 30, 1955), pp. 1D–3D; Miner, *Discovery!* p. 51.

32. Duncan and Scott, *History of Allen,* p. 58.

33. L. Wallace Duncan, comp., *History of Montgomery County, Kansas* (Iola, Kans.: L. Wallace Duncan, 1903), pp. 49, 147; Clark, *Towns and Minerals,* pp. 47–57.

34. Glenn A. Bradley, *The Story of the Santa Fe* (Boston: Richard G. Badger, 1920), p. 236; *Pioneer Edition* (Chanute, Kans.: Chanute Tribune, June 24, 1941), p. A5; T. F. Rager and John S. Gilmore, *History of Neosho and Wilson Counties, Kansas* (Fort Scott, Kans.: L. Wallace Duncan, 1902), p. 66.

35. Rager and Gilmore, *History of Neosho and Wilson,* p. 93; William W. Graves, *History of Neosho County* (St. Paul, Kans.: Journal Press, 1951), vol. 2, pp. 971–72.

36. Rager and Gilmore, *History of Neosho and Wilson,* p. 66; Graves, *History of Neosho County,* vol. 2, pp. 974, 1004–15; Harvey F. Nelson, "An Economic History of Chanute, Kansas" (master's thesis, University of Kansas, 1939), pp. 64–76; Miner, *Discovery!* pp. 85–86.

37. Genevieve L. Choguill and Harold S. Choguill, *A History of the Humboldt, Kansas, Community: 1855–1988* (North Newton, Kans.: Mennonite Press, 1988), pp. 36–40.

38. Miner, *Discovery!* pp. 51–52.

39. Ibid., Charles C. Drake, *"Who's Who?": A History of Kansas and Montgomery County* (Coffeyville, Kans.: Coffeyville Journal Press, 1943), p. 107; Clark, *Towns and Minerals,* p. 38.

40. Clark, *Towns and Minerals,* pp. 38–39; Duncan, *History of Montgomery,* p. 132.

41. *Watch Coffeyville Grow* (Coffeyville, Kans.: Coffeyville Daily Journal, April 6, 1907), pp. 3, 26, 29–37; Clark, *Towns and Minerals,* p. 37; Coffeyville at 100, Inc., *A History of Coffeyville* (Coffeyville, Kans.: Coffeyville Journal Press, 1969), pp. 35–38.

42. Duncan, *History of Montgomery,* p. 46; Southwest Directory and Publicity Company, comp., *Independence: The Heart of the Kansas Gas and Oil Field* (Independence, Kans.: H. W. Young and Son, 1907), pp. 7–17; Amy L. Rork, "Sense of Place in Montgomery County, Kansas: Perceptions of an Industrialized Rural Area" (master's thesis, University of Kansas, 1997), pp. 22–23.

43. Miner, *Discovery!* pp. 106–8.

44. Nixon, "Petroleum Industry," p. 384.

45. Ibid., p. 386.

46. *An Illustrated Hand Book Compiled from the Official Statistics, Descriptive of Butler County, Kansas* (El Dorado, Kans.: Daily and Weekly Republican, 1887), pp. 33–34; Jessie P. Stratford, *Butler County's Eighty Years: 1855–1935* (El Dorado, Kans.: Butler County News, 1935), p. 142.

47. Miner, *Discovery!* pp. 121–23.

48. Ibid., pp. 123–36.

49. Vol P. Mooney, *History of Butler County, Kansas* (Lawrence, Kans.: Standard Publishing Company, 1916), p. 284; Miner, *Discovery!* pp. 118, 130–31. Miner attributes the quotation to an unnamed contemporary observer.

50. Nixon, "Petroleum Industry," p. 387; Miner, *Discovery!* p. 147.

51. *Progress Issue* (El Dorado, Kans.: El Dorado Times, February 25, 1964), p. 11.

52. Ibid.; Miner, *Discovery!* pp. 138, 165; Stratford, *Butler County's Eighty Years,* pp. 91–95, 113–14.

53. Miner, *Discovery!* p. 74.

54. Curtis Hoover, "'Black Gold' Brings Millions in New Wealth to Kansas," *Kansas Business* 2 (June 1934): 10–11; *Mid-century Resources Edition, 1900–1950* (Arkansas City, Kans.: Arkansas City Daily Traveler, 1950).

55. Orsemus H. Bentley, ed., *History of Wichita and Sedgwick County, Kansas* (Chicago: C. F. Cooper and Company, 1910), vol. 1, pp. 1, 17, 25, and vol. 2, pp. 696–700.

56. Miner, *Discovery!* pp. 130, 138–39; H. Craig Miner, *Wichita: The Magic City* (Wichita: Wichita-Sedgwick County Historical Museum, 1988), pp. 122, 154.

57. Lockwood, Green and Company, *Industrial Survey of Wichita, Kansas* (Boston: Lockwood, Green and Company, 1927), p. 71; Burt Doze, "Courageous Oil Men Un-

latched the Door When Opportunity Knocked," in *Wichita People* (Wichita: Wichita Chamber of Commerce, 1946), pp. 73–76.

58. Miner, *Discovery!* p. 131; Miner, *Wichita: The Magic City,* pp. 154–55.

59. Nixon, "Petroleum Industry," pp. 389–405; Schoewe, "Geography of Kansas, Part IV," pp. 392–99.

60. Hoover, "'Black Gold, p. 10; "Petroleum—A Major Kansas Industry," *Kansas Business* 3 (April 1935): 8; "Oil—Kansas' Greatest Industry," *Kansas Business* 4 (July 1936): 13.

61. "Oil—Kansas' Greatest Industry," pp. 13–15; Raymond L. Flory, *Historical Atlas of McPherson County* (McPherson, Kans.: McPherson County Historical Society, 1983), p. 66; Miner, *Discovery!* p. 174.

62. Berton Roueche, "Profiles: Wheat Country," *New Yorker,* January 3, 1983, p. 42.

63. *Great Bend, Kansas: A Historical Portrait of the City* (Great Bend, Kans.: Centennial Book Committee, 1972), pp. 14, 41, 53.

64. *Russell Record,* May 10, 1951.

65. Jessie H. Rowland, *Pioneer Days in McPherson* (McPherson, Kans.: McPherson Junior Chamber of Commerce, 1947), p. 23; *Kansas Community Profile: McPherson* (Topeka: Kansas Department of Commerce, 1995); *Phillips County Review,* July 7, 1949.

66. "Development of World's Largest Known Gas Reserve Pushes Ahead," *Kansas Business Magazine* 18 (October 1950): 9, 73–74; Edith C. Thomson, *History of Stevens County Kansas* (n.p.: Edith C. Thomson, 1967), pp. 82–83.

67. Schoewe, "Geography of Kansas, Part IV," pp. 401–9; *Kansas Centennial Futurama Edition* (Liberal, Kans.: Southwest Daily Times, March 18, 1961), sec. 8, p. 1.

68. "Southwest Kansas after Petro-Chemical Industry," *Kansas Business Magazine* 22 (May 1954): 22; Schoewe, "Geography of Kansas, Part IV," p. 401.

69. Thomson, *History of Stevens,* pp. 79–85; *The History of Stevens County and Its People* (Hugoton, Kans.: Stevens County Historical Society, 1979), pp. 138–44.

70. *Kansas Centennial Futurama Edition,* sec. 8, p. 1; sec. 10, p. 1; Pauline Toland, ed., *Seward County Kansas* (Liberal, Kans.: Seward County Historical Society, 1979), pp. 80–82.

71. Agnesa Reeve, *Constant Frontier: The Continuing History of Finney County, Kansas* (Garden City, Kans.: Finney County Historical Society, 1996), p. 14; Cutler, *History of the State,* p. 1616; Anne M. Marvin, "The Fertile Domain: Irrigation as Adaptation in the Garden City, Kansas, Area" (Ph.D. diss., University of Kansas, 1985), pp. 45–59; James E. Sherow, *Watering the Valley: Development along the High Plains Arkansas River, 1870–1950* (Lawrence: University Press of Kansas, 1990), p. 83.

72. Sherow, *Watering the Valley,* pp. 80–82.

73. Ibid., pp. 86–88; Marvin, "Fertile Domain," pp. 51–53.

74. Sherow, *Watering the Valley,* pp. 88–91; Reeve, *Constant Frontier,* p. 28.

75. Reeve, *Constant Frontier,* pp. 32, 70, 92; Sherow, *Watering the Valley,* p. 91.

76. *Biennial Report, Kansas Secretary of State, 1959–1960* (Topeka: State Printer, 1960), pp. 920, 940, 961; *Sixteenth Biennial Report of the State Home for Feeble-Minded, 1910–1912* (Topeka: State Printer, 1912), pp. 115, 122–23; *Winfield State Hospital and Training Center* (Winfield, Kans.: Winfield State Hospital and Training Center, 1982), pp. 1–2; *Twenty-second Biennial Report of the State Orphans' Home, 1928–1930* (Topeka: State Printer, 1931), p. 30; Catherine Roe and Bill Roe, comps. and eds., *Atchison Centennial, June 20–26, 1854–1954* (Atchison, Kans.: Lockwood Company, 1954), p. 18; Mrs. Tommie J. Crispino, ed., *The Centennial Story of Parsons, Kansas* (Parsons, Kans.: n.p., 1971),

p. 67; *Fourth Biennial Report of the State Hospital for Epileptics, 1908–1910* (Topeka: State Printer, 1910), pp. 4, 43.

77. *Report of the Board of Commissioners of the State Industrial Reformatory, 1886* (Topeka: State Printer, 1887), p. 3; Pat Mitchell, *The Fair City: Hutchinson, Kansas* (Topeka: Josten's, 1982); *Third Biennial Report of the Kansas State Industrial Reformatory, 1898–1900* (Topeka: State Printer, 1900), pp. 5, 12–13.

78. Maureen Winter, ed., *Indians to Industry: A History of Hays and Ellis County* (Ellis, Kans.: Ellis County Star, 1967), pp. 69–70; History Book Committee, *At Home in Ellis County, Kansas 1867–1992* (Hays, Kans.: Ellis County Historical Society, 1991), vol. 1, pp. 45–46, 223; Helen P. Harris, "Agriculture and Fort Hays State University," *Kansas History* 9 (1986): 164–67.

79. Harris, "Agriculture and Fort Hays," pp. 164–67; Lyman D. Wooster, *Fort Hays Kansas State College: An Historical Story* (Hays, Kans.: Fort Hays Kansas State College, 1961), p. 194.

80. Hazel B. Baker, "Early History of Larned State Hospital," manuscript, n.d., located in the Kansas Collection at the University of Kansas, pp. 1–3; David K. Clapsaddle, *Larned State Hospital: The First Fifty Years* (Larned, Kans.: Larned Tiller and Toiler, 1980), p. 7.

81. Clapsaddle, *Larned State Hospital*, p. 7; *First Biennial Report of the Larned State Hospital, 1912–1914* (Topeka: State Printer, 1914), pp. 3–6; *Fourteenth Biennial Report of the Larned State Hospital, 1938–1940* (Topeka: State Printer, 1940), p. 9.

82. *First Biennial Report of the State Tubercular Sanatorium, 1912–1914* (Topeka: State Printer, 1914), pp. 3–8.

83. Ibid.; Norton Centennial Board, *From a Covered Wagon to Community Pride: Norton Centennial, 1872–1972* (Norton, Kans.: n.p., 1972), pp. 35–36; *Ninth Biennial Report of the State Tubercular Sanatorium, 1928–1930* (Topeka: State Printer, 1931), p. 23.

7. Urban Consolidation in a Railroad Mode, 1880–1950

1. Daniel Fitzgerald, *Ghost Towns of Kansas: A Traveler's Guide* (Lawrence: University Press of Kansas, 1988); Fitzgerald, *Faded Dreams: More Ghost Towns of Kansas* (Lawrence: University Press of Kansas, 1994).

2. Edward Vernon, ed., *Travelers' Official Railway Guide of the United States and Canada* (New York: National Railway Publication Company, 1868). Later editions were entitled *Travelers' Official Railway Guide for the United States and Canada; Travelers' Official Guide of the Railway and Steam Navigation Lines; The Official Guide of the Railway and Steam Navigation Lines of the United States, Canada, and Mexico;* and then *The Official Guide of the Railways and Steam Navigation Lines of the United States, Porto Rico, Canada, Mexico, and Cuba.*

3. V. V. Masterson, *The Katy Railroad and the Last Frontier* (Norman: University of Oklahoma Press, 1952), p. 30.

4. Officials of the Chicago, Rock Island and Pacific Railroad, instead of constructing their own tracks from Topeka into Kansas City, leased usage rights from the Union Pacific. This arrangement, which began in 1887, reduced traffic on the Rock Island's line between Topeka and Horton. See Imre E. Quastler, *The Railroads of Lawrence, Kansas, 1854–1900* (Lawrence, Kans.: Coronado Press, 1979), pp. 324–25.

5. Masterson, *Katy Railroad*, p. 237. For Fort Scott, see Edward F. Keuchel Jr., "The Railroad System of Fort Scott: A Study of the Impact of Railroad Development upon a Southeastern Kansas Community, 1867–1883" (master's thesis, University of Kansas, 1961).

6. Chauncy D. Harris, "A Functional Classification of Cities in the United States," *Geographical Review* 33 (1943): 86–99, quotation on p. 87. Important commentary on the Harris system includes John W. Alexander, "The Basic-Nonbasic Concept of Urban Economic Functions," *Economic Geography* 30 (1954): 246–61; Victor Roterus and Wesley Calef, "Notes on the Basic-Nonbasic Employment Ratio," *Economic Geography* 31 (1955): 17–20; Howard J. Nelson, "A Service Classification of American Cities," *Economic Geography* 31 (1955): 189–210; and Gunnar Alexandersson, *The Industrial Structure of American Cities* (Lincoln: University of Nebraska Press, 1956), pp. 14–20.

7. Harris typically compared sector percentages either to the total employment in manufacturing, retailing, and wholesaling or to the total employment of all gainful workers. Neither total was consistently available to me for Kansas towns, however, and so I adjusted his guidelines to use total urban populations as a base. Because state censuses from this period show approximately a quarter of local populations to be employed outside the home, I multiplied Harris's gainful-work percentages by .25 to obtain my threshold points.

8. *Gazetteer and Directory of the State of Kansas* (Lawrence, Kans.: Blackburn and Company, 1870); Robert L. Polk, comp., *Kansas State Gazetteer and Business Directory* (n.p.: R. L. Polk and Company, 1880); Polk, comp., *Kansas State Gazetteer and Business Directory* (n.p.: R. L. Polk and Company, 1891).

9. G. K. Renner, "The Kansas City Meat Packing Industry before 1900," *Missouri Historical Review* 55 (1960–1961): 18.

10. Ibid., pp. 20–21; Edwin D. Shutt, "The Saga of the Armour Family in Kansas City, 1870–1900," *Heritage of the Great Plains* 23 (fall 1990): 25–27.

11. Renner, "Kansas City Meat Packing," pp. 21–22; Charles N. Glaab, *Kansas City and the Railroads: Community Policy in the Growth of a Regional Metropolis* (Madison: State Historical Society of Wisconsin, 1962), pp. 168–69; Eva L. Atkinson, "Kansas City's Livestock Trade and Packing Industry, 1870–1914: A Study in Regional Growth" (Ph.D. diss., University of Kansas, 1971), pp. 122, 278–84.

12. Renner, "Kansas City Meat Packing," pp. 23–25.

13. Ibid., pp. 26–29.

14. Ibid., p. 27; Atkinson, "Kansas City's Livestock Trade," pp. 228–40.

15. Shutt, "Saga of the Armour Family," p. 28; William G. Cutler, ed., *History of the State of Kansas* (Chicago: A. T. Andreas, 1883), p. 1242; *Wyandotte County and Kansas City, Kansas: Historical and Biographical* (Chicago: Goodspeed Publishing Company, 1890), vol. 1, p. 407.

16. *Kansas City Star,* May 27, 1892.

17. Kansas City Kansas Mercantile Club, *The Story of Three Years' Progress in Kansas City, Kansas* (Kansas City: Meseraull, 1909); Perl W. Morgan, ed. and comp., *History of Wyandotte County, Kansas and Its People* (Chicago: Lewis Publishing Company, 1911), vol. 1, p. 472.

18. *Thirteenth Census of the United States Taken in the Year 1910,* vol. 9, *Manufactories, 1909, Reports by States* (Washington, D.C.: General Printing Office, 1912), p. 374.

19. Atkinson, "Kansas City's Livestock Trade," pp. 277, 292–95; Morgan, *History of Wyandotte,* pp. 477, 480; *Thirty-fifth Annual Report of the Department of Labor and Industry, 1920* (Topeka: State Printer, 1921), p. 18.

20. *Thirty-fifth Annual Report of the Department of Labor,* p. 18.

21. Cutler, *History of the State,* pp. 1244–45; Margaret Landis, *The Winding Valley and the Craggy Hillside: A History of the City of Rosedale, Kansas* (Kansas City: Margaret Landis, 1976), pp. 11–15.

22. Cutler, *History of the State,* pp. 1245–46; Edwin D. Shutt, "'Silver City': A History of the Argentine Community of Kansas City, Kansas" (master's thesis, Emporia State College, 1976), pp. 21–38.

23. Ibid., pp. 36–52.

24. Ibid., pp. 21–22, 63–96.

25. Carl Cabe, *Flour Milling,* Kansas Industry Series No. 2 (Lawrence: School of Business, University of Kansas, 1958), pp. 14–23.

26. Ibid., pp. 22–30, 41–51; Alice Lanterman, "The Development of Kansas City as a Grain and Milling Center," *Missouri Historical Review* 42 (1947–1948): 22–26; *Thirty-fifth Annual Report of the Department of Labor,* pp. 18–19.

27. Cabe, *Flour Milling,* pp. 18–31; "Kansas Mills Her Golden Grain to Lead Nation in Flour Production," *Kansas Business* 5 (June 1937): 4; Lanterman, "Development of Kansas City," p. 32.

28. *Facts concerning Kansas City Kansas* (Kansas City: Kansas City Kansas Chamber of Commerce, n.d.); Joseph H. McDowell, *Building a City: A Detailed History of Kansas City, Kansas* (Kansas City: Kansas City Kansan, 1969), p. 30; Larry K. Hancks and Meredith Roberts, *Roots: The Historic and Architectural Heritage of Kansas City, Kansas* (Kansas City: City of Kansas City, Kansas, 1976), p. 166.

29. McDowell, *Building a City,* pp. 27–33; "The Kansas Invasion," *Kansas Business Magazine* 9 (July 1941): 12.

30. "Kansas City, Kansas: Great Industrial Center," *Kansas Business* 6 (June 1938): 5, 19–22; "Kansas City Kansas Chamber Responsible for Huge Industrial Development," *Kansas Business Magazine* 9 (July 1941): 7, 12, 14; "General Motors Preparing Assembly Line at Fairfax," *Kansas Business Magazine* 14 (September 1946): 10; "Fairfax Industrial District: One of Finest in Nation," *Kansas Business Magazine* 15 (January 1947): 6, 64–66; "Sunshine Biscuit Is Bringing Most Modern Plant to Kansas," *Kansas Business Magazine* 16 (January 1948): 14; W. H. Radford, "Fairfax District Built Where Lewis and Clark Had Early Camp," *Kansas Business Magazine* 16 (June 1948): 8, 10–12.

31. Harold Crimmins, *A History of the Kansas Central Railway, 1871–1935,* Research Studies vol. 2, no. 4 (Emporia: Kansas State Teachers College, 1954), pp. 3–5.

32. William S. Burke and J. L. Rock, *The History of Leavenworth: The Metropolis of Kansas, and the Chief Commercial Center West of the Missouri River* (Leavenworth, Kans.: Leavenworth Board of Trade, 1880), pp. 21–22.

33. Frank M. Gable, "The Kansas Penitentiary," *Kansas Historical Collections, 1915–1918* 14 (1918): 406–7; Theodore L. Heim, "The Administration and Organization of the Kansas State Penitentiary" (master's thesis, University of Kansas, 1960), pp. 294–96; Harvey Hougen, "The Impact of Politics and Prison Industry on the General Management of the Kansas State Penitentiary, 1883–1909," *Kansas Historical Quarterly* 43 (1977): 298; Burke and Rock, *History of Leavenworth,* pp. 35–36; Alexander Caldwell, "Kansas Manufactures and Mines," *Kansas Historical Collections, 1883–1885* 3 (1886): 455.

34. *The Leading Industries of Leavenworth, Kansas* (Leavenworth, Kans.: Commercial and Manufacturing Publishing Company, 1883), p. 5; Richard L. Douglas, "A History of Manufactures in the Kansas District," *Kansas Historical Collections, 1909–1910* 11 (1910): 117.

35. Burke and Rock, *History of Leavenworth,* p. 37; J. H. Johnston III, *Leavenworth: Beginning to Bicentennial* (Leavenworth, Kans.: J. H. Johnston III, 1976), p. 87; Caldwell, "Kansas Manufactures," p. 456.

36. Lela Barnes, "The Leavenworth Board of Trade," *Kansas Historical Quarterly* 1

(1932): 360–78; Burke and Rock, *History of Leavenworth,* pp. 22–23; Caldwell, "Kansas Manufactures," p. 455.

37. *Advertising Leavenworth (Special Issue)* (Leavenworth, Kans.: Leavenworth Post, December 30, 1913), p. 21; Burke and Rock, *History of Leavenworth,* p. 38; "Leavenworth Was Center of State's Activities for Many Years," *Kansas Business* 4 (April 1936): 5.

38. Cutler, *History of the State,* pp. 434, 456; "Missouri Valley Steel Floats Biggest Barge Ever on River," *Kansas Business Magazine* 17 (August 1949): 18.

39. *Advertising Leavenworth,* p. 3; *Thirty-fifth Annual Report of the Department of Labor,* pp. 19–20; "Goodjohn Sash and Door Firm Rolling Out Lumber Products," *Kansas Business Magazine* 16 (October 1948): 22, 24; Gable, "Kansas Penitentiary," p. 431; Johnston, *Leavenworth,* p. 88.

40. Elvid Hunt and Walter E. Lorence, *History of Fort Leavenworth, 1827–1937,* rev. ed. (Fort Leavenworth, Kans.: Command and General Staff School Press, 1937), pp. 124–55.

41. Jesse A. Hall and Leroy T. Hand, *History of Leavenworth County, Kansas* (Topeka: Historical Publishing Company, 1921), pp. 235–36, 257; Johnston, *Leavenworth,* pp. 145–48; Hunt and Lorence, *History of Fort Leavenworth,* pp. 207–9.

42. James R. Shortridge, *Kaw Valley Landscapes: A Traveler's Guide to Northeastern Kansas,* rev. ed. (Lawrence: University Press of Kansas, 1988), pp. 33–41, 46–54.

43. William R. Petrowski, *The Kansas Pacific: A Study in Railroad Promotion* (New York: Arno Press, 1981), pp. 136–50; George L. Anderson, "Atchison, 1865–1886, Divided and Uncertain," *Kansas Historical Quarterly* 35 (1969): 30–45.

44. Anderson, "Atchison, 1865–1886," pp. 30–45.

45. Jay Gould did name Atchison as headquarters for one division of the Missouri Pacific.

46. Atchison Board of Trade, comp., *Atchison, the Railroad Centre of Kansas* (Atchison, Kans.: Board of Trade, 1874), pp. 5–14; Atchison Merchants' and Manufacturers' Bureau, *The Advantages of Atchison, Kansas, as a Commercial and Manufacturing Centre* (Atchison, Kans.: Merchants' and Manufacturers' Bureau, 1887), pp. 5–29.

47. Atchison Board of Trade, *Atchison, the Railroad Centre,* p. 6.

48. Atchison Merchants' and Manufacturers' Bureau, *Advantages of Atchison,* p. 12; Polk, *Kansas State Gazetteer* (1891); *Fiftieth Anniversary Edition* (Atchison, Kans.: Atchison Daily Globe, December 8, 1927), sec. 2, p. 8; "Atchison Hardware Firm in Outstanding 75-Year Growth," *Kansas Business Magazine* 14 (December 1946): 6–7.

49. Cutler, *History of the State,* p. 381; *Fiftieth Anniversary Edition,* sec. 4, p. 2; Catherine Roe and Bill Roe, comps. and eds., *Atchison Centennial, June 20–26, 1854–1954* (Atchison, Kans.: Lockwood Company, 1954), p. 22; "Kansas Grains Converted into Spirits/Alcohol at Atchison," *Kansas Business Magazine* 21 (July 1953): 8–9, 68.

50. Douglas, "History of Manufactures," pp. 119–20; Cutler, *History of the State,* p. 381; Roe and Roe, *Atchison Centennial,* p. 4B; "Atchison Plant Is Factor in National Transportation," *Kansas Business* 7 (June 1939): 10–11; "LFM Company at Atchison Expanding Foundry, Heavy Shop Facilities," *Kansas Business Magazine* 17 (July 1949): 8, 10, 52–53; "LFM Company Expands and Merges with Rockwell Manufacturing Company," *Kansas Business Magazine* 24 (February 1956): 7, 57–58.

51. Glenn D. Bradley, *The Story of the Santa Fe* (Boston: Richard G. Badger, 1920), pp. 54–61. For background of Holliday, see William E. Treadway, *Cyrus K. Holliday: A Documentary Biography* (Topeka: Kansas State Historical Society, 1979).

52. Frye W. Giles, *Thirty Years in Topeka: A Historical Sketch* (Topeka: G. W. Crane

and Company, 1886), pp. 121–23, 137–38; Roy D. Bird and Douglass W. Wallace, *Witness of the Times: A History of Shawnee County* (Topeka: Shawnee County Historical Society, 1976), pp. 88–89.

53. Quotations are from the *Topeka Commonwealth* as cited in Bird and Wallace, *Witness of the Times,* p. 89; Giles, *Thirty Years in Topeka,* pp. 138–39.

54. Bird and Wallace, *Witness of the Times,* p. 90; Bradley, *Story of the Santa Fe,* pp. 228, 235.

55. Giles, *Thirty Years in Topeka,* p. 105; *Gazetteer and Directory* (1870); Polk, *Kansas State Gazetteer* (1880); "Thompson Hardware Covers Five States; Plan Diamond Jubilee," *Kansas Business Magazine* 19 (April 1951): 8–9.

56. *Topeka and Its Advantages* (Topeka: Hall and O'Donald Litho Co., 1888), pp. 14, 17; *Report on the Manufactures of the United States at the Tenth Census* (Washington, D.C.: Government Printing Office, 1883), p. 243; Cabe, *Flour Milling,* p. 33.

57. *Topeka and Its Advantages,* p. 14; *Sixtieth Anniversary Issue* (Topeka: Topeka Daily Capital, July 16, 1939), p. C14; Bird and Wallace, *Witness of the Times,* p. 96; "Morrell's in Topeka for Ten Years," *Kansas Business Magazine* 9 (October 1941): 6; "Seymour Packing Company Puts Branches under Firm Name," *Kansas Business Magazine* 15 (September 1947): 6–7.

58. Quoted in Bird and Wallace, *Witness of the Times,* p. 87.

59. Ibid., p. 101; *Topeka and Its Advantages,* p. 20; Cutler, *History of the State,* p. 553.

60. *Topeka and Its Advantages,* p. 20; Bird and Wallace, *Witness of the Times,* p. 103.

61. "Topeka: Business, Industrial and Convention Center," *Kansas Business* 6 (January 1938): 27; *Topeka in the Valley of the Kaw: The Golden City of Quivira* (Topeka: Topeka Chamber of Commerce, 1929), p. 7; *Topeka, Kansas: A Thriving State Capital at the Nation's Crossroads* (Topeka: Topeka Chamber of Commerce, 1936); Bird and Wallace, *Witness of the Times,* pp. 103–4; Roy D. Bird, *Topeka: An Illustrated History of the Kansas Capital* (Topeka: Baranski Publishing Company, 1985), p. 63.

62. "Topeka: One of the Progressive Cities of the Nation," *Kansas Business* 4 (May 1936): 25; *Topeka in the Valley,* p. 4; *Topeka, Kansas: A Thriving State Capital.*

63. *New Horizons Issue* (Topeka: Topeka Daily Capital, March 6, 1958), p. 9; *Topeka in the Valley,* p. 5; "Topeka: One of the Progressive Cities," pp. 12–13.

64. Marco Morrow, "Modern Publishing Plant Spreads Fame of Kansas," *Kansas Business* 2 (February 1934): 8–9; "Topeka: One of the Progressive Cities," p. 14; "Topeka: Business, Industrial, and Convention Center," pp. 6, 16–18, 26.

65. *Sixtieth Anniversary Issue,* pp. A1–A6; Morrow, "Modern Publishing Plant," pp. 8–9; "Capper Farm Papers Give State Leading National Institution," *Kansas Business Magazine* 17 (March 1949): 8, 10, 12.

66. The slogan "Magic City" was coined by the longtime editor of the *Wichita Eagle* newspaper, Marshall Murdock. See H. Craig Miner, *Wichita: The Magic City* (Wichita: Wichita-Sedgwick County Historical Museum, 1988), p. 59.

67. *Wichita, Kansas* (Wichita: Wichita Board of Trade, 1888), p. 6.

68. Miner, *Wichita: The Magic City,* pp. 31–33.

69. Polk, *Kansas State Gazetteer* (1880, 1891); Jimmy M. Skaggs, "Wichita, Kansas: Economic Origins of Metropolitan Development, 1870–1960," in *Metropolitan Wichita: Past, Present and Future,* ed. Glenn W. Miller and Jimmy M. Skaggs (Lawrence: Regents Press of Kansas, 1978), p. 6.

70. Cutler, *History of the State,* p. 1393; Skaggs, "Wichita, Kansas," p. 8; Leslie Fitz, "The Development of the Milling Industry in Kansas," *Kansas Historical Collections,*

1911–1912 12 (1912): 59; "Kansas Mills Her Golden Grain," p. 20; Cabe, *Flour Milling,* p. 31; *Thirty-fifth Annual Report of the Department of Labor,* p. 29; Abraham E. Janzen, "The Wichita Grain Market" (master's thesis, University of Kansas, 1927).

71. *Thirty-fifth Annual Report of the Department of Labor,* p. 29; Miner, *Wichita: The Magic City,* pp. 75–77, 131; Skaggs, "Wichita, Kansas," p. 6; *Wichita, Kansas,* p. 6; Orsemus H. Bentley, ed., *History of Wichita and Sedgwick County, Kansas* (Chicago: C. F. Cooper and Company, 1910), vol. 2, p. 699.

72. Miner, *Wichita: The Magic City,* pp. 129–31; Bentley Barnabas, "Coleman Products Give Light and Warmth to Far Corners of Earth," *Kansas Business* 2 (November 1934): 9; "Anniversary: W. C. Coleman, 80 Years, the Coleman Company, 50 Years," *Kansas Business Magazine* 18 (April 1950): 8, 10, 42–46; *Thirty-fifth Annual Report of the Department of Labor,* p. 30.

73. Lockwood, Green and Company, *Industrial Survey of Wichita, Kansas* (Boston: Lockwood, Green and Company, 1927), pp. 16–17, 71–74; "Wichita: Metropolis of the Southwest," *Kansas Business* 6 (July 1938): 26. Most of the railroad employment in 1927 was at two repair shops, one owned by the Chicago, Rock Island and Pacific company, the other by a short-lived venture called the Kansas City, Mexico and Orient that was later absorbed into the Santa Fe system.

74. H. Craig Miner, *Discovery! Cycles of Change in the Kansas Oil and Gas Industry, 1860–1987* (Wichita: Kansas Independent Oil and Gas Association, 1987), pp. 130–39; Burt Doze, "Courageous Oil Men Unlatched the Door When Opportunity Knocked," in *Wichita People* (Wichita: Wichita Chamber of Commerce, 1946), pp. 73–76; "Cardwell Is a Big Name in Oil Equipment throughout the World," *Kansas Business Magazine* 9 (October 1951): 7, 74–78; "Koch Engineers Roam over World in Refinery Construction Work," *Kansas Business Magazine* 20 (October 1952): 10, 12, 14.

75. General accounts of the early Wichita aviation industry include Gerald O. Deneau, "Highlights of the Development of the Aircraft Industry in Kansas," *Transactions, Kansas Academy of Science* 71 (1968): 439–50; Sondra Van Meter, "The E. M. Laird Airplane Company: Cornerstone of the Wichita Aircraft Industry," *Kansas Historical Quarterly* 36 (1970): 341–54; H. Craig Miner, "A Roar from the Sky: Air-Mindedness in Wichita, 1908–1980," *Journal of the West* 30 (January 1991): 37–44; and Frank Joseph Rowe and Craig Miner, *Borne on the South Wind: A Century of Aviation in Kansas* (Wichita: Wichita Eagle and Beacon Publishing Company, 1994).

76. Rowe and Miner, *Borne on the South Wind.*

77. Miner, "Roar from the Sky," pp. 41–42; Andrew S. Swenson, "Airplane Activity Result of Years of Hard Work," *Kansas Business Magazine* 9 (August 1941): 7, 24–26.

78. Swenson, "Airplane Activity," p. 7; H. Craig Miner, *Kansas: The History of the Sunflower State, 1854–2000* (Lawrence: University Press of Kansas, 2002), pp. 303–13.

79. Deneau, "Highlights," pp. 445–46; Miner, "Roar from the Sky," pp. 42–43.

80. Miner, "Roar from the Sky," pp. 42–43; Julie Courtwright, "Want to Build a Miracle City? War Housing in Wichita," *Kansas History* 23 (2000): 218–39; Deneau, "Highlights," pp. 445–46; John Morrison, "Aircraft Factories Stabilize Post-war Wichita Employment," *Kansas Business Magazine* 17 (August 1949): 14, 16. The air force base changed its name from Wichita to McConnell in 1954.

81. "Salina" (Special Issue), *Kansas Business* 4 (June 1936): 9, 16.

82. Polk, *Kansas State Gazetteer* (1880, 1891); *Thirty-fifth Annual Report of the Department of Labor,* pp. 25–26; *Salina Facts* (Salina, Kans.: Salina Chamber of Commerce, 1930).

83. *Thirty-fifth Annual Report of the Department of Labor*, pp. 25–26; James C. Malin, *Winter Wheat in the Golden Belt of Kansas: A Study in Adaption to Subhumid Geographical Environment* (Lawrence: University of Kansas Press, 1944); Cutler, *History of the State*, p. 702; Cabe, *Flour Milling*, pp. 25–26, 31; Federal Writers' Project, Works Progress Administration, *A Guide to Salina, Kansas* (Salina, Kans.: Advertiser-Sun, 1939), pp. 11, 34–36.

84. Ruby P. Bramwell, *City on the Move: The Story of Salina* (Salina, Kans.: Survey Press, 1969), pp. 131–36; *Salina Facts*; Federal Writers' Project, *Guide to Salina*, pp. 8–9, 37–39.

85. Federal Writers' Project, *Guide to Salina*, pp. 8–9, 37–39; "Salina" (Special Issue); Harry Hughes and Helen C. Dingler, *From River Ferries to Interchanges: A Brief History of Saline County, Kansas from the 1850's to the 1980's* (Salina, Kans.: Harry Hughes and Helen C. Dingler, 1988), p. 137.

86. Pat Mitchell, *The Fair City: Hutchinson, Kansas* (Topeka: Josten's, 1982); "Remodeling the Nation with Kansas Fibre Products," *Kansas Business* 4 (March 1936): 12–13; "Fibre Products Company Is Reaching New Output Peaks," *Kansas Business Magazine* 15 (April 1947): 12.

87. See chapter 5 for details about the construction of and rationale for these railroad lines.

88. Cabe, *Flour Milling*, pp. 25–26; *Hutchinson, the Salt City in the Heart of the Great Kansas Wheat Belt* (Hutchinson, Kans.: News Company, 1908); Willard Welsh, *Hutchinson: A Prairie City in Kansas* (Wichita: McCormick-Armstrong Company, 1946), pp. 101–8.

89. *Hutchinson, the Salt City*; Welsh, *Hutchinson*, pp. 131–33, 141–43, 156; "Krause Corporation Builds World's Largest Disc Plow," *Kansas Business Magazine* 14 (September 1946): 6; "Krause into Million Dollar Building, Expansion Program," *Kansas Business Magazine* 16 (January 1948): 7, 63–65; "Dillon and Sons Expanding with New Warehouse, Food Stores," *Kansas Business Magazine* 15 (October 1947): 8.

90. Arman J. Habegger, "Out of the Mud: The Good Roads Movement in Kansas, 1900–1917" (master's thesis, University of Kansas, 1971), pp. 190–93; Beach Advertising Company, comp., *Hutchinson, Kansas* (Hutchinson, Kans.: Hutchinson Chamber of Commerce, 1951), pp. 14, 29; "Merger of Many Lines Makes Western Transit," *Kansas Business* 5 (October 1937): 15, 24; Welsh, *Hutchinson*, pp. 115–16.

91. The economic control of Kansas City over the state of Kansas was broken during the railroad period only in the far northwest. Spurs from the Chicago, Burlington and Quincy Railroad across southern Nebraska served four counties in the northern tier there, directing local trade into Omaha.

92. Janet P. Emery, *It Takes People to Make a Town: The Story of Concordia, Kansas: 1871–1971* (Salina, Kans.: Arrow Printing Company, 1970), p. 94.

93. *Dodge City's Diamond Jubilee* (Dodge City, Kans.: Dodge City Journal, May 23–25, 1947), pp. 31, 53.

94. *Thirty-fifth Annual Report of the Department of Labor*, p. 15.

95. *Pride, Progress Edition* (Pratt, Kans.: Pratt Daily Tribune, 1954), pp. 33–37.

96. *Pictorial Edition, Magazine Supplement* (McPherson, Kans.: McPherson Weekly Republican, March 1, 1901); Jessie H. Rowland, *Pioneer Days in McPherson* (McPherson, Kans.: McPherson Junior Chamber of Commerce, 1947), pp. 4, 23; Raymond L. Flory, ed., *Historical Atlas of McPherson County* (McPherson, Kans.: McPherson County Historical Society, 1983), pp. 50, 66.

97. *Great Bend, Kansas: A Historical Portrait of the City* (Great Bend, Kans.: Centennial Book Committee, 1972), p. 14; "Great Bend, the Oil Capital, Putting on Another Big

Show," *Kansas Business Magazine* 20 (October 1952): 70–71; "Oil Industry Is Making Great Bend a Boom Town," *Kansas Business Magazine* 22 (October 1954): 48; "Hays Oil Appreciation Days Honor Area Oil Men," *Kansas Business Magazine* 22 (October 1954): 52–53.

98. *Golden Jubilee Edition* (Belleville, Kans.: Belleville Telescope, July 1, 1937), p. 2B.

99. Norton Centennial Board, *From a Covered Wagon Ride to Community Pride: Norton Centennial, 1872–1972* (Norton, Kans.: n.p., 1972), p. 61; *Phillips County Review,* July 7, 1949; *Progress and Historical Edition* (Phillipsburg, Kans.: Phillips County Review, May 1952), sec. 1, pp. 1, 5.

100. Colby Chamber of Commerce, *Business and Industrial Opportunity in Colby, Kansas* (Colby, Kans.: Chamber of Commerce, 1946), p. 1; *Fiftieth Anniversary Edition* (Colby, Kans.: Colby Free Press–Tribune, October 4, 1939); *Thomas County: Yesterday and Today* 1 (January 1960): 17.

101. Joseph S. Vernon, *Dodge City and Ford County, Kansas: A History of the Old and a Story of the New* (Larned, Kans.: Tucker-Vernon Publishing Company, 1911), p. 17; *Thirty-fifth Annual Report of the Department of Labor,* p. 14; Carl F. Etrick, *Dodge City Semi-centennial Souvenir* (Dodge City, Kans.: Etrick Company, 1922); *Dodge City Greeter and Guide,* rev. ed., (Dodge City, Kans.: Strange and Hetzel Publishing Company, 1958), pp. 41, 49.

102. *Southwest Kansas Resource Edition* (Garden City, Kans.: Garden City Daily Telegram, June 12, 1940), sec. 1, p. 1; "Farmers Profit in Kansas Sugar Beet Industry," *Kansas Business* 5 (March 1937): 12–3; "Garden City Company Plans More Beet Sugar Production," *Kansas Business Magazine* 23 (June 1955): 42.

103. *Thirty-fifth Annual Report of the Department of Labor,* pp. 22, 27; *Fiftieth Anniversary Number* (Newton, Kans.: Newton Kansan, August 22, 1922), pp. 45, 52, 124–25; William T. Moran, *Santa Fe and the Chisholm Trail at Newton* (Newton, Kans.: Moran, 1970), pp. 16, 60, 72, 89; Rachel Waltner, *Brick and Mortar: A History of Newton, Kansas* (North Newton, Kans.: Mennonite Library and Archives, 1984), pp. 13, 16; Gwendoline Sanders and Paul Sanders, *The Sumner County Story* (North Newton, Kans.: Mennonite Press, 1966), p. 75.

104. Jessie P. Stratford, *Butler County's Eighty Years: 1855–1935* (El Dorado, Kans.: Butler County News, 1935), pp. 94–95, 113–14, 120; *Progress Issue* (El Dorado, Kans.: El Dorado Times, February 25, 1964), sec. I, pp. 1–9.

105. "Arkansas City" (Special Issue), *Kansas Business* 4 (February, 1936): 7; *Mid-century Resources Edition, 1900–1950* (Arkansas City, Kans.: Arkansas City Daily Traveler, 1950); James F. Clough, *The Story of Arkansas City* (Arkansas City, Kans.: n.p., 1961).

106. Eldie F. Caldwell, ed., *Illustrated Southern Kansas,* rev. ed. (Lawrence, Kans.: E. F. Caldwell, 1886), pp. 29, 33; Winfield Commercial Club, *Souvenir of Winfield, Cowley, Kansas, 1904* (Winfield, Kans.: n.p., 1904), pp. 3E, 6E; *Achievement Edition* (Winfield, Kans.: Winfield Daily Courier, February 25, 1952), p. 7I.

107. Cutler, *History of the State,* p. 712; *Special Centennial Edition* (Marysville, Kans.: Marysville Advocate, August 12, 1954).

108. Laura M. French, *History of Emporia and Lyon County* (Emporia, Kans.: Emporia Gazette, 1929), pp. 2, 30–38, 52, 56–66, 96; *Historical Booklet: Emporia, Kansas, Centennial Celebration* (Emporia, Kans.: Centennial Celebration Committee, 1957), pp. 5, 41–42, 64.

109. *Historical Booklet,* pp. 5, 41–42, 64; *Thirty-fifth Annual Report of the Department of Labor,* p. 14; French, *History of Emporia,* p. 58.

110. Quastler, *Railroads of Lawrence,* pp. 202–16; Kenneth A. Middleton, *The Indus-*

trial *History of a Midwestern Town,* Kansas Studies in Business No. 20 (Lawrence: School of Business, University of Kansas, 1941), pp. 25–53; David A. Dary, *Lawrence: Douglas County, Kansas: An Informal History* (Lawrence, Kans.: Allen Books, 1982), pp. 123–25; Brian Black, "Mastering the Kaw: The Bowersock Dam and the Development of Lawrence Industry," *Kansas History* 16 (1993): 262–75; *Thirty-fifth Annual Report of the Department of Labor,* p. 19.

111. *Thirty-fifth Annual Report of the Department of Labor,* p. 19; James R. Shortridge and Barbara G. Shortridge, "Yankee Town on the Kaw: A Geographical and Historical Perspective on Lawrence and Its Environs," in *Embattled Lawrence: Conflict and Community,* ed. Dennis Domer and Barbara Watkins (Lawrence: University of Kansas Continuing Education, 2001), pp. 5–19.

112. "A. L. Duckwall Company Adds Two New Stores in Expansion," *Kansas Business Magazine* 16 (August 1948): 16; "Duckwall's Observe 50th Year in New Offices and Warehouse," *Kansas Business Magazine* 19 (December 1951): 7, 52–53.

113. Charles M. Correll, "The First Century of Kansas State University," *Kansas Historical Quarterly* 28 (1962): 409–41.

114. James R. Shortridge, *Our Town on the Plains: J. J. Pennell's Photographs of Junction City, Kansas, 1893–1922* (Lawrence: University Press of Kansas, 2000), pp. 31–33, 94–127.

115. Charles A. Knouse, comp., *A Town between Two Rivers: Osawatomie, Kansas, 1854–1954* (Osawatomie, Kans.: Osage Valley Centennial, 1954), pp. 48–55; Berenice B. Wallace, comp., *The Story of Paola, Kansas, 1857–1950* (Paola, Kans.: n.p., n.d.), pp. 76–77; "Fluor Corporation Centers Metal Fabrication Work in Paola," *Kansas Business Magazine* 23 (October 1955): 7, 14, 16.

116. Oscar A. Copple and Joyce M. Hitchings, *History of Osage City, and Osage County* (n.p., 1970), pp. 60–63.

117. *Thirty-fifth Annual Report of the Department of Labor,* p. 23; "Bennett Creamery Company Is Typical Kansas Business," *Kansas Business* 7 (December 1939): 22–23.

118. *The Centennial Mirror, Special Edition* (Olathe, Kans.: Olathe Mirror, April 29, 1957), sec. 3, p. 7, and sec. 4, p. 1; *Olathe "The City Beautiful" Centennial, 1857–1957* (Olathe, Kans.: Olathe Centennial, 1957), pp. 11, 36; Shortridge, *Kaw Valley Landscapes,* pp. 195–97.

119. Elmer Coe, *Fort Scott as I Knew It* (Fort Scott, Kans.: Monitor Binding and Printing Company, 1940), pp. 47, 73–74, 85; *Thirty-fifth Annual Report of the Department of Labor,* p. 15.

120. *Thirty-fifth Annual Report of the Department of Labor,* p. 15; Coe, *Fort Scott,* pp. 12–14, 85–86; William G. Calhoun, comp., *Fort Scott: A Pictorial History* (Fort Scott, Kans.: Historic Preservation Association of Bourbon County, 1978), pp. 168–69; "Western Insurance Companies Observe 40th Year of Service," *Kansas Business Magazine* 18 (April 1950): 7, 53.

121. *Parsons: Its Past, Present and Future (Special Edition)* (Parsons, Kans.: Parsons Palladium, October 16, 1901), p. 9.

122. *Thirty-fifth Annual Report of the Department of Labor,* p. 24; *Down through the Years with Parsons Pioneer Firms (Special Edition)* (Parsons, Kans.: Parsons Sun, October 18, 1938), p. 6.

123. William E. Powell, "The Cherokee-Crawford Coal Field of Southeastern Kansas: A Study in Sequent Occupance," *Midwest Quarterly* 22 (1981): 123–24; *Pittsburg Diamond Jubilee, 1876–1951 (Official Program)* (Pittsburg, Kans.: n.p., 1951), p. 45; Federal Writ-

ers' Project, Works Progress Administration, *The Pittsburg Guide* (Pittsburg, Kans.: Pittsburg Chamber of Commerce, 1941), p. 1.

124. Pittsburg Chamber of Commerce, *Prosperous Pittsburg Pictorially Portrayed* (Pittsburg, Kans.: Pittsburg Publicity Company, 1915); *Pittsburg Diamond Jubilee*, pp. 32, 49; Pittsburg State University, *A Guide to Pittsburg State University* (Pittsburg, Kans.: The University, 1982); Federal Writer's Project, *Pittsburg Guide*, p. 21.

125. "Pittsburg Pottery Company Serves Wide Area with Pots and Artware," *Kansas Business Magazine* 21 (July 1953): 26; "Dickey Clay Manufacturing Is Expanding at Pittsburg," *Kansas Business Magazine* 16 (April 1948): 67; *Thirty-fifth Annual Report of the Department of Labor*, pp. 24–25; "McNally Pittsburg Corporation Expands Foundry Production and Offices," *Kansas Business Magazine* 19 (July 1951): 18, 20; "McNally Pittsburg Has New Machine Shop in Production," *Kansas Business Magazine* 24 (April 1956): 8–9, 69–70.

126. "Big Ammonium Plant to Kansas," *Kansas Business Magazine* 9 (August 1941): 8; "Spencer Chemicals Develop New Industry for Midwest," *Kansas Business Magazine* 15 (January 1947): 8–9, 62–63; "Spencer Chemical Company Will Make Million Dollar Expansion," *Kansas Business Magazine* 18 (March 1950): 26; "Growing Spencer Chemicals Stems from Area Planning," *Kansas Business Magazine* 21 (December 1953): 8–9, 57–59 (quotation on page 8); John A. Lawson, Sondra V. M. McCoy, Robert K. Ratzlaff, and Thomas R. Walther, *100 Years of Excellence: A History of the Pittsburg & Midway Coal Mining Company, 1885–1985* (Denver: Pittsburg & Midway Coal Mining Company, 1985), pp. 33–34, 44–45, 63.

127. "Eagle-Picher's Lead Smelter Spurs Big Tri-state Industry," *Kansas Business Magazine* 14 (October 1946): 8, 50; Claude H. Nichols, comp., *The Baxter Springs Story* (Baxter Springs, Kans.: Baxter Springs Centennial, 1958).

128. *Centennial Edition* (Iola, Kans.: Iola Register, May 30, 1955), pp. 1D–3D.

129. Ibid.; *Thirty-fifth Annual Report of the Department of Labor*, pp. 17–18.

130. *Thirty-fifth Annual Report of the Department of Labor*, pp. 17–18; Iola Chamber of Commerce, *The Story of Iola, Kansas: Crossroads of Mid-America* (Iola, Kans.: Iola Chamber of Commerce, 1959).

131. Genevieve L. Choguill and Harold S. Choguill, *A History of the Humboldt, Kansas, Community: 1855–1988* (North Newton, Kans.: Mennonite Press, 1988), pp. 36–37; "Monarch Cement Company Pushes Expansion Work at Humboldt," *Midwest Industry Magazine* 24 (August 1956): 9; Harvey F. Nelson, "An Economic History of Chanute, Kansas" (master's thesis, University of Kansas, 1939), pp. 70–72.

132. *Thirty-fifth Annual Report of the Department of Labor*, p. 23; William W. Graves, *History of Neosho County* (St. Paul, Kans.: Journal Press, 1951), vol. 2, pp. 971–73.

133. Ibid., pp. 974, 1013–15; *Thirty-fifth Annual Report of the Department of Labor*, p. 12; *Pioneer Edition* (Chanute, Kans.: Chanute Tribune, June 24, 1941), p. 11B.

134. *Pioneer Edition*, p. 5A; *Thirty-fifth Annual Report of the Department of Labor*, p. 12; Graves, *History of Neosho*, vol. 2, pp. 1005, 1015.

135. Charles C. Drake, *"Who's Who?": A History of Kansas and Montgomery County* (Coffeyville. Kans.: Coffeyville Journal Press, 1943), pp. 104–6; Coffeyville at 100, Inc., *A History of Coffeyville* (Coffeyville, Kans.: Coffeyville Journal Press, 1969), pp. 35–38.

136. *History of Coffeyville*, pp. 32, 38, 52; Drake, *"Who's Who?"* pp. 107–9; "50 Years in the Foundry Business," *Midwest Industry Magazine* 32 (October 1964): 12–13.

137. "Morrow Pattern Shop and Foundry One of Most Modern in Country," *Kansas Business Magazine* 19 (July 1951): 10; "Funk Manufacturing Comes Back with Power

Transmissions," *Midwest Industry Magazine* 33 (September 1965): 13, 56–59; "Dixon Company Has Grown to Be Coffeyville's Biggest Factory," *Kansas Business Magazine* 19 (September 1951): 24, 26.

138. Amy L. Rork, "Sense of Place in Montgomery County, Kansas: Perceptions of an Industrialized Rural Area" (master's thesis, University of Kansas, 1997), pp. 21–25.

8. Postindustrial Kansas I: The Interstate Cities since 1950

1. Daniel Bell, *The Coming of Post-industrial Society* (New York: Basic Books, 1973); David Harvey, *The Condition of Postmodernity: An Inquiry into the Origins of Cultural Change* (Cambridge, Mass.: Basil Blackwell, 1990); Frederic Jameson, *Postmodernism, or, The Cultural Logic of Late Capitalism* (Durham, N.C.: Duke University Press, 1991).

2. J. Neill Marshall and Peter A. Wood, *Services and Space: Key Aspects of Urban and Regional Development* (New York: Wiley, 1995); David Leonhardt, "Recession, and Then a Boom? Maybe Not This Time," *New York Times,* December 30, 2001, sec. 3, pp. 1, 10.

3. As I discussed in chapter 7, this classification is a modification of one created by Chauncy D. Harris, "A Functional Classification of Cities in the United States," *Geographical Review* 33 (1943): 86–99. Kansas censuses showed that approximately a quarter of local populations were employed outside the home in the period 1870–1950. This percentage had risen to half by the year 2000, necessitating the adjustments in thresholds noted in the text.

4. Arman J. Habegger, "Out of the Mud: The Good Roads Movement in Kansas, 1900–1917" (master's thesis, University of Kansas, 1971), pp. 44–47, 59, 145, 164–65; Sherry L. Schirmer and Theodore A. Wilson, *Milestones: A History of the Kansas Highway Commission and the Department of Transportation* (Topeka: Kansas Department of Transportation, 1986), pp. 1:10–1:12, 1:24; Paul S. Sutter, "Paved with Good Intentions: Good Roads, the Automobile, and the Rhetoric of Rural Improvement in the *Kansas Farmer*, 1890–1914," *Kansas History* 18 (1995): 284–99.

5. Schirmer and Wilson, *Milestones,* pp. 2:34, 3:9.

6. James D. Callahan, comp. and ed., *Jayhawk Editor: A Biography of A. Q. Miller, Senior* (Los Angeles: Sterling Press, 1955), pp. 76–91.

7. Habegger, "Out of the Mud," pp. 189–210; Schirmer and Wilson, *Milestones,* p. 2:20.

8. Ibid., 4:1–4:9, 4:18; *Topeka Daily Capital,* February 8, 1941; *Kansas City Times,* July 17, 1941; *Kansas City Star,* March 17, 1942.

9. Schirmer and Wilson, *Milestones,* pp. 4:3–4:4.

10. Ibid., pp. 4:5–4:9.

11. *Topeka Daily Capital,* August 18, 1912.

12. Schirmer and Wilson, *Milestones,* pp. 4:19–4:20.

13. Ibid., pp. 4:21–4:26.

14. *Lawrence Journal-World,* August 31, 1955; *Eighth Annual Report of the Kansas Turnpike Authority: 1960* (Topeka: State Printer, 1961), p. 15; "Sell $160,000,000 Bonds for Kansas Turnpike," *Kansas Business Magazine* 22 (October 1954): 72–73; *Third Annual Report of the Kansas Turnpike Authority: 1955* (Topeka: State Printer, 1956), p. 23.

15. Mark H. Rose, *Interstate Express Highway Politics, 1939–1989,* rev. ed. (Knoxville: University of Tennessee Press, 1990), pp. 41–94.

16. Ibid., pp. 96–99; Schirmer and Wilson, *Milestones,* p. 4:29.

17. Rose, *Interstate Express,* pp. 96–99.

18. *Highway Highlights* (State Highway Commission of Kansas) 17 (October 1956): 1–2; Schirmer and Wilson, *Milestones,* p. 4:36.

19. Schirmer and Wilson, *Milestones,* p. 4:36; *Topeka Daily Capital,* August 13, 1958.

20. *Topeka Daily Capital,* August 14, 1958; August 27, 1958; and November 20, 1958; *Topeka Journal,* July 22, 1959; *Kansas City Times,* December 15, 1959.

21. *Wichita Eagle and Beacon,* January 10, 1963.

22. *Kansas Highway Needs for 1975–1985* (Topeka: State Highway Commission, 1966), pp. 17–19; Schirmer and Wilson, *Milestones,* pp. 5:4–5:12.

23. *Rand McNally 2002 Commercial Atlas and Marketing Guide* (Chicago: Rand McNally, 2002), pp. 38–39; "Survey Defines Trade Areas," *Kansas Business News* 2 (August 1981): 14–15.

24. "Fairfax Industrial District Gains New Plants, Warehouses," *Kansas Business Magazine* 21 (January 1953): 7, 77; "Fairfax District Getting More Industrial Activity," *Kansas Business Magazine* 22 (November, 1954): 8–9, 60–61; "52 Autos an Hour, 16 Hours a Day," *Midwest Industry Magazine* 32 (September 1964): 14, 69–70.

25. "Town House Hotel Offers Latest Convention Facilities at K.C.," *Kansas Business Magazine* 19 (September 1951): 11, 57; "Fairfax Industrial District Gains," p. 7.

26. Michael J. Broadway and Terry Ward, "Recent Changes in the Structure and Location of the U.S. Meatpacking Industry," *Geography* 75 (1990): 76–79; Donald D. Stull and Michael J. Broadway, "The Effects of Restructuring on Beefpacking in Kansas," *Kansas Business Review* 14 (fall 1990): 10–16.

27. *Thirty-fifth Annual Report of the Department of Labor and Industry, 1920* (Topeka: State Printer, 1921), p. 18; *Kansas Community Profile: Kansas City* (Topeka: Kansas Department of Commerce and Housing, 1995); "Railroad Freight Yard Projects Bolster Industry of this Area," *Kansas Business Magazine* 17 (September 1949): 10–12; Edwin D. Shutt, "'Silver City': A History of the Argentine Community of Kansas City, Kansas" (master's thesis, Emporia State College, 1976), pp. 176–78.

28. "General Motors Puts New Installations in Kansas," *Kansas Business Magazine* 23 (May 1955): 7.

29. *Kansas Community Profile: Kansas City* (Topeka: Kansas Department of Commerce and Housing, 2002).

30. Dana Fields, "Kansas Speedway Puts Region on Development Fast Track," *Lawrence Journal-World,* August 27, 2000, pp. 1E, 3E; Shirley Christian, "City Pins Its Hopes on Kansas Speedway," *New York Times,* May 27, 2001, p. Y40.

31. Margaret Landis, *The Winding Valley and the Craggy Hillside: A History of the City of Rosedale, Kansas* (Kansas City: Margaret Landis, 1976), pp. 39–42; *Kansas Community Profile: Kansas City,* 2002.

32. *Kansas Community Profile: Kansas City,* 2002; James R. Shortridge, *Kaw Valley Landscapes: A Traveler's Guide to Northeastern Kansas,* rev. ed. (Lawrence: University Press of Kansas, 1988), pp. 24–25.

33. Fields, "Kansas Speedway," pp. 1E, 3E; Christian, "City Pins," p. Y40; "Kansas Speedway," *Business Facilities,* June, 2001, accessed at www.facilitycity.com/busfac/bf_01_06_special2.asp.

34. William S. Worley, *J. C. Nichols and the Shaping of Kansas City: Innovation in Planned Residential Communities* (Columbia: University of Missouri Press, 1990), pp. 77–174.

35. Ibid.; Elizabeth E. Barnes, *Historic Johnson County: A Bird's Eye View of the*

Development of the Area (Shawnee Mission, Kans.: Neff Printing Company, 1969), pp. 44–50.

36. Worley, *J. C. Nichols,* pp. 89–232; Joel Garreau, *Edge City: Life on the New Frontier* (New York: Doubleday, 1991).

37. Barnes, *Historic Johnson County,* p. 47.

38. Dan Bearth, "Business Moves to the Suburbs," *Kansas Business News* 2 (June 1981): 34–36; *Kansas Community Profile: Lenexa* (Topeka: Kansas Department of Commerce and Housing, 2002).

39. "World Trade/Kansas Style," *Kansas Business News* 3 (June 1982): 46–48; *Kansas Community Profile: Olathe* (Topeka: Kansas Department of Commerce and Housing, 2002).

40. *Kansas Community Profile: Olathe.*

41. Rachel Bolton, "Joe Dennis," *Kansas Business News* 3 (July 1982): 12–14; *Kansas Community Profile: Gardner* (Topeka: Kansas Department of Commerce and Housing, 2002).

42. Barnes, *Historic Johnson County,* pp. 45–46; Turner Lake, "Spotlight: Overland Park," *Kansas Business News* 5 (August 1984): 30, 32.

43. Lake, "Spotlight," pp. 29–30, 32, 34.

44. Ibid.; Bearth, "Business Moves," pp. 37–38; *Kansas Community Profile: Overland Park* (Topeka: Kansas Department of Commerce and Housing, 2002); "Sprint Headquarters to Display Company's Wealth," *Lawrence Journal-World,* December 8, 1997, p. 3D.

45. *Lawrence Journal-World,* January 29, 1996, p. 8B.

46. Mark Fagan, "Flour Power: Kansas Mills Lead the Nation in Flour Production," *Lawrence Journal-World,* May 7, 2000, pp. 1E–3E; Randy Brown, "Wichita Stockyards Gone . . . But Memories Live On," *Kansas Business News* 9 (January 1988): 16–18; *Kansas Community Profile: Wichita* (Topeka: Kansas Department of Commerce and Housing, 2002).

47. Susan Yerkes, "Low-Power TV: A New Broadcast Market Opens for Small Investors," *Kansas Business News* 3 (March 1982): 17; *Rand McNally 2002 Commercial Atlas,* pp. 38–39.

48. H. Craig Miner, *Wichita: The Magic City* (Wichita: Wichita-Sedgwick County Historical Museum, 1988), p. 193.

49. Ibid., pp. 193–95.

50. "Kansas Leading in Aircraft," *Kansas Business Magazine* 24 (March 1956): 8–12.

51. Ibid.; Dan Bearth, "Jerry Mallot," *Kansas Business News* 5 (October 1984): 24; Miner, *Wichita: The Magic City,* pp. 207–9.

52. Bearth, "Jerry Mallot," p. 24; *Kansas Community Profile: Wichita;* "Beech, Boeing Announce Plans for Major Expansions," *Kansas Business News* 4 (February 1983): 10–11.

53. Miner, *Wichita: The Magic City,* pp. 208–9; Leslie Wayne, "Brother versus Brother: Koch Family's Long Legal Feud Is Headed for a Jury," *New York Times,* April 28, 1998, pp. C1, C8; *Kansas Community Profile: Wichita.*

54. *Kansas Community Profile: Wichita;* "Frontier Chemical Completes Plant," *Midwest Industry Magazine* 24 (December 1956): 10–11; "Vulcan Chemical: Kansas Needs Them," *Kansas Business News* 9 (July 1988): 15–18.

55. Dan Bearth, "Coleman Sees Bright Lights Ahead," *Kansas Business News* 2 (July 1981): 32–37; *Kansas Community Profile: Wichita.*

56. Eileen O'Hara, "Garvey Plans Epic Center 'Landmark for Wichita,'" *Kansas Business News* 6 (August 1985): 14–15; Dan Bearth, "World's Largest Western Stores Blaze

Urban Trail," *Kansas Business News* 3 (November 1982): 16–20; Randy Brown, "Entrepreneurship: The Soul of Kansas," *Kansas Business News* 10 (April 1989): 9–11.

57. "Top Employers in Kansas" (Topeka: Kansas Department of Commerce and Housing, 2001), accessed at www.kdoch.state.ks.us/busdev/top_employers_2001.jsp.

58. "National Furniture's New Plant Has Open House at Wellington," *Kansas Business Magazine* 18 (August 1950): 7, 53.

59. *Kansas Community Profile: Wellington* (Topeka: Kansas Department of Commerce and Housing, 2002); *Kansas Community Profile: Augusta* (Topeka: Kansas Department of Commerce and Housing, 2002).

60. Dan Bearth, "Andover: The Promised Land," *Kansas Business News* 8 (April 1987): 56, 58–60.

61. *Kansas Community Profile: El Dorado* (Topeka: Kansas Department of Commerce and Housing, 2002); Jack M. Flint, Floyd Heir, and Carl L. Heinrich, *The Kansas Junior College* (Topeka: State Department of Public Instruction, 1968), p. 26; Thelma Helyar, ed., *Kansas Statistical Abstract, 2000* (Lawrence: Policy Research Institute, University of Kansas, 2001), p. 13-4.

62. "A Kansas Product with World Wide Distribution," *Kansas Business* 3 (January 1935): 12; "Kansas Firm Is Major Manufacturer of Coolers," *Midwest Industry Magazine* 34 (October 1966): 50, 52; "Winfield Is Proud of Plant Making School Color Crayons," *Kansas Business Magazine* 22 (January 1954): 10, 77–78.

63. Richard Archer, "Profile: Cowley County," *Kansas Business News* 3 (October 1982): 57; *Lawrence Journal-World,* January 24, March 6, October 1, and November 29, 1997; Traci Carl, "Main Street, Cowley County in Dire Straits," *Lawrence Journal-World,* February 2, 1997, p. B6.

64. Archer, "Profile: Cowley County," pp. 57–60; *Kansas Community Profile: Arkansas City* (Topeka: Kansas Department of Commerce and Housing, 2002); *Kansas Community Profile: Winfield* (Topeka: Kansas Department of Commerce and Housing, 2002).

65. "Other New Industry Follows Trailer Factories to Newton," *Kansas Business Magazine* 23 (December 1955): 7, 54; "Two New Factories Added to Newton Mobile Homes Center," *Midwest Industry Magazine* 24 (August 1956): 58; Dan Bearth, "Profile: Harvey County," *Kansas Business News* 4 (August 1983): 42–49.

66. Billy M. Jones, *Factory on the Plains: Lyle Yost and the Hesston Corporation* (Wichita: Center for Entrepreneurship, Wichita State University, 1987); "Hesston Company Expands Its Manufacture of Combine Parts," *Kansas Business Magazine* 21 (May 1953): 8, 88–89; "Hesston Reviews First 25 Years," *Midwest Industry Magazine* 40 (November 1972): 34; Dan Bearth, "They're Plowing New Ground at Hesston," *Kansas Business News* 3 (September 1982): 34–36.

67. *Kansas Community Profile: Hesston* (Topeka: Kansas Department of Commerce and Housing, 2002); *Kansas Community Profile: Newton* (Topeka: Kansas Department of Commerce and Housing, 2002).

68. *Kansas Community Profile: Topeka* (Topeka: Kansas Department of Commerce and Housing, 1986, 1994, 2002); "Southwestern Bell Moving Its Kansas Headquarters to Topeka," *Kansas Business Magazine* 18 (June 1950): 28.

69. *Kansas Community Profile: Topeka,* 2002; Dan Bearth, "Doug Wright," *Kansas Business News* 5 (October 1984): 27; *New Horizons Issue* (Topeka: Topeka Daily Capital, March 6, 1958), p. 9; *Seventy-fifth Anniversary Issue* (Topeka: Topeka Daily Capital, June 6, 1954), p. L12.

70. *Kansas Community Profile: Topeka,* 2002.

71. *Seventy-fifth Anniversary Issue,* p. N4.

72. Roy Bird, *Topeka: An Illustrated History of the Kansas Capital* (Topeka: Baranski Publishing Company, 1985), p. 133; Bearth, "Doug Wright," pp. 22, 27–30. The Interstate Commerce Commission rejected the merger of the Santa Fe with the Southern Pacific in 1986 after three years of combined operation.

73. Bearth, "Doug Wright," pp. 22, 27–35; Dan Bearth, "Capitol City Pops Lid Off Industrial Development," *Kansas Business News* 2 (December 1980): 32–35; Bearth, "Topeka Comes to Grips with Economic Hard Times," *Kansas Business News* 2 (September 1981): 32–37; Shortridge, *Kaw Valley Landscapes,* pp. 161–62.

74. Bearth, "Doug Wright," p. 30; "Huge Goodyear Plant to Topeka," *Kansas Business Magazine* 12 (August 1944): 8–9; "Latest in Cellophane," *Midwest Industry Magazine* 27 (August 1959): 14, 16; Richard Black, "Cellophane Plant Wraps Success," *Lawrence Journal-World,* May 3, 1998, p. E1; *Kansas Community Profile: Topeka,* 2002.

75. Shortridge, *Kaw Valley Landscapes,* pp. 161–62; *Kansas Community Profile: Topeka,* 2002; Michael Hooper, "President: Topeka Operation Viewed as Key Support Center," *Topeka Capital-Journal,* accessed at www.cjonline.com/webindepth/bnsf/stories/bus_bnsfprez.shtml; *Lawrence Journal-World,* June 14, 2002.

76. *Kansas Community Profile: Leavenworth* (Topeka: Kansas Department of Commerce and Housing, 2002); Leavenworth Area Development, "Major Employers—Leavenworth County, 2002," accessed at www.lvarea.com/data/majemp.htm.

77. Leavenworth Area Development, "Major Employers"; "Shipbuilding Hitting Lively Pace in Missouri Valley Yards," *Kansas Business Magazine* 20 (July 1952): 16; "Leavenworth Steel Has Complete Setup for Making Your Product," *Kansas Business Magazine* 21 (March 1953): 14.

78. Dan Bearth, "Profile: Leavenworth," *Kansas Business News* 4 (June 1983): 40–43.

79. Ibid.; Shortridge, *Kaw Valley Landscapes,* pp. 22–23, 26–27; Bill Foreman, "Leavenworth Invites Tourists to 'Do Some Time,'" *Lawrence Journal-World,* September 7, 1997, p. B5. Besides the federal penitentiary, the other three prisons in Leavenworth are the state corrections facility, the United States Disciplinary Barracks on Fort Leavenworth for military offenders, and, since the 1990s, a private institution owned by the Corrections Corporation of America that is used primarily to hold people awaiting trial or sentencing.

80. Bearth, "Profile: Leavenworth," p. 40; Shortridge, *Kaw Valley Landscapes,* pp. 46–53; J. H. Johnston III, *Leavenworth: Beginning to Bicentennial* (Leavenworth, Kans.: J. H. Johnston III, 1976), p. 32.

81. *Kansas Community Profile: Lawrence* (Topeka: Kansas Department of Commerce and Housing, 2002).

82. Ibid.; "Westvaco Chemical in Lawrence Doubles Capacity in Expansion," *Kansas Business Magazine* 21 (June 1953): 18, 41.

83. "Packer Plastics Takes Four Years to Double Facilities," *Midwest Industry Magazine* 40 (July 1972): 10, 12; Megan Neher, "Kmart Crew Celebrates Milestone," *Lawrence Journal-World,* November 17, 1996, p. E1; *Kansas Community Profile: Lawrence.*

84. James R. Shortridge and Barbara G. Shortridge, "Yankee Town on the Kaw: A Geographical and Historical Perspective on Lawrence and Its Environs," in *Embattled Lawrence: Conflict and Community,* ed. Dennis Domer and Barbara Watkins (Lawrence: University of Kansas Continuing Education, 2001), pp. 16–18.

85. Ibid.; John Walter, "The Lawrence Lifestyle: Good and Getting Better," *Kansas Business News* 1 (July 1980): 21–25; *Kansas Community Profile: Lawrence.*

86. For a penetrating analysis of attitudes and action in isolated county-seat towns during the middle and late twentieth century, see Kathleen Norris, *Dakota: A Spiritual Geography* (New York: Ticknor and Fields, 1993).

87. *Kansas Community Profile: Manhattan* (Topeka: Kansas Department of Commerce and Housing, 2002); *Kansas Community Profile: Junction City* (Topeka: Kansas Department of Commerce and Housing, 2002); *Lawrence Journal-World,* October 20, 1999.

88. *Kansas Community Profile: Manhattan;* Dan Bearth, "Profile: Manhattan," *Kansas Business News* 3 (June 1982): 40–45.

89. William Robbins, "Army's Plan to Expand Draws Neighbors' Ire," *New York Times,* June 30, 1989, p. 6; Dick Lipsey, "Fort Riley Backers Hold Breath for Base," *Lawrence Journal-World,* May 18, 1997, p. B10; *Kansas Community Profile: Junction City.*

90. Dan Bearth, "Profile: Junction City," *Kansas Business News* 4 (February 1983): 41–45; Don Terry, "Proud Base Faced with Notoriety by Association," *New York Times,* April 28, 1995, p. A11; John Milburn, "Communities Count on Military," *Lawrence Journal-World,* September 23, 2001, p. B4.

91. *Salina, Kansas Centennial, 1858–1958: Wagons to Wings* (Salina, Kans.: Salina Centennial, 1958); Ruby P. Bramwell, *City on the Move: The Story of Salina* (Salina, Kans.: Survey Press, 1969), p. 244.

92. Dan Bearth, "Salina's Big Leap Out," *Kansas Business News* 1 (November 1980): 34–39; *Kansas Community Profile: Salina* (Topeka: Kansas Department of Commerce and Housing, 1982, 1995).

93. *Kansas Community Profile: Salina,* 2002.

94. Bearth, "Salina's Big Leap Out," p. 35.

95. *Salina, Kansas Centennial;* Helyar, *Kansas Statistical Abstract, 2000,* pp. 6-26, 3-11; Linda Mowery-Denning, "Salina Hospital Merger Takes in Concordia," *Salina Journal,* August 5, 1995, p. 1.

96. Lyman D. Wooster, *Fort Hays Kansas State College: An Historical Story* (Hays, Kans.: Fort Hays Kansas State College, 1961), p. 194; History Book Committee, *At Home in Ellis County, Kansas, 1867–1992* (Hays, Kans.: Ellis County Historical Society, 1991), vol. 1, pp. 48–49; *Kansas Community Profile: Hays* (Topeka: Kansas Department of Commerce and Housing, 1982); Dan Bearth, "Profile: Hays," *Kansas Business News* 3 (April 1982): 36.

97. Bearth, "Profile: Hays," pp. 36–38.

98. Ted Blankenship, "Hays Hits Its Economic Stride 5 Years after Loss of Travenol," *Kansas Business News* 10 (March 1989): 32–34; Donald P. Schnacke, "Markets, Politics, Environmental Regulations Challenge Kansas Oil and Gas Industry," *Kansas Business News* 10 (September 1989): 11–12; *Kansas Community Profile: Hays,* 2002.

99. *Kansas Community Profile: Hays,* 2002; Bearth, "Profile: Hays," pp. 38–40; Helyar, *Kansas Statistical Abstract, 2000,* p. 3-14; Hays Medical Center, "Hospital Information," accessed at www.haysmed.com.

100. Book Committee, *They Came to Stay: Sherman County and Family History* (Goodland, Kans.: Sherman County Historical Society, 1981), vol. 3, pp. 318–20, 368; *Kansas Community Profile: Goodland* (Topeka: Kansas Department of Commerce and Housing, 1982).

101. "Matthew Harper Hamill Interview," typescript, 1994, located in the archives of the Prairie Museum of Art and History, Colby, Kansas.

102. *Wichita Eagle and Beacon,* February 1, 1963; Flint, Heir, and Heinrich, *Kansas Junior College,* p. 58.

103. *Kansas Community Profile: Russell* (Topeka: Kansas Department of Commerce and Housing, 2002).

104. *Lawrence Journal-World,* October 26, 2000, and June 18, 2001; Tim Unruh, "Russell on the Rebound," *Lawrence Journal-World,* December 2, 2002, p. B4.

105. Dan Bearth, "Robert Soelter," *Kansas Business News* 4 (December 1982): 19–22; *Kansas Community Profile: Abilene* (Topeka: Kansas Department of Commerce and Housing, 2002).

106. Roger Verdon, "Abilene's Greyhound Industry Searches for Recognition in Home State," *Kansas Business News* 2 (May 1981): 35–37.

107. *Lawrence Journal-World,* December 8, 1995.

108. Dan Bearth, "McPherson: Oasis of Energy," *Kansas Business News* 2 (April 1981): 28–35.

109. Ibid.; "Keith Swinehart," *Kansas Business News* 4 (March 1983): 16–17; *Kansas Community Profile: McPherson* (Topeka: Kansas Department of Commerce and Housing, 2002). The Johns-Manville plant is now a division of Schuller International Corporation and Sterling Drugs a part of Sanofi Winthrop. NCRA is an acronym formed from the company's former name, the National Co-operative Refinery Association.

110. "H. D. Lee Company Returns to Manufacture Popular Garments," *Kansas Business Magazine* 17 (December 1949): 8–9; Robert Burtch, "Manager Becomes Owner in Buy of Ottawa Truck," *Kansas Business News* 6 (March 1985): 8, 10; *Kansas Community Profile: Ottawa* (Topeka: Kansas Department of Commerce and Housing, 1982).

111. Ric Anderson, "Ottawa Basks in Turnaround," *Lawrence Journal-World,* December 17, 1995, pp. E1–E2; Anderson, "Ottawa Set Back by Firm's Closure," *Lawrence Journal-World,* February 20, 1997, p. B1; Mark Fagan, "Ottawa Draws Distribution Centers," *Lawrence Journal-World,* April 16, 2001, p. D6; *Kansas Community Profile: Ottawa,* 2002.

112. Dan Bearth, "Profile: Emporia," *Kansas Business News* 4 (December 1982): 37–40.

113. Ibid.; Stull and Broadway, "Effects of Restructuring," pp. 10–16; *Kansas Community Profile: Emporia* (Topeka: Kansas Department of Commerce and Housing, 1982, 2002).

114. *Kansas Community Profile: Emporia,* 2002; Bearth, "Profile: Emporia," pp. 37–40.

9. Postindustrial Kansas II: Life beyond the Exit Ramps

1. "Profile: Great Bend," *Kansas Business News* 3 (August 1982): 47; Donald P. Schnacke, "Markets, Politics, Environmental Regulations Challenge Kansas Oil and Gas Industry," *Kansas Business News* 10 (September 1989): 11.

2. Schnacke, "Markets, Politics," p. 11; "Oil Industry Is Making Great Bend a Boom Town," *Kansas Business Magazine* 22 (October 1954): 48; Margaret Shauers, "Good News? In Great Bend?" *Kansas Business News* 8 (July 1987): 34.

3. "Profile: Great Bend," pp. 47–49; *Kansas Community Profile: Great Bend* (Topeka: Kansas Department of Commerce and Housing, 2002); "Fuller Brush Starts Huge New Complex in Great Bend," *Midwest Industry Magazine* 39 (October 1971): 14; Margaret Shauers, "Fuller Brush: Largest Employer in Great Bend," *Kansas Business News* 8 (August 1987): 17–18.

4. *Kansas Community Profile: Great Bend;* "Profile: Great Bend," p. 47; *Rand McNally 2002 Commercial Atlas and Marketing Guide* (Chicago: Rand McNally, 2002), pp. 38–39.

5. "Kansas Brick and Tile, Hoisington, Is State's Newest, Most Modern," *Kansas Business Magazine* 23 (June 1955): 68–71; *Kansas Community Profile: Hoisington* (Topeka: Kansas Department of Commerce and Housing, 2002).

6. *Kansas Community Profile: Larned* (Topeka: Kansas Department of Commerce and Housing, 2002); "Doerr Metal Products Moves into New Factory at Larned," *Kansas*

Business Magazine 16 (October 1948): 48–49; "Doerr Metal Products Planning Another Major Plant Expansion," *Kansas Business Magazine* 21 (May 1953): 64–65; *Panorama of Progress, A Century of Living, Pawnee County, 1872–1972* (Larned, Kans.: Larned Tiller and Toiler, 1972).

7. Leroy Towns, "Hutchinson: The Growth Is Steady and Deliberate," *Kansas Business News* 1 (August 1980): 22–27.

8. Ibid.

9. Lillian Z. Martell, "Salt City Hits a Rough Spot," *Lawrence Journal-World,* April 25, 1999, pp. E1–E2; *Kansas Community Profile: Hutchinson* (Topeka: Kansas Department of Commerce and Housing, 2002).

10. *Kansas Community Profile: Hutchinson;* "Cessna Grows on Ground as Well as in the Air," *Midwest Industry Magazine* 34 (June 1966): 13, 39–41.

11. *Kansas Community Profile: Hutchinson.*

12. *Kansas Community Profile: Ellsworth* (Topeka: Kansas Department of Commerce and Housing, 2002); *Kansas Community Profile: Lyons* (Topeka: Kansas Department of Commerce and Housing, 2002).

13. Carl Manning, "Ellsworth Surprises State with Highest Increase in County Median Income," *Lawrence Journal-World,* August 12, 2002, pp. B1, B3; *Kansas Community Profile: Ellsworth;* Terry Rombeck, "Nuclear Storage Debate Hits Home in Lyons," *Lawrence Journal-World,* July 22, 2002, pp. A1–A5.

14. "Pratt's Industrial Growth Is Result of Community Selling," *Kansas Business Magazine* 21 (August 1953): 14–15; Tim Stucky, *Pratt, Kansas: A Centennial View, 1884–1984* (Pretty Prairie, Kans.: Prairie Publications, 1984), pp. 94–95, 104–7; *Kansas Community Profile: Pratt* (Topeka: Kansas Department of Commerce and Housing, 2002).

15. *Kansas Community Profile: Pratt.*

16. Agnesa Reeve, *Constant Frontier: The Continuing History of Finney County, Kansas* (Garden City, Kans.: Finney County Historical Society, 1996), pp. 117, 150–51, 158, 167, 210.

17. Ibid.; Leroy Towns, "Garden City: Optimism in a Land of Plenty," *Kansas Business News* 1 (June 1980): 19; David E. Kromm and Stephen E. White, "The High Plains Ogallala Region," in *Groundwater Exploitation in the High Plains,* ed. by David E. Kromm and Stephen E. White (Lawrence: University Press of Kansas, 1992), p. 15; Donald E. Green, "A History of Irrigation Technology Used to Exploit the Ogallala Aquifer," in Kromm and White, *Groundwater Exploitation,* pp. 28–43.

18. Donald D. Stull and Michael J. Broadway, "The Effects of Restructuring on Beefpacking in Kansas," *Kansas Business Review* 14 (fall 1990): 10–16.

19. Ibid., p. 11; Reeve, *Constant Frontier,* p. 188; Pauline Toland, ed., *Seward County Kansas* (Liberal, Kans.: Seward County Historical Society, 1979), pp. 158–59; *Kansas Community Profile: Liberal* (Topeka: Kansas Department of Commerce and Housing, 1982).

20. Stull and Broadway, "Effects of Restructuring," pp. 10–16.

21. "Liberal Has Wheat, Pancakes, but Big Talk Now Is Oil Boom," *Kansas Business Magazine* 21 (May 1953): 11; "Liberal Picks Up Tempo in Industry and Construction," *Kansas Business Magazine* 23 (June 1955): 22; *Kansas Centennial Futurama Edition* (Liberal, Kans.: Southwest Daily Times, March 18, 1961), sec. 8, p. 1; Toland, *Seward County,* p. 80.

22. "Liberal Picks Up Tempo," p. 22; "Tradewind Industries New Firm Making 'Wheat King' Truck Beds," *Kansas Business Magazine* 16 (December 1948): 11–12; "Beech Plant at Liberal Has Quality Production Record," *Kansas Business Magazine* 23 (June 1955): 7, 76–78.

23. Dan Bearth, "Profile: Liberal," *Kansas Business News* 5 (December 1983): 40–42,

44; *Kansas Community Profile: Liberal* (Topeka: Kansas Department of Commerce and Housing, 2002); Steve Brisendine, "Southwest Kansas Town a Microcosm of Hispanics' Gains in State," *Lawrence Journal-World,* May 17, 2001, p. B2.

24. Huber Self, *Environment and Man in Kansas: A Geographical Analysis* (Lawrence: Regents Press of Kansas, 1978), p. 105; Towns, "Garden City," p. 19.

25. *Kansas Community Profile: Garden City* (Topeka: Kansas Department of Commerce and Housing, 1982, 2002); Towns, "Garden City," pp. 18–20. Garden City's new economy and ethnic relationships were the subject of a major study in the 1980s sponsored by the Ford Foundation. A special issue of the journal *Urban Anthropology* edited by Donald D. Stull summarizes the results (vol. 19 [1990]: 303–427).

26. Towns, "Garden City," pp. 18–20; John W. Fountain, "Needy Workers Wait for a Kansas Plant to Reopen," *New York Times,* July 10, 2001, p. A8; Roxana Hegeman, "City Survives Despite ConAgra Fire," *Lawrence Journal-World,* January 1, 2002, p. B4.

27. Michael Paterniti, "Eating Jack Horner's Cow," *Esquire* 128 (November 1997): 90–97, 158, 160–63; Traci Carl, "Immigration Changes Face of Cowtowns," *Lawrence Journal-World,* July 6, 1997, pp. B1, B6; Betty Simecka, "Dodge City Revives Its Wild West Past," *Kansas Business News* 6 (May 1985): 52–53; Richard Archer, "Profile: Dodge City," *Kansas Business News* 4 (April 1983): 53–57.

28. Archer, "Profile: Dodge City," pp. 53–57; Self, *Environment and Man,* p. 105.

29. *Kansas Community Profile: Dodge City* (Topeka: Kansas Department of Commerce and Housing, 2002).

30. *Kansas Community Profile: Scott City* (Topeka: Kansas Department of Commerce and Housing, 2002).

31. *Kansas Community Profile: Hugoton* (Topeka: Kansas Department of Commerce and Housing, 2002); *Kansas Community Profile: Ulysses* (Topeka: Kansas Department of Commerce and Housing, 2002).

32. *Kansas Community Profile: Atchison* (Topeka: Kansas Department of Commerce and Housing, 2002); *Lawrence Journal-World,* February 5, 1997, and August 31, 2000.

33. Carol Beach, "Property Tax Rebate Attracts New Residents to Historic Town," *Kansas City Star,* August 20, 2000, pp. K1, K7.

34. *Kansas Community Profile: Holton* (Topeka: Kansas Department of Commerce and Housing, 2002); Randy Brown, "Clay Center: A Kansas Town Helping Itself," *Kansas Business News* 10 (June 1989): 18–19, 24; *Kansas Community Profile: Clay Center* (Topeka: Kansas Department of Commerce and Housing, 2002); Karen Kennedy, "Balderson Snow Plow Business Started from Blacksmith's Shop," *Kansas Business Magazine* 16 (March 1948): 10, 12, 14; Ron Welch, "Small Kansas Company Sells to Russia," *Kansas Business News* 9 (July 1988): 7, 14, 22–23; *Kansas Community Profile: Wamego* (Topeka: Kansas Department of Commerce and Housing, 2002).

35. History and Literature Club, *History of Horton and Surrounding Neighborhoods* (Horton, Kans.: Horton Pride, 1974), p. 19; *Kansas Community Profile: Horton* (Topeka: Kansas Department of Commerce and Housing, 2002); Ron Welch, "Horton Shapes Its Own Destiny," *Kansas Business News* 11 (January–February, 1990): 10–12.

36. *Kansas Community Profile: Sabetha* (Topeka: Kansas Department of Commerce and Housing, 2002).

37. *Kansas Community Profile: Hiawatha* (Topeka: Kansas Department of Commerce and Housing, 2002); *Lawrence Journal-World,* October 1, 1999.

38. *Kansas Community Profile: Marysville* (Topeka: Kansas Department of Commerce and Housing, 2002).

39. Ibid.; "Landoll Corporation: A Do-It-Yourself Firm," *Kansas Business News* 8 (August 1987): 43–46.

40. *Kansas Community Profile: Belleville* (Topeka: Kansas Department of Commerce and Housing, 1982, 1990, 2002).

41. Janet P. Emery, *It Takes People to Make a Town: The Story of Concordia, Kansas, 1871–1971* (Salina, Kans.: Arrow Printing Company, 1970), p. 100; *Kansas Community Profile: Concordia* (Topeka: Kansas Department of Commerce and Housing, 1982).

42. *Kansas Community Profile: Concordia,* 2002.

43. Leon Gennette, project director, *Cloud County History* (Concordia, Kans.: Cloud County Historical Society and Cloud County Genealogical Society, 1992), pp. 97–99.

44. *Kansas Community Profile: Beloit* (Topeka: Kansas Department of Commerce and Housing, 2002); John S. Morrell, "A History of Mitchell County" (mimeographed, n.d., copy located in Kansas Collection at the University of Kansas).

45. *Kansas Community Profile: Norton* (Topeka: Kansas Department of Commerce and Housing, 2002); *Kansas Community Profile: Phillipsburg* (Topeka: Kansas Department of Commerce and Housing, 2002).

46. *Kansas Community Profile: Phillipsburg;* Ron Welch, "The Kyle: 'A Mighty Good Little Road,'" *Kansas Business News* 9 (September 1988): 19–21.

47. *Kansas Community Profile: Louisburg* (Topeka: Kansas Department of Commerce and Housing, 2002); *Kansas Community Profile: Osawatomie* (Topeka: Kansas Department of Commerce and Housing, 2002); *Kansas Community Profile: Paola* (Topeka: Kansas Department of Commerce and Housing, 2002); *Miami County, Kansas 1987* (Paola, Kans.: Miami County Historical Society, 1987), p. 23.

48. Oscar A. Copple and Joyce M. Hitchings, *History of Osage City, and Osage County* (n.p.: 1970), p. 36; *Kansas Community Profile: Osage City* (Topeka: Kansas Department of Commerce and Housing, 1988, 2002); Jenny Upchurch, "Modular Homes Find Growing Acceptance," *Omaha World-Herald,* January 24, 1999, accessed at www.nutrendomaha.com/articles/12499.htm.

49. *Kansas Community Profile: Burlington* (Topeka: Kansas Department of Commerce and Housing, 2002); Wolf Creek Nuclear Operating Corporation, "About Wolf Creek," accessed at www.wcnoc.com/whynuclearfuel.cfm.

50. *Kansas Community Profile: Herington* (Topeka: Kansas Department of Commerce and Housing, 2002); "RITS News—ROCK 103" (a newsletter for enthusiasts of the Rock Island railroad system), July 1999, accessed at www.simpson.edu/~RITS/news/TheROCK/News103.html.

51. Dale E. Jones et al., *Religious Congregations and Membership in the United States* (Nashville, Tenn.: Glenmary Research Center, 2002), p. 23; *Kansas Community Profile: Hillsboro* (Topeka: Kansas Department of Commerce and Housing, 2002).

52. *Centennial Edition* (Iola, Kans.: Iola Register, May 30, 1955), pp. 1D–3D.

53. *Fort Scott* (Fort Scott, Kans.: City of Fort Scott, n.d.)

54. James McCain, "Burt Hoefs," *Kansas Business News* 3 (March 1982): 23–25; *Kansas Community Profile: Iola* (Topeka: Kansas Department of Commerce and Housing, 1982, 2002).

55. J. R. Turman, "Neodesha Example of Community Effort Toward Hiking Payrolls," *Kansas Business Magazine* 19 (June 1951): 38–40.

56. Ibid.; "City of Neodesha, Bankers Join Forces to Rescue Ailing Manufacturing Firm," *Kansas Business News* 3 (October 1982): 8–10; *Kansas Community Profile: Neodesha* (Topeka: Kansas Department of Commerce and Housing, 1982, 2002).

57. Lincoln Financial Group, "Our History," accessed at www.lincoln.lfg.com/who/history.htm; *Kansas Community Profile: Fort Scott* (Topeka: Kansas Department of Commerce and Housing, 2002).

58. *Kansas Community Profile: Fort Scott;* Ward/Kraft, Inc., "About Ward/Kraft," accessed at www.wardkraft.com/about.htm; "Dayco Corporation," in *Ozarks Labor Union Archives, Local 662* (Springfield, Mo.: Southwest Missouri State University, n.d.), accessed at www.library.smsu.edu/Meyer/SpecColl/olua16.html; Associated Wholesale Grocers, Inc., "History of AWG," accessed at www.awginc.com/about/history.htm.

59. *Fort Scott.*

60. Mid-America, Inc., *Preliminary Report Pertaining to Some of the Assets of Southeast Kansas* (Parsons, Kans.: Mid-America, Inc., n.d.). In 2000, Mid-America merged with a Pittsburg-based organization, the Southeast Kansas Economic Alliance, to form Southeast Kansas, Incorporated. See Trish Hollenbeck, "Mid-America, SEK Economic Alliance Join Forces," *Pittsburg Morning Sun,* October 30, 1999, accessed at www.morningsun.net/stories/103099/bus_1030990012.shtml.

61. Jack M. Flint, Floyd Heir, and Carl L. Heinrich, *The Kansas Junior College* (Topeka: State Department of Public Instruction, 1968), pp. 51–63; Thelma Helyar, ed., *Kansas Statistical Abstract, 2000* (Lawrence: Policy Research Institute, University of Kansas, 2001), p. 6-29.

62. Sherry L. Schirmer and Theodore A. Wilson, *Milestones: A History of the Kansas Highway Commission and the Department of Transportation* (Topeka: Kansas Department of Transportation, 1986), p. 5-27; Ted Blankenship, "No One Knows What Will Be in Governor's Road Program," *Kansas Business News* 8 (June 1987): 5–8; Bob Whittaker, "Whittaker Touts Southeast Kansas Highway Program," *Kansas Business News* 8 (June 1987): 18–19; Harold Campbell, "Enhancement of U.S. 400 Important to Area's Future," *Pittsburg Morning Sun,* March 10, 2000, accessed at www.morningsun.net/stories/031000/loc_0310000007.shtml.

63. *Kansas Community Profile: Fredonia* (Topeka: Kansas Department of Commerce and Housing, 2002).

64. *Kansas Community Profile: Chanute* (Topeka: Kansas Department of Commerce and Housing, 1982, 2002); *Kansas Community Profile: Parsons* (Topeka: Kansas Department of Commerce and Housing, 1987, 1993, 2002); Global Security.org, "Kansas Army Ammunition Plant," accessed at www.globalsecurity.org/military/facility/aap-kansas.htm; Mrs. Tommie J. Crispino, ed., *The Centennial Story of Parsons, Kansas* (Parsons, Kans.: n.p., 1971), p. 60.

65. *Kansas Community Profile: Independence* (Topeka: Kansas Department of Commerce and Housing, 1982, 2002); *Kansas Community Profile: Coffeyville* (Topeka: Kansas Department of Commerce and Housing, 2002).

66. *Kansas Community Profile: Chanute,* 1982; *Kansas Community Profile: Parsons,* 1982.

67. *Kansas Community Profile: Chanute,* 2002; interview with Anna Methvin, Chanute Chamber of Commerce, November 7, 2002.

68. *Kansas Community Profile: Parsons,* 2002; Campbell, "Enhancement of U.S. 400"; Crispino, *Centennial Story,* p. 1.

69. Amy L. Rork, "Sense of Place in Montgomery County, Kansas: Perceptions of an Industrialized Rural Area" (master's thesis, University of Kansas, 1997), pp. 54–57.

70. Ibid., pp. 57–61; Richard Brack, "Amazon.com Taps Site in Coffeyville," *Lawrence Journal-World,* April 14, 1999, p. D5; *Kansas Community Profile: Coffeyville.*

71. *Kansas Community Profile: Independence,* 1982, 2002; *Lawrence Journal-World,* January 8, 1999.

72. Rork, "Sense of Place," pp. 65–67; *Lawrence Journal-World,* July 3, 1995, and July 6, 1999; Traci Carl, "Southeast Kansas' Economy Takes Flight with Cessna Plant," *Lawrence Journal-World,* January 18, 1997, p. B8.

73. Ross Milloy, "Waste from Old Mines Leaves Piles of Problems, *New York Times,* July 21, 2000, p. A10.

74. Claude H. Nichols, comp., *The Baxter Springs Story* (Baxter Springs, Kans.: Baxter Springs Centennial, 1958); *Kansas Community Profile: Baxter Springs* (Topeka: Kansas Department of Commerce and Housing, 2002); *Kansas Community Profile: Galena* (Topeka: Kansas Department of Commerce and Housing, 2002); *Lawrence Journal-World,* June 14, 1995.

75. Dan Bearth, "Profile: Pittsburg," *Kansas Business News* 3 (February 1982): 30; *Kansas Community Profile: Columbus* (Topeka: Kansas Department of Commerce and Housing, 2002); *Kansas Community Profile: Girard* (Topeka: Kansas Department of Commerce and Housing, 2002).

76. Bearth, "Profile: Pittsburg," p. 29.

77. Ibid., pp. 29–36; *Kansas Community Profile: Pittsburg* (Topeka: Kansas Department of Commerce and Housing, 1982, 1987); *Kansas Community Profile: Columbus,* 1982, 1987; "Mac-Pitt Engineers Look to Other Basic Industries," *Midwest Industry Magazine* 34 (August 1966): 16, 66–67.

78. Interview with Jerry Lindberg, Pittsburg Industrial Development Office, December 19, 2002; *Kansas Community Profile: Pittsburg,* 2002; Pittsburg Chamber of Commerce, "Business and Industry (2002)," accessed at www.pittsburgkschamber.com/business.asp.

79. Ann D. Costantini and Misty Cox, *Pittsburg, Crawford County Visitor's Guide* (Pittsburg, Kans.: Crawford County Convention and Visitors Bureau, 2000); *Kansas Community Profile: Pittsburg,* 2002.

80. *Rand McNally 2002 Commercial Atlas,* pp. 38–39. Joplin officials began to expand a former community college into that city's own four-year institution in 1967. A decade later Missouri Southern College became state affiliated, but it still is much smaller than Pittsburg State University.

10. Conclusion

1. *Parsons: Its Past, Present and Future (Special Edition)* (Parsons, Kans.: Parsons Palladium, October 16, 1901), p. 1.

2. The St. Joseph and Denver City Railroad later became the St. Joseph and Grand Island. It is now the Northeast Kansas and Missouri Railroad. The St. Louis–San Francisco line through Fort Scott was known earlier as the Missouri River, Fort Scott and Gulf. It is now part of the Burlington Northern–Santa Fe system.

3. Location issues for colleges are discussed in Harry C. Humphreys, *The Factors Operating in the Location of State Normal Schools* (New York: Teachers College, Columbia University, 1923), p. 34, and in Paul V. Turner, *Campus: An American Planning Tradition* (Cambridge, Mass.: MIT Press, 1984), p. 23.

Bibliography

"A. L. Duckwall Company Adds Two New Stores in Expansion." *Kansas Business Magazine* 16 (August 1948): 16.
Abilene, Kansas: A Clean, Healthy Town. Abilene, Kans.: Abilene Commercial Club, 1915.
Achievement Edition. Winfield, Kans.: Winfield Daily Courier, February 25, 1952.
Adams, Franklin G. "The Capitals of Kansas." *Transactions of the Kansas State Historical Society* 8 (1903–1904): 331–51.
———, comp. *The Homestead Guide.* Waterville, Kans.: F. G. Adams, 1873.
Advertising Leavenworth (Special Issue). Leavenworth, Kans.: Leavenworth Post, December 30, 1913.
"Aircraft Instruments Made and Repaired by Garwin, Inc." *Kansas Business Magazine* 21 (December 1953): 11–12.
Alexander, John W. "The Basic-Nonbasic Concept of Urban Economic Functions." *Economic Geography* 30 (1954): 246–61.
Alexandersson, Gunnar. *The Industrial Structure of American Cities.* Lincoln: University of Nebraska Press, 1956.
All about NCRA. McPherson, Kans.: National Cooperative Refinery Association, 1976.
Allard, Mary M. *Holton, Kansas, 1856–1970.* Holton, Kans.: Bookman Club, 1970.
Allen, Joseph W. *Neodesha, Wilson County, Kansas, from the Wilderness Days to the Coming of the Railroads, 1865–1886.* Neodesha, Kans.: Joseph W. Allen, 1962.
"Allis-Chalmers in Topeka Growing in Facilities, Personnel and Products." *Midwest Industry Magazine* 42 (February 1974): 10, 12–13.
Allison, Nathaniel T. *History of Cherokee County, Kansas, and Representative Citizens.* Chicago: Biographical Publishing Company, 1904.
Alumbaugh, Jack. "SCKEDD: Looking Up the Road from South Central Kansas." *Kansas Business News* 10 (September 1989): 20–21.
"American Salt Company Expands Operations at Lyons Plant." *Kansas Business Magazine* 22 (September 1954): 22.
Ames, Charles E. *Pioneering the Union Pacific: A Reappraisal of the Building of the Railroad.* New York: Appleton-Century-Crofts, 1969.
"Ancient Art, Once Lost, Now a Major Kansas Industry." *Kansas Business* 2 (May 1934): 10–11.
Anderson, George L. "Atchison and the Central Branch Country, 1865–1874." *Kansas Historical Quarterly* 28 (1962): 1–24.
———. *Kansas West: An Epic of Western Railroad Building.* 1936. San Marino, Calif.: Golden West Books, 1963.
———. "Atchison, 1865–1886, Divided and Uncertain." *Kansas Historical Quarterly* 35 (1969): 30–45.
———. "A North-South Link: Missouri Pacific's Proposal: Union Pacific's Achievement, 1889–1910." *Kansas Quarterly* 2 (summer 1970): 88–96.

Anderson, Ric. "Ottawa Basks in Turnaround." *Lawrence Journal-World,* December 17, 1995, pp. E1–E2.

———. "Ottawa Set Back by Firm's Closure." *Lawrence Journal-World,* February 20, 1997, p. B1.

"Anniversary: W. C. Coleman, 80 Years, the Coleman Company, 50 Years." *Kansas Business Magazine* 18 (April 1950): 8, 10, 42–46.

Annual Report of the Department of Labor and Industry. Topeka: State Printer, various years.

Annual Report of the Kansas Turnpike Authority. Topeka: State Printer, various years.

Anthony, W. P. *Clay Center Illustrated.* Clay Center, Kans.: Clay Center Dispatch, 1901.

Anthony, the First Hundred Years, 1878–1978. Anthony, Kans.: Historical Museum of Anthony, 1978.

Archer, Richard. "Competition Heats Up for Kansas's Share of Military Contracts." *Kansas Business News* 3 (September 1982): 43–45.

———. "Profile: Cowley County." *Kansas Business News* 3 (October 1982): 57–60.

———. "Profile: Dodge City." *Kansas Business News* 4 (April 1983): 53–57.

———. "Military Aviation Gives Economy a Lift." *Kansas Business News* 5 (April 1984): 18–23.

"Arkansas City" (Special Issue). *Kansas Business* 4 (February 1936): 7.

Armitage, Merle. *Operations Santa Fe.* New York: Duell, Sloan and Pearce, 1948.

Associated Wholesale Grocers, Inc. "History of AWG." Accessed at www.awginc.com/about/history.htm.

Atchison Board of Trade, comp. *Atchison, the Railroad Centre of Kansas.* Atchison, Kans.: Board of Trade, 1874.

"Atchison Hardware Firm in Outstanding 75-Year Growth." *Kansas Business Magazine* 14 (December 1946): 6–7.

Atchison Merchants' and Manufacturers' Bureau. *The Advantages of Atchison, Kansas, as a Commercial and Manufacturing Centre.* Atchison, Kans.: Merchants' and Manufacturers' Bureau, 1887.

"Atchison Plant Is Factor in National Transportation." *Kansas Business* 7 (June 1939): 10–11.

The Atchison, Topeka and Santa Fe Railroad: Celebration of Its 80th Birthday (Special Issue). Atchison, Kans.: Atchison Daily Globe, September 17, 1940.

"Atchison's Fitz Overall Firm Spurs Production, Sales Plans." *Kansas Business Magazine* 18 (August 1950): 9, 49.

Atkinson, Eva L. "Kansas City's Livestock Trade and Packing Industry, 1870–1914: A Study in Regional Growth." Ph.D. diss., University of Kansas, 1971.

"Aviation Is Here to Stay." *Kansas Business News* 7 (March 1986): 12–14.

Bailles, Kendall. *From Hunting Ground to Suburb: A History of Merriam, Kansas.* Rev. ed. Merriam, Kans.: Merriam Chamber of Commerce, 1968.

Bain, William E. *Frisco Folks: Stories and Pictures of the Great Steam Days of the Frisco Road (St. Louis–San Francisco Railway Company).* Denver: Sage Books, 1961.

Baker, Hazel B. "Early History of Larned State Hospital." Manuscript, n.d. Located in the Kansas Collection at the University of Kansas.

Ballard, David E. "The First State Legislature." *Kansas Historical Collections, 1907–1908* 10 (1908): 232–79.

Barlow, Mary L., comp. *The Why of Fort Scott.* N.p., 1921.

Barnabas, Bentley. "Coleman Products Give Light and Warmth to Far Corners of Earth." *Kansas Business* 2 (November 1934): 9.

Barnes, Elizabeth E. *Historic Johnson County: A Bird's Eye View of the Development of the Area.* Shawnee Mission, Kans.: Neff Printing Company, 1969.

Barnes, Lela. "The Leavenworth Board of Trade, 1882–1892." *Kansas Historical Quarterly* 1 (1932): 360–78.

———, ed. "Letters of Cyrus Kurtz Holliday, 1854–1859." *Kansas Historical Quarterly* 6 (1937): 241–94.

———. "An Editor Looks at Early-Day Kansas: The Letters of Charles Monroe Chase." *Kansas Historical Quarterly* 26 (1960): 113–51, 267–301.

Barry, Louise. *The Beginning of the West: Annals of the Kansas Gateway to the American West, 1540–1854.* Topeka: Kansas State Historical Society, 1972.

Baruth, Wilma F. "Straight as the Crow Flies: Historical Geography of the Kansas City Southern Railway Company." Master's thesis, Kansas State University, 1986.

Bawden, William T. *A History of Kansas State Teachers College of Pittsburg, 1903–1941.* Pittsburg, Kans.: Kansas State Teachers College, 1952.

Beach, Carol. "Property Tax Rebate Attracts New Residents to Historic Town." *Kansas City Star,* August 20, 2000, pp. K1, K7.

Beach Advertising Co., comp. *Hutchinson, Kansas.* Hutchinson, Kans.: Hutchinson Chamber of Commerce, 1951.

Bearth, Dan. "Salina's Big Leap Out." *Kansas Business News* 1 (November 1980): 34–39.

———. "Capitol City Pops Lid Off Industrial Development." *Kansas Business News* 2 (December 1980): 32–35.

———. "Who'll Rule the Rails?" *Kansas Business News* 2 (December 1980): 20–31.

———. "Joe Pichler: A Dean Climbs Aboard Dillon's Rising Star." *Kansas Business News* 2 (January 1981): 30–33.

———. "McPherson: Oasis of Energy." *Kansas Business News* 2 (April 1981): 28–35.

———. "Annual Report: Kansas Public Companies." *Kansas Business News* 2 (June 1981): 23–32.

———. "Business Moves to the Suburbs." *Kansas Business News* 2 (June 1981): 34–39.

———. "Coleman Sees Bright Lights Ahead." *Kansas Business News* 2 (July 1981): 32–37.

———. "Hopes Rise on the High Plains." *Kansas Business News* 2 (August 1981): 24–28.

———. "Topeka Comes to Grips with Economic Hard Times." *Kansas Business News* 2 (September 1981): 32–37.

———. "Kansas General Aviation: The Future Days." *Kansas Business News* 2 (October 1981): 40–44.

———. "Atchison." *Kansas Business News* 3 (December 1981): 32–39.

———. "Profile: Pittsburg." *Kansas Business News* 3 (February 1982): 29–36.

———. "Profile: Hays." *Kansas Business News* 3 (April 1982): 36–41.

———. "Profile: Manhattan." *Kansas Business News* 3 (June 1982): 40–45.

———. "Boeing's Lionel Alford." *Kansas Business News* 3 (July 1982): 20–27.

———. "They're Plowing New Ground at Hesston." *Kansas Business News* 3 (September 1982): 34–36.

———. "'World's Largest Western Stores Blaze Urban Trail." *Kansas Business News* 3 (November 1982): 16–20.

———. "Profile: Emporia." *Kansas Business News* 4 (December 1982): 37–40.

———. "Robert Soelter." *Kansas Business News* 4 (December 1982): 19–22.

———. "Profile: Junction City." *Kansas Business News* 4 (February 1983): 41–45.
———. "Profile: Leavenworth." *Kansas Business News* 4 (June 1983): 40–43.
———. "Profile: Harvey County." *Kansas Business News* 4 (August 1983): 42–49.
———. "Profile: Southeast Kansas." *Kansas Business News* 4 (October 1983): 42, 44–48.
———. "Profile: Liberal." *Kansas Business News* 5 (December 1983): 40–42, 44.
———. "Kansas City, Kansas, Ushers in New Political and Business Climate." *Kansas Business News* 5 (January 1984): 42–45.
———. "Doug Wright." *Kansas Business News* 5 (October 1984): 22, 27–35.
———. "Jerry Mallot." *Kansas Business News* 5 (October 1984): 23–26.
———. "A Man of Ideas Brings Two Decades of Growth to Gott Corporation." *Kansas Business News* 6 (December 1984): 14–19.
———. "Andover: The Promised Land." *Kansas Business News* 8 (April 1987): 56, 58–60.
Beckman, Peter. "Atchison's First Railroad." *Kansas Historical Quarterly* 21 (1954): 153–66.
———. "The Overland Trade and Atchison's Beginnings." In *Territorial Kansas: Studies Commemorating the Centennial,* pp. 148–61. Lawrence: University of Kansas Publications, 1954.
"Beech, Boeing Announce Plans for Major Expansions." *Kansas Business News* 4 (February 1983): 10–11.
"Beech Plant at Liberal Has Quality Production Record." *Kansas Business Magazine* 23 (June 1955): 7, 76–78.
Belcher, Wyatt A. *The Economic Rivalry between St. Louis and Chicago, 1850–1880.* New York: Columbia University Press, 1947.
Bell, Daniel. *The Coming of Post-industrial Society.* New York: Basic Books, 1973.
"Bennett Creamery Company Is Typical Kansas Industry." *Kansas Business* 7 (December 1939): 22–23.
Benson, Janet E. "Good Neighbors: Ethnic Relations in Garden City Trailer Courts." *Urban Anthropology* 19 (1990): 361–86.
Bentley, Orsemus H., ed. *History of Wichita and Sedgwick County, Kansas.* 2 vols. Chicago: C. F. Cooper and Company, 1910.
Berry, Brian J. L. *A Geography of Market Centers and Retail Distribution.* Englewood Cliffs, N.J.: Prentice-Hall, 1967.
Betton, Frank H. "The Genesis of a State's Metropolis." *Kansas Historical Collections, 1901–1902* 7 (1902): 114–20.
Biennial Report of the Kansas Secretary of State. Topeka: State Printer, various years.
Biennial Report of the Kansas State Industrial Reformatory. Topeka: State Printer, various years.
Biennial Report of the Larned State Hospital. Topeka: State Printer, various years.
Biennial Report of the State Home for Feeble-Minded. Topeka: State Printer, various years.
Biennial Report of the State Hospital for Epileptics. Topeka: State Printer, various years.
Biennial Report of the State Orphans' Home. Topeka: State Printer, various years.
Biennial Report of the State Tubercular Sanatorium. Topeka: State Printer, various years.
"Big Ammonium Plant to Kansas." *Kansas Business Magazine* 9 (August 1941): 8.
"Big Rock Island Yard to Kansas." *Kansas Business Magazine* 16 (September 1948): 38–39.
"Bigness Has Advantages." *Midwest Industry Magazine* 32 (February 1964): 14, 16.
Biographical History of Barton County, Kansas. Great Bend, Kans.: Great Bend Tribune, 1912.

Bird, Roy D. *An Ethnic History of Shawnee County, Kansas.* Topeka: Topeka–Shawnee County Metropolitan Planning Commission, 1974.

———. *Topeka: An Illustrated History of the Kansas Capital.* Topeka: Baranski Publishing Company, 1985.

Bird, Roy D., and Douglass W. Wallace. *Witness of the Times: A History of Shawnee County.* Topeka: Shawnee County Historical Society, 1976.

Black, Brian. "Mastering the Kaw: The Bowersock Dam and the Development of Lawrence Industry." *Kansas History* 16 (1993): 262–75.

Black, Judie. "Archie Dykes." *Kansas Business News* 2 (November 1981): 24–29.

Black, Richard. "Cellophane Plant Wraps Success." *Lawrence Journal-World,* May 3, 1998, p. E1.

Blackmar, Frank W. "Penology in Kansas." *Kansas University Quarterly* 1 (April 1893): 156–60.

———. *The Life of Charles Robinson: The First Governor of Kansas.* Topeka: Crane and Company, 1902.

Blankenship, Ted. "No One Knows What Will Be in Governor's Road Program." *Kansas Business News* 8 (June 1987): 5–8.

———. "Beef May Be State's Top Industry." *Kansas Business News* 9 (January 1988): 25–27.

———. "Morrison Company Inc. and W. Dale Arnold." *Kansas Business News* 9 (July 1988): 8–9.

———. "There Are Real Problems, but Optimism, Too, in Health-Care Industry." *Kansas Business News* 9 (October 1988): 8–11.

———. "Wichitan Finds a Need, and Builds a Company." *Kansas Business News* 10 (January 1989): 14–16.

———. "Hays Hits Its Economic Stride 5 Years after Loss of Travenol." *Kansas Business News* 10 (March 1989): 32–34.

———. "Lyle Yost: The Company Builder." *Kansas Business News* 10 (June 1989): 20, 24.

Bolton, Rachel. "Joe Dennis." *Kansas Business News* 3 (July 1982): 12–14.

Book Committee. *The History of Greenwood County, Kansas.* 2 vols. Eureka, Kans.: Greenwood County Historical Society, 1986.

Book Committee. *New Branches from Old Trees: A New History of Wabaunsee County.* Alma, Kans.: Wabaunsee County Historical Society, 1976.

Book Committee. *They Came to Stay: Sherman County and Family History.* 3 vols. Goodland, Kans.: Sherman County Historical Society, 1980, 1981.

Borchert, John R. "American Metropolitan Evolution." *Geographical Review* 57 (1967): 301–32.

———. "America's Changing Metropolitan Regions." *Annals of the Association of American Geographers* 62 (1972): 352–73.

Bottom, Rachel. "Newsmakers: Joe Dennis." *Kansas Business News* 3 (July 1982): 12–14.

Boughton, Joseph S. *Lawrence, Kansas: A Good Place to Live.* Lawrence, Kans.: J. S. Boughton, 1904.

Bowen, William A. *The Willamette Valley: Migration and Settlement on the Oregon Frontier.* Seattle: University of Washington Press, 1978.

Bowers, D. N. *Seventy Years in Norton County, Kansas: 1872–1942.* Norton, Kans.: Norton County Champion, 1942.

Brack, Richard. "Amazon.com Taps Site in Coffeyville." *Lawrence Journal-World,* April 14, 1999, p. D5.

Bradley, Glenn D. *The Story of the Santa Fe.* Boston: Richard G. Badger, 1920.

Bramwell, Ruby P. *City on the Move: The Story of Salina.* Salina, Kans.: Survey Press, 1969.

Breckenridge, R. H. *An Industrial Survey of Pratt, Kansas.* Manhattan, Kans.: Engineering Experiment Station, Kansas State College, 1950.

Brigham, Lalla M. *The Story of Council Grove on the Santa Fe Trail.* Rev. ed. Council Grove, Kans.: Council Grove Republican, 1921.

Brisendine, Steve. "Southwest Kansas Town a Microcosm of Hispanics' Gains in State." *Lawrence Journal-World,* March 17, 2001, p. B2.

Broadway, Michael J. "The Origins and Determinants of Indochinese Secondary In-Migration in S. W. Kansas." *Heritage of the Great Plains* 20 (spring 1987): 19–29.

———. "Meatpacking and Its Social and Economic Consequences for Garden City, Kansas, in the 1980s." *Urban Anthropology* 19 (1990): 321–44.

———. "Beef Stew: Cattle, Immigrants, and Established Residents in a Kansas Beefpacking Town." In *Newcomers in the Workplace: New Immigrants and the Restructuring of the U. S. Economy,* edited by Louise Lamphere, Guillermo Grenier, and Alex Stepick, pp. 25-43. Philadelphia: Temple University Press, 1994.

Broadway, Michael J., and Donald D. Stull. "Rural Industrialization: The Example of Garden City, Kansas." *Kansas Business Review* 14, no. 4 (1991): 1–9.

Broadway, Michael J., and Terry Ward. "Recent Changes in the Structure and Location of the U.S. Meatpacking Industry." *Geography* 75 (1990): 76–79.

Bromberg, Nicolette. "The Fastest Route to Denver: The Development of the East-West Highway across Kansas." Master's thesis, University of Kansas, 1995.

Brown, A. Theodore. *Frontier Community: Kansas City to 1870.* Columbia: University of Missouri Press, 1963.

Brown, A. Theodore, and Lyle W. Dorsett. *K.C.: A History of Kansas City.* Boulder, Colo.: Pruett Publishing, 1978.

Brown, Charles. *Marysville as It Is.* Marysville, Kans.: Press of P. Springer, 1887.

Brown, Lawrence A., Rodrigo Sierra, Scott Digiacento, and W. Randy Smith. "Urban-System Evolution in Frontier Settings." *Geographical Review* 84 (1994): 249–65.

Brown, Randy. "Wichita Stockyards Gone . . . But Memories Live On." *Kansas Business News* 9 (January 1988): 16–18.

———. "What's on the Minds of Kansans? Highways, Highways, Highways." *Kansas Business News* 9 (October 1988): 30–34.

———. "Insurance Industry Thrives in Kansas." *Kansas Business News* 9 (November 1988): 8–10.

———. "Kansas Trucking Industry: Finding the Road to Profitability." *Kansas Business News* 10 (January 1989): 8–9.

———. "1989 Could Set Stage for Recovery." *Kansas Business News* 10 (February 1989): 28–30.

———. "Air Service in Kansas a Mixed Bag." *Kansas Business News* 10 (March 1989): 20–21, 26–28.

———. "Entrepreneurship: The Soul of Kansas." *Kansas Business News* 10 (April 1989): 9–11.

———. "Clay Center: A Kansas Town Helping Itself." *Kansas Business News* 10 (June 1989): 18–19, 24.

Brush, John E. "The Hierarchy of Central Places in Southwestern Wisconsin." *Geographical Review* 43 (1953): 380–402.

Bryant, Keith L., Jr. *Arthur E. Stilwell: Promoter with a Hunch.* Nashville, Tenn.: Vanderbilt University Press, 1971.
———. *History of the Atchison, Topeka and Santa Fe Railway.* New York: Macmillan, 1974.
Burgess, Ernest W., and J. J. Sippy. *Belleville Social Survey.* Belleville, Kans.: Telescope Print, 1914.
Burghardt, Andrew F. "The Location of River Towns in the Central Lowland of the United States." *Annals of the Association of American Geographers* 49 (1959): 305–23.
———. "A Hypothesis about Gateway Cities." *Annals of the Association of American Geographers* 61 (1971): 269–85.
Burke, William S., and J. L. Rock. *The History of Leavenworth: The Metropolis of Kansas, and the Chief Commercial Center West of the Missouri River.* Leavenworth, Kans.: Leavenworth Board of Trade, 1880.
Burns, Nancy. "The Collapse of Small Towns on the Great Plains: A Bibliography." *Emporia State Research Studies* 31 (summer 1982): 5–36.
Burr, Myron C., and Elizabeth Burr, eds. *The Kinsley-Edwards County Centennial, 1873–1973.* Kinsley, Kans.: Nolan Publishers, 1973.
Burtch, Robert. "Manager Becomes Owner in Buy of Ottawa Truck." *Kansas Business News* 6 (March 1985): 8, 10.
———. "Mobile Homes Seek New Foundation." *Kansas Business News* 6 (July 1985): 36–39.
Business Men's League. *Annual Review of Greater Kansas City.* Kansas City: Business Men's League, 1908.
Byers, Oliver P. "The Conception and Growth of a Kansas Railroad." *Kansas Historical Collections, 1911–1912* 12 (1912): 383–87.
———. "Early History of the El Paso Line of the Chicago, Rock Island and Pacific Railway." *Kansas Historical Collections, 1919–1922* 15 (1923): 573–78.
"Byers Portobase Corporation Is Booming Chanute Industry." *Kansas Business Magazine* 23 (October 1955): 18.
Cabe, Carl. *Flour Milling.* Kansas Industry Series No. 2. Lawrence: School of Business, University of Kansas, 1958.
Caldwell, Alexander. "Kansas Manufactures and Mines." *Kansas Historical Collections, 1883–1885* 3 (1886): 451–58.
Caldwell, Eldie F., ed. *Illustrated Southern Kansas.* Rev. ed. Lawrence, Kans.: E. F. Caldwell, 1886.
Calhoun, William G., comp. *Fort Scott: A Pictorial History.* Fort Scott, Kans.: Historic Preservation Association of Bourbon County, 1978.
Callahan, James D., comp. and ed. *Jayhawk Editor: A Biography of A. Q. Miller, Senior.* Los Angeles: Sterling Press, 1955.
Campa, Arthur L. "Immigrant Latinos and Resident Mexican Americans in Garden City, Kansas: Ethnicity and Ethnic Relations." *Urban Anthropology* 19 (1990): 345–60.
Campbell, Harold. "Enhancement of U.S. 400 Important to Area's Future." *Pittsburg Morning Sun*, March 10, 2000. Accessed at www.morningsun.net/stories/031000/loc_031000007.shtml.
"Capper Farm Papers Give State Leading National Institution." *Kansas Business Magazine* 17 (March 1949): 8, 10, 12.
"Cardwell Is a Big Name in Oil Equipment throughout the World." *Kansas Business Magazine* 19 (October 1951): 7, 74–78.
Carey, James C. *Kansas State University: The Quest for Identity.* Lawrence: Regents Press of Kansas, 1977.

Carl, Traci. "Southeast Kansas' Economy Takes Flight with Cessna Plant." *Lawrence Journal-World,* January 18, 1997, p. B8.
———. "Main Street, Cowley County in Dire Straits." *Lawrence Journal-World,* February 2, 1997, p. B6.
———. "Immigration Changes Face of Cowtowns." *Lawrence Journal-World,* July 6, 1997, pp. B1, B6.
Carter, Samuel. *Cowboy Capital of the World: The Saga of Dodge City.* Garden City, N.Y.: Doubleday, 1973.
Case, Nelson. *The Story of Little Town.* Oswego, Kans.: Oswego Independent, 1916.
Case, Theodore S., ed. *History of Kansas City, Missouri.* Syracuse, N.Y.: D. Mason and Company, 1888.
Castel, Albert. *A Frontier State at War: Kansas, 1861–1865.* Ithaca, N.Y.: Cornell University Press for the American Historical Association, 1958.
Centennial Edition. Fort Scott, Kans.: Fort Scott Tribune and Fort Scott Monitor, May 30, 1942.
Centennial Edition. Iola, Kans.: Iola Register, May 30, 1955.
Centennial Issue. Lawrence: Kansas Daily Tribune, July 4, 1876.
Centennial Leavenworth, 1854–1954. Leavenworth, Kans.: Historical Program Committee, 1954.
The Centennial Mirror, Special Edition. Olathe, Kans.: Olathe Mirror, April 29, 1957.
"Cessna Grows on Ground as Well as in the Air." *Midwest Industry Magazine* 34 (June 1966): 13, 39–41.
Charvat, Arthur. "Growth and Development of the Kansas City Stock Yards: A History, 1871–1947." Master's thesis, University of Kansas City, 1948.
Cherokee County, Kansas, in Pictures and Prose. N.p., 1923.
Cherryvale Centennial Committee. *Cherryvale Centennial, 1871–1971: A New Century Beckons.* N.p., 1971.
"Cherryvale Citizens Rally for New Factory Building." *Kansas Business Magazine* 16 (January 1948): 30, 32.
Choguill, Genevieve L., and Harold S. Choguill. *A History of the Humboldt, Kansas, Community: 1855–1988.* North Newton, Kans.: Mennonite Press, 1988.
Christaller, Walter. *Central Places in Southern Germany.* Trans. Carlisle W. Baskin. Englewood Cliffs, N.J.: Prentice-Hall, 1966.
Christian, Shirley. "City Pins Its Hopes on Kansas Speedway." *New York Times,* May 27, 2001, p. Y40.
"City of Neodesha, Bankers Join Forces to Rescue Ailing Manufacturing Firm." *Kansas Business News* 3 (October 1982): 8–10.
Clapsaddle, David K. *Larned State Hospital: The First Fifty Years.* Larned, Kans.: Larned Tiller and Toiler, 1980.
Clark, Carroll D., and Roy A. Roberts. *People of Kansas: A Demographic and Social Study.* Topeka: Kansas State Planning Board, 1936.
Clark, John G. *Towns and Minerals in Southeastern Kansas: A Study in Regional Industrialization, 1890–1930.* Special Distribution Publication No. 52. Lawrence: State Geological Survey of Kansas, 1970.
Clough, James F. *The Story of Arkansas City.* Arkansas City, Kans.: n.p., 1961.
"Coal Industry Prepared for Industrial Expansion." *Kansas Business* 6 (November 1938): 6, 22.

Coe, Elmer. *Fort Scott as I Knew It.* Fort Scott, Kans.: Monitor Binding and Printing Company, 1940.
Coffeyville at 100, Inc. *A History of Coffeyville.* Coffeyville, Kans.: Coffeyville Journal Press, 1969.
Colby Chamber of Commerce. *Business and Industrial Opportunity in Colby, Kansas.* Colby, Kans.: Chamber of Commerce, 1946.
Commercial Club of Arkansas City. *In and around Arkansas City, Kansas.* Arkansas City, Kans.: News Publishing Company, 1912.
Conknight, James D. "A Social History of Fort Scott, Kansas, at the Turn of the Century." Master's thesis, Pittsburg State University, 1973.
Connelley, William E. *The Life of Preston B. Plumb, 1837–1891.* Chicago: Browne and Howell Company, 1913.
———. *Kansas City, Kansas: Its Place in the History of the State!* Kansas City: Wyandotte County Historical Society, 1918.
———. *History of Kansas, State and People.* 5 vols. Chicago: American Historical Society, 1928.
Conzen, Michael P. "Capital Flows and the Developing Urban Hierarchy: State Bank Capital in Wisconsin, 1854–1895." *Economic Geography* 51 (1975): 321–38.
———. "A Transport Interpretation of the Growth of Urban Regions: An American Example." *Journal of Historical Geography* 1 (1975): 361–82.
———. "The Maturing Urban System in the United States, 1840–1910." *Annals of the Association of American Geographers* 67 (1977): 88–108.
Copley, Josiah. *Kansas and the Country Beyond on the Line of the Union Pacific Railway, Eastern Division.* Philadelphia: Lippincott, 1867.
Copple, Oscar A., and Joyce M. Hitchings. *History of Osage City, and Osage County.* N.p., 1970.
Correll, Charles M. "The First Century of Kansas State University." *Kansas Historical Quarterly* 28 (1962): 409–41.
"Corrugated Boxes through Rigid Tests in Love Plant." *Kansas Business Magazine* 24 (February 1956): 8–9.
Costantini, Ann D., and Misty Cox. *Pittsburg, Crawford County Visitor's Guide.* Pittsburg, Kans.: Crawford County Convention and Visitors Bureau, 2000.
Courtwright, Julie. "Want to Build a Miracle City? War Housing in Wichita." *Kansas History* 23 (2000): 218–39.
Cox, Thomas C. *Blacks in Topeka, Kansas, 1865–1915: A Social History.* Baton Rouge: Louisiana State University Press, 1982.
Crimmins, Harold. *A History of the Kansas Central Railway, 1871–1935.* Research Studies Vol. 2, No. 4. Emporia, Kans.: Kansas State Teachers College, 1954.
Crippin, Waldo R. "The Kansas Pacific: A Cross Section of an Age of Western Railroad Building." Master's thesis, University of Chicago, 1934.
Crisman, J. H., and W. W. Sargent. *Holton: The County Seat of Jackson County, Kansas.* N.p., 1888.
Crispino, Mrs. Tommie J., ed. *The Centennial Story of Parsons, Kansas.* Parsons, Kans.: N.p., 1971.
Crofoot, John, et al. "Growing from Within." *Kansas Business News* 6 (March 1985): 57–65.
Cronon, William. *Nature's Metropolis: Chicago and the Great West.* New York: Norton, 1991.

Crowther, Mary, and Mary Maley, eds. *As We Were: Pictorial History of Saline County.* Salina, Kans.: Saline County Historical Society, 1976.

Cruise, John D. "Early Days on the Union Pacific." *Kansas Historical Collections, 1909–1910* 11 (1910): 529–49.

Curl, Thelma J. "Promotional Efforts of the Kansas Pacific and Santa Fe to Settle Kansas." Master's thesis, University of Kansas, 1961.

Cushman, George L. "Abilene, First of the Kansas Cow Towns." *Kansas Historical Quarterly* 9 (1940): 240–58.

Cutler, William G., ed. *History of the State of Kansas.* Chicago: A. T. Andreas, 1883.

Cyr, John R. "Historic Landscapes of Cloud County, Kansas." Master's thesis, Kansas State University, 1981.

Dahman, James. "Southeast Kansas: Pittsburg Plant to Employ 750 Persons." *Kansas Business News* 10 (April 1989): 5, 13.

Dahms, F. A. "The Evolution of Settlement Systems: A Canadian Example, 1851–1970." *Journal of Urban History* 7 (1981): 169–204.

Darby, Harry, Jr. "State Highway Department Shows Progress." *Kansas Business* 2 (December 1934): 3-4.

"Darby Adds Stress Relieving Furnace and Pipe Fabrication." *Kansas Business Magazine* 21 (August 1953): 8–9, 61.

Dary, David A. *Lawrence, Douglas County, Kansas: An Informal History.* Lawrence, Kans.: Allen Books, 1982.

———. *The Santa Fe Trail: Its History, Legends, and Lore.* New York: Knopf, 2000.

"Dayco Corporation." In *Ozarks Labor Union Archives, Local 662.* Springfield, Mo.: Southwest Missouri State University, n.d. Accessed at www.library.smsu.edu/Meyer/SpecColl/olua16.html.

Deatherage, Charles P. *Early History of Greater Kansas City, Missouri and Kansas: The Prophetic City at the Mouth of the Kaw.* 3 vols. Kansas City: Interstate Publishing Company, 1927.

Debes, Leroy J. "The Impact of Industrialization upon Real Estate in Parsons, Kansas." Master's thesis, Kansas State University, 1973.

Deneau, Gerald O. "Highlights of the Development of the Aircraft Industry in Kansas." *Transactions, Kansas Academy of Science* 71 (1968): 439–50.

"Development of World's Largest Known Gas Reserve Pushes Ahead." *Kansas Business Magazine* 18 (October 1950): 9, 73–74.

Diamond Jubilee Edition. Winfield, Kans.: Winfield Daily Courier, February 16, 1948.

"Dickey Clay Manufacturing Is Expanding at Pittsburg." *Kansas Business Magazine* 16 (April 1948): 6–7.

"Dillon and Sons Expanding with New Warehouse, Food Stores." *Kansas Business Magazine* 15 (October 1947): 8.

Directory, Newton and Harvey County, Kansas, 1928–29. Newton, Kans.: Chamber of Commerce, 1928.

Directory of Kansas Manufactures, 1932. Topeka: Commission of Labor and Industry, 1932.

"Dixon Company Has Grown to Be Coffeyville's Biggest Factory." *Kansas Business Magazine* 19 (September 1951): 24, 26.

Dobak, William A. *Fort Riley and Its Neighbors: Military Money and Economic Growth, 1853–1895.* Norman: University of Oklahoma Press, 1998.

Dodge City Greeter and Guide. Rev. ed. Dodge City, Kans.: Strange and Hetzel, 1958.

Dodge City's Diamond Jubilee. Dodge City, Kans.: Dodge City Journal, May 23–25, 1947.

"Doerr Metal Products Moves into New Factory at Larned." *Kansas Business Magazine* 16 (October 1948): 48–49.

"Doerr Metal Products Planning Another Major Plant Expansion." *Kansas Business Magazine* 21 (May 1953): 64–65.

Douglas, Richard L. "A History of Manufactures in the Kansas District." *Kansas Historical Collections, 1909–1910* 11 (1910): 81–211.

Down through the Years with Parsons Pioneer Firms (Special Edition). Parsons, Kans.: Parsons Sun, October 18, 1938.

Drabenstott, Mark, Mark Henry, and Kristin Mitchell. "Where Have All the Packing Plants Gone? The New Meat Geography in Rural America." *Economic Review* 84, no. 3 (1999): 65–82.

Drake, Charles C. *"Who's Who?": A History of Kansas and Montgomery County*. Coffeyville, Kans.: Coffeyville Journal Press, 1943.

"Duckwall's Observe 50th Year in New Offices and Warehouse." *Kansas Business Magazine* 19 (December 1951): 7, 52–53.

Duncan, L. Wallace, comp. *History of Montgomery County, Kansas*. Iola, Kans.: L. Wallace Duncan, 1903.

Duncan, L. Wallace, and Charles F. Scott, eds. *History of Allen and Woodson Counties, Kansas*. Iola, Kans.: Iola Register, 1901.

Dykstra, Robert R. *The Cattle Towns*. New York: Atheneum, 1979.

"Eagle-Picher's Lead Smelter Spurs Big Tri-State Industry." *Kansas Business Magazine* 14 (October 1946): 8, 50.

Edson, Karen, ed. *Herington: A Century of Pride*. North Newton, Kans.: Mennonite Press, 1987.

Edwards, John B. *Early Days in Abilene*. 1896. Abilene, Kans.: C. W. Wheeler, 1940.

Ehernberger, James L., and Francis G. Gschwind. *Smoke above the Plains: Union Pacific, Kansas Division*. Calloway, Nebr.: E. and G. Publications, 1965.

Emery, Janet P. *It Takes People to Make a Town: The Story of Concordia, Kansas: 1871–1971*. Salina, Kans.: Arrow Printing Company, 1970.

Erickson, Ken C. "New Immigrants and the Social Service Agency: Changing Relations at SRS." *Urban Anthropology* 19 (1990): 387–408.

Etrick, Carl F. *Dodge City Semi-centennial Souvenir*. Dodge City, Kans.: Etrick Company, 1922.

"Exline Member of Small Club Boasting 100 Years of Existence." *Midwest Industry Magazine* 40 (July 1972): 14, 17.

Facts concerning Kansas City Kansas. Kansas City: Kansas City Kansas Chamber of Commerce, n.d.

Facts You Should Know about Herington, Kansas. Herington, Kans.: Herington Chamber of Commerce, n.d.

Fagan, Mark. "Flour Power: Kansas Mills Lead the Nation in Flour Production." *Lawrence Journal-World,* May 7, 2000, pp. 1E–3E.

———. "Ottawa Draws Distribution Centers." *Lawrence Journal-World,* April 16, 2001, p. D6.

"Fairfax District Getting More Industrial Activity." *Kansas Business Magazine* 22 (November 1954): 8–9, 60–61.

"Fairfax Gains New Industry." *Midwest Industry Magazine* 27 (January 1959): 42.

"Fairfax Industrial District Gains New Plants, Warehouses." *Kansas Business Magazine* 21 (January 1953): 7, 77.

"Fairfax Industrial District: One of Finest in Nation." *Kansas Business Magazine* 15 (January 1947): 6, 64–66.

"Family Business Is Big Business." *Kansas Business News* 6 (October 1985): 60.

Farber, Madeline K., ed. *Mulvane: City of the Valley.* Mulvane, Kans.: Mulvane Historical Society, 1977.

Farley, Alan W. "Samuel Hallett and the Union Pacific Railway Company in Kansas." *Kansas Historical Quarterly* 25 (1959): 1–16.

"Farmers Profit in Kansas Sugar Beet Industry." *Kansas Business* 5 (March 1937): 12–13.

Farnham, Wallace D. "The Pacific Railroad Act of 1862." *Nebraska History* 43 (1962): 141–67.

Faulk, Odie B. *Dodge City: The Most Western Town of All.* New York: Oxford University Press, 1977.

Federal Writers' Project, Works Progress Administration. *The Larned City Guide.* Larned, Kans.: Chamber of Commerce, 1938.

———. *A Guide to Salina, Kansas.* Salina, Kans.: Advertiser-Sun, 1939.

———. *Kansas: A Guide to the Sunflower State.* New York: Viking Press, 1939.

———. *A Guide to Fort Scott, Kansas.* Fort Scott, Kans.: Monitor Binding and Printing Company, 1940.

———. *A Guide to Hillsboro, Kansas.* Hillsboro, Kans.: Mennonite Brethren Publishing House, 1940.

———. *A Guide to Leavenworth, Kansas.* Leavenworth, Kans.: Leavenworth Chamber of Commerce, 1940.

———. *Missouri: A Guide to the "Show Me" State.* New York: Duell, Sloan and Pearce, 1941.

———. *The Pittsburg Guide.* Pittsburg, Kans.: Pittsburg Chamber of Commerce, 1941.

Feldman, Sidney P. "Kansas Meat Packing: Past and Future." *Kansas Business Review* 13 (February 1960): 10–11.

"Fibre Products Company Is Reaching New Output Peaks." *Kansas Business Magazine* 15 (April 1947): 12.

Fields, Dana. "Kansas Speedway Puts Region on Development Fast Track." *Lawrence Journal-World,* August 27, 2000, pp. 1E, 3E.

Fiftieth Anniversary Edition. Atchison, Kans.: Atchison Daily Globe, December 8, 1927.

Fiftieth Anniversary Edition. Colby, Kans.: Colby Free Press–Tribune, October 4, 1939.

Fiftieth Anniversary Edition. Herington, Kans.: Herington Times-Sun, August 24, 1937.

Fiftieth Anniversary Edition. Hoisington, Kans.: Hoisington Dispatch, March 9, 1939.

Fiftieth Anniversary Edition. Horton, Kans.: Tri-County News, 1936.

Fiftieth Anniversary Edition. Kingman, Kans.: Kingman Journal, March 31, 1939.

Fiftieth Anniversary Number. Newton, Kans.: Newton Kansan, August 22, 1922.

"50 Years in the Foundry Business." *Midwest Industry Magazine* 32 (October 1964): 12–13.

"52 Autos an Hour, 16 Hours a Day." *Midwest Industry Magazine* 32 (September 1964): 14, 69–70.

"Figure on the Future." *Midwest Industry Magazine* 36 (August 1968): 12.

First National Bank. *Goodland, Kansas.* Goodland, Kans.: The Bank, 1971.

Fitch, Don B. "Kansas General Aviation: The Early Days." *Kansas Business News* 2 (October 1981): 34–39.

Fitz, Leslie. "The Development of the Milling Industry in Kansas." *Kansas Historical Collections, 1911–1912* 12 (1912): 53–59.

Fitzgerald, Daniel. "Town Booming: An Economic History of Steamboat Towns along the Kansas-Missouri Border, 1840–1860." Master's thesis, University of Kansas, 1983.
———. "Kansas Was a Leader in Coal Mining." *Kansas Business News* 7 (May 1986): 48–49.
———. *Ghost Towns of Kansas: A Traveler's Guide.* Lawrence: University Press of Kansas, 1988.
———. *Faded Dreams: More Ghost Towns of Kansas.* Lawrence: University Press of Kansas, 1994.
Flint, Jack M., Floyd Heir, and Carl L. Heinrich. *The Kansas Junior College.* Topeka: State Department of Public Instruction, 1968.
Flory, Raymond L., ed. *McPherson at Fifty: A Kansas Community in the 1920s.* McPherson, Kans.: McPherson College, 1970.
———. *Historical Atlas of McPherson County.* McPherson, Kans.: McPherson County Historical Society, 1983.
Floyd, Edyth. "Kansas Is Potential Beet Sugar Center." *Kansas Business* 2 (August 1934): 5–6.
"Fluor Corporation Centers Metal Fabrication Work in Paola." *Kansas Business Magazine* 23 (October 1955): 7, 14, 16.
Foreman, Bill. "Leavenworth Invites Tourists to 'Do Some Time.'" *Lawrence Journal-World,* September 7, 1997, p. B5.
"Fort Riley—Cavalry Hub of the Nation." *Kansas Business Magazine* 9 (December 1941): 12, 26–27.
Fort Scott. Fort Scott, Kans.: City of Fort Scott, n.d.
Foulke, Steven V. "Shaping of Place: Mennonitism in South-Central Kansas." Ph.D. diss., University of Kansas, 1998.
Fountain, John W. "Needy Workers Wait for a Kansas Plant to Reopen." *New York Times,* July 10, 2001, p. A8.
"Frank Becker: A Power in Economic Development." *Kansas Business News* 8 (April 1987): 11–12.
Franzen, Susan L. *Behind the Facade of Fort Riley's Hometown.* Ames, Iowa: Pivot Press, 1998.
French, Laura M. *History of Emporia and Lyon County.* Emporia, Kans.: Emporia Gazette, 1929.
Frontenot, Gregory. "Junction City–Fort Riley: A Case of Symbiosis." In *The Martial Metropolis: U. S. Cities in War and Peace,* edited by Roger W. Lotchin, pp. 35–60. New York: Praeger, 1984.
"Frontier Chemical Completes Plant." *Midwest Industry Magazine* 24 (December 1956): 10–11.
"Fuller Brush Starts Huge New Complex in Great Bend." *Midwest Industry Magazine* 39 (October 1971): 14.
"Funk Manufacturing Comes Back with Power Transmissions." *Midwest Industry Magazine* 33 (September 1965): 13, 56–59.
Gable, Frank M. "The Kansas Penitentiary." *Kansas Historical Collections, 1915–1918* 14 (1918): 379–437.
Gaeddert, G. Raymond. *The Birth of Kansas.* Lawrence: University of Kansas Publications, 1940.
Gamble, Ralph C., Jr., and Scott McCubbin. "The Economic Location of Western Kansas." *Kansas Business Review* 19 (spring 1996): 21–29.

"Garden City and Southwest Get Set for Huge Stanolind Industry." *Kansas Business Magazine* 16 (April 1948): 26, 30, 32.

"Garden City Company Has Only Sugar Beet Refinery in Kansas." *Kansas Business Review* 18 (May 1950): 30.

"Garden City Company Plans More Beet Sugar Production." *Kansas Business Magazine* 23 (June 1955): 42.

Garreau, Joel. *Edge City: Life on the New Frontier.* New York: Doubleday, 1991.

Gates, Paul W. "The Railroads of Missouri, 1850–1870." *Missouri Historical Review* 26 (1932): 126–42.

———. *Fifty Million Acres: Conflicts over Kansas Land Policy, 1854–1890.* Ithaca, N.Y.: Cornell University Press, 1954.

Gaughan, Robert. "Kansas Cities Show Big Gains, Bright Prospects for '53." *Kansas Business Magazine* 21 (January 1953): 8–9, 74–77.

———. "Record Business Past Year, Prospects Bright for '54." *Kansas Business Magazine* 22 (January 1954): 8, 20, 22, 24, 26, 28.

———. "See Better Business in '55 as '54 Closes Strong." *Kansas Business Magazine* 23 (January 1955): 8, 18, 20, 22, 24, 28.

Gazetteer and Directory of the State of Kansas. Lawrence, Kans.: Blackburn and Company, 1870.

"General Motors Preparing Assembly Line at Fairfax." *Kansas Business Magazine* 14 (September 1946): 10.

"General Motors Puts New Installations in Kansas." *Kansas Business Magazine* 23 (May 1955): 7.

Gennette, Leon, project director. *Cloud County History.* Concordia, Kans.: Cloud County Historical Society and Cloud County Genealogical Society, 1992.

Georgia, A. J., et al. *A Twentieth Century History and Biographical Record of Crawford County, Kansas.* Chicago: Lewis Publishing Company, 1905.

Get Acquainted Edition. Emporia, Kans.: Emporia Gazette, June 13, 1942.

Gibson, Arrell M. *Wilderness Bonanza: The Tri-State District of Missouri, Kansas, and Oklahoma.* Norman: University of Oklahoma Press, 1972.

Giles, Frye W. *Thirty Years in Topeka: A Historical Sketch.* Topeka: G. W. Crane and Company, 1886.

"Girard Boasts of Largest Cheese Industry in Kansas." *Kansas Business Magazine* 16 (February 1948): 6–8.

Gish, Lowell. *The First 100 Years: A History of Osawatomie State Hospital.* Topeka: Boys' Industrial School, 1966.

———. *Reform at Osawatomie State Hospital: Treatment of the Mentally Ill, 1866–1970.* Lawrence: University Press of Kansas, 1972.

Glaab, Charles N. "Business Patterns in the Growth of a Midwestern City." *Business History Review* 33 (1959): 159–74.

———. *Kansas City and the Railroads: Community Policy in the Growth of a Regional Metropolis.* Madison: State Historical Society of Wisconsin, 1962.

Glass, Robert. "Age Structure and Economic Growth in Kansas." *Kansas Business Review* 6 (November–December 1982): 6–10.

Gleed, Charles S. "The First Kansas Railway." *Transactions of the Kansas State Historical Society, 1897–1900* 6 (1900): 357–59.

Glick, George W. "The Railroad Convention of 1860." *Transactions of the Kansas State Historical Society, 1905–1906* 9 (1906): 467–80.

Global Security.org. "Kansas Army Ammunition Plant." Accessed at www.globalsecurity.org/military/facility/aap-kansas.htm.
"GM Spends One Billion in Kansas City." *Midwest Industry Magazine* 26 (August 1958): 14–15, 24.
Golden Jubilee Edition. Belleville, Kans.: Belleville Telescope, July 1, 1937.
"Goodjohn Sash and Door Firm Rolling Out Lumber Products." *Kansas Business Magazine* 16 (October 1948): 22, 24.
Goodlander, Charles W. *Early Days of Fort Scott: Memoirs and Recollections.* Fort Scott, Kans.: Monitor Printing Company, 1900.
Goter, Dale. "Here Come the Shopping Malls." *Kansas Business News* 3 (July 1982): 6–7.
Gottschamer, John C. "The Relationship between Community Attitudes and Economic Growth and Development for Kansas Communities." Master's thesis, Kansas State University, 1974.
"Grain Products Is Unique Operation in West Kansas." *Kansas Business Magazine* 24 (May 1956): 46, 48.
Graves, Carl R. "Scientific Management and the Santa Fe Railway Shopmen of Topeka, Kansas, 1900–1925." Ph.D. diss., Harvard University, 1980.
Graves, William W. *History of Neosho County.* 2 vols. St. Paul, Kans.: Journal Press, 1949, 1951.
Great Bend, Kansas. Great Bend, Kans.: Great Bend Chamber of Commerce, N.d.
Great Bend, Kansas: A Historical Portrait of the City. Great Bend, Kans.: Centennial Book Committee, 1972.
"Great Bend, the Oil Capital, Putting on Another Big Show." *Kansas Business Magazine* 20 (October 1952): 70–71.
Greater Leavenworth Club. *Leavenworth, Kansas 1909: Her Resources and Advantages.* Leavenworth, Kans.: Greater Leavenworth Club, 1909.
Greeley, Horace. *An Overland Journey from New York to San Francisco in the Summer of 1859.* San Francisco: H. H. Bancroft and Co., 1860.
Green, Donald E. "A History of Irrigation Technology Used to Exploit the Ogallala Aquifer." In *Groundwater Exploitation in the High Plains,* edited by David E. Kromm and Stephen E. White, pp. 28–43. Lawrence: University Press of Kansas, 1992.
Greenbaum, Susan D. *The Afro-American Community in Kansas City, Kansas.* Kansas City: Brennan Printing Company, 1982.
Greene, Albert R. "United States Land-Offices in Kansas." *Transactions of the Kansas State Historical Society* 8 (1903–1904): 1–13.
———. "The Kansas River—Its Navigation." *Transactions of the Kansas State Historical Society* 9 (1905–1906): 317–58.
Greenwood, County, Kansas: Southeastern Kansas. N.p., n.d.
Grey, Mark A. "Immigrant Students in the Heartland: Ethnic Relations in Garden City, Kansas High School." *Urban Anthropology* 19 (1990): 409–32.
Griffin, Clifford S. *The University of Kansas: A History.* Lawrence: University Press of Kansas, 1974.
Groop, Richard E. "Small Town Population Change in Kansas: 1950–1970." Ph.D. diss., University of Kansas, 1976.
"Growing Spencer Chemicals Stems from Area Planning." *Kansas Business Magazine* 21 (December 1953): 8–9, 57–59.
"Guggenheim Buys Cherryvale Plant." *Kansas Business Magazine* 16 (January 1948): 28.

"H. D. Lee Company Returns to Manufacture Popular Garments." *Kansas Business Magazine* 17 (December 1949): 8–9.

Habegger, Arman J. "Out of the Mud: The Good Roads Movement in Kansas, 1900–1917." Master's thesis, University of Kansas, 1971.

Hale, Edward Everett. *Kanzas and Nebraska: The History, Geographical and Physical Characteristics, and Political Position of Those Territories; An Account of the Emigrant Aid Companies, and Directions to Emigrants.* Boston: Philips, Sampson and Co., 1854

Hall, Jesse A., and Leroy T. Hand. *History of Leavenworth County, Kansas.* Topeka: Historical Publishing Company, 1921.

Hamm, George A. "The Atchison Associates of the Santa Fe Railroad." *Kansas Historical Quarterly* 42 (1976): 353–65.

Hammer, Clint J. F. "History of Wyandotte County, Kansas." Master's thesis, Colorado State University, 1948.

Hancks, Larry K., and Meredith Roberts. *Roots: The Historic and Architectural Heritage of Kansas City, Kansas.* Kansas City: City of Kansas City, Kansas, 1976.

Hand-book of Concordia and Cloud County, Kansas. Chicago: C. S. Burch Publishing Co., 1888.

Harris, Chauncy D. "A Functional Classification of Cities in the United States." *Geographical Review* 33 (1943): 86–99.

Harris, Helen P. "Agriculture and Fort Hays State University." *Kansas History* 9 (1986): 164–74.

"Hartman Combines Foundry with Manufacturing." *Midwest Industry Magazine* 33 (May 1965): 12, 51–53.

Hartsock, D. Lane. "The Impact of the Railroads on Coal Mining in Osage County, 1869–1910." *Kansas Historical Quarterly* 37 (1971): 429–40.

Hartzler, F. E., et al. "What Future for Small Towns?" *Kansas Business News* 6 (September 1985): 15–20, 37–38.

Harvey, David. *The Condition of Postmodernity: An Inquiry into the Origins of Cultural Change.* Cambridge, Mass.: Basil Blackwell, 1991.

Haworth, B. Smith. *Ottawa University: Its History and Its Spirit.* Ottawa, Kans.: Ottawa University, 1957.

Haworth, Erasmus. *Special Report on Oil and Gas.* Topeka: State Printer, 1908.

Hayes, William E. *Iron Road to Empire: The History of 100 Years of Progress and Achievements of the Rock Island Lines.* New York: Simmons-Broadman, 1953.

Hays, Garry D., and Mark A. Plummer. "The Kansas Centennial: Politics and Business 100 Years Ago." *Kansas Business Review* 13 (December 1960): 2–11.

Hays Chamber of Commerce. *The Story of the Early Life of Fort Hays and of Hays City.* Hays, Kans.: Old Fort Hays Historical Association, 1959.

Hays City (Fort Hays) Kansas. Topeka: Crane and Company, 1920.

Hays Medical Center. "Hospital Information." Accessed at www.haysmed.com.

"Hays Oil Appreciation Days Honor Area Oil Men." *Kansas Business Magazine* 22 (October 1954): 52–53.

Haywood, C. Robert. *Trails South: The Wagon-Road Economy in the Dodge City-Panhandle Region.* Norman: University of Oklahoma Press, 1986.

———. *The Merchant Prince of Dodge City: The Life and Times of Robert M. Wright.* Norman: University of Oklahoma Press, 1998.

"He Bought a Good Idea and It Became a Business." *Kansas Business News* 8 (September 1987): 31–32.

Hegeman, Roxana. "City Survives Despite ConAgra Fire." *Lawrence Journal-World,* January 1, 2002, p. B4.

Heim, Theodore L. "The Administration and Organization of the Kansas State Penitentiary." Master's thesis, University of Kansas, 1960.

Helyar, Thelma, ed. *Kansas Statistical Abstract, 2000.* Lawrence: Policy Research Institute, University of Kansas, 2001.

Henderson, Harold J. "The Building of the First Kansas Railroad South of the Kaw River." *Kansas Historical Quarterly* 15 (1947): 225–39.

Hess, Mary. *Anatomy of a Town: Hesston, Kansas.* New York: Carlton Press, 1976.

"Hesston Company Expands Its Manufacture of Combine Parts." *Kansas Business Magazine* 21 (May 1953): 8, 88–89.

"Hesston Reviews First 25 Years." *Midwest Industry Magazine* 40 (November 1972): 34.

Highway Highlights (State Highway Commission of Kansas). Topeka: State Printer, various years.

Historical Booklet: Emporia, Kansas, Centennial Celebration. Emporia, Kans.: Centennial Celebration Committee, 1957.

Historical Centennial Edition. Atchison, Kans.: Atchison Daily Globe, June 20, 1954.

History and Literature Club. *History of Horton and Surrounding Neighborhoods.* Horton, Kans.: Horton Pride, 1974.

History Book Committee. *At Home in Ellis County, Kansas 1867–1992.* 2 vols. Hays, Kans.: Ellis County Historical Society, 1991.

History of Early Scott County. Scott City, Kans.: Scott County Historical Society, 1977.

The History of Stevens County and Its People. Hugoton, Kans.: Stevens County Historical Society, 1979.

Hollenbeck, Trish. "Mid-America, SEK Economic Alliance Join Forces." *Pittsburg Morning Sun,* October 30, 1999. Accessed at www.morningsun.net/stories/103099/bus_1030990012.shtml.

Hollibaugh, Mrs. E. F. *Biographical History of Cloud County, Kansas.* N.p., 1903.

Holliday, C. K. "The Cities of Kansas." *Kansas Historical Collections, 1883–1885* 3 (1886): 396–401.

Holloway, John N. *History of Kansas: From the First Exploration of the Mississippi Valley, to Its Admission into the Union.* Lafayette, Ind.: James, Emmons and Company, 1868.

Hooper, Michael. "President: Topeka Operation Viewed as Key Support Center." *Topeka Capital-Journal.* Accessed at www.cjonline.com/webindepth/bnsf/stories/bus_bnsfprez.shtml.

Hoover, Curtis. "'Black Gold' Brings Millions in New Wealth to Kansas." *Kansas Business* 2 (June 1934): 10–11.

Hope, Holly. *Garden City: Dreams in a Kansas Town.* Norman: University of Oklahoma Press, 1988.

Hosack, Robert E. "Chanute—The Birth of a Town." *Kansas Historical Quarterly* 41 (1975): 468–87.

Hotchkiss, Hubert G. *The Story of Oil and Gas in Kansas.* Wichita: Wichita Beacon, 1938.

Hougen, Harvey. "The Impact of Politics and Prison Industry on the General Management of the Kansas State Penitentiary, 1883–1909." *Kansas Historical Quarterly* 43 (1977): 297–318.

Howell, Frederick N. "Pittsburg, Kansas, and Its Industries." Master's thesis, University of Kansas, 1930.

———. "Some Phases of the Industrial History of Pittsburg, Kansas." *Kansas Historical Quarterly* 1 (1932): 273–94.

Hudson, John C. *Plains Country Towns.* Minneapolis: University of Minnesota Press, 1985.

"Huge Goodyear Plant to Topeka." *Kansas Business Magazine* 12 (August 1944): 8–9.

"Huge Plant Comes to Kansas." *Kansas Business Magazine* 18 (June 1950): 12–14.

Hughes, Harry, and Helen C. Dingler. *From River Ferries to Interchanges: A Brief History of Saline County, Kansas From the 1850's to the 1980's.* Salina, Kans.: Harry Hughes and Helen C. Dingler, 1988.

Hull, O. C. "Railroads in Kansas." *Kansas Historical Collections, 1911–1912* 12 (1912): 37–52.

Humphrey, James. "The Railroads of Kansas." *Kansas Historical Collections, 1883–1885* 3 (1886): 401–4.

———. "The Country West of Topeka Prior to 1865." *Kansas Historical Collections, 1886–1890* 4 (1890): 289–97.

Humphreys, Harry C. *The Factors Operating in the Location of State Normal Schools.* New York: Teachers College, Columbia University, 1923.

Hundredth Anniversary Edition. Leavenworth, Kans.: Leavenworth Times, March 7, 1957.

Hunt, Elvid, and Walter E. Lorence. *History of Fort Leavenworth, 1827–1937.* Rev. ed. Fort Leavenworth, Kans.: Command and General Staff School Press, 1937.

Hurley, L. M. *Newton, Kansas: #1 Santa Fe Rail Hub, 1871–1971.* Topeka: Josten's, 1985.

Hurt, R. Douglas. "Irrigation in the Kansas Plains Since 1930." *Red River Valley Historical Review* 4 (summer 1979): 64–72.

"Hutchinson Salt Industry Helps Keep Kansas on Map." *Kansas Business Magazine* 16 (August 1948): 20, 22, 24, 48–49.

Hutchinson, the Salt City in the Heart of the Great Kansas Wheat Belt. Hutchinson, Kans.: News Company, 1908.

Huttig, Jack. "There's Good News in the Kansas Aircraft Industry." *Kansas Business News* 10 (March 1989): 8–9.

An Illustrated Hand Book Compiled from the Official Statistics, Descriptive of Butler County, Kansas. El Dorado, Kans.: Daily and Weekly Republican, 1887.

The Industrial Directory of Kansas. Topeka: Department of Labor and Industry, 1919.

Ingalls, John J. *A Collection of the Writings of John James Ingalls.* Kansas City: Hudson Kimberly Publishing Co., 1902.

Ingalls, Sheffield. *History of Atchison County, Kansas.* Lawrence: Standard Publishing Company, 1916.

"Inspection Brings Them Back." *Midwest Industry Magazine* 29 (September 1961): 12–13.

Interesting Facts about Girard, Kansas. Girard, Kans.: Girard Press, 1930.

Iola Chamber of Commerce. *The Story of Iola, Kansas: Crossroads of Mid-America.* Iola, Kans.: Iola Chamber of Commerce, 1959.

"Irrigation Activity Looks Up throughout Southwestern Kansas." *Kansas Business Magazine* 21 (July 1953): 46–47.

"Irrigation Is Coming to Front in Western Kansas." *Kansas Business Magazine* 23 (June 1955): 8–10.

Iseley, Bliss. "Wichita, the Conference Center, Fast-Growing Industrial City." *Kansas Business Magazine* 20 (September 1952): 46–47.

Istas, Joan. "Positive Attitude, New Investment Stirs Economic Growth in Colby." *Kansas Business News* 3 (September 1982): 6–7.

"J. B. Ehrsam and Sons: A By-Word for Production." *Midwest Industry Magazine* 32 (July 1964): 14, 50.

Jackson, W. Turrentine. "The Army Engineers as Road Surveyors and Builders in Kansas and Nebraska, 1854–1858." *Kansas Historical Quarterly* 17 (1949): 37–59.

Jameson, Frederic. *Postmodernism, or, The Cultural Logic of Late Capitalism*. Durham, N.C.: Duke University Press, 1991.

January, Anna. *Historic Souvenir of Osawatomie and Environs*. Kansas City: R. P. Company, 1910.

Janzen, Abraham E. "The Wichita Grain Market." Master's thesis, University of Kansas, 1927.

"Jayhawk, Junior, Unique Hotel Is Completed in Topeka." *Kansas Business Magazine* 19 (January 1951): 58–59.

Jeffries, John B. "An Early History of Junction City, Kansas: The First Generation." Master's thesis, Kansas State University, 1963.

Jeffries, John B., and Irene Jeffries, comps. and eds. *Garden of Eden: A Pictorial History of Geary County Kansas*. Junction City, Kans.: Geary County Historical Society, 1978.

Jenks, George F. *A Kansas Atlas*. Topeka: Kansas Industrial Development Commission, 1952.

Jennings, A. Owen. *Topeka, Kansas: A Capital City*. Springfield, Mo.: A. O. Jennings, 1911.

Johnson, Arthur M., and Barry E. Supple. *Boston Capitalists and Western Railroads*. Cambridge, Mass.: Harvard University Press, 1967.

Johnson, Samuel A. *The Battle Cry of Freedom: The New England Emigrant Aid Company in the Kansas Crusade*. Lawrence: University of Kansas Press, 1954.

Johnston, J. H., III. *Leavenworth: Beginning to Bicentennial*. Leavenworth, Kans.: J. H. Johnston III, 1976.

Jones, Billy M. *Factory on the Plains: Lyle Yost and the Hesston Corporation*. Wichita: Center for Entrepreneurship, Wichita State University, 1987.

Jones, Carolyn. *The First One Hundred Years: A History of the City of Manhattan, Kansas, 1855–1955*. Manhattan, Kans.: Manhattan Centennial, 1955.

Jones, Dale E., Sherri Doty, Clifford Grammich, James E. Horsch, Richard Houseal, Mac Lynn, John P. Marcum, Kenneth M. Sanchagrin, and Richard H. Taylor. *Religious Congregations and Membership in the United States*. Nashville, Tenn.: Glenmary Research Center, 2002.

Kansas Anniversary Edition. Newton, Kans.: Newton Kansan, January 28, 1961.

"Kansas Brick and Tile, Hoisington, Is State's Newest, Most Modern." *Kansas Business Magazine* 23 (June 1955): 68–71.

Kansas Centennial, 1861–1961, Special Edition. Goodland, Kans.: Goodland Daily News, June 11, 1961.

Kansas Centennial Futurama Edition. Liberal, Kans.: Southwest Daily Times, March 18, 1961.

"Kansas City Kansas Chamber Responsible for Huge Industrial Development." *Kansas Business Magazine* 9 (July 1941): 7, 12, 14.

"Kansas City, Kansas: Great Industrial Center." *Kansas Business* 6 (June 1938): 5, 19–22.

Kansas City Kansas Mercantile Club. *The Story of Three Years' Progress in Kansas City, Kansas*. Kansas City: Meseraull, 1909.

"Kansas City Kansas" (Special Issue). *Kansas Business* 3 (December 1935): 5, 7, 12–14, 26.

Kansas City Stock Yards Commission. *75 Years of Kansas City Livestock Market History: 1871–1946*. Kansas City: Kansas City Stock Yards Commission, 1946.

Kansas Community Profile, various cities. Topeka: Kansas Department of Commerce and Housing, various years.

The Kansas Directory of Commerce. Wichita: Wichita Eagle and Beacon Publishing Company, various years.

"Kansas Firm Is Major Manufacturer of Coolers." *Midwest Industry Magazine* 34 (October 1966): 50, 52.

"Kansas Grains Converted into Spirits/Alcohol at Atchison." *Kansas Business Magazine* 21 (July 1953): 8–9, 68.

"Kansas Has Largest Overhaul Base in Air Transport Field." *Kansas Business Magazine* 15 (April 1947): 7, 49–50.

Kansas Highway Needs for 1975–1985. Topeka: State Highway Commission, 1966.

Kansas Highway Progress and Future Needs. Topeka: State Highway Commission, 1957.

"The Kansas Invasion." *Kansas Business Magazine* 9 (July 1941): 12.

"Kansas Leading in Aircraft." *Kansas Business Magazine* 24 (March 1956): 8–12.

"Kansas Mills Her Golden Grain to Lead Nation in Flour Production." *Kansas Business* 5 (June 1937): 4, 12, 19–20, 22.

"Kansas Natural Gas an Important Fuel for Homes and Industries." *Kansas Business* 4 (February 1936): 10–11, 18.

Kansas Pacific Railway Co. *Emigrant's Guide to the Kansas Pacific Railway Lands.* Lawrence: Land Department, Kansas Pacific Railway Co., 1871.

"Kansas-Packed TNT Popcorn Shipped All over the World." *Kansas Business Magazine* 22 (March 1954): 16, 18.

"A Kansas Product with World Wide Distribution." *Kansas Business* 3 (January 1935): 12.

Kansas Rail Report: 1988 Update. Topeka: Department of Transportation, 1988.

"Kansas Refineries Continue Growth, Improvement Programs." *Kansas Business Magazine* 18 (October 1950): 18.

"Kansas Refineries Fill Major Spot in Boosting Oil Industry." *Kansas Business Review* 17 (October 1949): 12–13, 55–56.

"Kansas Salt: An Essential Resource." *Kansas Business* 5 (November 1937): 8–10.

"Kansas Soya Products Has New Extraction Plant in Operation." *Kansas Business Magazine* 17 (June 1949): 7, 53–54.

"Kansas Speedway." *Business Facilities,* June, 2001. Accessed at www.facilitycity.com/busfac/bf_01_06_special2.asp.

"Kansas 'Spencer' Trademark on Trailers Scattered over Nation." *Kansas Business Magazine* 17 (January 1949): 10, 75.

Kansas Turnpike: Kansas City to Oklahoma Border: Preliminary Plans. Kansas City: Howard, Needles, Tammen and Burgendoff, 1954.

"Kansas Turnpike Opens after 22 Months Big Construction." *Midwest Industry Magazine* 24 (November 1956): 80–81.

"Keith Swinehart." *Kansas Business News* 4 (March 1983): 16–17.

Kellogg, James E. "The Decline of Railroad Passenger Travel in Kansas, 1950–1969." Master's thesis, University of Kansas, 1971.

Kellogg, Lyman B. "The Founding of the State Normal School." *Kansas Historical Collections, 1911–1912* 12 (1912): 88–98.

Kennedy, Karen. "Balderson Snow Plow Business Started from Blacksmith's Shop." *Kansas Business Magazine* 16 (March 1948): 10, 12, 14.

Kerr, John L. *The Story of a Western Power—The Missouri Pacific: An Outline History.* New York: Railway Research Society, 1928.

Kersey, Ralph T. *History of Finney County, Kansas.* 2 vols. Garden City, Kans.: Finney County Historical Society, 1950.

Keuchel, Edward F., Jr. "The Railroad System of Fort Scott: A Study of the Impact of Railroad Development upon a Southeastern Kansas Community, 1867–1883." Master's thesis, University of Kansas, 1961.

Kinchen, Oscar A. "Boom and Bust in Southwestern Kansas." *West Texas Historical Association Year Book* 24 (October 1948): 27–39.

Kingery, Cecil, ed. and comp. *Phillipsburg—Phillips County Centennial, 1872–1972.* N.p.: Phillipsburg Centennial Committee, 1972.

Kingman County Colonization Company. *Out There in Kansas: Kingman County.* Wichita: Eagle Press, 1901.

Kirkwood, Roger D. "Western Kansas Is Optimistic about Its Industrial Growth." *Kansas Business Magazine* 19 (May 1951): 6, 9–14, 20, 22, 24, 26, 36, 82, 85–86.

———. "Area Industry Looks Ahead to Another Big Year." *Kansas Business Magazine* 24 (January 1956): 8, 10, 12, 14, 16, 18.

Klein, Maury. *Union Pacific: The Birth of a Railroad, 1862–1893.* Garden City, N.Y.: Doubleday, 1987.

Knouse, Charles A., comp. *A Town between Two Rivers: Osawatomie, Kansas, 1854–1954.* Osawatomie, Kans.: Osage Valley Centennial, 1954.

Knudsen, Gwyneth M. "Convention Money: How Sweet It Is." *Kansas Business News* 9 (November 1988): 12–13.

"Koch Engineers Roam over World in Refining Construction Work." *Kansas Business Magazine* 20 (October 1952): 10, 12, 14.

"Krause Corporation Builds World's Largest Disc Plow." *Kansas Business Magazine* 14 (September 1946): 6.

"Krause into Million Dollar Building, Expansion Program." *Kansas Business Magazine* 16 (January 1948): 7, 63–65.

Krebs, Ollie, ed. *Horton, Kansas, 1886–1986.* Horton, Kans.: Horton Headlight, 1986.

Krider, Charles E., Don Eskew, and Steven Maynard-Moody. "Business Retention and Expansion in Kansas Medium-Sized Communities." *Kansas Business Review* 11 (summer 1988): 15–25.

Krider, Charles E., Genna M. Ott, and Patty Skalla. "Economic Development in Medium-Sized Kansas Towns." *Kansas Business Review* 12 (summer 1989): 10–16.

Kromm, David E., and Stephen E. White, eds. *Groundwater Exploitation in the High Plains.* Lawrence: University Press of Kansas, 1992.

———. "The High Plains Ogallala Region." In *Groundwater Exploitation in the High Plains,* edited by David E. Kromm and Stephen E. White, pp. 1–27. Lawrence: University Press of Kansas, 1992.

Laird, Judith F. "Argentine, Kansas: The Evolution of a Mexican-American Community, 1905–1940." Ph.D. diss., University of Kansas, 1975.

Lake, Turner. "Spotlight: Overland Park." *Kansas Business News* 5 (August 1984): 28–30, 32–34.

Lamphere, Louise, Guillermo Grenier, and Alex Stepick, eds. *Newcomers in the Workplace: New Immigrants and the Restructuring of the U.S. Economy.* Philadelphia: Temple University Press, 1994.

Landis, Margaret. *The Winding Valley and the Craggy Hillside: A History of the City of Rosedale, Kansas.* Kansas City: Margaret Landis, 1976.

"Landoll Corporation: A Do-It-Yourself Firm." *Kansas Business News* 8 (August 1987): 43–46.

Langsdorf, Edward, and Robert W. Richmond, eds. "Letters of Daniel R. Anthony, 1857–1862." *Kansas Historical Quarterly* 24 (1958): 6–30, 198–226, 351–70, 458–75.

Lanterman, Alice. "The Development of Kansas City as a Grain and Milling Center." *Missouri Historical Review* 42 (1947–1948): 20–33.

Lapham, Amos S. "Looking Backward." *Kansas Historical Collections, 1923–1925* 16 (1925): 504–14.

Larsen, Lawrence H. *The Urban West at the End of the Frontier.* Lawrence: Regents Press of Kansas, 1978.

"Latest in Cellophane." *Midwest Industry Magazine* 27 (August 1959): 14, 16.

Latrobe, Charles J. *The Rambler in North America, MDCCCXXXII–MDCCCXXXIII.* Vol. 1. New York: Harper and Brothers, 1835.

"Lawrence Canning Plant Does Volume Business." *Kansas Business* 3 (October 1935): 6.

Lawson, John A., Sondra V. M. McCoy, Robert K. Ratzlaff, and Thomas R. Walther. *100 Years of Excellence: A History of the Pittsburg & Midway Coal Mining Company, 1885–1985.* Denver: Pittsburg & Midway Coal Mining Company, 1985.

"Lead and Zinc from Part of the Unseen Empire of Kansas." *Kansas Business* 3 (August 1935): 8–9, 13.

The Leading Industries of Lawrence, Kansas. Lawrence: Commercial and Manufacturing Publishing Company, 1883.

The Leading Industries of Leavenworth, Kansas. Leavenworth, Kans.: Commercial and Manufacturing Publishing Company, 1883.

Leavenworth Area Development. "Major Employers—Leavenworth County, 2002." Accessed at www.lvarea.com/data/majemp.htm.

"Leavenworth" (Special Issue). *Kansas Business* 4 (April 1936): 5.

"Leavenworth Steel Has Complete Setup for Making Your Product." *Kansas Business Magazine* 21 (March 1953): 14.

"Leavenworth Was Center of State's Activities for Many Years." *Kansas Business* 4 (April 1936): 5.

Leonhardt, David. "Recession, Then a Boom? Maybe Not This Time." *New York Times,* December 30, 2001, sec. 3, pp. 1, 10.

Lewis, Tom. *Divided Highways: Building the Interstate Highways, Transforming American Life.* New York: Viking Penguin, 1997.

"LFM Company at Atchison Expanding Foundry, Heavy Shop Facilities." *Kansas Business Magazine* 17 (July 1949): 8, 10, 52–53.

"LFM Company Expands and Merges with Rockwell Manufacturing Company." *Kansas Business Magazine* 24 (February 1956): 7, 57–58.

"Liability Crisis Devastates General Aircraft Industry." *Kansas Business News* 8 (November 1987): 28–30.

"Liberal Benefiting from Oil-Gas Discoveries." *Kansas Business Magazine* 22 (October 1954): 22, 24.

"Liberal Has Wheat, Pancakes, but Big Talk Now Is Oil Boom." *Kansas Business Magazine* 21 (May 1953): 11.

"Liberal Picks Up Tempo in Industry and Construction." *Kansas Business Magazine* 23 (June 1955): 22.

Liberman, Harold M. "Early History of the Kansas School for the Deaf, 1861–1873." Master's thesis, University of Kansas, 1966.

Lincoln Financial Group. "Our History." Accessed at www.lincoln.lfg.com/who/history.htm.

Lipsey, Dick. "Fort Riley Backers Hold Breath for Base." *Lawrence Journal-World,* May 18, 1997, p. B10.

"Little Cash, Much Know-How Pay Off." *Midwest Industry Magazine* 32 (May 1964): 24–27, 58–59.
Lockwood, Green and Company. *Industrial Survey of Wichita, Kansas.* Boston: Lockwood, Green and Company, 1927.
Long, Elmer E. *Pittsburg, Kansas, 1902: Its Business and Its Beauties as Seen through the Camera.* Pittsburg, Kans.: Means-Moore Company, 1902.
"Low-Power TV." *Kansas Business News* 3 (March 1982): 17.
Lowe, Jesse H. *Pioneer History of Kingman County, Kansas.* N.p., n.d.
Lowther, Charles C. *Dodge City, Kansas.* Philadelphia: Dorrance and Company, 1940.
"Luxra Only Water Heater Manufacturer in Kansas." *Kansas Business Magazine* 23 (February 1955): 8, 53.
"Lyons Manufacturing Sold to Oklahoma City Company." *Midwest Industry Magazine* 37 (July 1969): 18–19.
"Mac-Pitt Engineers Look to Other Basic Industries." *Midwest Industry Magazine* 34 (August 1966): 16, 66–67.
MacVicar, Peter. *A Historical Sketch of Washburn College.* Burlington, Kans.: Republican-Patriot Printers, 1886.
Madden, John L. "The Kansas Economy in Historical Perspective, 1860–1900." Ph.D. diss., Kansas State University, 1968.
Mahoney, Timothy R. *River Towns in the Great West: The Structure of Provincial Urbanization in the American Midwest, 1820–1870.* Cambridge, Engl.: Cambridge University Press, 1990.
"Makers of Jayhawk Boxes Had First Paper Mill in the West." *Kansas Business Magazine* 18 (July 1950): 7, 59–62.
Malin, James C. *Winter Wheat in the Golden Belt of Kansas: A Study in Adaption to Subhumid Geographical Environment.* Lawrence: University of Kansas Press, 1944.
———. *The Grassland of North America: Prolegomena to Its History.* Lawrence, Kans.: James C. Malin, 1947.
———. *The Nebraska Question, 1852–1854.* Lawrence, Kans.: James C. Malin, 1953.
Malott, Deane W. *On Growing Up in Abilene, Kansas, 1898–1916.* Abilene, Kans.: Dickinson County Historical Society, 1992.
"Mangelsdorf Expands to Boost East Kansas Popcorn Industry." *Kansas Business Magazine* 16 (October 1948): 18, 20.
Manning, Carl. "Ellsworth Surprises State with Highest Increase in County Median Income." *Lawrence Journal-World,* August 12, 2002, pp. B1, B3.
Manufactures of the United States in 1860 Compiled from the Original Returns of the Eighth Census. Washington, D.C.: Government Printing Office, 1865.
Marshall, J. Neill, and Peter A. Wood. *Services and Space: Key Aspects of Urban and Regional Development.* New York: Wiley, 1995.
Marshall County, a County of Wealth, Beauty, and Historic Interest (Special Issue). Topeka: Kansas Commercial News, January–February 1902.
Martell, Lillian Z. "Salt City Hits a Rough Spot." *Lawrence Journal-World,* April 25, 1999, pp. E1–E2.
Martin, George W. "The Territorial and Military Combine at Fort Riley." *Transactions of the Kansas State Historical Society* 7 (1901–1902): 361–90.
———. "The Boundary Lines of Kansas." *Kansas Historical Collections, 1909–1910* 11 (1910): 53–74.
———. "A Chapter from the Archives." *Kansas Historical Collections, 1911–1912* 12 (1912): 359–75.

Martin, Jennie. *A Brief History of the Early Days of Ellis, Kansas.* N.p., 1903.
Marvin, Anne M. "The Fertile Domain: Irrigation as Adaptation in the Garden City, Kansas, Area." Ph.D. diss., University of Kansas, 1985.
Masterson, V. V. *The Katy Railroad and the Last Frontier.* Norman: University of Oklahoma Press, 1952.
"Matthew Harper Hamill Interview." Typescript, 1994. Located in the archives of the Prairie Museum of Art and History, Colby, Kansas.
McCain, James. "Burt Hoefs." *Kansas Business News* 3 (March 1982): 23–25.
McCandless, Perry. *A History of Missouri.* Vol. 2, *1820–1860.* Columbia: University of Missouri Press, 1972.
McCoy, Joseph G. *Historic Sketches of the Cattle Trade of the West and Southwest.* Glendale, Calif.: Arthur H. Clark, 1940.
McCoy, Sondra V. M. "The Patriarch of Abilene: Cleyson L. Brown and the United Empire, 1898–1935." *Kansas History* 5 (1982): 107–19.
McCoy, Sondra V. M., and Jan Hults. *1001 Kansas Place Names.* Lawrence: University Press of Kansas, 1989.
McDowell, Joseph H. *Building a City: A Detailed History of Kansas City, Kansas.* Kansas City: Kansas City Kansan, 1969.
McKenna, Joseph M. *The Topeka Metropolitan Area: Its Political Units and Characteristics.* Lawrence: Governmental Research Center, University of Kansas, 1962.
———. *The Growth and Problems of Metropolitan Wyandotte County.* Lawrence: Governmental Research Center, University of Kansas, 1963.
McLean, Robert A., Anthony L. Redwood, and Morris M. Kleiner. "Factors in Firms' Decisions to Locate or Expand in Kansas." *Kansas Business Review* 7 (fall 1983): 13–18.
McNally, Edward T., comp. *Pittsburg Almanac, 1876–1976.* Pittsburg, Kans.: E. T. McNally, 1986.
"McNally Pittsburg Corporation Expands Foundry Production and Offices." *Kansas Business Magazine* 19 (July 1951): 18, 20.
"McNally Pittsburg Has New Machine Shop in Production." *Kansas Business Magazine* 24 (April 1956): 8–9, 69–70.
McNamara, John M. *Three Years on the Kansas Border.* New York: Miller, Orton, and Mulligan, 1856.
"McPherson Flour Mill Example of 'Value Added.'" *Kansas Business News* 8 (April 1987): 36, 38.
"Meat Packing Industry Blazed Early Trails to Prosperity." *Kansas Business* 5 (November 1937): 6–7, 23–24.
"Meat Packing of Prime Importance to Kansas." *Kansas Business* 3 (February 1935): 8–9.
Meinig, Donald W. "American Wests: Preface to a Geographical Interpretation." *Annals of the Association of American Geographers* 62 (1972): 159–84.
Mellinger, Gwyneth. "Northcentral Kansas Economy Rebounds from Agricultural Crisis." *Kansas Business News* 10 (March 1989): 14–15.
———. "Kansas: A Good Place to Live." *Kansas Business News* 10 (April 1989): 14–15.
"Merger of Many Lines Makes Western Transit." *Kansas Business* 5 (October 1937): 15, 24.
Meyer, David R. "A Dynamic Model of the Integration of Frontier Urban Places into the United States System of Cities." *Economic Geography* 56 (1980): 120–40.
———. "The Rise of the Industrial Metropolis: The Myth and the Reality." *Social Forces* 68 (1990): 731–52.

Miami County, Kansas 1987. Paola, Kans.: Miami County Historical Society, 1987.
Mickey, F. S., ed. *Souvenir Book of Olathe and Johnson County, Kansas.* Olathe, Kans.: Olathe Mirror, n.d.
Mid-America, Inc. *Preliminary Report Pertaining to Some of the Assets of Southeast Kansas.* Parsons, Kans.: Mid-America, n.d.
Mid-Century Resources Edition, 1900–1950. Arkansas City, Kans.: Arkansas City Daily Traveler, 1950.
Middleton, Kenneth A. *The Industrial History of a Midwestern Town.* Kansas Studies in Business No. 20. Lawrence: School of Business, University of Kansas, 1941.
Milburn, John. "Communities Count on Military." *Lawrence Journal-World,* September 23, 2001, p. B4.
Miller, Glenn W., and Jimmy M. Skaggs, eds. *Metropolitan Wichita: Past, Present and Future.* Lawrence: Regents Press of Kansas, 1978.
Miller, W. H. *The History of Kansas City.* Kansas City: Birdsall and Miller, 1881.
Milloy, Ross. "Waste from Old Mines Leaves Piles of Problems." *New York Times,* July 21, 2000, p. A10.
"Millstone or Milestone." *Midwest Industry Magazine* 28 (July 1960): 12–13.
Miner, H. Craig. "Border Frontier: The Missouri River, Fort Scott & Gulf Railroad in the Cherokee Neutral Lands, 1868–1870." *Kansas Historical Quarterly* 35 (1969): 105–29.
———. *The St. Louis–San Francisco Transcontinental Railroad: The Thirty-fifth Parallel Project, 1853–1890.* Lawrence: University Press of Kansas, 1972.
———. "The Kansas and Neosho Valley: Kansas City's Drive for the Gulf." *Journal of the West* 17 (October 1978): 75–85.
———. *Wichita: The Early Years, 1865–80.* Lincoln: University of Nebraska Press, 1982.
———. *West of Wichita: Settling the High Plains of Kansas, 1865–1890.* Lawrence: University Press of Kansas, 1986.
———. *Discovery! Cycles of Change in the Kansas Oil and Gas Industry, 1860–1987.* Wichita: Kansas Independent Oil and Gas Association, 1987.
———. *Wichita: The Magic City.* Wichita: Wichita-Sedgwick County Historical Museum, 1988.
———. "A Roar from the Sky: Air-Mindedness in Wichita, 1908–1950." *Journal of the West* 30 (January 1991): 37–44.
———. *Kansas: The History of the Sunflower State, 1854–2000.* Lawrence: University Press of Kansas, 2002.
Miner, H. Craig, and William E. Unrau. *The End of Indian Kansas: A Study of Cultural Revolution, 1854–1871.* Lawrence: Regents Press of Kansas, 1978.
Missouri Pacific Railway Company. *The Empire That Missouri Pacific Serves!* St. Louis: Van Hoffman Press, n.d.
"Missouri Valley Steel Floats Biggest Barge Ever on River." *Kansas Business Magazine* 17 (August 1949): 18.
Mitchell, Pat. *The Fair City: Hutchinson, Kansas.* Topeka: Josten's, 1982.
Mitchell, Robert D., and Warren R. Hofstra. "How Do Settlement Systems Evolve? The Virginia Backcountry during the Eighteenth Century." *Journal of Historical Geography* 21 (1995): 123–47.
"Monarch Cement Company Pushes Expansion Work at Humboldt." *Midwest Industry Magazine* 24 (August 1956): 9.
Moon, Henry. *The Interstate Highway System.* Washington, D.C.: Association of American Geographers, 1994.

Mooney, Vol P. *History of Butler County, Kansas.* Lawrence, Kans.: Standard Publishing Company, 1916.

Moore, Henry M. *Early History of Leavenworth City and County.* Leavenworth, Kans.: Samuel Dodsworth Book Co., 1906.

Moran, William T. *Santa Fe and the Chisholm Trail at Newton.* Newton, Kans.: Moran, 1970.

Morgan, Perl W., ed. and comp. *History of Wyandotte County, Kansas and Its People.* 2 vols. Chicago: Lewis Publishing Company, 1911.

Morrell, John S. "A History of Mitchell County." Mimeographed, n.d. Copy located in Kansas Collection at the University of Kansas.

"Morrell's in Topeka for Ten Years." *Kansas Business Magazine* 9 (October 1941): 6.

Morrison, John. "Aircraft Factories Stabilize Post-war Wichita Employment." *Kansas Business Magazine* 17 (August 1949): 14, 16.

Morrow, Marco. "Modern Publishing Plant Spreads Fame of Kansas." *Kansas Business* 2 (February 1934): 8–9.

"Morrow Pattern Shop and Foundry One of Most Modern in Country." *Kansas Business Magazine* 19 (July 1951): 10.

Mowery-Denning, Linda. "Salina Hospital Merger Takes in Concordia." *Salina Journal,* August 5, 1995, p. 1.

Muller, Edward K. "Selective Urban Growth in the Middle Ohio Valley, 1800–1860." *Geographical Review* 66 (1976): 178–99.

———. "Regional Urbanization and the Selective Growth of Towns in North American Regions." *Journal of Historical Geography* 3 (1977): 21–39.

———. "From Waterfront to Metropolitan Region: The Geographical Development of American Cities." In *American Urbanism: A Historiographical Review,* edited by Howard Gillette, Jr. and Zane L. Miller, pp. 105–33. New York: Greenwood Press, 1987.

Napier, Rita. "Squatter City: The Construction of a New Community in the American West, 1854–1861." Ph.D. diss., American University, 1976.

"National Battery Firm Picks Kansas for Plant Expansions." *Kansas Business Magazine* 16 (May 1948): 11–12.

"National Brand of Bedding Made in Topeka." *Kansas Business* 3 (March 1935): 10.

"National Furniture's New Plant Has Open House at Wellington." *Kansas Business Magazine* 18 (August 1950): 7, 53.

"National Gypsum Adding Big Plant." *Kansas Business Magazine* 19 (January 1951): 20.

"National Gypsum Company Completing Medicine Lodge Wallboard Plant." *Kansas Business Magazine* 19 (May 1951): 7, 83–85.

"Need Brings Rapid Growth." *Midwest Industry Magazine* 36 (November 1968): 12–13, 16.

Neff, Phillip, and Robert M. Williams. *The Industrial Development of Kansas City.* Kansas City: Federal Reserve Bank, 1954.

Neher, Megan. "Kmart Crew Celebrates Milestone." *Lawrence Journal-World,* November 17, 1996, p. E1.

Nelson, Harvey F. "An Economic History of Chanute, Kansas." Master's thesis, University of Kansas, 1939.

Nelson, Howard J. "A Service Classification of American Cities." *Economic Geography* 31 (1955): 189–210.

New Horizons Issue. Topeka: Topeka Daily Capital, March 6, 1958.

"New Kansas Company." *Kansas Business* 6 (February 1938): 6.

"New Marley Company Facility Features Employee Comfort, Mechanical Handling Systems." *Midwest Industry Magazine* 37 (September 1969): 14, 61.

"New Spencer Company Research Center." *Kansas Business Magazine* 24 (January 1956): 9.
New Year's Edition. Belleville, Kans.: Belleville Telescope, December 29, 1905.
Nichols, Alice. *Bleeding Kansas*. New York: Oxford University Press, 1954.
Nichols, Claude H., comp. *The Baxter Springs Story*. Baxter Springs, Kans.: Baxter Springs Centennial, 1958.
Nickel, Janet, and Carleen Hill. "Analysis of the Impact of the Aviation Industry on the Kansas Economy." *Kansas Business Review* 17 (fall 1993): 20–27.
Nimz, Dale. "Building the 'Historic City': Significant Houses in East Lawrence." Master's thesis, George Washington University, 1985.
"1988 a Big Year in Southwest Kansas." *Kansas Business News* 10 (February 1989): 5.
Nixon, Earl K. "The Petroleum Industry of Kansas." *Transactions of the Kansas Academy of Science* 51 (1948): 369–424.
Norris, Kathleen. *Dakota: A Spiritual Geography*. New York: Ticknor and Fields, 1993.
North Lawrence Civic Association. "Early History of North Lawrence." Mimeographed, 1930. Copy located in the Kansas Collection at the University of Kansas.
Norton Centennial Board. *From a Covered Wagon Ride to Community Pride: Norton Centennial, 1872–1972*. Norton, Kans.: n.p., 1972.
"Norton Is Having a Building Boom." *Kansas Business Magazine* 20 (May 1952): 66–67.
Nyquist, Edna E. *Pioneer Life and Lore of McPherson County, Kansas*. McPherson, Kans.: Democrat-Opinion Press, 1932.
Official State Atlas of Kansas. Philadelphia: L. H. Everts and Company, 1887.
O'Hara, Eileen. "Garvey Plans Epic Center 'Landmark for Wichita.'" *Kansas Business News* 6 (August 1985): 14–15.
"Oil Industry Is Making Great Bend a Boom Town." *Kansas Business Magazine* 22 (October 1954): 48.
"Oil—Kansas' Greatest Industry." *Kansas Business* 4 (July 1936): 13–15, 24–25.
Olathe "The City Beautiful" Centennial, 1857–1957. Olathe, Kans.: Olathe Centennial, 1957.
Oliva, Leo E. *Fort Hays: Keeping Peace on the Plains*. Topeka: Kansas State Historical Society, 1980.
———. *Fort Larned: Guardian of the Santa Fe Trail*. Topeka: Kansas State Historical Society, 1982.
———. *Fort Scott: Courage and Conflict on the Border*. Topeka: Kansas State Historical Society, 1984.
———. *Fort Wallace: Sentinel on the Smoky Hill Trail*. Topeka: Kansas State Historical Society, 1998.
Olmstead, Robert P. *Rock Island Reflections*. Woodridge, Ill.: McMillan Publications, 1982.
Oswego Centennial, Inc. *Oswego Centennial, 1867–1967*. Oswego, Kans.: Oswego Independent, 1967.
Oswego Chamber of Commerce. *Oswego, Kansas: Live, Trade, Play in Oswego*. Oswego, Kans.: n.p., 1955.
"Other New Industry Follows Trailer Factories to Newton." *Kansas Business Magazine* 23 (December 1955): 7, 54.
"Ottawa Steel Products Firm into 1948 Expansion Program." *Kansas Business Magazine* 16 (April 1948): 8–9.
Overton, Richard C. *Burlington West: A Colonization History of the Burlington Railroad*. Cambridge, Mass.: Harvard University Press, 1941.
"Pacific Intermountain Opens a New Kansas City Truck Terminal." *Kansas Business Magazine* 21 (March 1953): 56–58.

"Packer Plastics Takes Four Years to Double Facilities." *Midwest Industry Magazine* 40 (July 1972): 10, 12.

Panorama of Progress: A Century of Living, Pawnee County, 1872–1972. Larned, Kans.: Larned Tiller and Toiler, 1972.

Parsons: Its Past, Present and Future (Special Edition). Parsons, Kans.: Parsons Palladium, October 16, 1901.

Parsons, Labette County, Kansas: Years from 1869 to 1895. Parsons, Kans.: Bell Bookcraft Shop, [1901?].

Paterniti, Michael. "Eating Jack Horner's Cow." *Esquire* 128 (November 1997): 90–97, 158, 160–63.

Peier, J. Dale. "Change Is Constant as Rural Communities Struggle to Hold On." *Kansas Business News* 6 (September 1985): 40–41.

Pens-Lens View Company. *Pens-Lens Views of the Galena-Empire Mining Camp.* Carthage, Mo.: Pens-Lens View Company, 1899.

"Petroleum—A Major Kansas Industry." *Kansas Business* 3 (April 1935): 8–10.

Petrowski, William R. *The Kansas Pacific: A Study in Railroad Promotion.* New York: Arno Press, 1981.

"Phillips Lube Manufacturing Centralized in New K. C. Plant." *Kansas Business Magazine* 17 (June 1949): 8–9, 52–53.

Pictorial Edition, Magazine Supplement. McPherson, Kans.: McPherson Weekly Republican, March 1, 1901.

Pictorial Historical Edition. Atchison, Kans.: Atchison Daily Globe, July 16, 1894.

Pin-ups from Wellington (Special Edition). Wellington, Kans.: Wellington Monitor-Press, June 5, 1944.

Pioneer Edition. Chanute, Kans.: Chanute Tribune, June 24, 1941.

"Pipe Lines Important Factor in State's Oil Production." *Kansas Business Magazine* 20 (December 1952): 16, 18, 20.

Pittsburg Chamber of Commerce. *Prosperous Pittsburg Pictorially Portrayed.* Pittsburg, Kans.: Pittsburg Publicity Company, 1915.

———. "Business and Industry (2002)." Accessed at www.pittsburgkschamber.com/business.asp.

Pittsburg Diamond Jubilee, 1876–1951 (Official Program). Pittsburg, Kans.: n.p., 1951.

"Pittsburg Pottery Company Serves Wide Area with Pots and Artware." *Kansas Business Magazine* 21 (July 1953): 26.

"Pittsburg" (Special Issue). *Kansas Business* 4 (January 1936): 9, 11, 14–16.

Pittsburg State University. *A Guide to Pittsburg State University.* Pittsburg, Kans.: The University, 1982.

Plummer, Mark A. *Frontier Governor: Samuel J. Crawford of Kansas.* Lawrence: University Press of Kansas, 1971.

Polk, Robert L., comp. *Kansas State Gazetteer and Business Directory.* N.p.: R. L. Polk and Company, 1880, 1891.

Popplewell, Frank S. "St. Joseph, Missouri, as a Center of the Cattle Trade." *Missouri Historical Review* 32 (1937–1938): 443–57.

Porter, Don, comp. *100 Years of History Significant to Kansas: 1854–1954.* Lawrence: World Company, 1954.

Portrait and Biographical Record of Dickinson, Saline, McPherson, and Marion Counties. Chicago: Chapman Bros., 1893.

"Post-war Air Transport and Kansas." *Kansas Business Magazine* 11 (August 1943): 5–6.

Powell, William E. "European Settlement in the Cherokee-Crawford Coal Field of Southeastern Kansas." *Kansas Historical Quarterly* 41 (1975): 150–65.

———. "Coal and Pioneer Settlement in Southeasternmost Kansas." *Ecumene* 9 (1977): 6–16.

———. "The Cherokee-Crawford Coal Field of Southeastern Kansas: A Study in Sequent Occupance." *Midwest Quarterly* 22 (1981): 113–25.

"Pratt's Industrial Growth Is Result of Community Selling." *Kansas Business Magazine* 21 (August 1953): 14–15.

Pred, Allan R. *Urban Growth and the Circulation of Information: The United States System of Cities, 1790–1810.* Cambridge, Mass.: Harvard University Press, 1973.

———. *Urban Growth and City-Systems in the United States, 1840–1860.* Cambridge, Mass.: Harvard University Press, 1980.

Pretzer, Don D. "The Organization of County Elevators in Kansas." Master's thesis, Kansas State University, 1970.

Pride, Woodbury F. *The History of Fort Riley.* Topeka: Capper, 1926.

Pride, Progress Edition. Pratt, Kans.: Pratt Daily Tribune, 1954, 1961.

Proclamation Edition. Chanute, Kans.: Chanute Daily Tribune, February 1, 1904.

"Profile: Emporia." *Kansas Business News* 4 (December 1982): 37–40.

"Profile: Great Bend." *Kansas Business News* 3 (August 1982): 47–49.

Progress and Historical Edition. Phillipsburg, Kans.: Phillips County Review, May, 1952.

Progress Issue. El Dorado, Kans.: El Dorado Times, February 25, 1964.

Quastler, Imre E. *The Railroads of Lawrence, Kansas, 1854–1900.* Lawrence, Kans.: Coronado Press, 1979.

———. "Charting a Course: Lawrence, Kansas, and Its Railroad Strategy, 1854–1872." *Kansas History* 18 (1995): 18–33.

Radford, W. H. "Fairfax District Built Where Lewis and Clark Had Early Camp." *Kansas Business Magazine* 16 (June 1948): 8, 10–12.

Rager, T. F., and John S. Gilmore. *History of Neosho and Wilson Counties, Kansas.* Fort Scott, Kans.: L. Wallace Duncan, 1902.

"Rail Car Building Makes Santa Fe a Prominent Kansas Manufacturer." *Kansas Business Magazine* 17 (April 1949): 8, 10.

"Rail Fight Brews in N.W. Kansas." *Kansas Business News* 2 (October 1981): 10, 46.

"Railroad Freight Yard Projects Bolster Industry of This Area." *Kansas Business Review* 17 (September 1949): 10–12.

Rand McNally 2002 Commercial Atlas and Marketing Guide. Chicago: Rand McNally and Company, 2002.

Ratzlaff, Robert K., and Thomas R. Walther. "Crawford County: From Coal to Soybeans, 1900–1941." *Heritage of Kansas* 11 (spring 1978): 1–15.

"Real Estate Hot Spots." *Kansas Business News* 5 (April 1984): 30–38.

Redwood, Anthony, David L. Petree, and Gary L. Albrecht. "Long-Term Structural Changes in the Kansas Economy." *Kansas Business Review* 8 (winter 1984–1985): 1–9.

Reeve, Agnesa. *Constant Frontier: The Continuing History of Finney County, Kansas.* Garden City, Kans.: Finney County Historical Society, 1996.

Reim, Virginia B. "The Kansas Pacific." Master's thesis, University of Kansas, 1920.

"Remodeling the Nation with Kansas Fibre Products." *Kansas Business* 4 (March 1936): 12–13.

Renner, G. K. "The Kansas City Meat Packing Industry before 1900." *Missouri Historical Review* 55 (1960–1961): 18–29.

Report of the Board of Commissioners of the State Industrial Reformatory, 1886. Topeka: State Printer, 1887.

Report on the Manufactures of the United States at the Tenth Census. Washington, D.C.: Government Printing Office, 1883.

Reps, John W. *Cities of the American West: A History of Frontier Urban Planning.* Princeton, N.J.: Princeton University Press, 1979.

———. *The Forgotten Frontier: Urban Planning in the American West before 1890.* Columbia: University of Missouri Press, 1991.

Rice, Harvey D. *Reminiscences.* Topeka: n.p., 1894.

Richards, Rex. "Kansas City's Industrial Muscle." *Kansas Business News* 10 (May 1989): 4.

Richardson, Albert D. *Beyond the Mississippi.* Hartford, Conn.: American Publishing Co., 1867.

Richardson, Elmo R., and Alan W. Farley. *John Palmer Usher: Lincoln's Secretary of the Interior.* Lawrence: University of Kansas Press, 1960.

Richmond, Robert W. "The First Capitol of Kansas." *Kansas Historical Quarterly* 21 (1955): 321–25.

Riegel, Robert E. "The Missouri Pacific Railroad to 1879." *Missouri Historical Review* 18 (1923–1924): 3–26.

———. "The Missouri Pacific Railroad, 1879–1900." *Missouri Historical Review* 18 (1923–1924): 173–96.

Riley, Darrell G. "Community in Transition: Kansas City's Economy in the 1870s." Master's thesis, University of Kansas City, 1988.

Rinehart, Mrs. Bennett. *Blaze Marks on the Border: The Story of Arkansas City, Kansas, Founded 1870–1871.* North Newton, Kans.: Mennonite Press, 1970.

"RITS News—ROCK 103" (a newsletter for enthusiasts of the Rock Island railroad system), July 1999. Accessed at www.simpson.edu/~RITS/news/TheROCK/News103.html.

Robbins, William. "Army's Plan to Expand Draws Neighbors' Ire." *New York Times,* June 30, 1989, p. 6.

Robinson, Charles. *The Kansas Conflict.* New York: Harper Brothers, 1892.

"Rock Island System Observes 100 Years; Is Big Business in State." *Kansas Business Magazine* 20 (November 1952): 32, 34.

Roe, Catherine, and Bill Roe, comps. and eds. *Atchison Centennial, June 20–26, 1854–1954.* Atchison, Kans.: Lockwood Company, 1954.

Rohrer, Susan K. "Unintended Consequences: The Argentine Neighborhood That Refused to Die." Master's thesis, University of Kansas, 1986.

Rombeck, Terry. "Nuclear Storage Debate Hits Home in Lyons." *Lawrence Journal-World,* July 22, 2002, pp. A1–A5.

Root, Frank A., and W. E. Connelley. *The Overland Stage to California.* Topeka: F. A. Root and W. E. Connelley, 1901.

Root, George A., and Russell K. Hickman. "Pike's Peak Express Companies." *Kansas Historical Quarterly* 13 (1944–1945): 163–95, 211–42, 485–526; 14 (1946): 36–92.

Rork, Amy L. "Sense of Place in Montgomery County, Kansas: Perceptions of an Industrialized Rural Area." Master's thesis, University of Kansas, 1997.

Rose, Mark H. *Interstate Express Highway Politics, 1939–1989.* Rev. ed. Knoxville: University of Tennessee Press, 1990.

Roterus, Victor, and Wesley Calef. "Notes on the Basic-Nonbasic Employment Ratio." *Economic Geography* 31 (1955): 17–20.

Roueche, Berton. "Profiles: Wheat Country." *New Yorker,* January 3, 1983, pp. 37–40, 42–44, 46, 48–50.
Rowe, Frank Joseph, and Craig Miner. *Borne on the South Wind: A Century of Aviation in Wichita.* Wichita: Wichita Eagle and Beacon Publishing Company, 1994.
Rowland, Jessie H. *Pioneer Days in McPherson.* McPherson, Kans.: McPherson Junior Chamber of Commerce, 1947.
Ruley, A. N. *A. N. Ruley's History of Brown County.* Hiawatha, Kans.: Hiawatha World, 1930.
Sagaser, A. Bower. "Building the Main Line of the Missouri Pacific through Kansas." *Kansas Historical Quarterly* 21 (1955): 326–30.
Salina Facts. Salina, Kans.: Salina Chamber of Commerce, 1930.
Salina, Kansas Centennial, 1858–1958: Wagons to Wings. Salina, Kans.: Salina Centennial, 1958.
"Salina" (Special Issue). *Kansas Business* 4 (June 1936): 9, 16–17.
"Salvation for the Small Business." *Midwest Industry Magazine* 32 (February 1964): 12–13, 63.
Sanborn, Gina, with David Burress, Pat Oslund, and Anthony Redwood. "The Importance of the Service Sector: The Case of Johnson County." *Kansas Business Review* 12 (fall 1988): 1–10.
Sanders, Gwendoline, and Paul Sanders. *The Sumner County Story.* North Newton, Kans.: Mennonite Press, 1966.
———. *The Harper County Story.* North Newton, Kans.: Mennonite Press, 1968.
Santa Fe 80th Anniversary Special Issue. Topeka: Topeka Daily Capital, September 18, 1940.
Savage, I. O. *A History of Republic County, Kansas.* Beloit, Kans.: Jones and Chubbic, 1901.
Scenes Near Baxter Springs. Baxter Springs, Kans.: Baxter Springs Investment Company, 1888.
"Schafer Plow Company of Pratt, One of the Larger One-Way Builders." *Kansas Business Magazine* 18 (May 1950): 22, 24, 28.
Schamp, Dick. *"Post Rock to Moon Rock": A Brief History of Russell.* N.p.: Russell County Historical Society, 1971.
"Schimmel Hotel Company Contracts to Operate New K.C. Project." *Kansas Business Review* 18 (July 1950): 42–43.
Schirmer, Sherry L., and Theodore A. Wilson. *Milestones: A History of the Kansas Highway Commission and the Department of Transportation.* Topeka: Kansas Department of Transportation, 1986.
Schnacke, Donald P. "Markets, Politics, Environmental Regulations Challenge Kansas Oil and Gas Industry." *Kansas Business News* 10 (September 1989): 11–12.
Schnell, J. C., and P. E. McLean. "Why the Cities Grew: An Historiographical Essay on Western Urban Growth, 1850–1880." *Bulletin, Missouri Historical Society* 28 (1972): 162–77.
Schoewe, Walter H. "The Geography of Kansas, Part IV, Economic Geography: Mineral Resources." *Transactions, Kansas Academy of Science* 61 (1958): 359–468.
Schroeder, Walter A. "The Presettlement Prairie in the Kansas City Region (Jackson County, Missouri)." *Missouri Prairie Journal* 7, no. 2 (December 1985): 3–12.
Schruben, Francis W. *Wea Creek to El Dorado: Oil in Kansas, 1860–1920.* Columbia: University of Missouri Press, 1972.
Scott, Angelo. "How Natural Gas Came to Kansas." *Kansas Historical Quarterly* 21 (1954): 233–46.

Scott, Charles F. "The Discovery and Development of Natural Gas in Kansas." *Kansas Historical Collections, 1901–1902* 7 (1902): 126–28.

Self, Huber. *Environment and Man in Kansas: A Geographical Analysis.* Lawrence: Regents Press of Kansas, 1978.

"Sell $160,000,000 Bonds for Kansas Turnpike." *Kansas Business Magazine* 22 (October 1954): 72–73.

Serda, Daniel. *Boston Investors and the Early Development of Kansas City, Missouri.* Kansas City: Midwest Research Institute, 1992.

———. *A Blow to the Spirit: The Kaw River Flood of 1951 in Perspective.* Kansas City: Midwest Research Institute, 1993.

Settle, Raymond W., and Mary L. Settle. *War Drums and Wagon Wheels: The Story of Russell, Majors, and Waddell.* Lincoln: University of Nebraska Press, 1966.

Seventieth Anniversary Edition. Marysville, Kans.: Marshall County News, February 27, 1931.

Seventy-fifth Anniversary Edition. Hugoton, Kans.: Hugoton Hermes, August 17, 1961.

Seventy-fifth Anniversary Edition. Leavenworth, Kans.: Leavenworth Times, March 6, 1932.

Seventy-fifth Anniversary Issue. Topeka: Topeka Daily Capital, June 6, 1954.

"Seymour Packing Absorbs Three Other Firms in Expansion Move." *Kansas Business Magazine* 18 (September 1950): 16.

"Seymour Packing Company Puts Branches under Firm Name." *Kansas Business Magazine* 15 (1947): 6–7.

"Shallow Water Refining Company Is Completing Expansion Project." *Kansas Business Magazine* 18 (May 1950): 36.

Shauers, Margaret. "Good News? In Great Bend?" *Kansas Business News* 8 (July 1987): 34–38.

———. "Fuller Brush: Largest Employer in Great Bend." *Kansas Business News* 8 (August 1987): 17–18.

Sheridan, Anne Mosher. "The Development of Nucleated Settlement Systems: The Case of the Upper Mississippi Valley, 1800–1860." Master's thesis, Pennsylvania State University, 1983.

Sheridan, Richard. *Economic Development in Southcentral Kansas, Part 1A: An Economic History, 1500–1900.* Lawrence: School of Business, University of Kansas, 1956.

Sherman, William T. *Home Letters of General Sherman.* Edited by M. A. De Wolfe Howe. New York: Scribner's, 1909.

Sherow, James E. "Rural Town Origins in Southwest Reno County." *Kansas History* 3 (1980): 99–111.

———. *Watering the Valley: Development along the High Plains Arkansas River, 1870–1950.* Lawrence: University Press of Kansas, 1990.

Sherr, Lawrence A., and Alan J. Timblick. "The Impact of the Aircraft Industry on the Economy of Kansas." *Kansas Business Review* 19 (November 1966): 1–3.

Sherwood, Leon A. *The Spirit of Independence.* Independence, Kans.: Tribune Print, 1970.

"Shipbuilding Hitting Lively Pace in Missouri Valley Yards." *Kansas Business Magazine* 20 (July 1952): 16.

Shortridge, James R. "The Post Office Frontier in Kansas." *Journal of the West* 13 (July 1974): 83–97.

———. *Kaw Valley Landscapes: A Traveler's Guide to Northeastern Kansas.* Rev. ed. Lawrence: University Press of Kansas, 1988.

———. *Peopling the Plains: Who Settled Where in Frontier Kansas.* Lawrence: University Press of Kansas, 1995.
———. *Our Town on the Plains: J. J. Pennell's Photographs of Junction City, Kansas, 1893–1922.* Lawrence: University Press of Kansas, 2000.
Shortridge, James R., and Barbara G. Shortridge. "Yankee Town on the Kaw: A Geographical and Historical Perspective on Lawrence and Its Environs." In *Embattled Lawrence: Conflict and Community,* edited by Dennis Domer and Barbara Watkins, pp. 5–19. Lawrence: University of Kansas Continuing Education, 2001.
Shutt, Edwin D. "'Silver City': A History of the Argentine Community of Kansas City, Kansas." Master's thesis, Emporia State College, 1976.
———. "The Saga of the Armour Family in Kansas City, 1870–1900." *Heritage of the Great Plains* 23 (fall 1990): 25–42.
Sibert, Ed. "The Abandonment of Railroad Trackage in Kansas." *Railroad History Monograph* 9 (January 1980): 1–10.
Siler, Zoe M. *This Is Our Town: Cherryvale, Kansas.* Cherryvale, Kans.: Zoe M. Siler, 1961.
Simecka, Betty. "Dodge City Revives Its Wild Wild West." *Kansas Business News* 6 (May 1985): 52–53.
Simmons, Donald H., ed. *Centennial History of Argentine, Kansas City, Kansas, 1880–1980.* Kansas City: Simmons Funeral Home, 1980.
Sixtieth Anniversary Issue. Topeka: Topeka Daily Capital, July 16, 1939.
Skaggs, Jimmy D. *The Cattle-Trailing Industry.* Lawrence: University Press of Kansas, 1973.
———. *Prime Cut: Livestock Raising and Meatpacking in the United States, 1607–1983.* College Station: Texas A&M University Press, 1986.
Slagg, Winifred N. *Riley County Kansas: A Story of Early Settlements, Rich Valleys, Azure Skies and Sunflowers.* Manhattan, Kans.: Winifred N. Slagg, 1968.
"Small Town Firm, Big Time Business." *Midwest Industry Magazine* 28 (September 1960): 14, 16.
Smith, Henry Nash. *Virgin Land: The American West as Symbol and Myth.* Cambridge, Mass.: Harvard University Press, 1950.
Snell, Joseph. "Farm Problems Lead to Progress." *Kansas Business News* 7 (January 1986): 53.
Snell, Joseph W., and Robert W. Richmond. "When the Union and Kansas Pacific Built through Kansas." *Kansas Historical Quarterly* 32 (1966): 161–86, 334–52.
Snell, Joseph W., and Don W. Wilson. "The Birth of the Atchison, Topeka and Santa Fe Railroad." *Kansas Historical Quarterly* 34 (1968): 113–42, 325–56.
"Snowden-Mize Drug Company Has Served Midwest 37 Years." *Kansas Business Magazine* 16 (May 1948): 10, 58–59.
Socolofsky, Homer E. *Kansas Governors.* Lawrence: University Press of Kansas, 1990.
"Sometimes the Tails Don't Move and the Skies Can Be Less Than Friendly." *Kansas Business News* 8 (July 1987): 8–12.
Sorkin, Pamela J., and Murry L. Sorkin, eds. *Sorkin's Directory of Business and Government.* St. Louis: Sorkin's Directories, various years.
Southwest Directory and Publicity Company, comp. *Independence: The Heart of the Kansas Gas and Oil Field.* Independence, Kans.: H. W. Young and Son, 1907.
"Southwest Kansas after Petro-Chemical Industry." *Kansas Business Magazine* 22 (May 1954): 22.
Southwest Kansas Resource Edition. Garden City, Kans.: Garden City Daily Telegram, June 12, 1940.

"Southwestern Bell Moving Its Kansas Headquarters to Topeka." *Kansas Business Magazine* 18 (June 1950): 28.
Special Centennial Edition. Holton, Kans.: Holton Recorder, June 9, 1955.
Special Centennial Edition. Marysville, Kans.: Marysville Advocate, August 12, 1954.
Special Edition. Lawrence: Lawrence Daily Journal, January, 1880.
Special Historical Edition. Holton, Kans.: Holton Recorder, March 14, 1935.
Special Issue. Lawrence: Lawrence Daily World. February, 1903.
Special Issue—Advancing Eureka. Eureka, Kans.: Democratic Messenger, June 29, 1899.
Special Issue to Dedicate Gas Company Building. Phillipsburg, Kans.: Phillips County Review, July 7, 1949.
Special Seventy-fifth Anniversary Issue. Lawrence: Lawrence Journal-World, October 10, 1929.
Speer, John. *Life of Gen. James H. Lane.* Garden City, Kans.: John Speer, 1896.
"Spencer Chemical Company Expands Production of Nitrogen Products." *Kansas Business Magazine* 17 (May 1949): 9.
"Spencer Chemical Company Will Make Million Dollar Expansion." *Kansas Business Magazine* 18 (March 1950): 26.
"Spencer Chemicals Develop New Industry for Midwest." *Kansas Business Magazine* 15 (January 1947): 8–9, 62–63.
"Sprint Headquarters to Display Company's Wealth." *Lawrence Journal-World,* December 8, 1997, p. 3D.
"Star Manufacturing Starts Its New Production Line in Kansas." *Kansas Business Review* 17 (September 1949): 14, 16.
"State Backs Rail Loan, Studies Merger." *Kansas Business News* 5 (July 1984): 12–14.
"State Lawmakers View and Praise Industrial Progress at Fairfax." *Kansas Business Magazine* 19 (February 1951): 8, 51–52.
Stone, Irene G. "The Lead and Zinc Field of Kansas." *Kansas Historical Collections, 1901–1902* 7 (1902): 243–60.
Stotler, Jacob. *Annals of Emporia and Lyon County: Historical Incidents of the First Quarter of a Century, 1857 to 1882.* Emporia, Kans.: n.p., 1898.
Stoughton, P. D. *Greenwood County, Kansas—Let Me Show You.* Madison, Kans.: P. D. Stoughton, 1910.
Stratford, Jessie P. *Butler County's Eighty Years: 1855–1935.* El Dorado, Kans.: Butler County News, 1935.
Stucky, Tim. *Kingman: A Centennial View.* Kingman, Kans.: Golden Valley Parks and Recreation Foundation, 1983.
———. *Pratt, Kansas: A Centennial View, 1884–1984.* Pretty Prairie, Kans.: Prairie Publications, 1984.
Stull, Donald D. "'I Come to the Garden': Changing Ethnic Relations in Garden City, Kansas." *Urban Anthropology* 19 (1990): 303–20.
———. "Knock 'em Dead: Work on the Killfloor of a Modern Beefpacking Plant." In *Newcomers in the Workplace: New Immigrants and the Restructuring of the U. S. Economy,* edited by Louise Lamphere, Guillermo Grenier, and Alex Stepick, pp. 44–77. Philadelphia: Temple University Press, 1994.
———, ed. "When the Packers Came to Town: Changing Ethnic Relations in Garden City, Kansas." *Urban Anthropology* 19 (1990): 303–427.
Stull, Donald D., and Michael J. Broadway. "The Effects of Restructuring on Beefpacking in Kansas." *Kansas Business Review* 14 (fall 1990): 10–16.

Stull, Donald D., Michael J. Broadway, and K. C. Erickson. "The Price of a Good Steak: Beef Packing and Its Consequences for Garden City, Kansas." In *Structuring Diversity: Ethnographic Perspectives on the New Immigration,* edited by Louise Lamphere, pp. 35–64. Chicago: University of Chicago Press, 1992.

Stull, Donald D., Michael J. Broadway, and David Griffith, eds. *Any Way You Cut It: Meat Processing and Small-Town America.* Lawrence: University Press of Kansas, 1995.

"Sunshine Biscuit Is Bringing Most Modern Plant to Kansas." *Kansas Business Magazine* 16 (January 1948): 14.

"Survey Defines Trade Areas." *Kansas Business News* 2 (August 1981): 14–15.

Sutherland, James, comp. *Atchison City Directory and Business Mirror for 1859–60.* St. Louis: Sutherland and McEvoy, 1859.

———. *Atchison City Directory and Business Mirror for 1860–61.* Indianapolis: James Sutherland, 1860.

———. *Lawrence City Directory and Business Mirror for 1860–61.* Indianapolis: James Sutherland, 1860.

———. *Leavenworth City Directory and Business Mirror for 1860–61.* Leavenworth, Kans.: James Sutherland, 1860.

Sutter, Paul S. "Paved with Good Intentions: Good Roads, the Automobile, and the Rhetoric of Rural Improvement in the *Kansas Farmer,* 1890–1914." *Kansas History* 18 (1995): 284–99.

"Sutton Corporation Extends Line of Air Conditioner Products." *Kansas Business Magazine* 23 (March 1955): 8–9.

Swan, Robert A., Jr. *The Ethnic Heritage of Topeka, Kansas: Immigrant Beginnings.* Topeka: n.p., 1974.

Swenson, Andrew S. "Airplane Activity Result of Years of Hard Work." *Kansas Business Magazine* 9 (August 1941): 7, 24–26.

Taft, Robert. "Kansas and the Nation's Salt." *Transactions of the Kansas Academy of Science* 49 (1946): 223–72.

Taylor, David G. "The Economic Development of Leavenworth, Kansas, 1854–1870." Master's thesis, University of Kansas, 1966.

———. "The Business and Political Career of Thomas Ewing, Jr.: A Study of Frustration." Ph.D. diss., University of Kansas, 1970.

———. "Boom Town Leavenworth: The Failure of a Dream." *Kansas Historical Quarterly* 38 (1972): 389–415.

———. "Thomas Ewing, Jr., and the Origins of the Kansas Pacific Railway Company." *Kansas Historical Quarterly* 42 (1976): 155–79.

Taylor, Loren L., coord. *A Short Ethnic History of Wyandotte County.* Kansas City: Kansas City Kansas Ethnic Council, n.d.

"Tele-Marketing Creates New Jobs in Kansas." *Kansas Business News* 9 (August 1988): 18.

Terry, Don. "Proud Base Faced with Notoriety by Association." *New York Times,* April 28, 1995, p. A11.

Third Industrial Edition. Independence, Kans.: South Kansas Tribune, December 30, 1903.

Thirteenth Census of the United States Taken in the Year 1910. Vol. 9, *Manufactories, 1909, Reports by States.* Washington, D.C.: General Printing Office, 1912.

Thomas County: Yesterday and Today. Twelve monthly issues bound as a single volume. Colby, Kans.: Prairie Printers, 1960.

Thompson, Bill. "State Strives to Improve Business Climate." *Kansas Business News* 7 (March 1986): B5–B6.

"Thompson Hardware Covers Five States; Plan Diamond Jubilee." *Kansas Business Magazine* 19 (April 1951): 8–9.

"Thompson-Hayward Chemical Company Expanding Kansas Operations." *Kansas Business Magazine* 16 (August 1948): 11–12.

Thomson, Edith C. *History of Stevens County Kansas.* N.p.: Edith C. Thomson, 1967.

"Three Generations Represented in Silverdale Stone Company." *Kansas Business Magazine* 19 (March 1951): 8, 49–51.

"To Unite for Statewide Road Improvement." *Kansas Business* 5 (August–September 1937): 8–9, 22–23.

Toland, Pauline, ed. *Seward County Kansas.* Liberal, Kans.: Seward County Historical Society, 1979.

"Top Employers in Kansas." Topeka: Kansas Department of Commerce and Housing, 2002. Accessed at www.kdoch.state.ks.us/busdev/top_employers_2001.jsp.

"Top Ten Insurance Companies Remain Healthy in Premium Writing Field." *Kansas Business News* 1 (August 1980): 14–16.

Topeka and its Advantages. Topeka: Hall and O'Donald Litho Co., 1888.

"Topeka: Business, Industrial and Convention Center." *Kansas Business* 6 (January 1938): 6, 10, 16–18, 26–29.

Topeka: Guide to the Capital City of Kansas. Topeka: Topeka Junior Chamber of Commerce, 1957.

Topeka in the Valley of the Kaw: The Golden City of Quivira. Topeka: Topeka Chamber of Commerce, 1929.

Topeka, Kansas: A Thriving State Capital at the Nation's Crossroads. Topeka: Topeka Chamber of Commerce, 1936.

"Topeka: One of the Progressive Cities of the Nation." *Kansas Business* 4 (May 1936): 12–14, 25–26.

"Town House Hotel Offers Latest Convention Facilities at K.C." *Kansas Business Magazine* 19 (September 1951): 11, 57.

Towns, Leroy. "For Air Midwest, the Sky's Not the Limit." *Kansas Business News* 1 (December 1979): 17–19.

———. "Garden City: Optimism in a Land of Plenty." *Kansas Business News* 1 (June 1980): 18–23.

———. "Mark It 'Made in Kansas' and Sell It to the World." *Kansas Business News* 1 (June 1980): 12–16.

———. "Hutchinson: The Growth Is Steady and Deliberate." *Kansas Business News* 1 (August 1980): 22–27.

Towns, Leroy, and Ron Welch. "Getting around Kansas." *Kansas Business News* 1 (December 1979): 12–16.

"Tradewind Industries New Firm Making 'Wheat King' Truck Beds." *Kansas Business Magazine* 16 (December 1948): 11–12.

Treadway, William E. *Cyrus K. Holliday: A Documentary Biography.* Topeka: Kansas State Historical Society, 1979.

Trottman, Nelson. *History of the Union Pacific: A Financial and Economic Survey.* New York: Ronald Press, 1923.

Turk, Eleanor L. "The Germans of Atchison, 1854–1859: Development of an Ethnic Community." *Kansas History* 2 (1979): 146–56.

Turman, J. R. "Neodesha Example of Community Effort toward Hiking Payrolls." *Kansas Business Magazine* 19 (June 1951): 38–40.

Turner, Paul V. *Campus: An American Planning Tradition.* Cambridge, Mass.: MIT Press, 1984.

Turner, Ruth. *Weir, Kansas, 1776–1976, Bicentennial Salute.* Weir, Kans.: Weir Bicentennial Committee, 1976.

"2 Hospital Tables." *Kansas Business News* 7 (July 1986): 22–23.

"Two New Brick Plants Soon by Kansas Firms." *Kansas Business Magazine* 22 (August 1954): 24.

"Two New Factories Added to Newton Mobile Homes Center." *Midwest Industry Magazine* 24 (August 1956): 58.

Tyler, Orville Z. *The History of Fort Leavenworth, 1937–1951.* Fort Leavenworth, Kans.: Command and General Staff College, 1952.

Union Pacific Railroad. *Farmers without Land Should Take the Central Branch Union Pacific Railroad for the Great Homestead Area of Northwestern Kansas.* Atchison, Kans.: Union Pacific Railroad, 1879.

———. *The Golden Belt Lands along the Line of the Kansas Division of the Union Pacific Railway.* Kansas City: Union Pacific Railroad, 1884.

———. *Kansas: Its Resources and Attractions.* Rev. ed. Omaha: Union Pacific Railroad, 1903.

"United Brick and Tile Doubles Output at Expanded Weir Plant." *Kansas Business Magazine* 16 (October 1948): 14.

Unruh, Tim. "Russell on the Rebound." *Lawrence Journal-World,* December 2, 2002, p. B4.

Upchurch, Jenny. "Modular Homes Find Growing Acceptance." *Omaha World-Herald,* January 24, 1999. Accessed at www.nutrendomaha.com/articles/12499.htm.

"Val-Agri." *Kansas Business News* 4 (May 1983): 16–17.

Van Meter, Sondra. "The E. M. Laird Airplane Company: Cornerstone of the Wichita Aircraft Industry." *Kansas Historical Quarterly* 36 (1970): 341–54.

Van Meter, Sondra, and Jan Hults. *1001 Kansas Place Names.* Lawrence: University Press of Kansas, 1989.

Vance, James E., Jr. *The Merchant's World: The Geography of Wholesaling.* Englewood Cliffs, N.J.: Prentice-Hall, 1970.

Vasser, James, and David C. Darling. "Kansas's County Pull-Factor Analysis: 1982–1987." *Kansas Business Review* 11 (fall 1987): 17–20.

Verckler, Stewart P. *Cowtown-Abilene: The Story of Abilene, Kansas, 1867–1975.* New York: Carlton Press, 1961.

Verdon, Roger. "Abilene's Greyhound Industry Searches for Recognition in Home State." *Kansas Business News* 2 (May 1981): 35–37.

Vernon, Edward, ed. *Travelers' Official Railway Guide of the United States and Canada.* Later (with different editors) entitled *Travelers' Official Railway Guide for the United States and Canada, Travelers' Official Guide of the Railway and Steam Navigation Lines, The Official Guide of the Railway and Steam Navigation Lines of the United States, Canada, and Mexico,* and then *The Official Guide of the Railways and Steam Navigation Lines of the United States, Porto Rico, Canada, Mexico, and Cuba.* New York: National Railway Publication Company, various years.

Vernon, Joseph S. *Dodge City and Ford County, Kansas: A History of the Old and a Story of the New.* Larned, Kans.: Tucker-Vernon Publishing Company, 1911.

Vestal, Stanley. *Queen of Cow Towns: Dodge City.* New York: Harper, 1952.

Villard, Henry. *Memoirs of Henry Villard.* 2 vols. Boston: Houghton Mifflin, 1904.

Vincent, Frank. "History of Salt Discovery and Production in Kansas, 1887–1915." *Kansas Historical Collections, 1915–1918* 14 (1918): 358–78.

Voss, Stuart. F. "Town Growth in Central Missouri, 1815–1880: An Urban Chaparral." *Missouri Historical Review* 64 (1969–1970): 64–80, 197–217, 322–50.

"Vulcan Chemical: Kansas Needs Them." *Kansas Business News* 9 (July 1988): 15–18.

Wade, Richard C. *The Urban Frontier: The Rise of Western Cities, 1790–1830.* Cambridge, Mass.: Harvard University Press, 1959.

Wallace, Berenice B., comp. *The Story of Paola, Kansas, 1857–1950.* Paola, Kans.: n.p., n.d.

Walter, John. "The Lawrence Lifestyle: Good and Getting Better." *Kansas Business News* 1 (July 1980): 21–25.

Walter, Sara, and Iona Spencer. *Territorial Days: A Story of Historic Lecompton, Kansas.* Lecompton, Kans.: Lecompton Historical Society: 1980.

Waltner, John D. "The Process of Civilization of the Kansas Frontier: Newton, Kansas, 1871–1873." Master's thesis, University of Kansas, 1971.

Waltner, Rachel. *Brick and Mortar: A History of Newton, Kansas.* North Newton, Kans.: Mennonite Library and Archives, 1984.

Ward/Kraft. Inc. "About Ward/Kraft." Accessed at www.wardkraft.com/about.htm.

Warne, Clinton. "The Development of the Kansas Motel Industry and the Kansas Highway System." *Kansas Business Review* 12 (February 1959, sec. 2): 1–4.

Watch Coffeyville Grow. Coffeyville, Kans.: Coffeyville Daily Journal, April 6, 1907.

Waters, L. L. *Steel Trails to Santa Fe.* Lawrence: University of Kansas Press, 1950.

Wayne, Leslie. "Brother versus Brother: Koch Family's Long Legal Feud Is Headed for a Jury." *New York Times,* April 28, 1998, pp. C1, C8.

Weichselbaum, Theodore. "Statement of Theodore Weichselbaum of Ogden, Riley County, July 17, 1908." *Kansas Historical Collections, 1909–1910* 11 (1910): 561–71.

Welch, Ron. "The Push Is On for More Conventions." *Kansas Business News* 1 (February 1980): 12–19.

———. "The $1 Billion Stop and Look Vacation Game." *Kansas Business News* 1 (May 1980): 20–25.

———. "Cities Learn Industries Don't Always Deliver." *Kansas Business News* 1 (September 1980): 20–21.

———. "Menninger Moves to Fortify Its Future." *Kansas Business News* 2 (February 1981): 39–41.

———. "Ah! Kansas Comes Alive with Logo." *Kansas Business News* 6 (October 1985): 13–14.

———. "Kansas: Hub of the Cattle Feeding Industry." *Kansas Business News* 9 (January 1988): 12–13.

———. "Small Kansas Company Sells to Russia." *Kansas Business News* 9 (July 1988): 7, 14, 22–23.

———. "The Kyle: 'A Mighty Good Little Road.'" *Kansas Business News* 9 (September 1988): 19–21.

———. "Kansas Cities Learn to Help Themselves." *Kansas Business News* 9 (October 1988): 5, 13, 26–27, 33–34.

———. "Kansas Tourism." *Kansas Business News* 10 (June 1989): 14–15.

———. "It's an Age of Specialized Health Care." *Kansas Business News* 10 (August 1989): 12–13.

———. "Northeast Kansas: Economic Development in High Gear." *Kansas Business News* 10 (August 1989): 18–19.

———. "Horton Shapes Its Own Destiny." *Kansas Business News* 11 (January–February 1990): 10–12.

Welcome to Fort Riley. Fort Riley, Kans.: United States Army, 1974.
Wellemeyer, John F., and Earl Walker. *The Public Junior College in Kansas.* Topeka: Kansas Association of Public Junior Colleges, 1937.
Wellington Centennial Book Committee. *Trails to Turnpikes: Wellington Centennial, 1871–1971.* Wellington, Kans.: n.p., 1971.
Wells, Eugene T. "St. Louis and Cities West, 1820–1880: A Study in Historical Geography." Ph.D. diss., University of Kansas, 1951.
Welsh, Willard. *Hutchinson: A Prairie City in Kansas.* Wichita: McCormick-Armstrong Company, 1946.
Wenger, V. L. *A History of Railroad Construction and Abandonment within the State of Kansas.* N.p.: Kansas Corporation Commission, 1972.
Wenzl, Timothy F. *Discovering Dodge City's Landmarks.* Spearville, Kans.: Spearville News, 1980.
"Western Bindery Products Company Binds Industrial Record Forms." *Kansas Business Magazine* 21 (April 1953): 22, 24.
"Western Insurance Companies Observe 40th Year of Service." *Kansas Business Magazine* 18 (April 1950): 7, 53.
"Westvaco Chemical in Lawrence Doubles Capacity in Expansion." *Kansas Business Magazine* 21 (June 1953): 18, 41.
Whebell, Charles F. J. "Corridors: A Theory of Urban Systems." *Annals of the Association of American Geographers* 59 (1969): 1–26.
Whitney, Carrie W. *Kansas City, Missouri: Its History and Its People 1800–1908.* Chicago: S. J. Clarke Publishing Co., 1908.
Whittaker, Bob. "Whittaker Touts Southeast Kansas Highway Program." *Kansas Business News* 8 (June 1987): 18–19.
"Wichita: A City of Progress and Industry." *Kansas Business Magazine* 9 (August 1941): 5, 7, 24–26, 28–29, 32–34.
"Wichita Citizens Recreating Old Cow Town Setting of 1872." *Kansas Business Magazine* 20 (August 1952): 56–57.
"Wichita Has Nation's Largest Broom Corn Market." *Kansas Business* 5 (March 1937): 11–12.
Wichita, Kansas. Wichita: Wichita Board of Trade, 1888.
"Wichita: Metropolis of the Southwest." *Kansas Business* 6 (July 1938): 5, 23–24, 26, 28–30.
Wichita People. Wichita: Wichita Chamber of Commerce, 1946.
"Wichita" (Special Issue). *Kansas Business* 3 (November 1935): 5, 7, 9, 11–14, 19, 22.
Wiebe, Raymond F. *Hillsboro, Kansas: The City on the Prairie.* Hillsboro, Kans.: Centennial Commission and Hillsboro Historical Society, 1985.
"Will the Bread-Basket Belt Ever Lead in Shoe Manufacturing?" *Kansas Business* 5 (March 1937): 10, 13.
"William Leipold." *Kansas Business News* 3 (March 1982): 16.
Williams, Walter. *History of Northwest Missouri.* Chicago: Lewis Publishing Company, 1915.
Wilson, Don W. *Governor Charles Robinson of Kansas.* Lawrence: Regents Press of Kansas, 1975.
Winfield Commercial Club. *Souvenir of Winfield, Cowley County, Kansas, 1904.* Winfield, Kans.: N.p., 1904.
"Winfield Is Proud of Plant Making School Color Crayons." *Kansas Business Magazine* 22 (January 1954): 10, 77–78.

Winfield State Hospital and Training Center. Winfield, Kans.: Winfield State Hospital and Training Center, 1982.

Wingo, Wayne C. "A History of Thomas County, Kansas, 1885–1964." Master's thesis, Fort Hays State College, 1964.

Winter, Maureen, ed. *Indians to Industry: A History of Hays and Ellis County.* Ellis, Kans.: Ellis County Star, 1967.

Wolf, Frank E. "Railway Development in Kansas." Master's thesis, University of Kansas, 1917.

Wolf Creek Nuclear Operating Corporation, "About Wolf Creek." Accessed at www.wcnoc.com/whynuclearfuel.cfm.

Wooster, Lyman D. *Fort Hays Kansas State College: An Historical Story.* Hays, Kans.: Fort Hays Kansas State College, 1961.

"World Trade/Kansas Style." *Kansas Business News* 3 (June, 1982): 46–48.

Worley, William S. *J. C. Nichols and the Shaping of Kansas City: Innovation in Planned Residential Communities.* Columbia: University of Missouri Press, 1990.

Wright, Robert M. *Dodge City, the Cowboy Capital, and the Great Southwest.* Wichita: Wichita Eagle Press, 1913.

Wyandotte County and Kansas City, Kansas: Historical and Biographical. Chicago: Goodspeed Publishing Company, 1890.

"Wyatt Manufacturing Company Expanding Plant; Into Export." *Kansas Business Magazine* 16 (August 1948): 8, 49.

Wyckoff, William K. "Revising the Meyer Model: Denver and the National Urban System, 1859–1879." *Urban Geography* 9 (1988): 1–18.

Wyman, Walker D. "Freighting: A Big Business on the Santa Fe Trail." *Kansas Historical Quarterly* 1 (1931–1932): 17–27.

———. "Kansas City, Mo.: A Famous Freighter Capital." *Kansas Historical Quarterly* 6 (1937): 3–13.

———. "Atchison: A Great Frontier Depot." *Kansas Historical Quarterly* 11 (1942): 297–308.

Yerkes, Susan. "Low-Power TV: A New Broadcast Market Opens for Small Investors." *Kansas Business News* 3 (March 1982): 17.

Yesterday, Today and Tomorrow in Historic Fort Scott, Kansas. Fort Scott, Kans.: Fort Scott Chamber of Commerce, n.d.

Yost, Nellie S. *Medicine Lodge: The Story of a Kansas Frontier Town.* Chicago: Swallow Press, 1970.

You and Your Interstate Highways. Topeka: State Highway Commission, 1961.

Young and Company. *Business and Professional Directory.* Kansas City: Young and Company, 1898.

Young, Frederic R. *Dodge City: Up through a Century in Story and Pictures.* Dodge City, Kans.: Boot Hill Publications, 1972.

Index

Abernathy Furniture Company, 233–34
Abilene, Kans., 129, 271, 331–32
Acers, Nelson, 183
Acers, Roswell W., 183
Acme Foundry and Machine Company, 190, 280
Adams, Charles F., 222
Airosol Corporation, 359–60
Airplane industry, 250–54, 306–09, 311–14, 340, 379
Allen, Martin, 205
Alliance Insurance Company, 261, 333
Altoona, Kans., 187
Amazon.com, 364–65
Andover, Kans., 312
Anthony, Daniel, 62
Anthony, George T., 235
Argentine, Kans., 150–51, 226–28
Arkansas City, Kans.
    before 1900, 153–55, 401–02n. 11
    between 1900 and 1950, 266
    oil industry in, 195
    since 1950, 313–14
Arkansas Valley Branch (Union Pacific Railroad), 156
Arkansas Valley Town Company (Atchison, Topeka and Santa Fe Railroad), 138
Armour, Philip, 222
Armour, Simeon, 222
Armour Packing Company, 222–25, 298
Armstrong, Silas, 91
Army towns, 215–21, 271, 288, 319–21, 324–25
Ash Grove Cement Company, 188, 278–79, 362
Atchison, David. R., 63
Atchison, Kans.
    before 1870, 62–67
    between 1870 and 1950, 236–40
    orphan's home in, 204
    railroad promotion in, 99–101, 108–09, 115
    railroads in, 112, 150, 162–65, 204
    since 1950, 349–49
Atchison and St. Joseph Railroad, 66, 99
Atchison and Pike's Peak Railroad, 108
Atchison Castings Corporation, 239, 348
Atchison, Topeka and Santa Fe Railroad
    in Argentine, 226–28, 297
    beginnings of, 112
    building across Kansas, 137–49, 153–54, 177, 401–02n. 11
    building into Kansas City, 150, 211
    in Chanute, 279
    in Emporia, 269
    near Hutchinson, 258
    promotion of cities by, 96, 372–73
    south of Wichita, 265–66
    in Topeka, 240–41, 243–45, 318–19
Atlantic and Pacific Highway, 259
Augusta, Kans., 192–95, 266, 311–12

Baden, John P., 267
Baker University, 378
Balderson, Neil, 350
Baldwin City, Kans., 93–94, 379
Bank of Horton, 350
Barton, Edward, 182
Baxter Springs, Kans., 120–21, 180, 276–77, 365–66
Beech, Walter, 196, 251–52
Beech Aircraft Corporation, 251–54, 308–09, 344–45
Bellemont, Kans., 54
Belleville, Kans., 167, 262, 289, 352–53
Beloit, Kans., 160–61, 204, 262, 352–54
Bennett Creamery, 273
Benton, Thomas Hart, 41, 52–53
Bethany College, 378–79
Bethel College, 265, 379

471

Betts Baking Company, 259
Bicentennial Center, 317–18, 326–27
Binney and Smith Company, 313
Blanchard, Ben, 181
Blickerstaff, Raymond E., 263
Blish Mize Hardware Company, 238
Blue Mont Central College, 87
Boeing Airplane Company, 253–54, 308–09
Boot Hill Museum, 264, 346–47
Border-tier railroad. *See* Kansas and Neosho Valley Railroad; Missouri River, Fort Scott and Gulf Railroad
Bosland, Kans., 137
Bowman, George C., 243
Boyd, McDill, 354
Branscomb, Charles, 70
Brookover, Earl, 343
Brookville, Kans., 134–35, 157
Brookville Hotel, 332
Brown, William P., 189
Buchanan, James, 60–61
Burlingame, Kans., 138
Burlington, Kans., 356
Burns, John T., 94
Butler County Community College, 313
Butterfield, David A., 128
Butterfield Overland Despatch, 128

Caldwell, Alexander, 232
Caldwell, Kans., 153–54, 169
California, migrants to, 45, 49, 53–54
Caney, Kans., 187, 195
Capper, Arthur, 246, 269
Capper Publications, 246, 317
Carbon Coal Company, 138
Carbondale, Kans., 112, 138
Cardwell Manufacturing Company, 251
Carey, Emerson, 182, 258
Carney, Dan, 310
Carney, Frank, 310
Cathedral of the Immaculate Conception, 57–58
Cattle trade, 4, 110, 135–37, 140–44
 in Abilene, 136–37
 in Baxter Springs, 120–21
 in Bosland, 137
 in Caldwell, 153–54
 in Coffeyville, 122
 in Dodge City, 147
 in Ellsworth, 137
 in Newton, 140
 in Wichita, 141, 144
Central Fibre Products Company, 258
Central Overland, California and Pike's Peak Express Company, 66
Central-place theory, 5, 389n. 3
Cessna, Clyde, 196, 251
Cessna Aircraft Company, 251–53, 308–09, 314, 365
Challis, Luther C., 112
Chanute, Kans.
 before 1890, 119–20, 123
 between 1910 and 1950, 278–79
 oil and gas boom in, 187–88
 since 1950, 362–64
Chanute, Octavius, 119–20, 122
Chanute Refining Company, 188, 279
Cherokee, Kans., 177
Cherokee Nation, 118
Cherryvale, Kans., 187, 359
Chetopa, Kans., 2, 123–34, 398n. 65
Cheyenne Bottoms, 338
Chicago, Rock Island and Pacific Railroad, 164–72, 350, 356
Chicago Zinc and Mining Company, 180
Clay Center, Kans., 158, 270, 349
Clifton, Kans., 158–59
Clothing industry, 277–79, 333, 358–60, 366
Clyde, Kans., 159–60
Coal
 in Cherokee and Crawford Counties, 174–80, 275
 in Fort Scott area, 125–26, 174
 in Leavenworth, 232–33
 in Osage County, 112, 137–38, 272
Coates, Kersey, 113, 116, 227
Coffeyville, Kans.
 before 1890, 122, 401–02n. 11
 between 1910 and 1950, 279–80
 oil and gas boom in, 189–90
 since 1950, 362, 364–65
Coffeyville Vitrified Brick and Tile Company, 189, 280
Colby, Kans., 158, 167, 263–64, 328–30, 378
Colby Community College, 329–30
Coleman, William C., 250, 310

Coleman Company, 250, 310
College of Emporia, 268
College towns, 215–21, 269–71, 288, 321–24, 327–30, 334–35, 378–79
Columbus, Kans., 151–52, 366
Concordia, Kans., 159–60, 260–62, 352–53
Congleton, Thomas, 305
Consolidated Barb Wire Company, 270
Consolidated Kansas City Smelting and Refining Company, 227
Coolidge, Kans., 148, 260
Corporate Woods, 305
Council Grove, Kans., 4, 113, 163, 168–69
Crawford, George A., 125–26
Cray, Cloud L., 238
Cudahy, John, 195

Dawson, H. D., 299
Dayco Corporation, 360
Day, J. Robert, 310
Defense Highway Act of 1941, 289–90
Delco Remy battery plant, 298–99
Dennis, Joseph, 304
Derby, A. L., 194, 250
Derby Oil Company, 194, 196, 309
Devlin, Tom, 310
Didde, Carl, 335
Didde Corporation, 335
Dillon, John S., 259
Dixon Industries, 281
Dodds, John F., 138
Dodge City, Kans.
 before 1900, 3, 146–47
 between 1900 and 1950, 264, 270
 since 1950, 342–47, 379
Dold, Jacob, 195
Doniphan, Kans., 54–55
Drover's Cottage, 136
Duckwall, Alva L., 271, 331
Duckwall/Alco Stores, 271, 331

Eagle-Picher Company, 276, 366
Eastin, Lucian, 57, 59
Eaton Corporation, 340
Eisenhower, Dwight, 331
Eisenhower Center, 331
El Dorado, Kans., 192–96, 266, 312–13
El Dorado State Park, 313

Elkhart, Kans., 172
Ellis, Kans., 2, 134–35
Ellsworth, Kans., 130, 134, 156, 341
Elwood, Kans., 54, 99
Elwood and Marysville Railroad, 99
Emery, Philip, 93–94
Empire City, Kans., 179
Empire Gas and Fuel Company, 193–94, 196
Emporia, Kans.
 before 1870, 88–89, 139
 between 1870 and 1950, 268–69
 quest for university in, 87–89
 railroads in, 112–13, 139, 150
 since 1950, 334–35, 378
Emporia State University, 268, 334–35
Englewood, Kans., 172
Erie, Kans., 120
Eskridge, Charles, 88–89
Evcon Industries, 310
Ewing, Hugh, 60, 99
Ewing, Thomas, 60, 99
Ewing, William N., 227
Excel Corporation, 344, 346

Fairfax Industrial District, 229–31, 296–97
Farmland Industries Refinery, 363
Federal Aid Highway Act of 1944, 290–91
Feedlots, 338, 342–45, 347–48
Fitch, Howard A., 227–28
Flair-Fold Corporation, 351
Flood of 1951, 296–98
Flour milling
 in Atchison, 238
 in Dodge City, 264
 in Hutchinson, 258
 in Kansas City, 228–29
 in Leavenworth, 234
 in Salina, 256–57
 in Topeka, 242
 in Wichita, 248–49, 306
Fluor Corporation, 272, 355
Foley, Dan, 310
Foley, Robin, 310
Forbes Air Force Base, 316–19
Fordism, 282
Fort Fletcher, 131
Fort Harker, 129–30
Fort Hays, 131–32, 205

Fort Hays State University, 205–06, 328
Fort Leavenworth, 42–47, 50, 56, 58–63, 75, 235, 320–21
Fort Leavenworth-Fort Gibson Military Road, 59, 120
Fort Leavenworth-Fort Kearny Military Road, 46, 59, 63–64, 161
Fort Leavenworth-Fort Riley Military Road, 59
Fort Riley, 59, 76, 157–58, 271–72, 324–25
Fort Scott, Kans.
  before 1870, 125–27, 174, 398n. 65
  between 1870 and 1950, 273–74
  potential of its site, 4, 70
  since 1950, 360–61
Fort Scott Foundry and Machine Shops, 125–26
Fort Wallace, 133
Fred Harvey restaurants, 212, 265, 314
Fredonia, Kans., 187, 362
Free-State Hotel, 71, 81
Freighting. *See* Military freighting
Frontenac, Kans., 177, 180, 368
Fuller Brush Company, 337
Furniture industry, 234

Galena, Kans., 179–80, 276–77, 365–66
Galey, James, 184
Garden City, Kans., 148, 201–03, 264, 342–46, 379
Garden City Company, 264
Gardner, Kans., 303–04
Garrity, Mary Carol, 349
Garvey, Ray H., 249, 310
Garvey, Willard, 310
Gates Rubber Corporation, 358
Girard, Kans., 366, 398n. 65
Glass industry, 280
Globe Oil and Refining Company, 197–98, 261
Goodlander, Charles W., 125–26
Goodland, Kans., 167–68, 263, 328–30
Goodrich, Jack L., 294
Gould, Charles N., 192
Gould, Jay, 162–65, 182, 237, 255
Graham, Ethan L., 280
Graham, Glenn, 280
Graves, William P., 257

Graves Truck Line, 257, 326
Great Bend, Kans., 145–47, 198, 262, 336–37
Great Western Manufacturing Company, 234, 320
Great Western Stove Company, 234
Great Western Sugar Company, 328
Growth rates of cities, 209–10, 283–88, 357–58, 377–78
Guffey, James, 184

Hagaman, James M., 160
Haldex Brake Products Corporation, 358
Hale, Edward Everett, 70
Hallett, Samuel, 102–04, 396n. 13
Hannibal and St. Joseph Railroad, 50–51, 54, 56, 65, 97–99, 108, 115–17
Harris, Chauncy D., 214
Haskell Indian Nations University, 321
Hay and Forage Industries, 315
Hays, Kans.
  before 1900, 132
  between 1900 and 1950, 198, 205–06, 262
  since 1950, 327–28, 378
Hays Medical Center, 328
H. D. Lee Mercantile Company, 256
Helmers, Henry J., 234
Helmers Manufacturing Company, 234
Henry, Theodore C., 256
Herington, Kans., 168–69, 356
Herington, Monroe D., 168–69
Hesston, Kans., 315
Hesston Manufacturing Company, 315
Hiawatha, Kans., 165, 268, 351
Highways, 12, 257–59, 262–63, 288–95, 299, 302–03, 305, 323, 361–62, 373–74
Hill, Burton, 243
Hill Packing Company, 243, 317
Hill's Pet Products, 243, 317
Hillsboro, Kans., 356–57
Hoisington, Andrew J., 164
Hoisington, Kans., 164, 337–38
Holliday, Cyrus K., 73–82, 112, 139, 240
Holton, Kans., 349
Horton, Kans., 165–66, 350
Hospital/health center cities, 215–21, 288, 326, 328, 342, 368, 378–79
Hotel Lassen, 251, 307
Houston, Samuel D., 160

Hugoton, Kans., 200–01, 347–48
Hugoton natural gas field, 199–201, 265, 376
Humboldt, Kans., 188–89, 278
Hutchinson, Clinton C., 145
Hutchinson, Kans.
  before 1880, 145
  between 1880 and 1950, 169–70, 257–59
  reformatory in, 204–05
  salt industry in, 181–83
  since 1950, 338–41
Hyde, Albert A., 250
Hyer, Charles H., 273

Ideal Truck Lines, 263
Independence, Kans., 121, 190, 281, 362, 364–65
Independence, Mo., 47–48, 50
Indian lands, as impediments to railroads, 51–52
Insurance industry, 245, 317
Interstate Highway Act of 1956, 292–93
Interstate highways, 12, 293–95, 299, 302–03, 305, 323, 374
Iola, Kans., 183, 185–87, 277–78, 358
Iowa Beef Processors (IBP), 335, 344–45
Irrigation, 201–03, 328, 343, 347
Iuka, Kans., 170

Jacob Dold and Sons, 249
Jayhawk Ordnance Works, 276
John Morrell Company, 313–14
Johnson County Cooperative Association, 273
Johnson, Thomas, 76
Joplin, Mo., 173, 175–76, 180, 368
Joplin Railroad, 175, 177
Joy, James F., 116–17, 119–21, 174
J. S. Dillon and Sons, 259, 339
Junction City, Kans.
  before 1870, 105
  between 1870 and 1950, 271–72
  as a possible cattle town, 136
  railroads in, 105–07, 112–13, 134, 156–57
  since 1950, 324–25
Junction City and Fort Kearny Branch (Union Pacific Railroad), 156–59, 167

Kan-Build, Incorporated, 356
Kanopolis, Kans., 182

Kanotex Refining Company, 195, 266, 313–14
Kansas and Neosho Valley Railroad, 113–14, 126
Kansas Central Railroad, 231
Kansas City, Kans., 221–31, 295–301, 379. *See also* Wyandotte, Kans.
Kansas City, Mo., 113–17, 150, 211–13, 379. *See also* Town of Kansas
Kansas City International Airport, 320
Kansas City Southern Railroad, 178–79, 275, 367
Kansas City Stock Yards Company, 222–23
Kansas City Structural Steel Company, 228
Kansas Co-op Refining Company, 188
Kansas Farm Bureau, 324
Kansas Gas and Gasoline Company, 195
Kansas Industrial Development Commission, 252, 273, 276
Kansas Natural Gas Company, 190–91
Kansas-Nebraska Natural Gas Company, 199, 263, 354
Kansas Ordnance Works, 276, 362
Kansas Pacific Railroad. *See* Union Pacific Railroad
Kansas Rolling Mills, 227
Kansas Speedway, 301
Kansas State Fair, 259, 339–40
Kansas State Industrial Reformatory, 204–05
Kansas State Insane Asylum, 90–94, 206
Kansas State Penitentiary, 84–86, 231–32
Kansas State Sanatorium, 206–07, 354
Kansas State School for the Blind, 93–94
Kansas State School for the Deaf, 93–94, 273
Kansas State Teachers College, 268
Kansas State University, 271
Kansas Turnpike, 291–93, 299
Kansas Wagon Manufacturing Company, 232
Kansas Wesleyan College, 158, 257, 326
Kenny, Chauncey S., 207
Kickapoo City, Kans., 62–63
King, Edward, 303
King, Zenas, 241
King of the Road Corporation, 330
King Radio Corporation, 303
King Wrought Iron Bridge Manufactory and Iron Works, 241

Kinsley, Kans., 145–46
Kinsley cutoff (Atchison, Topeka and Santa Fe Railroad), 169–70, 258
KN Energy, 199, 263, 354
Koch, Fred C., 251, 309–10
Koch Industries, 251, 309
Krause, Henry, 259
Kyle Railroad, 354

Laird, Emil M., 196, 251
Lakin, David L., 140
Landoll, Don, 351
Landoll Corporation, 351
Lane, James, 81, 104, 110–12
Lanyon, Robert, 176, 179, 186
Larned, Kans., 145–46, 206, 338
Lawrence, Amos A., 70, 72
Lawrence, Kans.
  before 1870, 71–72
  between 1870 and 1950, 269–70
  quest for capital in, 75, 80–83
  quest for university in, 72, 87–89
  railroads in, 104–05, 110–12, 115
  since 1950, 319, 321–23, 378
Lawrence Paper Company, 270
Lawrence University, 72
Lead and zinc mining, 175–80, 276–77, 365–66
Lear, William, 309
Lear, Incorporated, 309
Leavenworth, Henry, 42
Leavenworth, Kans.
  before 1870, 2, 4, 56–62, 66–67
  between 1870 and 1950, 231–36
  quest for prison in, 84–86
  railroads in, 99–103, 111–12, 115–16
  since 1950, 319–21
Leavenworth and Pike's Peak Express Company, 61
Leavenworth, Lawrence and Galveston Railroad, 110–12, 117–19, 121–22, 150, 153, 272
Leavenworth, Pawnee and Western Railroad, 82, 100–02
Leawood, Kans., 302
Lecompte, Samuel, 78–79
Lecompton, Kans., 78–80
Lehigh Portland Cement Company, 278, 358

Lenexa, Kans., 302–03
Liberal, Kans., 171–72, 201, 264–65, 342–45, 379
Lindsborg, Kans., 378–79
Locomotive Finished Material Company, 239
Louisburg, Kans., 355
Low, Marcus A., 164–72
Lyons, Kans., 182, 198, 341

Manhattan, Kans., 86–88, 271, 324–25, 378
Manufacturing cities, 215–21, 287
Manypenny, George, 52
Marshall, Jack, 303
Marymount College, 257
Marysville, Kans., 161, 268, 351, 378
McCall Pattern Company, 324
McConnell Air Force Base, 308
McCoy, John C., 48, 55
McCoy, Joseph G., 135–36, 221
McNally, Thomas, 177
McNally Pittsburg Incorporated, 177, 179, 276, 367–68
McPherson, Kans., 199, 261–62, 332–33
Mead, James R., 142–44
Meade, Kans., 170–71
Meatpacking. *See* packinghouses
Menninger Foundation, 319
Mercantile model of urban growth, 6–8, 66–67, 254–55, 369–75, 379–80
Meridian Highway, 257, 262, 289
Messner, Roe, 312
Meyer, August R., 227
Meyer, David R., 7–8
Mid-America, Incorporated, 361
Midland Brake Corporation, 358
Midwest Grain Products, 238, 348
Midwest Solvents Company, 238, 348
Miege, Jean Baptiste, 57–58, 225
Military freighting, 55–56, 58–60, 63–64
Miller, A. Q., 289
Mills, William M., 184
Mining as a factor in city growth, 11, 173–201. *See also* specific minerals
Mining towns, 215–21, 287–88, 375–77
Minneola, Kans., 80
Mission Hills, Kans., 301–02
Missouri, Kansas and Texas Railroad, 118–19, 122–26, 211, 213, 274, 362, 372–73

Missouri Pacific Railroad
  building across Kansas, 162–65
  near Coffeyville, 190, 280
  near Fort Scott, 274
  near Hutchinson, 182
  near Iola, 186
  in Missouri, 115
  in Osawatomie, 163, 272, 355
  near Pittsburg, 177
  near Wichita, 248
Missouri River, Fort Scott and Gulf Railroad, 114, 117–21, 126, 174–75, 177, 179–80, 429n. 2
Missouri River Railroad, 162
Missouri Valley Steel Company, 234, 320
Mobile-home manufacturing industry, 314–15, 330
Moellendick, Jacob M., 196, 251
Moffett, E. R., 175
Monarch Portland Cement Company, 188, 278
Monfort Packing Company, 343, 345
Mormons, 45–46, 55, 61, 64
Mormon War, 61, 65
Morrow, Ivan, 280
Morse, L. V., 222
Morton, Joy, 182
Mount Carmel Regional Medical Center, 368
Mulberry, Kans., 180
Muller, Edward K., 7

National Beef Packing Company, 344–45
National Co-operative Refinery Association (NCRA), 198–99, 262, 333
National Furniture Company, 311
National Land Company (Union Pacific Railroad), 128, 131, 137
National Refinery Company, 280
Natural gas industry, 183–90, 196–201
Neodesha, Kans., 152–53, 184–85, 187, 279, 359–60
Nesch and Moore Brick Company, 177
New Century Airport, 303–04
New England Emigrant Aid Company, 69–73
Newton, Kans.
  before 1880, 2, 139–43, 152, 400n. 106
  between 1880 and 1950, 265
  since 1950, 314–15

Nichols, Jesse C., 273, 301–02
Nickerson, Kans., 143, 145
Northern Natural Gas Company, 200–01, 348
North Newton, Kans., 265, 379
Northrup, Hiram M., 48, 92
Norton, Kans., 206–07, 263, 354
Nu-Wa Industries, 363–64

Oakes, Oliver, 158
Oak Tree Inn, 351
Official Guide of the Railways, 210–14
Oil industry
  in southeastern Kansas, 183–90
  in south-central Kansas, 191–96, 250–51, 312
  in western Kansas, 196–201, 261–62, 330, 336–37, 342
Olathe, Kans., 93–94, 114, 119, 272–73, 303–04
Olathe Naval Air Station, 273, 289, 303–04
Oregon, migrants to, 44–45, 47
Osage City, Kans., 138, 163, 272, 356
Osage Products Company, 356
Osawatomie, Kans., 163, 272, 355
Oswego, Kans., 151–52
Ottawa, Kans., 112, 119, 163, 271–72, 294–95, 333–34
"Ottawa-South" highway, 294–95
Ottawa Truck Corporation, 333
Ottawa University, 272
Overland Park, Kans., 304–05
Ozark Smelting and Mining Company, 190, 280

Pacific Railroad Act of 1862, 100–01, 108, 161
Pacific Railroad of Missouri, 115. *See also* Missouri Pacific Railroad
Packinghouses
  in Emporia, 334–35
  in Hutchinson, 183
  in Kansas City, 149, 221–26, 296
  in St. Joseph, 46
  in southwestern Kansas, 343–46
  in Topeka, 242–43, 317
  in Weston, 43
  in Wichita, 195, 249, 306
Panhandle Eastern Pipe Line Company, 200–01, 344

Paola, Kans., 92, 114, 163, 272, 355
Parker, Kans., 122
Parsons, Kans., 123–25, 204, 274–75, 362–64, 373
Parsons, Levi, 113, 122–24, 126, 212–13, 398n. 65
Pawnee, Kans., 76–78, 394n. 11
Peet Brothers Soap Company, 226
Penrose, Spencer, 203
Perry, John D., 103, 106
Peter, Thomas J., 138–46
Phillips, William A., 129
Phillipsburg, Kans., 167, 199, 263, 354
Phillipsburg Cooperative Refinery, 198–99
Pielsticker, Fred A., 194
Pike's Peak gold rush, 61–62, 65
Pike's Peak Highway, 262
Pipelines, 191, 194, 200–01
Pittsburg, Kans., 175–80, 275–76, 365–68, 378
Pittsburg Pottery Company, 177, 275, 367
Pittsburg State University, 178–79, 275, 367–68
Pizzuti, Tony, 350
Plankinton, John, 222
Platte Country (northwestern Missouri), 42–43
Plumb, Preston, 88, 112, 129, 139, 268
Pomeroy, Samuel C., 64–65, 75, 99, 101, 108, 110, 112, 160, 240
Postindustrial Kansas, 282–368
Prairie Oil and Gas Company, 190, 281
Prairie Village, Kans., 302
Pratt, Kans., 170, 198, 261, 341–42, 378
Pratt Regional Medical Center, 342
Printing industry, 245–46, 317, 335, 366
Procter and Gamble soap factory, 226, 296–97
Producers' Packing Company, 343, 345
Public institutions as a factor in city growth, 4, 10, 204–05, 375
Pusch, Charles F., 161

Railroad promotion, 96–172
Railroads as factors in city growth, 10–11, 49–54, 96, 210–214, 371–72. *See also specific railroad companies and specific cities*
Railroad towns, 215–21, 286–87, 374

Reeder, Andrew H., 58, 75–79, 394n. 22
Reliance Manufacturing Company, 359
Rice, Oscar, 274
Richards, Joseph H., 274
Richardson, Francis, 131
Rivers as transportation routes, 41–67
Robinson, Charles, 70, 72–73, 81, 83, 85, 87
Rock Island Highway, 289
Roedelheimer, Adam, 132
Rosedale, Kans., 226–28, 300
Roseport, Kans., 54
Roubidoux, Joseph, 44
Rowena Hotel, 79
Russell, Edward, 88, 395n. 45
Russell, Kans., 131, 198–99, 262, 330–31
Russell, Majors, and Waddell, 58, 60–61
Russell Stover Company, 331–33, 358

Sabetha, Kans., 350–51
St. Benedict's College, 65
St. John's College, 267, 313
St. Joseph, Mo.
  before 1870, 44–46, 49–50, 53–54, 67, 99
  later railroads in, 161, 165
St. Joseph and Grand Island Railroad, 161, 429n. 2
St. Joseph and Iowa Railroad, 165
St. Louis, entrepreneurial thinking in, 51, 101–02, 116, 127
St. Louis and Pittsburg Zinc Company, 176
St. Louis and San Francisco Railway, 150–53, 177, 179, 248, 274
St. Mary College, 235, 320
St. Mary of the Plains College, 264, 346
Salina, Kans.
  before 1870, 129–30, 136
  between 1870 and 1950, 255–57
  later railroads in, 158, 163–64
  since 1950, 325–26
Salina and Southwestern Railroad, 158
Salina, Lincoln and Western Railroad, 158
Salina Regional Medical Center, 326
salt industry, 181–83, 257–58, 339, 341
Santa Fe, trade with, 41–42, 46–48, 55
S. C. Edgar Zinc Company, 187
Schilling Air Force Base, 325, 327
Scott, Lucien, 232
Scott City, Kans., 347

Seaboard Corporation, 341
Seaton, John, 239
Sergeant, John B., 175
Seventh Street Trafficway (Kansas City, Kans.), 230, 299
Seymour, Thomas F., 243
Seymour Packing Company, 243
Sharp, George. 314
Shawnee Indian mission, 75–76, 78
Sherman, William T., 60–61, 235
Shoemaker, Robert M., 127–28
Sinclair, Harry, 281
Sinclair Oil Company (ARCO), 190, 281, 365
Sisters of St. Joseph, 159–60, 353
Skelly, William G., 194
Skelly Refinery, 194, 266, 312
Smith, Don, 329
Smith, Olaf C., 227–28
Smith, Ray, 338
Smoky Hill Trail, 128
Soda ash, 182–83, 258
Solomon Valley Branch (Union Pacific Railroad), 161, 402n. 24
Solvay Process Company, 258
Southwestern College, 155, 267
Spencer, Kenneth A., 276
Spencer Chemical Company, 276–77, 367
Sprint Corporation, 305
Standard Oil Refinery, 184–85, 188, 279, 359
Stanley, Guy E., 229–30
Star Highway, 263
State capital, selection of, 75–83
State university, selection of site for, 84–89
Steamboats, 49–51, 72, 97
Stearman, Lloyd, 196, 251
Sternberg Museum of Natural History, 328–29
Stevens, Robert S., 112–13, 122, 124–25
Stilwell, Arthur, 178
Stilwell Hotel, 178
Stone, A. B., 227
Strang, William, 304
Strong, William B., 154
Sugar beets, 202–04, 328, 333
Sunflower Manufacturing Company, 354
Sunflower Ordnance Works, 273
Sunflower Village, 273
Superior Industries, 368

Swift and Company, 224
Swinehart, Keith, 333
Swink, George W., 203

Tabor College, 356
Target distribution center, 319
Taylor, Albert R., 205
Taylor Forge, 355
Tecumseh, Kans., 74, 79
Thayer, Eli, 69
Tony's Pizza Service, 325–26
Topeka, Kans.
  before 1870, 73–75
  between 1870 and 1950, 240–46
  quest for capital in, 75, 80–83
  railroad convention in, 98–99, 110
  railroads in, 104–05, 112, 115, 166, 168
  since 1950, 316–19
Topeka cutoff (Union Pacific Railroad), 161, 268
Topeka Daily Capital, 246
Topeka Expocentre, 318
Topeka Iron and Steel Company, 241
Tourism industry, 328–29, 331, 361
Town House hotel, 296
Town of Kansas, 48–49, 55. See also Kansas City, Mo.
Trade, as a factor in city growth, 3–4, 9–10, 41, 67. See also Mercantile model of urban growth
Tradewind Industries, 344
Travenol Laboratories, 327–28
Treece, Kans., 180
Turner, Kans., 299–300
Turner frontier theory, 5
Tuthill, James G., 181

Ulysses, Kans., 347–48
Union Pacific Railroad
  branch lines from, 155–61
  building across Kansas, 127–37
  Central Branch of, 108–09, 159–60, 162, 167
  Eastern Division of, 51, 97, 102–07
  in Hiawatha and Marysville, 268, 351
  in Kansas City, 226, 230, 297
  promotion of cities by, 96, 371–72
  Southern Branch of, 112–13

INDEX 479

United Iron Works, 276, 367
United States Sugar Company, 202–03
University of Kansas, 89, 270, 321
University of Kansas Medical Center, 300
University of Territorial Kansas, 85
Updegraff, William W., 92
U. S. Energy Partners, 330–31

Valu Merchandisers, 360
Vance, James E., 6
Van Horn, Robert, 117
Vickers Petroleum Company, 194
Vulcan Chemical Company, 309–10

Wade, Richard C., 5
Wallace, Kans., 134
W. A. L. Thompson Hardware Company, 242
Wamego, Kans., 134–35, 349–50
Ward/Kraft Corporation, 360
Ware, Eugene F., 126–27
Warner Manufacturing Company, 273
Washburn University, 86, 240, 393n. 17
Waterville, Kans., 109
Watson Wholesale Grocery Company, 256
Webb, William E., 132
Weir City, Kans., 180
Wellington, Kans., 153–54, 169, 265, 311–12
Western Insurance Company, 274, 360
Western Shale Products Company, 274
Western Transit Company, 259
Weston, Mo., 43–46, 50, 55–56, 62
Westport, Mo., 48–49, 55
Whebell, Charles F. J., 6
White Eagle Refinery, 194, 266
Wholesaling cities, 215–21, 288, 370–72
Wholesaling industry
  in Atchison, 238
  in Hutchinson, 258–59
  in Salina, 256
  in Topeka, 242
  in western Kansas, 260
  in Wichita, 247–48
Wichita, Kans.
  before 1880, 2–3, 141–44

between 1880 and 1950, 247–54
impact of oil industry on, 195–96
later railroads in, 150–52, 169
since 1950, 305–11, 379
Wichita and Southwestern Railroad, 142–43
Wichita Eagle, 306
Wichita Natural Gas Company, 192, 195
Wichita State University, 310
Williams, John M. S., 70
Willson, Edward P., 234
Winfield, Kans.
  before 1900, 152, 154–55, 204
  between 1900 and 1950, 195, 266–67
  since 1950, 313–14
Witt, Omer F., 366
Wolf Creek Nuclear Reactor, 356
Wolfe, Gurden, 314–15
Wolfe, Norman, 315
Wolff, Charles, 243
Wolff Packing Company, 243
Woodlands Race Track, 300
Woodson, Daniel, 78–79
World War II, 252–54, 273, 276–77, 289–90, 307–08, 310
Worrell, Squire, 202
Wright, Robert M., 147
W. S. Dickey Clay Manufacturing Company, 177, 276
Wulf, H. F. G., 278
Wyandot Indians, 91–92
Wyandotte, Kans., 90–93, 102–05. *See also* Kansas City, Kans.
Wyandotte constitutional convention, 81–82
Wyckoff, William K., 7–8

Yellow Freight Systems, 366
Yellow Transit Company, 277, 366
Yost, Lyle E., 315
Yucca Soap Company, 250

Zinc smelting, 175–77, 179–80, 275